Multicritical Phenomena

NATO ASI Series

Advanced Science Institutes Series

A series presenting the results of activities sponsored by the NATO Science Committee, which aims at the dissemination of advanced scientific and technological knowledge, with a view to strengthening links between scientific communities.

The series is published by an international board of publishers in conjunction with the NATO Scientific Affairs Division

A	**Life Sciences**	Plenum Publishing Corporation
B	**Physics**	New York and London
C	**Mathematical and Physical Sciences**	D. Reidel Publishing Company Fordrecht Boston, and Lancaster
D	**Behavioral and Social Sciences**	Martinus Nijhoff Publishers
E	**Engineering and Materials Sciences**	The Hague, Boston, and Lancaster
F	**Computer and Systems Sciences**	Springer-Verlag
G	**Ecological Sciences**	Berlin, Heidelberg, New York, and Tokyo

Recent Volumes in this Series

Volume 98 —Quantum Metrology and Fundamental Physical Constants
edited by Paul H. Cutler and A. A. Lucas

Volume 99 —Techniques and Concepts in High-Energy Physics II
edited by Thomas Ferbel

Volume 100—Advances in Superconductivity
edited by B. Deaver and John Ruvalds

Volume 101—Atomic and Molecular Physics of Controlled Thermonuclear Fusion
edited by Charles J. Joachain and Douglass E. Post

Volume 102—Magnetic Monopoles
edited by Richard A. Carrigan, Jr., and W. Peter Trower

Volume 103—Fundamental Processes in Energetic Atomic Collisions
edited by H. O. Lutz, J. S. Briggs, and H. Kleinpoppen

Volume 104—Short-Distance Phenomena in Nuclear Physics
edited by David H. Boal and Richard M. Woloshyn

Volume 105—Laser Applications in Chemistry
edited by K. L. Kompa and J. Wanner

Volume 106—Multicritical Phenomena
edited by R. Pynn and A. Skjeltorp

Series B: Physics

Multicritical Phenomena

Edited by

R. Pynn

Institut Laue-Langevin
Grenoble, France

and

A. Skjeltorp

Institute for Energy Technology
Kjeller, Norway

Plenum Press
New York and London
Published in Cooperation with NATO Scientific Affairs Division

Proceedings of a NATO Advanced Study Institute on
Multicritical Phenomena,
held April 10-21, 1983,
in Geilo, Norway

Library of Congress Cataloging in Publication Data

NATO Advanced Study Institute on Multicritical Phenomena (1983: Geilo, Norway)
Multicritical phenomena.

(NATO ASI series. Series B, Physics; v. 106)
"Proceedings of a NATO Advanced Study Institute on Multicritical Phenomena, held April 10-21, 1983, in Geilo, Norway"—T.p. verso.
Includes bibliographical references and index.
1. Critical phenomena (Physics)—Congresses. 2. Phase transformations (Statistical physics)—Congresses. 3. Spin glasses—Congresses. I. Pynn, R. II. Skjeltorp, A. III. Title. IV. Series.
QC173.4.C74N38 1983 530.4 83-27076

ISBN-13: 978-1-4612-9698-0 e-ISBN-13: 978-1-4613-2741-7
DOI: 10.1007/978-1-4613-2741-7

Softcover reprint of the hardcover 1st edition 1984
A Division of Plenum Publishing Corporation
233 Spring Street, New York, N.Y. 10013

PREFACE

This book comprises the Proceedings of a NATO Advanced Study Institute on Multicritical Phenomena held in Geilo, Norway, between 10-21 April 1983. This school was the seventh to be held in Geilo, on various aspects of phase transitions. In spite of its apparently restrictive title the school was planned as a forum for the discussion of phase transitions and instabilities in systems, with competing interactions and competing order parameters. Thus, in addition to the canonical multicritical points, subjects were diverse as critical phenomena in random magnetic systems and routes to chaos were discussed.

The subject matter of the school is naturally divided into a series of categories which to some extent, reflect the historical development of interest in competing phenomena at phase transitions. Multicritical points in equilibrium systems, defined phenomenologically as points of sudden change of behaviour on an otherwise smooth phase boundary, were the first topics of the school. The theoretical consensus which has emerged during the past decade, largely as a result of calculations with the renormalisation group, was reviewed in some detail. The results presented, however, apply only to pure systems (in which dirt and other manifestations of reality are irrelevant) in that small realm close to the phase transition known as "asymtopia." The experimental difficulties in reaching this ideal world are many and even the elegant experiments reported at this school can only probe into some of the multitude of theoretical predictions which the concepts of universality and the renormalisation group produce so readily. In the field of multicritical dynamics, a complex and lengthy calculation has provided results away from the asymptotic regime. However, the detailed experimental test of this theoretical tour de force is still lacking.

For both experimental and theoretical reasons, magnetic systems have been more popular than their structural conterparts among students of phase transitions. Elegant examples presented at this school which confirm this preference, were the measurements of tricritical phenomena in DAG, the bicritical points observed in various three-dimensional antiferromagnets and the Lifshitz point of MnP. However, a number of new results obtained by various means on structural transitions in stressed perovskites also attracted attention. The newest result to be presented at the meeting, a quantitative theoretical and experimental description of a multicritical point in stressed $LaAlO_3$, was an excellent example of results which can be obtained with such systems. In spite of the dichotomy between the "magnetic" and "structural" communities it is now clear that both groups are speaking the same language. This was certainly not the case at the first Geilo meeting in 1971.

Good-natured dissention between theorists and experimentalists was a recurring theme of this meeting. Nowhere was this more evident than in the discussion of random systems. Theoretical arguments for (different values of) the lower marginal dimensionality of Ising systems in random fields abound and none is yet in agreement with the relatively meagre experimental data. Experiments confirm some predictions, such as the Lorentzian-squared shape of the structure factor in randomly diluted antiferromagnets, but are in radical disagreement with current theory on such elementary points as the existence of multiple phase transitions in systems with competing anisotropies. Nevertheless the theoretical understanding of the types of randomness which may cause departure from pure behaviour is an encouraging sign of progress.

The percolation problem, which aroused little excitement, four years ago at the 1979 Geilo school, reappeared this year as an unsolved riddle. Although the model of one-dimensional links within an infinite cluster has been used with some success to describe experimental data, it is theoretically inconsistent in some situations. Modified models involving "links and blobs" or even a fractal structure which is self-similar under dilation have been proposed but none is yet fully understood nor have they faced the test of experimental vindication.

Competition between interactions leads in spin glasses to frustration and a diversity of almost degenerate ground states. The mean field theory which, until recently, was in a somewhat confused state, predicts a phase transition to a frozen state at low temperatures. However theoretical arguments insist that the lower marginal dimensionality for spin glasses is four and that no transition should exist in our three-dimensional world. Certainly, the results, presented in Geilo, of Monte-Carlo calculations for Ising spins in two dimensions show no indication of a sharp spin-glass transition but favour a model of progressive freezing of the spin configuration. The apparent existence of a transition in many experiments and its similarity to the mean-field predictions thus remains an enigma.

A model of competing interactions much discussed at the meeting was the Anisotropic Next Nearest Neighbour Ising (or ANNNI) model. Although this model cannot be solved rigorously, mean-field and low-temperature expansion techniques have been used to demonstrate a variety of solutions. These include incommensurate and commensurate phases, pinned and unpinned solitons and chaotic metastable states. Close to the ordering transition, where the mean-field theory of the ANNNI model can be solved exactly, an infinite (Cantor) set of commensurate phases occurs and forms an incomplete Devil's staircase. Qualitative features of this model resemble phenomena observed in systems as disparate as thiourea, manganese phosphide and cerium antimonide, all of which were discussed at the Study Institute.

The concept of numerically iterated maps as a method for investigating the structure of incommensurate phases was described and shown to yield a variety of states including regular soliton lattices and randomly pinned, non-interacting solitons in a spatially chaotic regime. A numerical study of the circle map (which describes a system with two competing frequencies) was shown to generate an incomplete Devil's staircase and it was conjectured that the fractal dimension of the latter may be universal for a certain class of maps. Similar mapping techniques were described in detail in lectures on temporal chaos and on the routes by which such chaotic states can be reached.

In the last lecture of the Study Institute participants learnt that a multicritical point may be observed in driven hydrodynamic systems and that non-classical exponents have been observed in one such system. The latter observation, which caused some consternation among theorists, struck a chord among those who had attended the first Geilo school in 1971. Then, the report (by the same group) of a non-classical value of the order-parameter exponent β for the structural transformation of $SrTiO_3$ was greeted with more than a little suspicion!

This short preface cannot, of course, do justice to the multitude of topics covered at the Study Institute and included in these proceedings. Rather we hope to have provided a (subjective) flavour. A strong impression left by this school, is that subjects once thought to be well understood, are not and that one may look forward to a period of activity and clarification.

May, 1983

R.A. Cowley
R. Pynn
A.T. Skjeltorp
H. Thomas
W.P. Wolf

CONTENTS

Multicriticality: A theoretical Introduction 1
M.E. Fisher (Invited Lecturer)

Partial Differential Approximants for Multicritical Singularities: An Overview 7
D.F. Styer

Experimental Studies of Magnetic Tricritical Points: Problems and Progress. 13
W.P. Wolf (Invited Lecturer)

Experimental Studies of Bicritical Points in 3D Antiferromagnets 35
Y. Shapira (Invited Lecturer)

The Experimental Evidence for a Lifshitz Point in MnP 53
Y. Shapira (Invited Lecturer)

Nature of the Smectic-A-Smectic-C Transition near a Nematic-Smectic-A-Smectic-C Multicritical Point 73
C.C. Huang and S.C. Lien

Dynamics near Multicritical Points. 81
V. Dohm (Invited Lecturer)

Critical Dynamics of Sound. 113
K. Fossheim (Invited Lecturer) and J.O. Fossum

Multicritical Phenomena at Structural Phase Transitions . . . 129
K. Fossheim (Invited Lecturer)

Bi- and Tetra-critical Behaviour of Uniaxially Stressed $LaAlO_3$ 143
K.A. Müller (Invited Lecturer), W. Berlinger, J.E. Drumheller and J.G. Bednorz

Fluctuation Driven Multicritical Points 155
A. Aharony (Invited Lecturer) and D. Blankschtein

A New Multicritical Point in A Seven Dimensional Parameter Space 165
S. Galam

Multicritical Points in Incommensurate Systems 171
P. Toledano

Neutron Investigation of Modulated Structures 177
R. Currat (Invited Lecturer)

Nature of The Incommensurate Phase in Rb_2ZnCl_4 201
S.R. Andrews and H. Mashiyama

New Phenomena in Incommensurate Barium Sodium Niobate: Memory and Relaxation Effects 207
G. Errandonéa, A. Litzler, H. Savary, J-C. Tolédano, J. Schnek and J. Aubrée

Critical Fluctuations and Magnetic Ordering in Praseodymium Metal 213
W.G. Stirling and K.A. McEwen

Soft Mode and Central Peak as A Consequence of Competing Interactions 219
P.A. Lindgård

Criticality and Crossover in Structural Phase Transitions. . 221
P.D. Beale

Anomalous Vibrational Absorption Near the Antiferroelectric Phase Transition in Alkali Cyanides 227
I. Vilfan, R. Pirc and F. Lüty

A Plenitude of Commensurate Phase in Simple Models 233
M.E. Fisher (Invited Lecturer)

Commensurability, Chaos and The Devil's Staircase 237
P. Bak (Invited Lecturer) and M.H. Jensen

The Complete Devil's Staircase and Universality of Mode Locking Structures in Discrete Mappings 265
M.H. Jensen, P. Bak and T. Bohr

Low Temperature Analysis of the ANNNI Model in A Field . . . 271
J. Smith and J. Yeomans

Field Behaviour of Spin Models with Competing Interactions 277
S.R. Salinas

Phase Transitions in Crystals with Competing Interactions 283
T. Janssen and J.A. Tjon

Commensurate Melting and Domain Walls in Surface Phases 289
M.E. Fisher (Invited Lecturer)

The Role of Domain Walls in The Critical Behaviour of the Three-Dimensional p-State Asymmetric Clock Model. 293
J. Yeomans

Interfacial Adsorption in Multi-State Models 301
W. Selke

Vortex Pinning in Thin Superfluid Films 307
J.L. McCauley Jr.

Theory of Critical Phenomena in Random Systems 309
A. Aharony (Invited Lecturer)

Critical Behaviour of Disordered Ising Systems: Supression of The Specific Heat Divergence at T_c 329
G. Jug

Disorder, Random Fields and Competing Interactions in Antiferromagnets 333
R.A. Cowley (Invited Lecturer), R.J. Birgeneau, G. Shirane and H. Yoshizawa

The Origin and Influences of The Spin-Glass Problem 363
B.R. Coles (Invited Lecturer)

Phase (?) Transitions in Systems with Spin Glass and Long-Range Order Regimes 373
B.R. Coles (Invited Lecturer)

Time Effects of The Field Cooled Magnetization of A Au(Fe) Spin Glass 379
P. Nordblad, P. Svedlindh, L. Lundgren and O. Beckman

Spin Glasses: Recent Theoretical Developments 383
A.P. Young (Invited Lecturer)

Critical and Multicritical Points, Lines and Surfaces of Spin Glasses in The Presence of Magnetic Fields and Anisotropy 399
D. Sherrington

Some Renormalization Group Results on the Quenched Gauge Field Model of The Spin Glass 405
R.K. Ritala

Frustration, Landau Levels and Superconducting Diamagnetism on Networks 409
G. Toulouse (Invited Lecturer)

Introduction to Chaos: Phenomenon, Structure of Spectra and Diffusion. 423
S. Grossmann (Invited Lecturer) and S. Thomae

Multicritical Behaviour of A Nonequilibrium System 451
T. Riste (Invited Lecturer) and K. Otnes

Participants 461

Index 467

MULTICRITICALITY: A THEORETICAL INTRODUCTION

Michael E. Fisher

Baker Laboratory, Cornell University, Ithaca

New York 14853, U.S.A.

BRIEF OVERVIEW

To understand multicriticality the first step is to understand ordinary critical behavior and, in particular, the theory of scaling. Ferromagnetic criticality provides the simplest example which suffices to define the critical exponents α, β, γ, etc.; the concept of universality classes depending on spatial dimensionality, d, and spin dimensionality, n; the basic nature of scaling and consequent exponent relations; the universality of the scaling functions and amplitude ratios. Study of fluid criticality reveals the importance and the significance of scaling axes and the linear scaling fields, say, $\tilde{t}$ and $\tilde{h}$, which are combinations of the physical fields $t = (T-T_c)/T_c$ and $h = (\mu-\mu_c)/k_BT$ [1,2] : the primary scaling axis is obvious from the orientation of the first order phase boundary; the secondary axis is less obvious but its slope leads, for example, to a $|t|^{1-\alpha}$ singularity in the diameter of the coexistence curve[1]. Then it is important to recognize the existence of <u>nonlinear</u> scaling fields, which can be represented by multiplying $\tilde{t}$ and $\tilde{h}$ by analytic functions of t and h. These enter even when symmetry dictates the linear scaling axes and they lead, for example, to $|t|^{1-\alpha}$ additive terms in the initial susceptibility of a ferromagnet[3]. Similarly, in the case of a lambda or critical line, they lead to the prediction that the specific heat singularity of an <u>antiferromagnet</u> in zero field should match that in $\partial(\chi T)/\partial T$, where χ is the differential susceptibility[4] . Finally, renormalization group theory[5], leads to singular correction factors of the form $(1 + c_1 t^\theta + \ldots)$ with, typically, $\theta \simeq 1/2 < 1$.

A <u>multicritical point</u> may be defined phenomenologically as a point of sudden change of behavior on an otherwise smooth, universal line of critical points : see Ref. 6, which reviews the theory of multicritical transitions and focuses on the spin-flop bicritical point. Thus at a tricritical point the continuous, critical transition abruptly becomes first order[6] : tricriticality is observed in antiferromagnets[7], in helium three-four mixtures[8], in the equilibrium polymerization of sulphur in solvents such as triphenylmethane[9], in binary alloys such as Fe + Aℓ, and,especially, in ternary and quaternary fluid mixtures[10]. A bicritical point, may, similarly, be characterized as the meeting of two separate critical lines corresponding to two, distinct, competing order parameters[6], say $\Psi_{\parallel}$ and $\Psi_{\perp}$. Bicritical points have been observed most clearly in weakly anisotropic antiferromagnets such as $GdA\ell O_3$ and MnF_2 in carefully oriented magnetic fields[11]; but they are also seen under anisotropic mechanical stress in structural phase transitions in perovskites such as $SrTiO_3$ [12]. If one of the ordered phases is spatially modulated, one may have a simple Lifshitz point[13] as recently observed[14] in MnP, or the somewhat analogous NAC (nematic-smectic A - smectic C) multicritical point as seen in liquid crystal mixtures[15]. As other variables are introduced, tetracriticality may be observed in conjunction with bicriticality and degenerate bicritical points arise in systems of high symmetry[6,11(b)].

A zeroth order but informative 'classical' theory of multicriticality is based on a phenomenological expansion of the free energy or, better, a minimization of an effective Landau-Ginzburg-Wilson Hamiltonian. This theory illustrates scaling[6] and predicts correctly the basic number of relevant variables (or "fields") which are four in number for tricriticality but 7 (or 8) for bicriticality. In many cases the overall topology and rough shape of the phase diagram is also given correctly[6]. However, renormalization group theory[6,16-19], exact solutions[20], and series expansion analyses[21,22] are required to determine the borderline dimensionality, $d_>$, below which the classical theory fails, and then to determine critical exponents, particularly the appropriate multicritical crossover exponents, ϕ, to check multicritical universality, to find amplitude ratios, effective exponents (which may crossover very slowly[23]), scaling functions, etc. Scaling alone leads to considerable insight provided the need for determining the correct linear scaling axes is remembered : this is of especial importance in bicriticality[1,6,11]. Scaling can also be used to provide valuable criteria, within the context of a given system or model, for the relevance, marginality, or irrelevance of a given perturbation (dipolar, cubic[6], chiral[24], etc.) which is suspected of causing multicritical crossover to new behavior[25].

For bicritical points the upper borderline dimensionality is found to be $d_> = 4$ and for $n_{\parallel} + n_{\perp} < n^*(d)$, with[6] $n^*(3) \simeq 3.13$, the asymptotic behavior is controlled by standard n-vector criticality[6,17,18] with $\phi \simeq 1.18$ and 1.25 for $(n_{\parallel} + n_{\perp}) = 2$ or 3, respectively[21,22]. (Here $n_{\parallel}$ and $n_{\perp}$ specify the number of components of the two competing order paramaters). A Lifshitz point at which m spatial components $\nabla_\mu \psi$, in the LGW Hamiltonian vanish (i.e. which has m 'soft' directions) has[13] a borderline dimensionality $d_> = d + \frac{1}{2} m$. One sees this from the behavior of the leading (zeroth) order renormalization group recursion relations if one allows for distinct spatial rescaling factors, b_μ and b_λ, for the m 'soft' directions and (d-m) 'hard' directions.

Standard tricritical points lead to a borderline dimensionality[16,19] $d_> = 3$, which indicates classical tricritical exponents for real bulk systems. This prediction accords well with experiments[7]; nevertheless marked discrepancies with the classical predictions for amplitude ratios are observed[20]. These deviations are associated with a <u>marginal variable</u>, z, which may be identified[20] in terms of the range, R_0, of the forces as $z \propto 1/R_0^d$. (When $z \to 0$ one knows rigorously that classical theory becomes exact for all d). In renormalization group theory[16,25] the marginal variable gives rise to complicated logarithmic factors, $(1 + c \ln|t|)^\mu$, with various exponents μ depending on n; but these have <u>not</u> been detected experimentally[7]. Rather one finds[20] the data can be well represented by <u>nonuniversal</u> scaling functions which are, however, parametrized by the <u>single</u> variable z. In the spherical model limit, $n \to \infty$, this conclusion is exact theoretically[20]. (Furthermore for $3 < d < 4$ the parameter z provides a dramatic example of a <u>dangerous irrelevant</u> variable[20]). For realistic values, $n = 1$ and $n = 2$, however, more theoretical effort is required to understand existing and future tricritical experiments !

ACKNOWLEDGEMENTS

The support of the National Science Foundation, in part through the Materials Science Center at Cornell University, and travel, living, and accommodation expenses from the NATO Advanced Study Institute are gratefully acknowledged. Special thanks are due to the Organizing Committee, Roger Pynn, Arne T. Skjeltorp and Gerd Jarrett whose hard work made the Institute physically possible, scientifically rewarding, and in all ways most enjoyable.

REFERENCES

1. M.E. Fisher, Phys. Rev. Lett. 34:1634 (1975).
2. M.E. Fisher, Proc. Nobel Symp. 24, "Collective Properties of Physical Systems", Eds. B. Lundqvist and S. Lundqvist (Academic Press, New York 1974) pp. 16-37.
3. A. Aharony and M.E. Fisher, Phys. Rev. B27:4394 (1983).
4. M.E. Fisher, Phil. Mag. 7:1731 (1962); W.P. Wolf and A.F.G. Wyatt, Phys. Rev. Lett. 13:368 (1964).
5. See e.g. M.E. Fisher, Revs. Mod. Phys. 46:597 (1974).

6. M.E. Fisher, AIP Conf. Proc. No. 24, "Magnetism and Magnetic Materials, 1974" (AIP, 1975), pp. 273-280 : Note Eq. (5.2) should read $\Delta_{\parallel} = \Delta_{\perp} \simeq 1.65$.
7. See e.g. N. Giordano and W.P. Wolf, Phys. Rev. Lett. 35:799 (1975) and the lectures by W.P. Wolf in these Proceedings.
8. See e.g. G. Ahlers and D. Greywall, Phys. Rev. Lett. 29:849 (1972); E.K. Riedel, H. Meyer and R.P. Behringer, J. Low Temp. Phys. 22:369 (1976); etc.
9. See J.C. Wheeler and P.Pfeuty, Phys. Rev. Lett. 46:1409 (1981), J. Chem. Phys. 74:6415 (1981).
10. See e.g. J.C. Lang and B. Widom, Physica 81A:190 (1975).
11. See e.g.(a) H, Rohrer, Phys. Rev. Lett. 34:1638 (1975) and (b) the review lecture by Y. Shapira in these Proceedings.
12. See e.g. S. Stokka and K. Fossheim, Phys. Rev. B25:4896 (1982) and the Lectures by K. Fossheim and by K.A. Müller in these Proceedings.
13. R.M. Hornreich, M. Luban and S. Shtrikman, Phys. Rev. Lett. 35:1678 (1975) and see the review : R.M. Hornreich, J. Magn. Magnetic Matls. 15-18:387 (1980).
14. C.C. Becerra, Y. Shapira, N.F. Oliveira, Jr. and T.S. Chang, Phys. Rev. Lett. 44:1692 (1980); see also the Lectures by Y. Shapira in these Proceedings.
15. By D.L. Johnson and coworkers e.g. D.L. Johnson et al. Phys. Rev. B16:470 (1977); R. de Hoff et al, Phys. Rev. Lett. 47:664 (1981); D. Brisbin, D.L. Johnson, H. Fellner, M.E. Neubert (Kent State Univ. preprint Feb. 1983).
16. E.K. Riedel and F.J. Wegner, Phys. Rev. Lett. 29:349 (1972).
17. M.E. Fisher and D.R. Nelson, Phys. Rev. Lett. 32:1350 (1974).
18. D.R. Nelson, J.M. Kosterlitz and M.E. Fisher, Phys. Rev. Lett. 33:813 (1974); Phys. Rev. B 13:412 (1976); E. Domany, D.R. Nelson and M.E. Fisher, Phys. B15:3493 (1977); etc.
19. D.R. Nelson and M.E. Fisher, Phys. Rev. B 11:1030; 12:263 (1975).
20. See e.g. M.E. Fisher and S. Sarbach, Phys. Rev. Lett. 41:1127 (1978); S. Sarbach and M.E. Fisher, Phys. Rev. B 18:2350 (1978); 20:2797 (1979).
21. P. Pfeuty, D. Jasnow and M.E. Fisher, Phys. Rev. B10:2088 (1974).
22. M.E. Fisher and R.M. Kerr, Phys. Rev. Lett. 39:667 (1977); M.E. Fisher, H. Au-Yang and J.-H. Chen, J. Phys. C 13:L459 (1980). See also the seminar on partial differential approximants by D.F. Styer in these Proceedings.
23. See e.g. P. Seglar and M.E. Fisher, J. Phys. C13:6613 (1980).
24. For chiral crossover in the melting of commensurate adsorbed surface phases see D.A. Huse and M.E. Fisher, Phys. Rev. Lett. 49:793 (1982) and M.E. Fisher below in these Proceedings.

25. See the discussion by D.A. Huse and M.E. Fisher, J. Phys. C 15:L585 (1982) which also illustrates the form ordinary and multicritical scaling must take when logarithmic specific heat singularities are present and when marginal variables and possibly, continuously-variable (non-universal) exponents are involved.
26. M.J. Stephen, E. Abrahams, and J.P. Straley, Phys. Rev. B12:256 (1975); M.J. Stephen, Phys. Rev. B12:1015 (1975).

EDITORS' NOTE

At the Advanced Study Institute, copies of references 2 and 6 from the foregoing list were made available to participants. Reference 6, in particular, contains material which both introduces these Proceedings and sets them in the correct context.

PARTIAL DIFFERENTIAL APPROXIMANTS

FOR MULTICRITICAL SINGULARITIES: AN OVERVIEW

Daniel F. Styer

Baker Laboratory
Cornell University
Ithaca, New York 14853 U.S.A.

ABSTRACT

The partial differential approximant (or PDA) is a two variable generalization of the Dlog Padé approximant which permits effective series analysis of multicritical behavior. This brief review surveys the motivation of PDAs, their applications and their matematical properties. Full literature references are included.

In any reliable approximation scheme, the approximant must be capable of exhibiting the same behavior as is displayed by the function to be approximated. For a thermodynamic function near a multicritical point, this behavior is generally given by an appropriate scaling form: thus in a system described by two variables, x and y, a thermodynamic function f(x,y) near a multicritical point (x_c,y_c) is expected to vary as

$$f(x,y) \approx |\Delta\tilde{x}|^{-\gamma} Z(\Delta\tilde{y}/|\Delta\tilde{x}|^{\phi}) + B(x,y) \quad , \tag{1}$$

where γ and ϕ are possibly nonintegral *multicritical exponents* and B(x,y) is a non-singular (or less singular) *background* function. The *scaling variables* are defined by

$$\Delta\tilde{x} = (x-x_c) - (y-y_c)/e_1 \quad , \quad \Delta\tilde{y} = (y-y_c) - e_2(x-x_c) \quad , \tag{2}$$

while the *scaling function* Z(z) is a single variable function which may itself display singularities. Effective approximants for multicritical functions such as f(x,y) must, first, be capable of displaying scaling behavior, and, second, be readily calculable from the information known concerning the function. This information is, typically, a finite number of terms $f_{i,i'}$ in the power series expansion

$$f(x,y) = \sum_{i,i'=0}^{\infty} f_{i,i'} x^{i} y^{i'} . \tag{3}$$

A third criterion, desirable but not essential, is that the approximant should permit easy estimation of the important multicritical parameters, namely the singular point (x_c, y_c), the exponents γ and ϕ, and the scaling axis slopes e_1 and e_2, without requiring computation of the approximant in the less interesting regions away from the multicritical point.

All three of these requirements are met by the method of *partial differential approximants,* proposed by Fisher (1977 a,b) and elaborated by Fisher and Au-Yang (1979). To understand this method, note that in general, all solutions of the partial differential equation

$$U(x,y) + P(x,y)F(x,y) = Q(x,y)\frac{\partial F(x,y)}{\partial x} + R(x,y)\frac{\partial F(x,y)}{\partial y} , \tag{4}$$

where U,P.Q and R are smooth functions, will near a common zero (x_c, y_c) of Q and R, take on precisely the scaling form (1)! In fact the exponents γ and ϕ, the slopes e_1 and e_2, and the background $B(x_c, y_c)$ of these solutions are dictated solely by the functions ^{c},P,Q and R: only the "function of integration" Z(z) depends upon the boundary conditions imposed on (4). [Explicit formulae for these multicritical parameters are given in the review by Fisher and Chen (1982) .]

Although this observation contains the germ of a useful approximation method, we must still discover how to select the so-called *defining functions* U,P,Q and R. Ideally, these would be chosen so that the power series solution of (4) reproduces the known coefficients $f_{i,i'}$ in the power series for the function, f(x,y). [Note that appropriate boundary conditions must be imposed on (4): see the general discussion by Fisher and Styer (1982) § 3, and the specific example of Fisher and Chen (1982) § 11.] Although this process is conceptually appealing, it is difficult to reduce to an algorithm. In practical calculations, the defining functions are instead taken as two-variable *polynomials* of selected "shape",* the coefficients of which are calculated by demanding that, when F(x,y) is replaced by f(x,y) in (4), both sides of the equation match to as high an order as

*The "order" or "shape" of a two-variable polynomial is described by a *label set* of integer pairs, such as $\underset{\sim}{J} = \{(j,j')\}$. A polynomial $U_{\underset{\sim}{J}}(x,y)$ of shape $\underset{\sim}{J}$ is given by $\Sigma u_{j,j'} x^{j} y^{j'}$ with $(j,j') \in \underset{\sim}{J}$.

possible. This demand leads merely to a set of linear equations for the coefficients of the defining polynomials in terms of the known coefficients $f_{i,i'}$. When this easily implemented, practical algorithm produces an approximant whose power series does indeed match the known coefficients $f_{i,i'}$, then the approximant is termed *faithful*. Conditions under which approximants are faithful, as well as a complete definition and a discussion of the existence and uniqueness of PDAs, have been presented by Fisher and Styer (1982).

In summary, one calculates a PDA by first selecting polynomial shapes $\underset{\sim}{J}, \underset{\sim}{L}, \underset{\sim}{M}$ and $\underset{\sim}{N}$ and then solving simultaneous linear equations to find the defining polynomials $U_{\underset{\sim}{J}}, P_{\underset{\sim}{L}}, Q_{\underset{\sim}{M}}$ and $R_{\underset{\sim}{N}}$. Once the polynomials are known, one may calculate the approximant F(x,y) by solving the *defining equation* (4) subject to some appropriate boundary condition; however this is frequently unnecessary because the most important multicritical parameters can be deduced directly from the defining polynomials alone without integrating the defining equation.

It is clear from the above that a PDA is capable, in principle, of approximating multicritical functions. This promise has been confirmed by several applications to significant problems. For example, Stilck and Salinas (1981) *tested* the PDA approach by constructing approximants to known functions and comparing the known and predicted multicritical behavior, finding rather good agreement. They also *applied* the method to dimensional crossover (from d=2 to d=3) in the Ising model and confirmed a scaling hypothesis for the tetracritical point. Earlier, Fisher and Kerr (1977) had studied the classical anisotropic Heisenberg model, which displays bicritical crossover from Ising-like to XY-like behavior as coupling constants in the Hamiltonian are varied. Their results for the bicritical location and exponents were fully consistent with those of a single-variable analysis of the same model (Pfeuty, Jasnow and Fisher 1974). However the PDA method can go beyond such results and actually estimate the singular scaling function Z(z), which is essentially inaccessible to single-variable methods: this extension, which requires integrating the defining equation (4), was performed for the anisotropic Heisenberg model by Fisher, Chen and Au-Yang (1980) (see also Fisher and Chen 1982).

Finally, note that the scaling form (1) describes *multicritical behavior* when $\phi > 0$, but it describes *corrections to scaling* when $\phi < 0$, and thus PDAs may be used to study ordinary critical phenomena. This was done by Chen, Fisher and Nickel (1982), who studied two models which interpolate smoothly between the Ising and Gaussian models. For the three-dimensional Ising model they estimated the susceptibility exponent as $\gamma = 1.2385 \pm .0015$, and the corrections-to-scaling exponent as $\theta = -\phi = 0.54 \pm .05$. These results agree with, for example, Zinn-Justin's (1981) single variable analysis, but again the PDA technique has the potential to probe more deeply and investigate the scaling function as well as the exponents.

Because partial differential approximants have been successful in these practical applications, the study of their mathematical properties has become an interesting problem. An important result in this area would be a convergence theorem, but a proof promises to be difficult and has not yet been attempted. However, valuable theorems concerning the *invariance* of approximants under a change of variable have been established (Styer and Fisher 1983, Styer 1983). Because invariant approximants are believed to be superior to non-invariant approximants, these results have direct applications to the practical problems encountered when approximating real functions.

In conclusion, partial differential approximants have been shown to be effective tools in the theoretical study of tetracritical and bicritical points and of corrections-to-scaling. One may hope that they will be equally successful in investigating other aspects of crossover and multicriticality, such as tricritical and Lifshitz points.

ACKNOWLEDGMENTS

I am indebted to Professor Michael E. Fisher for help and advice concerning all aspects of the work reviewed here. The Graduate School of Cornell University and NATO generously supported my attendance at the Advanced Study Institute. Finally I am grateful for the support of the National Science Foundation through the Materials Science Center at Cornell University.

REFERENCES

Chen, J.-H., Fisher, M.E., and Nickel, B.G., 1982, Phys. Rev. Lett., 48: 630-634.

Fisher, M.E. 1977a, in: "Statistical Mechanics and Statistical Methods in Theory and Application", U. Landman, ed., Plenum Press, New York.

Fisher, M.E. 1977b, Physica B, 86-88: 590-592.

Fisher, M.E., and Au-Yang, H., 1979, J. Phys. A, 12: 1677-1692; 13: 1517.

Fisher, M.E. and Chen, J.-H., 1982 in: "Phase Transitions: Cargèse 1980", M. Lévy, J.C. Le Guillou, and J. Zinn-Justin, eds., Plenum Press, New York.

Fisher, M.E., Chen, J.-H., and Au-Yang, H., 1980, J. Phys. C, 13: L459-464.
Fisher, M.E., and Kerr, R.M., 1977, Phys. Rev. Lett., 39: 667-670.
Fisher, M.E., and Styer, D.F., 1982, Proc. Roy. Soc. A, 384: 259-287.
Pfeuty, P., Jasnow, D., and Fisher, M.E., 1974, Phys. Rev. B, 10: 2088-2112.
Stilck, J.F., and Salinas, S.R., 1981, J. Phys. A, 14: 2027-2046.
Styer, D.F., 1983, submitted to Proc. Roy. Soc. A.
Styer, D.F., and Fisher, M.E., 1983, Proc. Roy. Soc. A, [in press].
Zinn-Justin, J., 1981, J. Physique, 42: 783-792.

EXPERIMENTAL STUDIES OF MAGNETIC TRICRITICAL POINTS: PROBLEMS AND PROGRESS

Werner P. Wolf

Yale University
Becton Center, P.O. Box 2157
New Haven, CT 06520 USA

INTRODUCTION

Experimental and theoretical studies of tricritical phenomena have constituted an active field of research over the past decade. Early experiments on $^3He-^4He$ mixtures[1] and metamagnets[2-5] were interpreted by Griffiths[6] in terms of a new type of critical phenomenon involving the simultaneous criticality of two different densities, and a number of theoretical predictions were subsequently made concerning the nature of the singularities which might be expected at such a point.[7-21]

Experimental efforts to verify these predictions have encountered a number of difficulties, several of which are intrinsic to the nature of critical phenomena in general, but exacerbated by the strong singularity which characterizes tricritical behavior.

In this paper we shall review the nature of these difficulties and the progress which has been made to overcome them. We shall concentrate on the work which has been done on magnetic systems since these have received the most attention and also because magnetic systems afford the most direct access to the thermodynamic fields which describe the phase diagrams. Summaries of the experimental work on magnetic systems prior to 1977 have been published in references 22 and 23. The only other extensively studied tricritical system is the mixture of $^3He-^4He$ and an excellent review of this work has been given in reference 24.

EXPERIMENTAL PROBLEMS

Tricritical Exponents: Possible Confusions

The thermodynamic singularities which occur at or near a tricritical point are described, as usual, by characteristic exponents whose asymptotic values are predicted by theory.[7,9,11] For the experimentalist, there are three potential sources of confusion in comparing measurements with the theory: 1) one must be careful to distinguish between the ordering density and the nonordering density [M_s (or $M^\dagger$) the staggered magnetization and M the uniform magnetization in the case of a metamagnet]; 2) different paths in the field-temperature plane will have different exponents. In particular, paths along the antiferromagnetic-paramagnetic phase boundary will have different exponents from all other paths, which will have the same exponents irrespective of path direction; 3) different authors have used different notations to designate the same exponent at various times, and there is at present no single accepted notation. Reviews of the various notations have been given in references 10, 23, and 24. Some of the exponents commonly used in magnetic studies are summarized in Fig. 1.

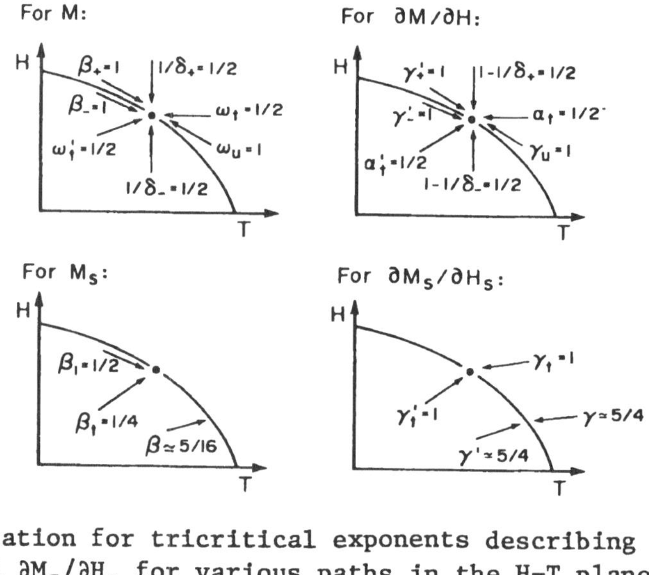

Fig. 1. Notation for tricritical exponents describing M, $\partial M/\partial H$, Ms, and $\partial M_s/\partial H_s$ for various paths in the H-T plane. Some equivalent definitions include for M: $(\beta_+ \equiv \beta_u^+ \equiv \beta_0 \equiv \omega_u') = (\beta_- \equiv \beta_u^-) = \omega_u = 1$ for paths parallel to the phase boundary and $1/\delta_+ = 1/\delta_- \equiv 1/\delta_{nt} \equiv 1/\delta_0 = \beta_{nt} = (\omega_t \equiv \omega_t) = \omega_t' = 1/2$ for paths not parallel to the phase boundary. For $\partial M/\partial H$: $(\gamma_+' \equiv \lambda'^+_u) = (\gamma_-' \ \lambda'^-_u) = (\gamma_u \equiv \gamma_0 \equiv \lambda_u) = 1$ parallel to the phase boundary and $(\alpha_t \equiv \lambda_t) = (\alpha_t' \equiv \lambda_t') = (1-1/\delta_+) = (1-1/\delta_-) = 1/2$ for paths not parallel to the phase boundary. For M_s, $\beta_1 \equiv \beta_{tr}$ (see Refs. 10 and 24).

Theory also predicts that tricritical points should have logarithmic correction terms modifying the usual exponent singularities,[14,20] but so far there has been no convincing experimental evidence for any such effects.[25,26]

Resolution Limits

It is not difficult to understand why it should be hard to detect logarithmic correction terms or, indeed, determine the tricritical exponents themselves with any accuracy. In common with other critical points, one would expect true asymptotic behavior only rather close to the tricritical point, where finite experimental resolution limits the range over which measurements can be made. The theory generally makes no predictions for the range over which asymptotic behavior is to be expected and the experimentalist is thus faced with the problem of deciding which part of the observed behavior is affected by neither the deviations due to higher order correction terms nor those due to finite experimental resolution limits. This is a problem which affects all kinds of critical point studies but there are several factors which make it particularly severe in the case of tricritical points. These include the following:

a) Tricritical points involve first order transitions. These are intrinsically susceptible to hysteresis effects and the formation of domains in the regions of coexistence. Consequently, experimental results can be quite sensitive to such factors as the speed and the precise path along which the phase transition is approached. Recent work has shown[27,28] that sweep speeds as low as 0.3 Oe/sec (3×10^{-5} T/sec) may still be too fast to give true equilibrium behavior (see Fig. 2). It seems clear that great care must be exercised to eliminate such effects.

b) First order transitions are also very susceptible to "rounding" effects which can arise from various sample imperfections, deviations from stoichiometry, from strains or from small cracks or other irregularities. All of these tend to smear the actual discontinuity which corresponds to a first order transition and make it hard to distinguish a small discontinuity just below the tricritical point from a second order inflection just above the tricritical point.

Some indication of these difficulties is illustrated by the values which have been reported for the "tricritical temperature" of $FeCl_2$, one of the two metamagnetic materials on which most of the tricritical point studies have been reported.[2,25,30-33,35] Table 1 summarizes the results which have been quoted in the literature and it is clear that there are serious difficulties, at least for this particular material. In part, this may be due to the special properties of the Fe^{2+} ion which can readily be oxidized to Fe^{3+}, but the problem of sample perfection is clearly one to which more attention must be given in the future.

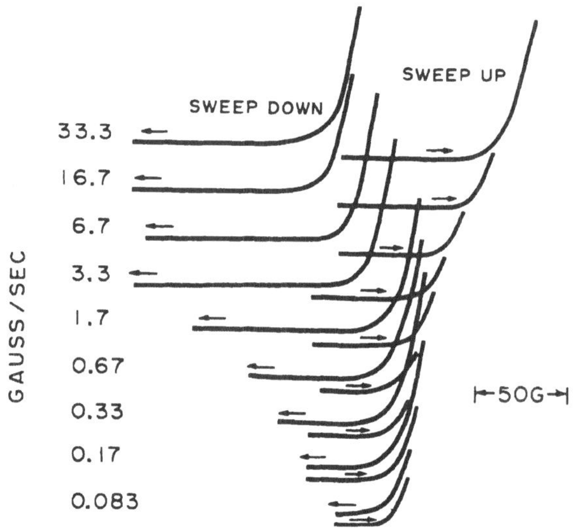

Fig. 2. Example of sweep speed dependent hysteresis effects at a first order metamagnetic transition. The curves show the onset of light scattering at the lower phase boundary of $Dy_3Al_5O_{12}$ at T=1.4K with $H||[110]$. Note the measurable hysteresis for sweep speeds as slow as 0.33 Gauss/sec. (Ref. 28).

Table 1. Apparent Tricritical Temperature of $FeCl_2$

Experimental Study	"T_t" (K)
Jacobs and Lawrence[2] (1967)	$\approx$ 20.4
Vettier et al.[30] (1973)	20.3 ± 0.1
Griffin et al.[31] (1974)	21.5 ± 0.1
Birgeneau et al.[32] (1974)	21.15 ± 0.1
Griffin and Schnatterly[33] (1974)	20.79 ± 0.11
Dillon et al.[35] (1978)	21.60 ± 0.02
Salamon and Shang[25] (1980)	20.54

c) In magnetic materials, there is an additional problem arising from the long range nature of magnetic dipole interactions. As a result, magnetic properties become shape dependent and an appropriate demagnetizing correction must be applied before the intrinsic properties can be extracted from experimental data. These corrections can be quite large. Figure 3 shows the magnetization of $Dy_3Al_5O_{12}$ (DAG), the second of two extensively studied metamagnets, as a function of both the applied field (H_a) and the internal field (H_i), obtained from H_a using the standard demagnetizing correction

$$H_i = H_a - NM \quad , \qquad (1)$$

where M is the magnetization (here measured using the Faraday rotation) and N is a demagnetizing factor estimated from the sample shape. It can be seen that H_i can differ from H_a by more than a factor of two, so that extreme care must be applied in the analysis of high resolution experiments.

The problem is further complicated by the fact that Eq. (1) applies strictly only for uniform ellipsoids, in that N becomes a function of the magnetic equation of state and position for all other sample shapes. Ignoring this fact and assuming N to be a constant can lead to significant errors.[28,36] Figure 4 shows, for example, some magnetization results for a disc shaped sample of $Dy_5Al_5O_{12}$, plotted as a function of "H_i," where "H_i" is given by Eq. (1) with a constant N, chosen to make the initial rise at the phase transition vertical. It seems clear that the resulting curve (a) is unphysical and that the observed variation must result from nonlinear, inhomogeneous demagnetizing effects. This is confirmed by a simple magnetostatic model calculation[28] (curve b) based on the assumed equation of state shown as the broken curve (c) in Fig. 4.

This and similar calculations show that demagnetizing effects in nonellipsoidal samples can be quite significant, especially in regions where $\partial M/\partial H_i$ is large. Unfortunately this is just in the vicinity of both first and second order phase transitions and, thus, of importance in the analysis of tricritical point behavior. The simple solution is, of course, to use ellipsoidal (or spherical) samples but, in practice this is sometimes not possible either because the material cannot be shaped, as in the case of $FeCl_2$, or because the experimental technique calls for planar surfaces, as in the case of magneto-optical methods. The alternative approach is then to use very thin samples which may approximate reasonably well to an infinite plate with a uniform demagnetizing factor. In practice, this limit is not always achieved and significant errors may then result.

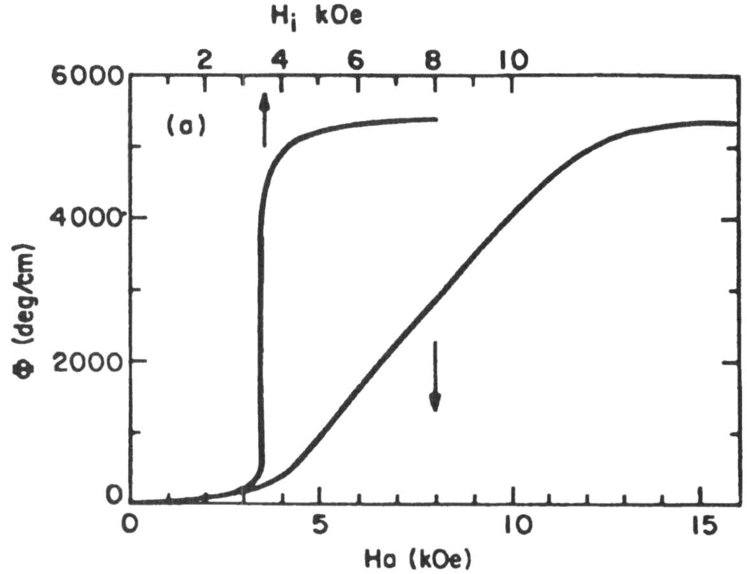

Fig. 3. Example of demagnetizing effect on the measurement of magnetization, using Faraday rotation. Lower curve shows rotation, Φ, for a thin plate of $Dy_3Al_5O_{12}$, magnetized perpendicular to the plate at T=1.4K, as a function of the applied field H_a. Upper curve shows the same data plotted as a function of the internal field $H_i=H_a-4\pi M$, where $M=c\Phi$ and c is a normalization constant. Note that the two field scales are the same, so that the demagnetizing correction is more than 50% of H_a. (J. F. Dillon Jr., private communication).

d) Demagnetizing effects can also affect measurements in another way. If a sample contains any cracks, voids, scratches, or other irregularities, these will distort the internal field throughout the entire sample, in a way which is quite difficult to calculate. A simple estimate of the effect of spherical cavity in a linear medium, for which $M = \chi H$, shows that the change of internal field can be quite large. The change of field at $\vec{r}$ due to a cavity field of radius a is given by

$$\Delta\vec{H} = -[\frac{\vec{\mu}}{r^3} - \frac{3(\vec{\mu}\cdot\vec{r})}{r^5}\vec{r}] \quad , \tag{2}$$

where for large χ^{28} $\mu \sim a^3H$. Thus, $\Delta H/H \sim (a/r)^3$, so that close to the cavity $\Delta H/H \sim 1$! In high resolution experiments, where fields are measured to parts in 10^4 or better, such effects can clearly be very serious, even if there are only few cavities or cracks.

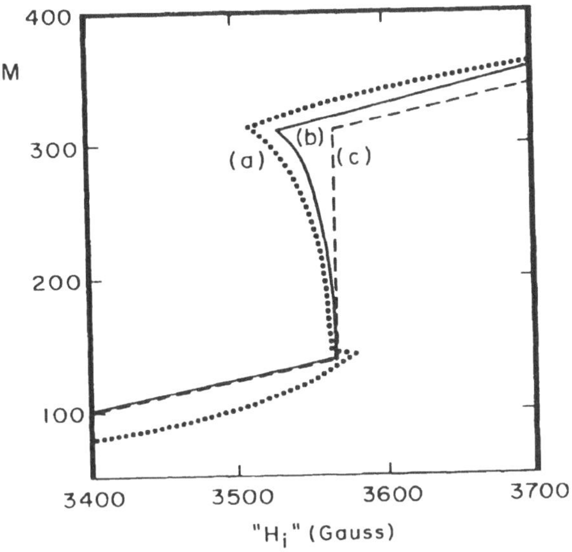

Fig. 4. Effect of inhomogeneous demagnetizing fields on the analysis of high resolution magnetization measurements. The points show experimental data (curve a) for a thin (1:6) disc of $Dy_3Al_5O_{12}$ at T=1.56K and H||[110], plotted as a function of "H_i"=H_a-N_eM, with a constant demagnetizing factor N_e=10.64, chosen so as to make the initial slope at the phase transition vertical. The solid curve (b) is the result of a magnetostatic model calculation based on an assumed equation of state, shown as curve c. (After Refs. 36 and 28).

There is, at present, no direct evdience that perturbations of this kind are responsible for any of the unexplained results which have been observed (see below), but certainly seems quite likely. Further studies on sample perfection are clearly needed.

e) Even in a perfect sample there will be problems with inhomogeneous internal fields whenever there are domains, as there will be at all first order transitions. Domains in metamagnets have been observed optically in both $FeCl_2$[37] and $Dy_3Al_5O_{12}$[38] and they have been found to be typically in the 10-100 μm size range. In macroscopic measurements, the effect of domains will be averaged to a large extent, but it is clear that the contributions from domain walls will not be completely negligible, especially close to phase boundaries. In experiments using more localized probes, such as laser[33,35] or neutron beams[32] there may be additional complications due to the fact that only a few domains contribute to the observed signal, so that the averaging may be incomplete.

In any case, domains constitute an unwelcome but inevitable complication which accompany all magnetic first order transitions and it seems clear that they may and do affect high resolution measurements near tricritical points.

Data Analysis

It is evident that experimental data obtained in the vicinity of multicritical points must be treated with even more care than usual critical point measurements, which are also known to be susceptible to serious systematic errors. So far, there have been relatively few detailed studies of tricritical points (see below) and none of these have employed systematic data analyses such as varying the fitting ranges, calculating χ^2 residues or checking several samples.[39] The present state of progress must, therefore, be regarded as strictly preliminary, even though a number of interesting comparisons between theory and experiment have been reported. There is no doubt that more detailed experiments are still needed and one may hope that improvements in sample preparation, measuring techniques, and data analysis will provide critical tests for such predictions as logarithmic corrections and higher order corrections to scaling.

EXPERIMENTAL RESULTS

With the above limitations in mind, it is nevertheless of some interest to examine the results which have been obtained so far and to compare the tricritical exponents which have been estimated with the predictions of the theory. The most extensive experiments have been made on three magnetic systems: $FeCl_2$,[2,25,26,29-35,37] $Dy_3Al_5O_{12}$ (DAG),[3,4,17,40-45] $CsCoCl_3 \cdot 2D_2O$,[46] and on liquid ^3He-^4He mixtures.[1,24] There have also been a few computer model studies using Monte Carlo methods.[47-49]

Ferrous Chloride: $FeCl_2$

This is frequently cited as the protypical metamagnet, with layers of ferromagnetically aligned spins (S = 1) coupled antiferromagnetically. The crystal structure is rhombohedral ($R\bar{3}m$) and there is a strong crystal field anisotropy favoring the crystal c-axis. The Néel temperature is about 23.6 K and the field required to destroy antiferromagnetism[2] is only 10.6 kG (1.06 T) at T = 0 K. Unfortunately, the layered crystal structure also leads to highly anisotropic mechanical properties, which make it impossible to shape crystals other than as thin plates which readily cleave into even thinner flakes. Moreover, the material is extremely hygroscopic and unprotected samples are rapidly decomposed by moisture in the air. Also, as mentioned previously, the Fe^{2+} ions can be partially oxidized to Fe^{3+}, causing a random disruption of the lattice which may affect the critical properties significantly.

For all of these reasons it is difficult to obtain reproducible high resolution data for $FeCl_2$, but a number of quite successful experiments have been reported. Several of these have used optical techniques to determine the phase boundaries[29,31,33,35] and results of the most recent and most accurate of these,[35] are shown as the dots in Fig. 5. For temperatures below the tricritical point, light scattering was used to detect the presence of domains in the mixed phase region corresponding to the first order transition, and the Faraday rotation (Φ) was used to measure the magnetization along the phase boundaries. Above the tricritical temperature the Faraday effect alone was used to locate the second order phase boundary.

The three exponents characterizing the approach of the magnetization to its tricritical value, β_+, β_-, and ω_u, are shown in Table 2. Also shown is the exponent β_u which describes the vanishing of the magnetization discontinuity as $T \to T_t$ from below. Neglecting logarithmic correction factors, all of these exponents are predicted to have the classical value of 1, and it can be seen that the experiments are quite consistent with this prediction. The only significant discrepancy occurs for ω_u corresponding to $T > T_t$ where the phase boundary is particularly hard to locate. Figure 5 also shows some earlier optical measurements[33] which, apart from some small shifts and changes of scale, are quite consistent with the later data.

However, there is a third series of experiments, using neutron scattering to locate the phase boundaries,[32] which are not at first sight consistent with the optical measurements. The difference is significant only very close to the tricritical point and it arises, basically, because the neutron experiments used an independent method (a change in the critical scattering intensity) for determining T_t. With this as a constraint, one is forced to conclude that the lower phase boundary is strongly curved upwards, as shown in Fig. 5. The corresponding exponent is then found to be close to 0.3. The upper phase boundary is essentially consistent with the optical measurements and a value of β_+ close to 1.

This marked difference has been explained by Birgeneau and Berker[26] in terms of a crossover from tricritical to critical behavior, resulting from a higher impurity content of Fe^{3+} in the neutron sample. They postulated that this would destroy the antiferromagnetic long range order in the presence of a field, so that the observed first order transition ends not at a tricritical point but, in fact, at an ordinary critical point. An analogous effect had previously been observed in $Dy_3Al_5O_{12}$ (DAG) for fields along [111], where the applied field induces a staggered magnetic field and thus wipes out the second order phase boundary (see below).

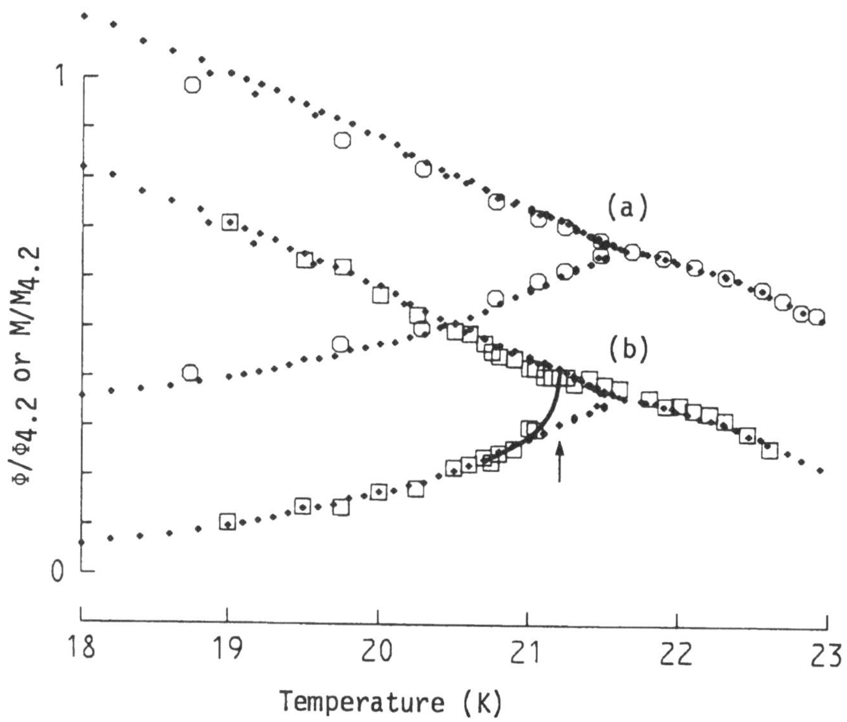

Fig. 5. Phase diagram of $FeCl_2$ in the M-T plane. The results of optical measurements[35,33] are shown as • and ○. The neutron results[32] are shown as □ . The three sets of data have been subjected to small shifts and changes of scale to facilitate comparisons. The arrow indicates the location of T_t inferred from the critical scattering. The scale on the ordinate is the magnetization M (or Faraday rotation Φ) relative to the value at T=0K.

While this explanation is not unreasonable, it should be noted that there is also another possible explanation which could be relevant in this case. The analysis of the neutron scattering experiment rests essentially on a single determination of the tricritical temperature based on the temperature dependence of the critical scattering along the observed phase boundary. It is hard to see how such a measurement could lead to a significantly incorrect determination of the tricritical point but, it must be pointed out, there is no theory for the variation of critical neutron scattering in this region and an intuitive analysis could, perhaps, be wrong.

Table 2. Tricritical Exponents: Theory and Experiment.

Exponent[10]	Theory[7,11]	DAG[45a,17]	$FeCl_2$[35,32]	$CsCoCl_3 \cdot 2D_2O$[46]	^3He-^4He[24]
β_+	1	∿ 1	1.00 ± 0.02	0.7 ± 0.3	
β_-	1	∿ 1	1.00 ± 0.07	0.7 ± 0.3	
β_u	1	0.98 ± 0.05	1.00 ± 0.08	0.7 ± 0.2	1.00 ± 0.05
ω_u	1	∿ 1	1.29 ± 0.16		∿ 1
γ_u	1	1.01 ± 0.07			1.05 ± 0.07
δ_+	2	2.12 ± 0.24		2.9 ± 0.4	2.00 ± 0.08
δ_-	2	2.14 ± 0.26			
ϕ	2	1.95 ± 0.11		2.0 ± 0.2	2.08 ± 0.06[17]
β_1	1/2		0.19 ± 0.02*	0.30 ± 0.10	
β_t	1/4			0.15 ± 0.02	

*This value is subject to a systematic uncertainty since it was derived from the experiment[32] which also gave the anomalous result for β_-.

In any case, it would seem clear that the detailed behavior of $FeCl_2$ near the tricritical point still needs further experiments, specifically to clarify the role of impurities and the details of neutron scattering in this region. In this connection it should be noted that there is one more recent experiment on $FeCl_2$, using light scattering and specific heat measurements, which claims to show curvature of the phase boundaries arising from logarithmic corrections.[25] However, as has been pointed out by Birgeneau and Berker,[26] the samples used in these experiments showed evidence of even larger amounts of Fe^{3+} impurity, so that the conclusions must be treated with caution.

Dysprosium Aluminum Garnet (DAG): $Dy_3Al_5O_{12}$

This is a very different metamagnetic material from $FeCl_2$. It is cubic (Ia3d; O_h^{10}) with excellent chemical and mechanical properties. It orders at about 2.5 K and the field to destroy the antiferromagnetic state is about 4 kG (0.4 T) at T = 0 K. All of this makes it very attractive for detailed studies of field induced phase transitions[41] but there are a number of complications. First, the cubic structure implies three pairs of mutually perpendicular sublattices which will react differently to fields applied in different directions. For this reason, almost all of the early experiments were performed with $H||[111]$, making equal angles with the three pairs of sublattices.

A series of ever more detailed experiments[3,4,40,41,43,45] showed clearly that the field induced phase transition is first order at low temperatures but that it ends not in a tricritical point, but in an ordinary critical point, above which there is no phase transition in the presence of a field.

An explanation for this rather unexpected result was produced by Blume et al.[44] who pointed out that the particular symmetry of the garnet structure allows terms in the free energy which couple the antiferromagnetic order parameter (the staggered magnetization M_s) and the applied field through a term of the form $aM_sH_xH_yH_z$ where H_x, H_y, and H_z are the components along the cubic axes. Such a term has exactly the same effect as that due to a staggered field, $-M_sH_s$, so that in any finite field there is always an induced nonzero M_s and no second order phase boundary where $M_s \to 0$.

The critical point which ends the first order phase boundary thus corresponds to a "wing" critical point and one would expect the usual exponent $\beta \approx 0.32$ for an Ising model. The experiment[45d] gives $\beta = 0.30 \pm 0.02$.

The theory of Blume et al. also predicts, unambiguously, that the "induced staggered field" effect should vanish for applied field orientations for which one (or more) of the components vanish, so that one would then expect true tricritical behavior.†

Magnetization measurements with H||[110], using a novel hysteresis method to locate the phase boundaries for $T < T_t$, confirmed this expectation.[45a] Figure 6 shows the phase diagram in the M-T which is clearly consistent with a linear approach of all three phase boundaries as T_t is approached. The fitted exponent describing the approach of the first order discontinuity to zero is $\beta_u = 0.98 \pm 0.05$. Figure 6 also shows that the upper phase boundary is kinked at T_t, in contrast to the prediction of mean field theory but in agreement with various model calculations.[13,15,18,20]

Measurements have also been made of the magnetization along the tricritical isotherm and of the susceptibility along the path $M = M_t$, $T > T_t$. The corresponding exponents are $\delta_\pm$ and γ_u and the results are shown in Table 2. It can be seen that the agreement with values based on renormalization group theory is very good.

Table 2 also shows the experimental and theoretical values for the crossover exponent ϕ. This is difficult to measure directly since the theoretical value is $\phi = 2$, which is the same as the nonsingular second order term describing the shape of the H-T phase boundary near T_t. Giordano has proposed[17] an ingeneous method for separating the singular and nonsingular contributions and the experimental value shown in Table 2 is based on his method of analysis. It can be seen that the agreement is very good.

There have been two attempts to test scaling in DAG. An earlier analysis[42] was not successful because it failed to take into account the effects of the induced staggered field, which had not been recognized at that time. Some results of the later analysis are shown in Fig. 7, in which the variables are the reduced magnetization $m = (M - M_t)/M_t$ and the scaling field $g = h - pt$, where p is the slope of the H-T phase boundary at T_t and h and t are respectively $(H - H_t)/H_t$ and $(T_t - T)/T_t$.

It would appear that the data scale quite well, using the classical exponents without any logarithmic correction factors. However, it should be noted that the regions of field and temperature for

---

†Fisher and Saarbach[18] have pointed out that there are actually two types of "staggered fields" which couple to the order parameters. The one which is usually ignored, which they denote as H_3, must also be zero for a true tricritical point. It is not completely clear whether the condition $H_xH_yH_z = 0$ also ensures that the induced $H_3 = 0$, but it seems reasonable and there is no experimental evidence to the contrary. If $H_3 \neq 0$, the first order phase boundary would terminate in a critical end point and not a tricritical point.

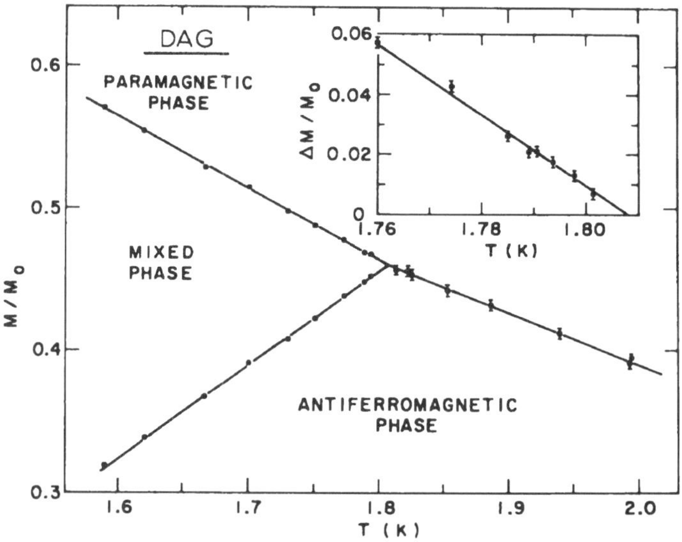

Fig. 6. Phase diagram of $Dy_3Al_5O_{12}$ (DAG) in the M–T plane for fields in the [110] direction. The inset shows the variation of the first order magnetization discontinuity ΔM as $T \to T_t$. The corresponding tricritical experiment is $\beta_u = 0.98 \pm 0.05$. (After Ref. 45a).

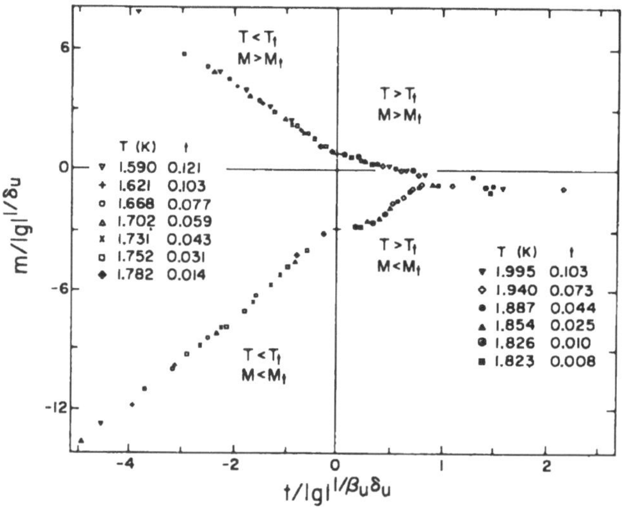

Fig. 7. Tricritical scaling plot for $Dy_3Al_5O_{12}$ (DAG) with H||[110]. The scaling variables m and g are defined in the text. The exponents were fixed to have their theoretical values $\beta_u = 1$ and $\delta_u = 2$. [Using the best experimental values ($\beta_u = 0.98$ and $\delta_u = 2.13$) would give essentially the same results.] Note the change of scale for $t > 0$ ($T > T_t$). (After Ref. 45c).

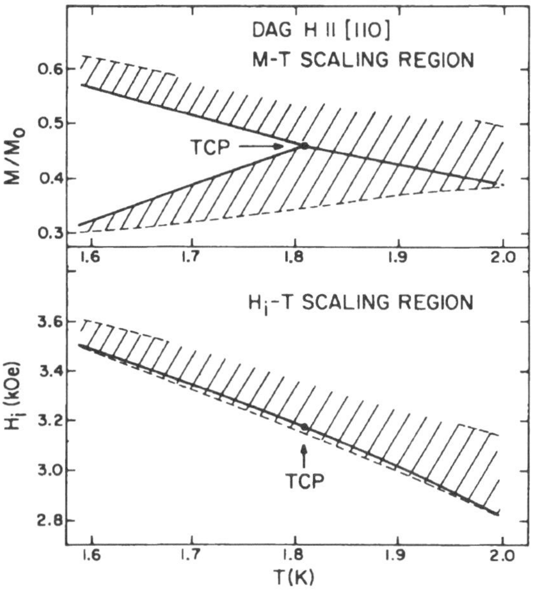

Fig. 8. Tricritical scaling regions for DAG with H||[110] in the M-T and H_i-T planes. The shaded regions show the ranges where scaling was found within the experimental uncertainties. The broken lines indicate where the scaling limits were reached; where there are no broken lines, the data did not extend to the edge of the scaling region. Note the very narrow region in the H_i-T plane for $H<H_c(T)$. (After Ref. 45c).

which the scaling holds are really quite restricted, as can be seen in Fig. 8. In particular for $H < H_t$, the scaling region is quite limited, although it should be remarked that the region appears much larger when plotted in the corresponding M-T plane. This emphasizes again the strong divergence of the singularity at the tricritical point. Attempts were made[27] to enlarge the scaling region by choosing various linear combinations of h and t for the second scaling axis, in place of the variable t, but no significant improvement could be obtained. It would appear that nonlinear scaling fields[50] are needed to describe the equation of state over a wider region.

All of the tricritical point measurements which have been reported for DAG relate to the magnetization and, so far, there have been none for the order parameter, M_s.

Earlier neutron scattering experiments around the critical point[51] and also at very low temperatures[52] showed that quantitative measurements are hard in DAG bacause Dy has a rather high absorption cross section and also because the high degree of perfection of the crystals may lead to significant extinction effects. However, it would seem well worthwhile to attempt to overcome these since the variation of M_s is an essential component in the complete description of the tricritical point.

Cesium Cobalt Chloride: $CsCoCl_3 \cdot 2D_2O$

This is a four sublattice metamagnet with a somewhat unusual structure. Strongly coupled linear chains of Co^{2+} ions are antiferromagnetically aligned, with a small canting angle which gives each chain a net moment. Neighboring chains are weakly coupled also antiferromagnetically and this coupling can be broken by an applied field. The Néel temperature is about 3.4 K and the critical field at T = 0 K is 2.85 kG (0.285 T). The magnetic space group was reported to be[46] $P_{2b}cca'$.

Measurements in the vicinity of the tricritical point were made using neutron scattering[46] and the variations of both M and M_s were studied as a function of H and T. The results for the corresponding exponents are summarized in Table 2. It can be seen that the agreement with theory is generally not very good, even allowing for the relatively large error limits quoted. The reason for this is not clear at this time. It has been suggested[25] that the discrepancies might be due to logarithmic correction factors, but systematic experimental errors cannot be ruled out. In particular, it is difficult to shape samples of this material, so that demagnetizing corrections are hard to estimate. Also, it is possible that disorder due to incomplete deuteration of the water of crystallization may introduce random staggered field effects which could be significant.[26] In any case, it is clear that the tricritical behavior of this material is not yet understood completely.

In spite of these quantitative difficulties, there is one aspect of the results for $CsCoCl_3 \cdot 2D_2O$ which is particularly interesting and which has not been studied for either of the two more extensively investigated systems discussed above. Crossover effects from critical to tricritical behavior are, of course, to be expected as the tricritical point is approached and the form of the crossover is predicted by scaling theory.[7,8,13] Appropriate scaling fields may be defined as $\mu_1 = [H - H_c(T)]/H_t$ and $\mu_2 = |T - T_t|/T_t - C\mu_1$, where $H_c(T)$ describes the phase boundary. The constant C may be chosen arbitrarily and in the analysis shown below, it was taken equal to 1. Other values would give qualitatively similar results. The corresponding scaling functions are $M^* = [M - M_c(T)]/\mu_2^{(1-\alpha_t)\phi}$ and $M_{st}^* = M_s/\mu_2^{\beta_t\phi}$. Log-log plots of these as functions of μ_1/μ_2^{ϕ} are shown in Fig. 9. As expected, data for different fields and temperatures collapse on distinct curves which clearly display crossover. For

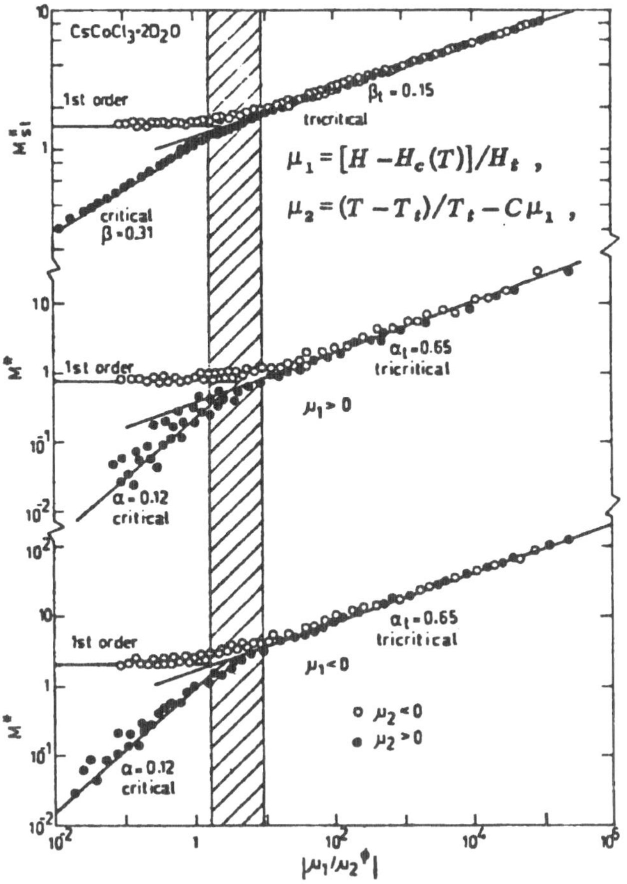

Fig. 9. Tricritical-Critical crossover in $CsCoCl_3.2D_2O$. Variation of the scaled staggered magnetization M^*_{st} and scaled uniform magnetization M^* as a function of the scaling variable μ_1/μ_2^ϕ (See text). The asymptotic slopes correspond, respectively, to the exponents β and β_t and α and α_t for $1 \gg \mu_1/\mu_2^\phi \gg 1$ (After Ref. 46b).

small values of μ_1/μ_2^ϕ the curves become linear with shapes corresponding to the critical exponents for an Ising-like system, while for high values the shapes lead to the anomalous "tricritical" exponents previously noted in Table 2. The crossover region occurs for $\mu_1/\mu_2^\phi \sim 1 - 10$ as might be expected.

Further work on this material would, clearly, be of interest. The crucial step will be to remove or reduce the inhomogeneous demagnetizing effects or other "rounding" mechanisms which currently appear to limit the experimental resolution.

Mixtures of Liquid ^3He-^4He

This system is really outside the scope of this paper since the phase transition is not primarily due to magnetic effects. However, there are a number of striking similarities with the magnetic systems which should be noted.

The "nonordering" variable corresponding to the magnetization is the concentration of one isotope relative to the other. Conventionally[24] one writes $x = {}^3He/({}^3He + {}^4He)$. The variable corresponding to an applied field is $\Delta = \mu_3 - \mu_4$, where μ_3 and μ_4 are the chemical potentials of the two isotopes. Neither of these is directly accessible, but one can infer variations with Δ from measurements of the vapor pressure.[24] The experiments are, thus, less direct than those with magnetic materials but from a careful analysis of extensive thermodynamic data one can extract many of the same tricritical exponents. The most detailed work has been done by Meyer and his collaborators and an excellent summary together with references to earlier papers is given in reference 24.

Table 2 lists the tricritical exponents which have been measured for ^3He-^4He, where we have changed the notation to be consistent with that used commonly for magnetic systems.[10] It can be seen that the exponents which have been measured agree very well both with the theory and with the corresponding values for the metamagnets. So far, no method for measuring the order parameter has been found for liquid He.

The data for ^3He-^4He were also tested for scaling and plots very similar to Fig. 7 were found. One remarkable feature of these results was the finding that the ranges of Δ and T over which the data scaled were also very similar to those for the metamagnet. In particular the very narrow scaling range in the ordered phase of the H_i-T plot ($H_i < H_c(T)$ in Fig. 8) was again found in the Δ-T plot for ^3He-^4He. This reemphasizes that nonlinear scaling fields are needed for a more complete description of the tricritical region.

Computer Models

One way of avoiding some of the problems inherent in experiments on physical systems is to study computer models. These can be defined precisely in terms of appropriate microscopic interactions and their thermodynamic behavior can be studied using Monte Carlo techniques.[47-49] One advantage of such an approach is the fact that it is possible to study both ordering and nonordering densities at any external field. There are no complications from such extraneous effects as demagnetizing corrections, impurities or, as in the case of liquid systems, gravitational effects.

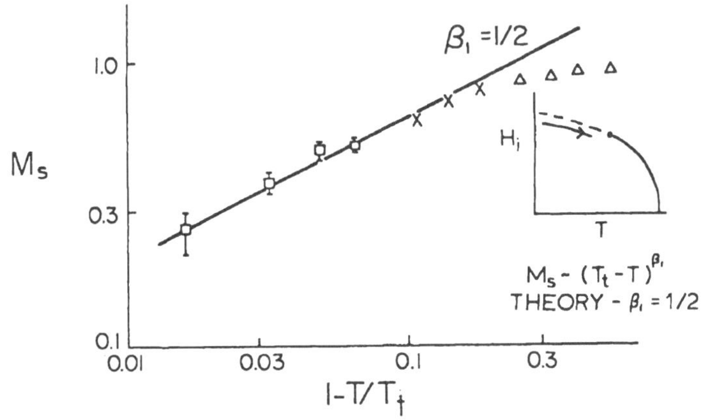

Fig. 10. Variation of staggered magnetization M_s, with temperature obtained from computer model study. The inset shows the path chosen for the calculation. The solid line indicates the slope β_1=1/2 predicted by renormalization group theory (After Ref. 48).

On the other hand, there are limitations set by the finite size of the models which can be studied and this can lead to serious errors, if the extrapolation to the thermodynamic limit is not carried out properly. One must also be careful to ensure true equilibrium, which may be reached only slowly, especially near first order transitions. One can thus find hysteresis effects[49] very much like those which are observed in real physical systems, and there will always be some errors due to the finite number of iterations which can be made for any given point.

With present day computers, the overall uncertainties in such studies are generally comparable with those inherent in magnetic or liquid He experiments. Some typical results[48] are shown in Fig. 10, in which the variation of the log of the staggered magnetization, M_s, is plotted as a function of the log of the reduced temperature for a path along the first order phase boundary for $T < T_t$. It can be seen that the results are consistent with the theoretical slope corresponding to the exponent $\beta_1 = 1/2$, but the errors become large for $(T_t - T)/T_t \sim 0.02$. One may confidently expect that the accuracy of such studies will improve as more powerful computers become available, but it must be recognized that significant increases in both speed and size will be needed to provide data which will answer such delicate questions as the importance of logarithmic corrections.

CONCLUSIONS

It is clear that tricritical points constitute an active field of research in which much remains to be done.

On the theoretical side, it would be useful to have quantitative information on such factors as the corrections to asymptotic behavior, the form of the scaling functions and the values of the amplitudes which multiply the singular terms in the free energy. It would also be of great interest to have some more detailed quantitative predictions for the effects of impurities.

On the experimental side, the immediate problems are more mundane. Additional experiments are needed on the three magnetic systems which have been studied in some detail up to now, both to resolve some unexplained ambiguities and to extend the results to quantities which have not been measured before. Among the most serious of the unresolved questions are the effects of impurities and inhomogeneous demagnetizing corrections. Both of these call for better samples. It may be that the answer can be found in studying completely new metamagnetic materials from which it is intrinsically easier to prepare more perfect samples. A number of potential candidates have been discussed by Stryjewski and Giordano.[23]

On the other hand, there is clearly also scope for improving and extending the measurements on the materials which have been studied up to now. At the same time, it must be noted that, with only a few exceptions, the results which have been reported for metamagnetic materials are in good agreement with the predictions of theory, as can be seen in Table 2. The fact that it is possible to study both the ordering and nonordering densities makes magnetic systems very attractive for tricritical point studies and one can certainly expect further results in the near future.

ACKNOWLEDGMENTS

It is a pleasure to acknowledge a number of informative and stimulating discussions with N. Giordano, M. E. Fisher, and R. J. Birgeneau and I would like to thank J. Brug for help with the preparation of this paper and making available some of his unpublished results. This work was supported in part by NSF Grant Number DMR-8216222.

REFERENCES

1. E. H. Graf, D. M. Lee, and J. D. Reppy, Phys. Rev. Lett. 19, 417 (1967).
2. I. S. Jacobs and P. E. Lawrence, Phys. Rev. 164, 866 (1967).
3. M. Ball, W. P. Wolf, and A. F. G. Wyatt, Phys. Lett. 10, 7 (1964).
4. B. F. Keen, D. P. Landau, and W. P. Wolf, J. Appl. Phys. 38, 967 (1967).
5. V. A. Schmidt and S. A. Friedberg, Phys. Rev. B 1, 2250 (1970).
6. R. B. Griffiths, Phys. Rev. Lett. 24, 715 (1970).
7. E. K. Riedel and F. J. Wegener, Phys. Rev. Lett. 29, 349 (1972); Phys. Rev. B 7, 248 (1973); ibid B 9, 294 (1974).
8. A. Hankey, H. E. Stanley, and T. S. Chang, Phys. Rev. Lett. 29, 278 (1972).
9. R. Bausch, Z. Phys. 254, 81 (1972).
10. R. B. Griffiths, Phys. Rev. B 7, 545 (1973), and references cited therein.
11. M. E. Fisher and D. R. Nelson, Phys. Rev. Lett. 32, 1350 (1974); Phys. Rev. B 11, 1030 (1975); ibid B 12, 263 (1975).
12. M. E. Fisher, AIP Conf. Proc. 24, 273 (1975), and references cited therein.
13. D. R. Nelson and J. Rudnick, Phys. Rev. Lett. 35, 178 (1975); Phys. Rev. B 13, 2208 (1976).
14. M. J. Stephen, E. Abrahams, and J. P. Straley, Phys. Rev. B 12, 256 (1975), and references cited therein.
15. V. J. Emery, Phys. Rev. B 11,3397 (1975).
16. D. R. Nelson, AIP Conf. Proc. 29, 450 (1976).
17. N. Giordano, Phys. Rev. B 14, 2927 (1976).
18. M. E. Fisher and S. Saarbach, Phys. Rev. Lett. 41, 1127 (1978); J. Appl. Phys. 49, 1350 (1978).
19. S. Fishman and A. Aharony, J. Phys. C L729 (1979).
20. M. J. Stephen, J. Phys. C 13, L83 (1980).
21. I. D. Lawrie, J. Phys. C 13, 2739 (1980).
22. W. P. Wolf, Physica 86-88B, 550 (1977).
23. E. Stryjewski and N. Giordano, Adv. Phys. 26, 487 (1977).
24. E. K. Riedel, H. Meyer, and R. P. Behringer, J. Low Temp. Phys. 22, 369 (1976), and references cited therein.
25. M. B. Salamon and H. T. Shang, Phys. Rev. Lett. 44, 879 (1980); Phys. Rev. B 22, 4401 (1980).
26. R. J. Birgeneau and A. N. Berker, Phys. Rev. B 26, 3751 (1982).
27. N. Giordano, Thesis, Yale University, 1977 (unpublished).
28. J. Brug, Thesis, Yale University, 1984 (unpublished).
29. D. H. Robbins and P. Day, Chem. Phys. Lett. 19, 529 (1973).
30. C. Vettier, H. L. Alberts, and D. Bloch, Phys. Rev. Lett. 31, 1414 (1973).
31. J. A. Griffin, S. E. Schnatterly, Y. Farge, M. Regis, and M. P. Fontana, Phys. Rev. B 10, 1960 (1974).
32. R. J. Birgeneau, G. Shirane, M. Blume, and W. C. Koehler, Phys. Rev. Lett. 33, 1098 (1974); AIP Conf. Proc. 24, 258 (1975).

33. J. A. Griffin and S. E. Schnatterly, Phys. Rev. Lett. 33, 1576 (1974); AIP Conf. Proc. 24, 195 (1975).
34. J. F. Dillon, E. Yi Chen, H. J. Guggenheim, and R. Alben, Phys. Rev. B 15, 1422 (1977).
35. J. F. Dillon, E. Yi Chen, and H. J. Guggenheim, Phys. Rev. B 18, 377 (1978).
36. J. Brug, G. Domann, and W. P. Wolf, J. Mag. and Mag. Mat. 31-34, 585 (1983).
37. J. F. Dillon, Jr., E. Yi Chen, and H. J. Guggenheim, Solid State Commun. 16, 371 (1975).
38. J. F. Dillon, Jr., E. Yi Chen, N. Giordano, and W. P. Wolf, Phys. Rev. Lett. 33, 98 (1974).
39. See, for example, J. A. Griffin, M. Huster, and R. J. Folweiler, Phys. Rev. B 22, 4370 (1980).
40. D. P. Landau, B. E. Keen, B. Schneider, and W. P. Wolf, Phys. Rev. B 3, 2310 (1971).
41. W. P. Wolf, B. Schneider, D. P. Landau, and B. E. Keen, Phys. Rev. B 5, 4472 (1972), and references contained therein.
42. G. F. Tuthill, F. Harbus, and H. E. Stanley, Phys. Rev. Lett. 31, 527 (1973).
43. A. T. Skjeltorp, R. Alben, and W. P. Wolf, AIP Conf. Proc. 18, 770 (1974).
44. M. Blume, L. M. Corliss, J. M. Hastings, and E. Schiller, Phys. Rev. Lett. 32, 544 (1974); also, R. Alben, M. Blume, L. M. Corliss, and J. M. Hastings, Phys. Rev. B 11, 295 (1975).
45. N. Giordano and W. P. Wolf, (a) Phys. Rev. Lett. 35, 799 (1975); (b) ibid 39, 342 (1977); (c) AIP Conf. Proc. 29, 459 (1975); (d) Physica 86-88B, 593 (1977); (e) Phys. Rev. B 21, 2008 (1980).
46. A. L. M. Bongaarts, W. J. M. de Jonge, and P. Van der Leeden, (a) Phys. Rev. Lett. 37, 1007 (1976); (b) Phys. Rev. B 15, 3424 (1977), and references contained therein.
47. D. P. Landau, Phys. Rev. Lett. 28, 449 (1972).
48. D. P. Landau, Phys. Rev. B 14, 4054 (1976).
49. A. K. Jain and D. P. Landau, Phys. Rev. B 22, 445 (1980).
50. A. Aharony and M. E. Fisher, Phys. Rev. B 27, 4394 (1983).
51. J. C. Norvell, W. P. Wolf, L. M. Corliss, J. M. Hastings, and R. Nathans, Phys. Rev. 186, 557 (1969).
52. M. Steiner and N. Giordano, Phys. Rev. B 25, 6886 (1982).

EXPERIMENTAL STUDIES OF BICRITICAL POINTS IN 3D ANTIFERROMAGNETS

Y. Shapira

Francis Bitter National Magnet Laboratory
Massachusetts Institute of Technology
Cambridge, MA 02139, USA

INTRODUCTION

The theoretical work of Fisher, Nelson, and Kosterlitz[1-3] on the bicritical point (BCP) of a 3d antiferromagnet has stimulated many experimental investigations. Some of these experiments will be reviewed here. Recent works on random fields are not included. A short review of works up to 1976 was given by Wolf.[4]

THEORETICAL BACKGROUND

A. Crossover Due to a Change of the Spin Dimensionality

Critical points are divided into universality classes (UC's).[5,6] In magnetic materials, the UC is often (but not always[7]) determined by two parameters: (i) the lattice dimensionality d, and (ii) the spin dimensionality (or number of components of the order parameter) n. Here we shall be concerned only with d=3. The most common values of n are: n=1 (Ising) which corresponds physically to a magnetic material with a single easy axis; n=2(XY) which corresponds to an easy plane; and n=3 (Heisenberg) which corresponds to full isotropy (in 3d).

An interesting situation arises when a small perturbation $\mathcal{H}_1$ breaks the symmetry of the original Hamiltonian $\mathcal{H}_0$, and thereby causes a "crossover" from the critical behavior of the original UC to that of a new UC. In what follows we shall be concerned with the crossover associated with a change of n. This crossover was discussed by Pfeuty, Jasnow and Fisher, who used extended scaling and high-temperature series for a ferromagnet.[8] Renormalization-group treatments of this problem were reviewed by Aharony.[7]

The perturbation which lowers the spin dimensionality is usually represented by g. In discussions of the BCP, the order-disorder line $T_c(g)$ consists of two branches (g>0 and g<0) which meet at the BCP (g=0). The crossover exponent ϕ_g depends on the spin dimensionality n at g=0. High-temperature series give: $\phi_g=1.25\pm0.015$ for n=3, and $\phi_g=1.175\pm0.015$ for n=2.[8]

B. Low-Anisotropy Easy-Axis Antiferromagnets

Most experimental studies of BCP's were carried out on low-anisotropy antiferromagnets. The reason for this is that in many antiferromagnets the spin dimensionality can be changed easily by applying a magnetic field $\vec{H}$. In essence, this comes about because the magnetic field produces an effective anisotropy which prefers the plane perpendicular to it. That is, configurations with the staggered magnetization, $\vec{M}_{st}$, perpendicular to $\vec{H}$ are preferred.

$\vec{H}$ paralle to the easy axis. Consider a low anisotropy easy-axis antiferromagnet, with H parallel to the easy axis. The phase diagram obtained from mean-field theory (MFT)[9] is shown in Fig. 1(a). There are two ordered phases: (1) the antiferromagnetic (AF) phase, with $\vec{M}_{st}\parallel\vec{H}$, and (2) the spin-flop (SF) phase, with $\vec{M}_{st}\perp\vec{H}$. The paramagnetic (disordered) phase is designated by P. The AF and SF phases are separated by a line $H_{sf}^2(T)$ of first-order spin-flop transitions. The lines $T_c^{\parallel}$ and $T_c^{\perp}$ are lines of second-order transitions. The three phase boundaries meet at the BCP ($T=T_b$, $H^2=H_b^2$). The Néel temperature is T_N.

The BCP may be viewed as that point on the order-disorder line where the effective easy-plane anisotropy due to $\vec{H}$ just balances the intrinsic easy-axis anisotropy. From a modern viewpoint, transitions on the line $T_c^{\parallel}$ are Ising-like ($n_{\parallel}=1$). Transitions on the line $T_c^{\perp}$ are either Ising ($n_{\perp}=1$) or XY ($n_{\perp}=2$), depending on whether or not there is a preferred direction in the perpendicular plane. At the BCP, $n=n_{\parallel}+n_{\perp}$.

Figure 1(b) is a sketch of the phase diagram predicted by Fisher, Nelson and Kosterlitz (FNK).[1-3] The phase boundaries near the BCP are affected by the crossover due to the change in n. Thus, the λ line (consisting of the lines $T_c^{\parallel}$ and $T_c^{\perp}$) has a notch or "umbilicus". This umbilicus is not present in the mean-field phase diagram.

The scaling axes at the BCP were discussed by Fisher.[2] They are shown in Fig. 1(b) as $\tilde{g}=0$ and $\tilde{t}=0$. The scaling variables are:

$$\tilde{t} = t + q(H^2 - H_b^2), \qquad \tilde{g} = H^2 - H_b^2 - pt, \tag{1}$$

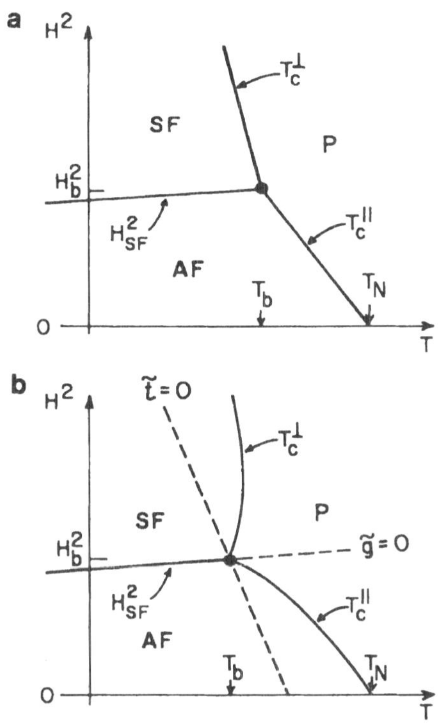

Fig. 1. Schematic phase diagram for a low-anisotropy easy-axis antiferromagnet in a magnetic field $\vec{H}$ parallel to the easy axis. (a) Mean-field theory. (b) Fisher-Nelson-Kosterlitz (FNK) theory.

where $t = (T - T_b)/T_b$, and $p = T_b[d(H_{sf}^2)/dT]_b$. The parameter q is not accurately given by the theory, but Fisher gives the estimate

$$q \cong -\left(\frac{n+2}{3nT_b}\right)[dT_c^{\parallel}/d(H^2)]_{H=0} . \tag{2}$$

The shapes of the lines $T_c^{\parallel}$ and $T_c^{\perp}$ near the BCP are predicted to obey the equation

$$\tilde{g}/\tilde{t}^{\phi} = w_{\perp}, - w_{\parallel} , \tag{3}$$

where $\phi = \phi_g(n)$, and $w_{\perp,\parallel}$ are nonuniversal constants. The ratio $Q(n) = w_{\perp}/w_{\parallel}$ is, however, universal. Thus, Q(2)=1 by symmetry, and Q(3)= 2.34±0.08.[10] [An earlier estimate, Q(3)=2.51,[2] was used in the interpretation of some experimental data.]

Other predictions for the BCP will be discussed later in connection with the experimental data.

Tetracritical point. In Fig. 1 the AF-SF transitions are abrupt (first-order). This corresponds to the most common experimental situation. However, there are causes in which the AF and SF phases are separated by an "intermediate phase" (IN) in which the aorientation of $\vec{M}_{st}$ changes gradually from parallel to perpendicular. This was discussed in detail by Bruce and Aharony[12] who considered the effects of a cubic anisotropy. This anisotropy can be of two types: (1) the cube edges are preferred (v<0), or (2) the cube diagonals are preferred (v>0). In case (1), and assuming a weak cubic anisotropy, one obtains the phase diagram in Fig. 1(b). The critical lines are then still described by Eq. (3). Case (2) leads to Fig. 2. The two lines which surround the IN phase are critical lines. The multicritical point where all four critical lines meet is a tetracritical point. The P-AF and P-SF lines are described by Eq. (3), i.e., the shift exponent ψ is equal to ϕ_g. However, the two critical lines which surround the IN phase are governed by a shift exponent $\psi = \phi_g + |\phi_v|$, where ϕ_v is a crossover exponent associated with the cubic anisotropy. In the cases considered here, ϕ_v is negative and is numerically small, e.g., $\phi_v \cong -0.03$ for d=n=3.[13]

Skew magnetic field. In general, $\vec{H}$ may make a finite angle with the easy axis, so that there is a perpendicular component $H_{\perp}$ in addition to $H_{\parallel}$. The expected phase diagram in the $T\text{-}H_{\parallel} - H_{\perp}$ space was discussed in several papers.[2,3,14-16] It was assumed that the AF-SF transition in the $T\text{-}H_{\parallel}$ plane was abrupt, as in Fig. 1.

Figure 3(a), taken from a PRL by Rohrer and Gerber,[15] shows the expected phase diagram for an orthorhombic material when $H_{\perp}$ points along the medium anisotropy axis (i.e., preferred axis in the perpendicular plane). The phase diagram when $H_{\perp}$ is along the hard

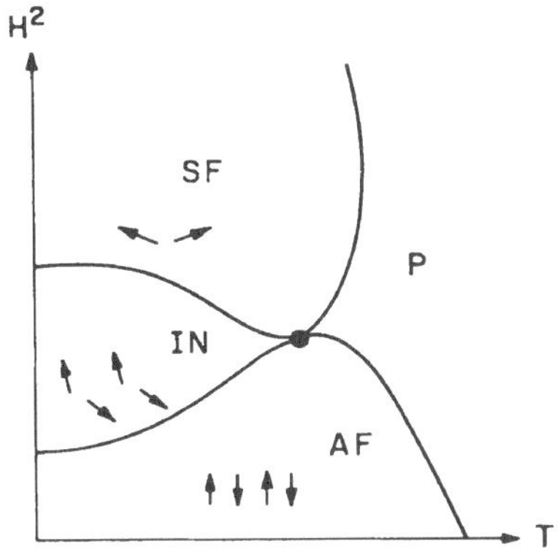

Fig. 2. Schematic phase diagram for an easy-axis antiferromagnet with a cubic anisotropy of the type v>0.

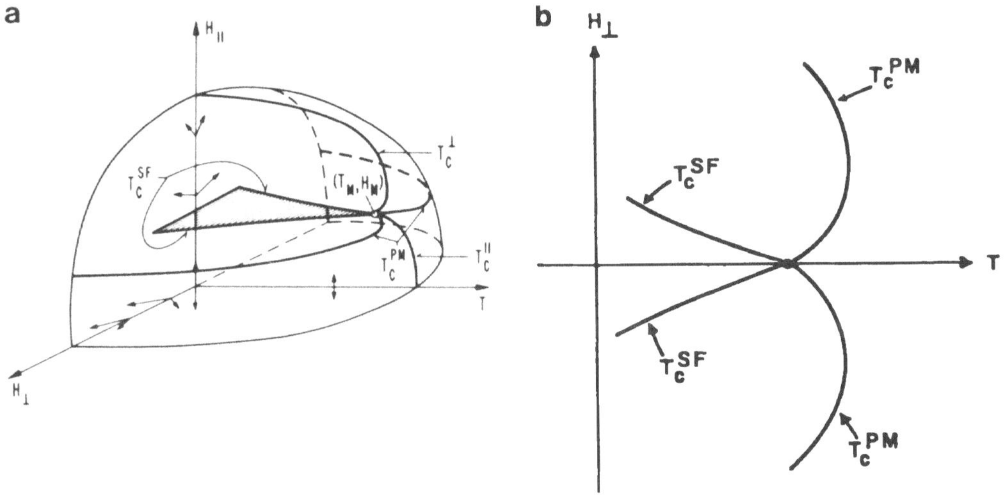

Fig. 3. (a) Phase diagram of an orthorhombic antiferromagnet in the space of $H_\parallel, H_\perp$, and T. Here, $H_\parallel$ is parallel to the easy axis, and $H_\perp$ is parallel to the axis of medium anisotropy. The shaded area is the shelf of first-order spin-flop transitions. (After Rohrer and Gerber, Ref. 15). (b) The phase diagram in the plane which passes through the multicritical point (T_M, H_M) and which is parallel to the $H_\perp$ and $\tilde{g}=0$ axes.

axis is quite different,[14] and will not be discussed here. The phase diagram for a material with no anisotropy in the perpendicular plane is qualitatively similar to Fig. 3(a), except that the line $T_c^{\perp}$ lies in a "furrow".[2] This furrow results from the crossover between XY symmetry for $H_{\perp}=0$ to the Ising symmetry for $H_{\perp}\neq 0$.

The shaded area in Fig. 3(a) represents a "shelf" of first-order spin-flop transitions. The shelf terminates on two critical lines, designated as T_c^{SF}. The width of the shelf decreases as T increases, and it vanishes at the multicritical point (T_M,H_M). The width at T=0 was discussed by Chepurnykh,[17] and by Rohrer and Thomas.[18]

Figure 3(b) shows the phase diagram in a plane which passes through the multicritical point and which is parallel to the $H_{\perp}$ and $\tilde{g}=0$ axes. (The latter axis is usually nearly parallel to the $\bar{T}$ axis.) In this plane the multicritical point is a tetracritical point. The lines T_c^{PM}, for small $H_{\perp}$, are given by $t \propto |H_{\perp}|^{1/\phi}$ (Refs. 2,3). The lines T_c^{SF} are expected to be governed by a similar expression, but with ϕ replaced by $\phi + |\phi_v|$.[15,16] It should be noted that in the $T\text{-}H_{\parallel}$ plane, the multicritical point is still a bicritical point.

C. Degenerate and Virtual Bicritical Points

The BCP for an easy-axis antiferromagnet with no anisotropy in the perpendicular plane is shown (again) in Fig. 4(a). The field is along the easy axis. Figure 4(b) shows the situation when the axial anisotropy vanishes, i.e., a fully isotropic antiferromagnet. The BCP now occurs at H=0, and is called a "degenerate BCP".[2] The phase boundary at low H is given by[2,3]

$$T_c(H) - T_N = aH^{2/\phi} - bH^2 , \qquad (4)$$

where $\phi = \phi_g(n=3)$, and $\underline{a}$ and $\underline{b}$ are positive constants. The term $aH^{2/\phi}$ represents the effect of the crossover from Heisenberg to XY symmetries. This term dominates at low H, and leads to an initial increase of T_c with increasing H. The bow-shaped curve described by Eq. (4) is in striking contrast to the mean-field prediction

$$T_c(H) - T_N = -BH^2 \qquad \text{(MFT)} . \qquad (5)$$

The term $-bH^2$ in Eq. (6) is expected to be comparable to $-BH^2$.

An analogous situation occurs in an easy-plane antiferromagnet for which the transition at H=0 has XY character. When $\vec{H}$ is applied in the easy plane, it forces $\vec{M}_{st}$ to lie perpendicular to it. The transitions at $H\neq 0$ have Ising character. The phase boundary at low H is given by Eq. (4), but with $\phi=\phi_g(n=2)$.

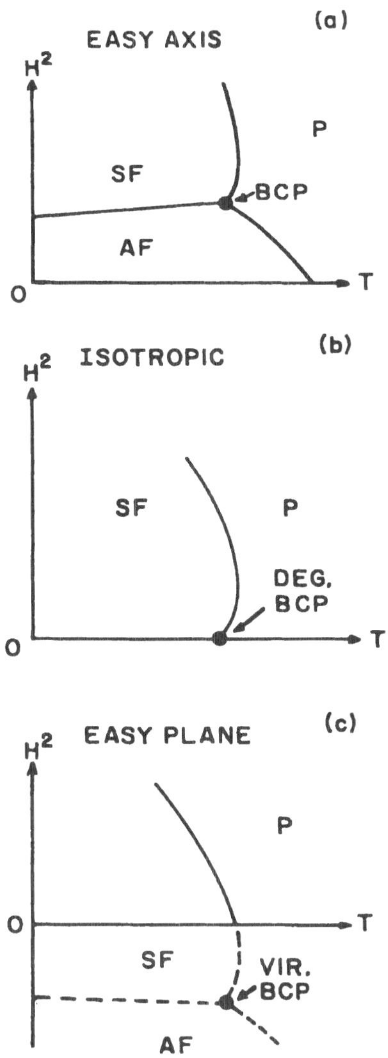

Fig. 4. Phase diagrams of low-anisotropy antiferromagnets with cylindrical symmetry about an axis and with $\vec{H}$ parallel to that axis. (a) Easy-axis, i.e., anisotropy favors the axis. (b) Isotropic, i.e., no anisotropy at all. (c) Easy-plane, i.e., anisotropy favors the plane perpendicular to the axis. These three cases correspond to ordinary BCP, degenerate BCP, and virtual BCP, respectively.

The "virtual BCP" was introduced in order to account for the behavior of an easy-plane antiferromagnet when $\vec{H}$ is perpendicular to the easy plane.[19] The transitions at both zero and finite H then have XY character, so that, strictly, there is no change in symmetry. Nevertheless, deviations from mean-field behavior occur at low H if the intrinsic easy-plane anisotropy is small. In Ref. 19 it is shown that the case of a weak easy-plane anisotropy can be treated by analogy to the easy-axis case. The analogy leads to a BCP which is (mathematically) at $H^2<0$. The phase diagram corresponding to this virtual BCP is shown in Fig. 4(c). Only the region $H^2 \geqslant 0$ can be observed; all transitions at $H^2<0$ are virtual. The expression for T_c in the observable region of $H^2 \geqslant 0$ is

$$T_c(H) - T_N = a[(H^2 + h_0{}^2)^{1/\phi} - h_0{}^{2/\phi}] - bH^2 \quad , \tag{6}$$

where $\phi = \phi_g(n=3)$, and $-h_0{}^2$ is the value of H^2 at the virtual BCP.

EXPERIMENTAL STUDIES OF PHASE DIAGRAMS

A. Introductory Remarks

Phase diagrams of low-anisotropy antiferromagnets were measured since the mid 50's. Prior to the work of FNK, most results were interpreted using MFT. The FNK predictions led to a renewed experimental activity. The new generation of experiments started with Rohrer's discovery of the umbilicus in $GdAlO_3$.[20] Many of the new experiments were carried out with a much higher precision than in typical early studies. This was dictated by two facts: the umbilicus is usually quite small, and its shape is sensitive to field misalignment. I shall make only two comments regarding the experiments which predated the FNK theory: (1) Early studies of both $GdAlO_3$ (Ref. 21) and MnF_2 (Ref. 9) did not reveal the umbilicus. The observations of the umbilicus in both of these materials (discussed below) occurred only after the theorists said it should be there! (2) In some materials an umbilicus was observed even before the FNK theory was advanced. Examples are the data in Refs. 22 and 23. However, it does not appear that these observations were considered at the time to be an indication of a fundamental flaw in MFT.

BCP's were investigated by susceptibility, ultrasonic, thermal-expansion, magnetostriction, NMR and neutron-diffraction techniques. Accurate field alignment is important in measurements of the phase boundaries near the BCP's of easy-axis antiferromagnets. In the case of MnF_2, King and Rohrer aligned the field to within 10^{-4} deg from the easy axis.[24] However, adequate results were obtained in some materials with alignments of 10^{-2} - 10^{-1} deg. In general, the misalignment should be reduced to a level where the effect of the perpendicular field component $H_\perp$ is not the limiting factor in the precision for the phase boundaries. In many realistic situations, the mosaic spread inside the sample, and demagnetizing effects in non-ellipsoidal samples, place a limit on the achievable alignment. One advantage of degenerate BCP's (H_b=0) is that the data are not sensitive to field alignment. Another advantage is that the expression for the phase boundary, Eq. (4), contains fewer parameters than the expression for the ordinary BCP which involves ϕ, T_b, H_b, p, q, $w_\perp$ and $w_\parallel$. In fitting data for an ordinary BCP, one often takes p and H_b from data for the spin-flop line. Sometimes T_b is also known with sufficient accuracy. To reduce the number of parameters further, one sometimes assumes that $w_\perp/w_\parallel$, or ϕ, or q, are correctly given by the theory. The number of parameters which are allowed to vary in the fit usually depends on the precision of the data.

B. BCP's with n=2 in Easy-Axis Antiferromagnets

$GdAlO_3$. $GdAlO_3$ is an orthorhombic antiferromagnet with T_N= 3.878 K. The phase boundaries near the BCP, and for $\vec{H}$ parallel to the easy axis, were determined by Rohrer.[20] The measurements were repeated, with a better field alignment, by Rohrer and Gerber.[15] Figure 5(a) is taken from the latter work. The umbilicus is very clear. Fits to Eq. (3) gave ϕ=1.17±0.02 and Q = $w_\perp/w_\parallel$ = 0.9±0.2, in good agreement with the theoretical values ϕ(n=2) = 1.175 and Q(n=2) = 1. The orientation of the $\tilde{t}$=0 scaling axis (related to the parameter q) was found to agree with Fisher's estimate, Eq. (2).[20]

$NiCl_2 6H_2O$. This material is a monoclinic antiferromagnet with T_N=5.34K. The phase boundaries were measured by Oliveira et al.[25] The umbilicus was observed. The values Q=1.06±0.22, and ϕ=1.29±.07 were obtained. The result for q was consistent with Fisher's estimate.

$CsMnBr_3 2D_2O$. CMB is an orthorhombic antiferromagnet with T_N=6.30 K. Its bicritical point was investigated by Basten et al.[26] Although CMB is a pseudo-1d antiferromagnet, a crossover to a 3d critical behavior is expected as the critical lines are approached. Thus, Basten et al. fitted the phase diagram near the BCP (which showed a pronounced umbilicus) to the FNK theory for 3d. In this fit, Q=1 was assumed. The result ϕ=1.22±0.06 is in agreement with the theoretical value for n=2. However, the value for q differs considerably from Fisher's estimate. This was attributed to the fact that Fisher's estimate is based on a mean-field type approach which is not expected to be valid for a pseudo-1d material.

C. BCP's with n=3 in Easy-Axis Antiferromagnets

MnF_2. MnF_2 is a tetragonal antiferromagnet, with the easy axis parallel to c and with T_N=67.3 K. The phase diagram near the BCP, and for $\vec{H}\parallel$c, was measured by King and Rohrer,[24] and by Shapira and Becerra.[27] Both data show the umbilicus and are consistent with

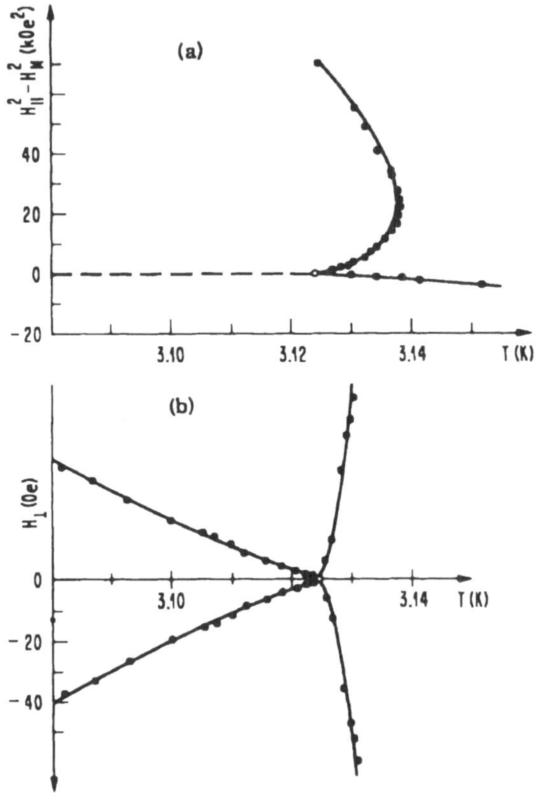

Fig. 5. Phase transitions near the multicritical point of $GdAlO_3$, marked by an open circle. The solid lines are best fits to the data. (a) Bicritical behavior in the $H_{\parallel}$-T plane. The broken line is the spin-flop line. (b) Tetracritical behavior in a plane which passes through the multicritical point and which is parallel to the $H_{\perp}$-T plane. (After Rohrer and Gerber, Ref. 15).

each other. However, because the data of King and Rohrer are more precise, I shall concentrate on their results.

A fit of the King-Rohrer data to Eq. (3) gave $\phi=1.279\pm0.031$, which agrees with the prediction $\phi(n=3) = 1.25$. The value for q was only slightly lower than Fisher's estimate. However, the value $Q=1.56\pm0.35$ obtained from the fit was significantly lower than the predicted universal ratio $Q(3)=2.34\pm0.08$. This discrepancy was discussed in detail by Fisher et al.[10]

A puzzling feature of the data of King and Rohrer is the appearance of an anomalous susceptibility peak near the $\tilde{g}=0$ line at T slightly above T_b, i.e., in the paramagnetic phase. They also state that such a peak is observed in $GdAlO_3$ when $\vec{H}$ is accurately aligned along the easy axis. The origin of the anomalous susceptibility peak is unknown.

Cr_2O_3. This easy-axis antiferromagnet has trigonal crystal structure, and T_N=307.3 K. The phase boundaries for $\vec{H}$ parallel to the easy axis were determined by Shapira and Becerra.[28] The umbilicus was observed clearly. A fit to Eq. (3), in which Q was held fixed at the then-accepted theoretical value $Q=2.51$, gave $\phi=1.22\pm0.15$. This value is consistent with theory. However, the parameter q was smaller by a factor of 3 than Fisher's estimate. A fit in which ϕ was held fixed at 1.25 led to the same conclusion.

The discrepancy with Fisher's estimate was explained by noting that the derivation of Eq. (2) involves two steps: (1) a mean-field type approach (supplemented by renormalization-group results) which leads to a relation between q and the mean-field slope $[dT_c^{\parallel}/d(H^2)]_{MF}$, and (2) equating the mean-field slope to the measured slope near T_N. Shapira and Becerra argue that the second step is inappropriate when the uniaxial anisotropy is extremely small. In that case the BCP is very close to the Néel point, so that the shape of the boundary $T_c^{\parallel}$ near T_N is influenced by the BCP, and is not given by MFT. In Cr_2O_3, $(T_N-T_b)/T_b \sim 10^{-3}$.

D. Degenerate BCP's

$RbMnF_3$. $RbMnF_3$ is a cubic antiferromagnet with T_N=83 K. The cubic anisotropy field is $H_a \sim 4$ Oe at T=0 compared to the exchange field $H_E \cong 8\times10^5$ Oe. This makes the material a nearly ideal isotropic Heisenberg antiferromagnet. (The effect of the cubic anisotropy is discussed later.)

The order-disorder phase boundary $T_c(H)$ was determined by Shapira and Becerra.[29] The results are shown in Fig. 6. Note that the ordinate is H, and not H^2. The phase boundary is bow-shaped, as predicted by FNK. A fit to Eq. (4), supplemented by a conservative error analysis, gives $\phi=1.258\pm0.08$. The same phase boundary was determined later by Shapira and Oliveira, using another technique.[30] The data were very close to those in Fig. 6. The results for two samples gave $\phi=1.278\pm0.026$, and $\phi=1.274\pm0.045$. All the experimental values for ϕ are close to the prediction $\phi(n=3)=1.25$. The experimental values for the coefficient b in Eq. (4) agreed with an independent estimate (analogous to Fisher's estimate of q).

$KNiF_3$. This material is a cubic antiferromagnet with $T_N=246.4$ K and $H_a/H_E \cong 7\times10^{-5}$. The phase boundary in fields up to 180 kOe was measured by Becerra et al.,[31] and by Shapira and Oliveira.[32] An initial increase of T_c with increasing H was observed, in qualitative agreement with the FNK theory. An accurate determination of ϕ was not possible.

$CsMnF_3$. This material is an easy-plane hexagonal antiferromagnet with $T_N=51.4$ K. The hexagonal anisotropy in the easy plane is very small, $H_a/H_E < 10^{-5}$. The transition at T_N is expected to be XY-like. The phase diagram when $\vec{H}$ is in the easy plane should correspond to that near a degenerate BCP with n=2. Measurements[19] show that the phase boundary is bow-shaped, as predicted by FNK. The crossover exponent, obtained from fits of data for two samples to Eq. (4), is $\phi=1.185\pm0.03$ and 1.184 ± 0.025. This agrees with the predicted value $\phi(n=2) = 1.175$.

E. Virtual BCP

Figure 7 shows the phase boundary of $CsMnF_3$ when $\vec{H}$ is perpendicular to the easy plane.[19] Note that T_c is not linear in H^2. This was attributed to the "existence" of a virtual BCP (with n=3) at $H^2<0$. Several facts were cited in support of this interpretation: (i) The phase boundary is well described by Eq. (6), with $\phi=1.25$. The solid line in Fig. 7 is a three-parameter fit to this equation. (ii) The values of two of the parameters (h_0 and b) obtained from this fit agree with independent estimates which do not involve the data for this phase boundary.

F. Tetracritical Behavior

I shall distinguish between: (i) a tetracritical behavior which occurs when there is an intermediate phase (IN) between the AF and SF phases (Fig. 2), and (ii) a tetracritical behavior which occurs when a perpendicular field component $H_\perp$ is superimposed on a parallel field $H_\parallel=H_b$ (Fig. 3b).

The existence of an intermediate phase in $CoBr_2\cdot6(0.48D_2O, 0.52H_2O)$ was first reported by Basten et al.[33] This was confirmed by a recent neutron study of the critical scattering near the two boundaries of the IN phase.[34] Basten et al. stated that the IN phase persisted in temperatures up to a tetracritical temperature. However, a phase diagram was not presented.

As already noted, $RbMnF_3$ is a cubic antiferromagnet. The weak cubic anisotropy prefers the alignment of $\vec{M}_{st}$ along the body diagonals. Therefore, a tetracritical behavior associated with an intermediate phase is expected when $\vec{H}$ is parallel to [100]. However, because there is no uniaxial anisotropy, the tetracritical point occurs at H=0, and only two of the four critical lines may be observed. (The other two lines are at $H^2<0$. By applying a suitable stress along [100] it should be possible to move these two lines into the observable region of $H^2 \geqslant 0$.) The P-SF line was already shown (Fig. 6). The second line (IN-SF) was studied by the present author.[35] The results were in qualitative agreement with the predictions of Bruce and Aharony,[12] but the data were not accurate enough for determining the shift exponent ψ.

The second type of tetracritical behavior ($H_{\|}=H_b$, $H_{\perp}\neq 0$) was studied in $GdAlO_3$ by Rohrer and Gerber.[15] Their results for the phase diagram in the $H_{\perp}$-T plane are shown in Fig. 5(b). A fit of the two lines above the multicritical temperature gave $\phi=1.19\pm0.02$, in good agreement with the expected value. The other two critical lines (associated with the "shelf") gave a shift exponent $\psi = \phi + |\phi_v| = 1.19\pm0.02$. Thus, ϕ_v is quite small, as expected.

Measurements in the $H_{\perp}$-T plane (with $H_{\|}=H_b$) were also made in Cr_2O_3, but only the two critical lines above the multicritical temperature were studied.[28] The results were consistent with theoretical predictions.

CRITICAL EXPONENTS AND THE STAGGERED MAGNETIZATION

The theory for the BCP leads to many predictions besides the shapes of the phase boundaries.[1,2,36] Only a few experimental tests of these additional predictions were made thus far.

From generalized scaling it follows that the magnetization discontinuity ΔM across the spin-flop line should behave like $|t|^{\tilde{\beta}}$, where $\tilde{\beta} = 2-\alpha-\phi$.[1,2] This relation was tested in $GdAlO_3$ by Rohrer et al.[15,20] Their final value, $\tilde{\beta}=0.88\pm0.02$ is in agreement with the predicted value $\tilde{\beta}(n=2) = 0.85\pm0.04$.

The T-dependence of the usual susceptibility $(\partial M_{\|}/\partial H_{\|})$ along the line $\tilde{g}=0$ should diverge as $t^{-\tilde{\gamma}}$, where $\tilde{\gamma} = 2\phi + \alpha - 2$.[1,2] Measurements in $GdAlO_3$, by Rohrer,[20] gave $\tilde{\gamma}=0.15$, which is smaller than the predicted value 0.33. It was suggested that the discrepancy was due to a residual field misalignment.

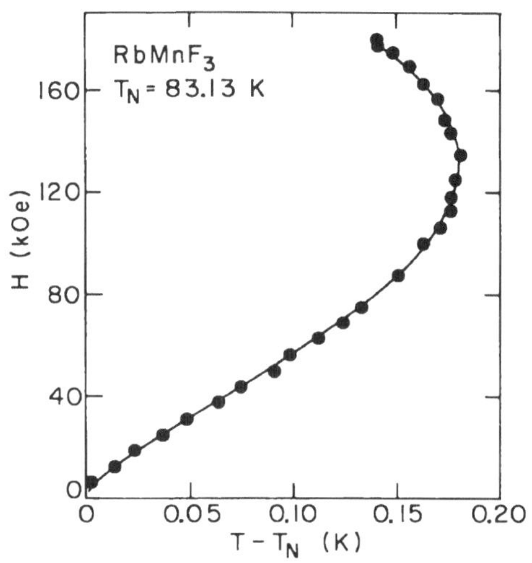

Fig. 6. The phase boundary between the paramagnetic phase and the ordered phase of $RbMnF_3$. Note that the ordinate is H. (After Shapira and Becerra, Ref. 29).

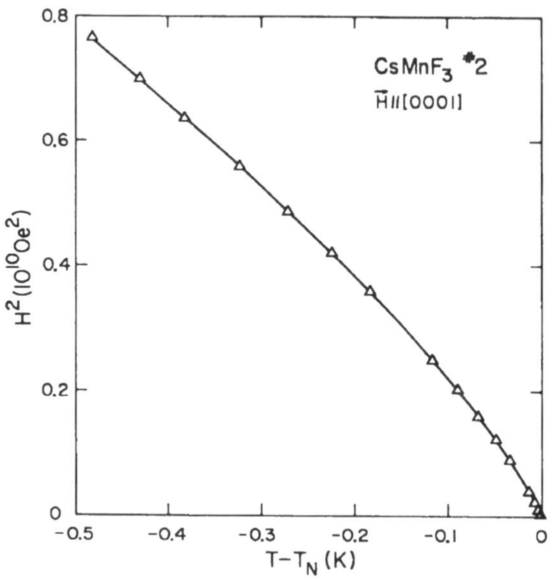

Fig. 7. Phase boundary of $CsMnF_3$ for $\vec{H}$ perpendicular to the easy plane. The solid line is a fit to Eq. (6), with ϕ held fixed at 1.25. (After Shapira, Oliveira, and Chang, Ref. 19).

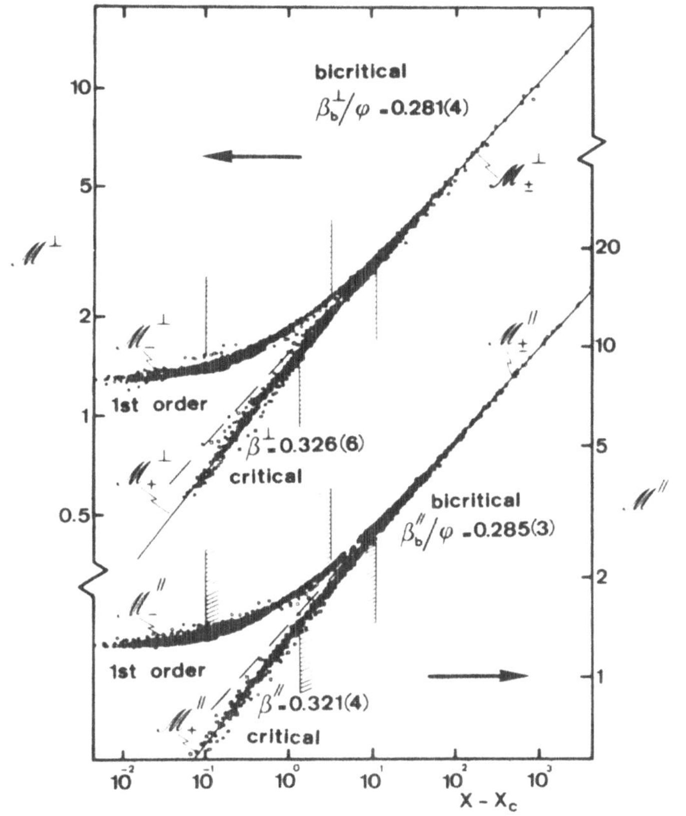

Fig. 8. Optimum accumulation of $M_{st}^{\parallel}$ and $M_{st}^{\perp}$ data, scaled according to Eq. (7) to yield the scaling functions $\mathcal{M}^{\parallel}$ and $\mathcal{M}^{\perp}$. Either of these two scaling functions consists of two branches: $\mathcal{M}_{-}$ and $\mathcal{M}_{+}$. x_c=0 for $\mathcal{M}_{-}$, and x_c=1 for $\mathcal{M}_{+}$. [After Basten, Frikkee and de Jonge, Phys Rev B 22, 1429 (1980)].

A detailed test of the extended scaling assumption was performed by Basten et al. who measured the staggered magnetization of $CsMnBr_3 2D_2O$ using neutron scattering.[26] As noted, although this is a pseudo-1d material, its behavior near the BCP (n=2) can be interpreted using the theory for 3d. From extended scaling it follows that

$$M_{st}^{\parallel}(\tilde{g},\tilde{t}) = |\tilde{t}|^{\beta_b} \, M_{\pm}^{\parallel}\left(\frac{\tilde{g}}{|\tilde{t}|^{\phi}}\right) , \tag{7}$$

where $M_{\pm}^{\parallel}(x)$ is a scaling function with two branches, for $\tilde{t}>0$ and $\tilde{t}<0$. The subscript b in β_b is to emphasize that it refers to the BCP. The variable x is normalized so that x=1 on the Ising line $T_c^{\parallel}$. A relation similar to Eq. (7) also holds for $M_{st}^{\perp}$.

According to Eq. (7), $M_{st}^{\parallel}/|\tilde{t}|^{\beta_b}$ should depend only on the ratio $\tilde{g}/|\tilde{t}|^{\phi}$. Such "data collapsing" is shown in Fig. 8, which is taken from Ref. 26. Here, ϕ=1.20 and β_b=0.34 were used. They are close to the predicted values ϕ=1.175 and β_b=0.346, respectively.

It is shown in Ref. 26 that the scaling function $M_{\pm}^{\parallel}(x)$ should have the following asymptotic behaviors

$$M_{\pm}^{\parallel}(x) \propto x^{\beta_b/\phi} \qquad \text{for} \quad x \to \infty \tag{8a}$$

$$M_{+}^{\parallel}(x) \propto (x-1)^{\beta} \qquad \text{for} \quad x \to 1 \tag{8b}$$

$$M_{-}^{\parallel}(x) = \text{const.} \qquad \text{for} \quad x \to 0 \, . \tag{8c}$$

These three equations correspond, respectively, to the regions near the $\tilde{t}$=0 axis, near the line $T_c^{\parallel}$, and near the spin-flop line. The exponent β in Eq. (8b) refers to the Ising transitions on the $T_c^{\parallel}$ line (to be distinguished from β_b). Similar expressions hold for $M_{\pm}^{\perp}$. The data of Fig. 8 show the behaviors expected from Eqs (8). The values for β and β_b/ϕ which are indicated in Fig. 8 are close to those predicted by the theory.

ACKNOWLEDGEMENTS

Much of my knowledge of bicritical points was gained while working together with C.C. Becerra, N.F. Oliveira, Jr., and T.S. Chang. The Francis Bitter National Magnet Laboratory is supported by the National Science Foundation.

REFERENCES

1. M.E. Fisher and D.R. Nelson, Phys. Rev. Lett. 32, 1350 (1974).
2. M.E. Fisher, AIP Conf. Proc. 24, 273 (1975); Phys. Rev. Lett. 34, 1634 (1975).
3. J.M. Kosterlitz, D.R. Nelson, and M.E. Fisher, Phys. Rev. B 13, 412 (1976).
4. W.P. Wolf, Physica 86-88B, 550 (1977).
5. L.P. Kadanoff, in "Critical Phenomena, Proceedings of the International School of Physics 'Enrico Fermi', Course LI" (Academic Press, New York, 1971).
6. H.E. Stanley, T.S. Chang, F. Harbus, and L.L. Liu, in "Local Properties at Phase Transitions, Proceedings of the International School of Physics 'Enrico Fermi', Course LIX" (North Holland, Amsterdam, 1976).
7. A. Aharony, in "Phase Transitions and Critical Phenomena", edited by C. Domb and M.S. Green (Academic Press, London, 1976), Vol. 6.
8. P. Pfeuty, D. Jasnow, and M.E. Fisher, Phys. Rev. B 10, 2088 (1974). See also, S. Singh and D. Jasnow, Phys. Rev. B 12, 493 (1975), and P.R. Gerber and M.E. Fisher, Phys. Rev. B 13, 5042 (1976).
9. See, for example, Y. Shapira and S. Foner, Phys. Rev. B 1, 3083 (1970).

.0. M.E. Fisher, J.H. Chen, and H. Au-Yang, J. Phys. C 13, L459 (1980). See also, Ref. 11.

.1. A.D. Bruce, J. Phys. C 8, 2992 (1975).

.2. A.D. Bruce and A. Aharony, Phys. Rev. B 11, 478 (1975). I only quote their results for $n<n_c(d=3)$, because $n_c(3)$ is believed to be higher than 3 [see E. Domany and E.K. Riedel, J. Appl. Phys. 50, 1804 (1979)].

3. D. Mukamel, M.E. Fisher, and E. Domany, Phys. Rev. Lett. 37, 565 (1976).

4. H. Rohrer, B. Derighetti, and Ch. Gerber, Physica 86-88B, 597 (1977).

5. H. Rohrer and Ch. Gerber, Phys. Rev. Lett. 38, 909 (1977), and J. Appl. Phys. 49, 1341 (1978).

6. E. Domany and M.E. Fisher, Phys. Rev. B 15, 3510 (1977).

7. C.K. Chepurnykh, Soviet Physics-Solid State 10, 1517 (1968).

8. H. Rohrer and H. Thomas, J. Appl. Phys. 40, 1025 (1969).

9. Y. Shapira, N.F. Oliveira, Jr., and T.S. Chang, Phys. Rev. Lett. 42, 1292 (1979), and Phys. Rev. B 21, 1271 (1980). See also, N.F. Oliveira, Jr. and Y. Shapira, J. Appl. Phys. 50, 1790 (1979).

0. H. Rohrer, Phys. Rev. Lett. 34, 1638 (1975).

1. K.W. Blazey, H. Rohrer, and R. Webster, Phys. Rev. B 4, 2287 (1971).

2. W. van der Lugt and N.J. Poulis, Physica 26, 917 (1960).

3. G.J. Butterworth and J.A. Woollam, Phys. Lett. 29A, 259 (1969).

24. A.R. King and H. Rohrer, Phys. Rev. B 19, 5864 (1979), and AIP Conf. Proc. 29, 420 (1976).
25. N.F. Oliveira, Jr., A. Paduan Filho, S.R. Salinas, and C.C. Becerra, Phys. Rev. B 18, 6165 (1978); N.F. Oliveira, Jr., A. Paduan Filho and S.R. Salinas, Phys. Lett. A 55, 293 (1975).
26. J.A.J. Basten, E. Frikkee, and W.J.M. deJonge, Phys. Lett. A 68, 385 (1978); Phys. Rev. Lett. 42, 897 (1979); Phys. Rev. B 22, 1429 (1980).
27. Y. Shapira and C.C. Becerra, Phys. Lett. A 57, 483 (1976), Errata, ibid, 58, 493 (1976).
28. Y. Shapira and C.C. Becerra, Phys. Rev. B 16, 4920 (1977); Phys. Lett. A 59, 75 (1976).
29. Y. Shapira and C.C. Becerra, Phys. Rev. Lett. 38, 358 (1977), Errata, ibid, 38, 733 (1977).
30. Y. Shapira and N.F. Oliveira, Jr., Phys. Rev. B 17, 4432 (1978).
31. C.C. Becerra, Y. Shapira, and N.F. Oliveira, Jr., Phys. Rev. B 18, 5060 (1978).
32. Y. Shapira and N.F. Oliveira, Jr., J. Appl. Phys. 49, 1374 (1978).
33. J.A.J. Basten, W.J.M. deJonge, and E. Frikkee, Phys. Rev. B 21, 4090 (1980).
34. J.P.M. Smeets, E. Frikkee, and W.J.M. deJonge, Phys. Rev. Lett. 49, 1515 (1982).
35. Y. Shapira, J. Appl. Phys. 52, 1926 (1981).
36. See E. Domany, D.R. Nelson, and M.E. Fisher, Phys. Rev. B 15, 3493 (1977), and references therein.

THE EXPERIMENTAL EVIDENCE FOR A LIFSHITZ POINT IN MnP

Y. Shapira

Francis Bitter National Magnet Laboratory
Massachusetts Institute of Technology
Cambridge, MA 02139, USA

I. INTRODUCTION

In 1975 Hornreich, Luban and Shtrikman introduced a new multicritical point, which they called a Lifshitz point (LP).[1] Since then several experimental attempts to observe the LP in a real material were made. These include: unsuccessful searches in magnet systems,[2,3] a study of the structural transition of $RbCaF_3$ which suggested the existence of a LP,[4] and experiments on liquid crystals.[5-7] In this lecture I shall review several studies of MnP, which strongly suggest the existence of a LP. This review is largely based on works done in collaboration with C.C. Becerra and N.F. Oliveira, Jr. (University of São Paulo), R.M. Moon and J.W. Cable (Oak Ridge National Laboratory), and T.S. Chang (MIT).[8-12]

II. THE LIFSHITZ POINT

The theory of the LP is reviewed in Ref. 13. Here I shall concentrate mainly on theoretical results which are relevant to MnP. I shall start by introducing the LP via two related simple models: the R-S model and the ANNNI model.

A. The LP in The R-S and ANNNI Models

Magnetic systems exhibit a variety of ordered phases. Among them are: (1) the ferromagnetic phase, with a magnetization uniform in space, and (2) various helicoidal (modulated) phases in which the magnetization is periodic in space. Examples of the latter are sinusoidal, spiral, cone, and fan phases.

Early works[14-16] showed that one mechanism which can lead to helicoidal phases is a competition between exchange interactions of a given spin with various neighbors. The R-S model[17] is a particular model for such a competition. In this model the spins are located on a simple cubic lattice. The exchange interactions in the x-y plane are between nearest-neighbors only, with an exchange constant $J_0>0$ (i.e., ferromagnetic). Along the z-direction there are two competing exchange interactions: $J_1>0$ between nearest neighbors, and $J_2<0$ between next-nearest neighbors.[18] The exchange interactions may be of the Ising, XY, or Heisenberg types. A special case of the R-S model is the ANNNI model,[19] with Ising interactions, and with $J_0=J_1$.

To introduce the LP it is convenient to consider some mean-field (MF) results for the R-S model[17]. the q-dependent susceptibility in the paramagnetic (para) phase is given by

$$\chi(\vec{q}) = A/[kT - B\tilde{J}(\vec{q})] \,, \qquad \text{MF} \qquad (1)$$

where A and B are constants which depend on the spin S, and $\tilde{J}(\vec{q})$ is the Fourier transform of $J(\vec{r})$. Note that $\tilde{J}(\vec{q}) = \tilde{J}(-\vec{q})$. As T decreases, an order-disorder transition will occur when the first divergence in $\chi(\vec{q})$ appears. If the highest value of $\tilde{J}(\vec{q})$ is at $\vec{q}=\pm\vec{q}_0$ then $kT_c = B\tilde{J}(\vec{q}_0)$. The ordered phase just below T_c is characterized by $\vec{q}_0$. In the present model the value of q_0 depends on $\kappa \equiv J_2/J_1 = -|J_2/J_1|$.[20] If $\kappa > -0.25$ then $q_0=0$ and the ordered phase is ferromagnetic. For $\kappa < -0.25$, q_0 is nonzero, is parallel to $\hat{z}$, and is given by

$$\cos(q_0 a) = -1/(4\kappa) \,, \qquad \text{MF} \qquad (2)$$

where <u>a</u> is the lattice parameter. The ordered phase just below T_c is then helicoidal.

We now turn to the phase diagram. To be specific, we assume that J_0 and J_1 are fixed, but that $J_2=\kappa J_1$ can vary. The phase diagram is sketched in Fig. 1. The order-disorder boundary $T_c(\kappa)$ consists of two segments: (1) the para-ferro segment $T_\lambda(\kappa)$ for $\kappa > -0.25$, and (2) the para-helicoidal segment $T_\lambda^*(\kappa)$ for $\kappa < -0.25$. The point where the two segments meet, at $\kappa_L = -0.25$, is a LP. The boundary $T_1(\kappa)$ separates the two ordered phases.

From Eq. (2) it follows that q_0 approaches zero continuously as the LP is approached along the line T_λ^*. Near the LP, $q_0 \propto (\kappa_L-\kappa)^{1/2}$. From Eq. (1) it follows that the uniform susceptibility $\chi(q=0)$ diverges on the line T_λ. On the para-helicoidal line T_λ^*, $\chi(q=0)$ is finite, but it increases as $(\kappa_L-\kappa)^{-2}$ as $\kappa \to \kappa_L$. An interesting feature of the LP is that for $\kappa=\kappa_L$ the line shape of $\chi(q_z)$ vs q_z is not a Lorentzian.[17]

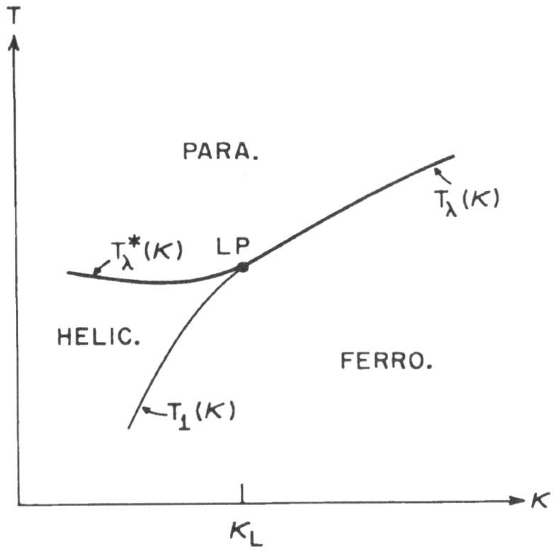

Fig. 1. Schematic for the phase diagram near a Lifshitz point. Here, T is the temperature, and $\kappa=J_2/J_1$ is the ratio of the exchange constants in the R-S (or ANNNI) model.

Going beyond the MF approximation, the ANNNI model was attacked by high-T series,[17] and by other methods.[19] The high-T series give $\kappa_L = -0.27$, compared to the mean-field value -0.25. On the T_λ^* line and close to the LP, these series give

$$q_0 \propto (\kappa_L - \kappa)^{\beta_k} , \tag{3}$$

where $\beta_k = 0.5 \pm 0.15$.

An essential property of the LP is that q_0 approaches zero continuously as the LP is approached along the line T_λ^*. The ANNNI model, with two competing exchange interactions, has this property. However, when there are more than two competing exchange interactions, it is possible sometimes for another type of a critical point to occur.[21] At this critical point, para, ferro and helicoidal phases meet, but q_0 jumps discontinuously from a finite value to zero. In the MF approximation such a critical point can arise when the function $\tilde{J}(\vec{q})$ has three local maxima: one at $\vec{q}=0$, and two others at $\vec{q} = \pm q_1 \hat{z}$, say. The relative heights of these maxima change when the exchange constants change. When the maximum at q=0 is the highest, the transition is to a ferro phase. When the other pair of maxima become the highest (with q_1 remaining nonzero), the transition is to a helicoidal phase with a finite q. In this case, the para-ferro-helicoidal point is not a LP. Thus, not every para-ferro-helicoidal point is a LP.

In many discussions of the LP it is imagined that the exchange interactions can be changed by changing a parameter P, such as pressure or material composition. In such discussions, P plays the role of κ. The phase diagram is then plotted in the P-T plane, and the LP is at (P_L,T_L).

B. A More General Treatment of the LP

A theory of Hornreich et al.[1,13] is not based on a specific model of competing exchange interactions. Instead, it starts from a Landau-type free energy density F. For an isotropic system and for a scalar order parameter M, F has the form

$$F = a_2M^2 + a_4M^4 + a_6M^6 + \ldots + c_1(\nabla M)^2 + c_2(\nabla^2 M)^2 + \ldots \quad (4)$$

The novel feature (compared to a simple critical point) is that c_1 is assumed to vary continuously as a function of a parameter P. In particular, c_1 changes its sign at $P=P_L$. The coefficient c_2 is assumed to be positive. When $c_1(P)$ is positive, the ordered phase is ferromagnetic. For a negative $c_1(P)$, the ordered phase is helicoidal. The LP occurs at $P=P_L$, where $c_1=0$. Thus, the hallmark of the LP is the vanishing of the term $c_1(\nabla M)^2$ in the free energy. The renormalization-group counterpart of this condition is discussed in Ref. 1 (see also Ref. 22).

Equation (4) can be generalized as follows: (i) In a model for an anisotropic crystal, the term $c_1(\nabla M)^2$ in Eq. (4) is replaced by a sum of $c_i(\partial M/\partial x_i)^2$, where $i=1,2,\ldots d$ (d is the lattice dimensionality). At a LP, one or more of the c_i's must vanish. The number of c_i's which vanish is called m. Clearly, $m \leqslant d$. (ii) The scalar M in Eq. (4) can be replaced by a vector with n components.

The universality class (UC) of a LP is characterized by d,n, and m. A discussion of the UC's of all Lifshitz points which can occur in various 3d crystal structures was given by Hornreich.[23] For orthorhombic crystals, the only allowed UC is n=m=1 (and d=3, of course). Because MnP has an orthorhombic structure, we shall be concerned primarily with this UC. The ANNNI model, in 3d, also belongs to this UC.

The upper critical dimension for a LP with m=1 is $d_u=4.5$.[1] Thus, $\varepsilon=1.5$ for a LP with d=3, n=m=1. The critical exponents for this LP, to first order in ε, are given in Ref. 1 (most of them in a footnote). Later, we will need the exponent β_k, defined by the behavior of q_0 on the line T^*_λ, i.e.,

$$q_0 = \text{const.}\ |P - P_L|^{\beta_k} . \quad (5)$$

This is analogous to Eq. (3) for the ANNNI model. To first order in

ε, the exponent β_k retains its mean-field value 0.5. A second-order expansion in ε gives β_k=0.54 for d=3, n=m=1.[24]

Several treatments of the phase diagram for d=3, n=m=1 were given. They include: results based on a Landau-type theory,[25] high-T series calculations,[17] Monte-Carlo simulations,[19] and scaling theory supplemented by RG calculations.[26] The two segments of the λ line, T_λ^* and T_λ, are expected to be tangent to each other at the LP. Also, at the LP the λ line is expected to have an inflection point.[17] The crossover exponent ϕ, which governs the shape of the λ line near the LP, will be discussed later. The only detailed treatment of the ferro-helicoidal line $T_1(P)$ is a Landau-type treatment.[25] It indicates that the transitions on this line are of first order, and that the line is tangent to the λ line at the LP. It should be mentioned, however, that Selke and Fisher (Ref. 19, 1980 paper) have suggested that the transitions on the line T_1 may not be of first order.

In the preceding discussion we have considered only "ferromagnetic LP's" at which ferro, helicoidal, and para phases meet. More general types of LP's are discussed in Refs. 1,13. Still other types of LP's are discussed in Refs. 21,22. These generalizations will not be needed here.

III. SOME PROPERTIES OF MnP

A. Global Phase Diagram for $\vec{H} \parallel \hat{b}$

MnP is a metal with orthorhombic structure (a>b>c). The magnetic properties of single crystals were first measured by Huber and Ridgley.[27] Later, the static magnetic properties were investigated in Japan by Komatsubara et al. (e.g., Refs. 28,29), and by others.[30] Several neutron-diffraction studies were performed in the sixties.[31-33]

The early studies showed that MnP has the following properties. At H=0, MnP becomes ferromagnetic at T_C=291 K. The magnetic moment in the ferro phase is parallel to c. The b and a directions are the intermediate and hard axes, respectively. Still at H=0, a first-order transition from the ferro phase to a "screw phase" occurs as MnP is cooled through $T_\alpha \cong 47$ K. (The value of T_α depends slightly on purity.[34]) The screw phase is a spiral, with $\vec{q} \parallel a$, and with the moments rotating in the cb plane (easy-medium plane). The period is roughly 9 lattice spacings.[31,32] Finer details of the screw phase were investigated recently.[35,36]

We shall be primarily concerned with the phase diagram for $\vec{H} \parallel b$. The global features of this phase diagram were established by

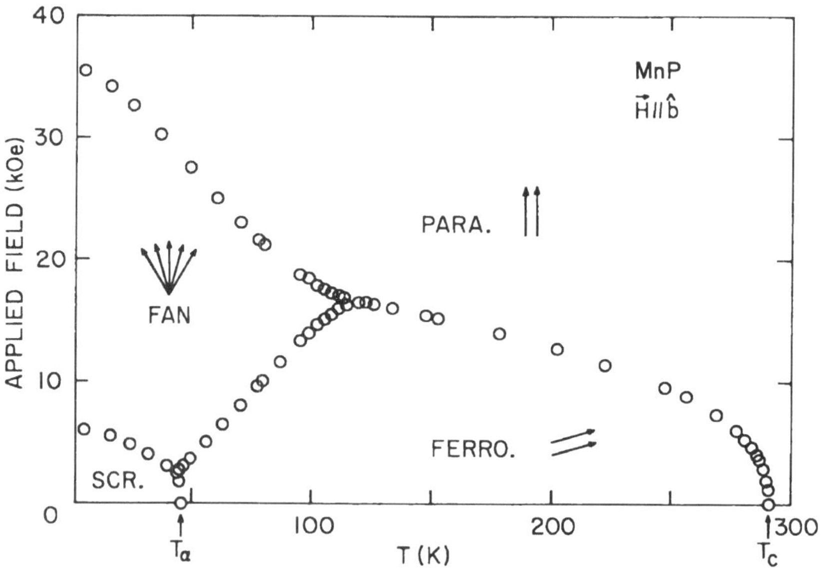

Fig. 2. Global phase diagram of MnP, for $\vec{H}$ parallel to b. The screw phase is designated as SCR. (Ref. 9).

Komatsubara et al.[28,29] Figure 2 shows more recent results.[9] In addition to the ferro and screw phases there is also a "fan phase". This phase is helicoidal, with $\vec{q} \parallel \underline{a}$.[33] The local moment oscillates about the b direction, remaining always in the bc plane. This is in contrast to the screw phase in which the moment undergoes a full rotation in the bc plane. The screw-ferro and screw-fan transitions are of first order, and the screw-ferro-fan triple point is an ordinary triple point. Of greater interest here is the other triple point, i.e., the para-ferro-fan (PFF) triple point. Because the fan phase is helicoidal, the PFF triple point satisfied one of the prerequisites for a LP. This fact, first pointed out by C.C. Becerra, was the stimulus for the recent investigations of MnP.

The para-ferro line $H_\lambda(T)$ and the para-fan line $H^*_\lambda(T)$ are lines of second-order transitions. [The notation $H_\lambda(T)$ and $H^*_\lambda(T)$ is purposely chosen to be similar to the notation $T_\lambda(\kappa)$ and $T^*_\lambda(\kappa)$ in Sec. II.] The para-ferro transition may be interpreted physically as follows. At $H>H_\lambda(T)$ the magnetic moment is parallel to b. When H is lowered through H_λ, the magnetic moment develops a uniform component along c. As H decreases further, the c component grows while the b component decreases. At H=0, the moment is a parallel to c. (Because the ±c directions are equivalent, there are two types of domains in the ferro phase.) The order parameter associated with the para-ferro transition is the c component of the magnetization, so that n=1. In the case of the para-fan transition, the c component of the magnetization develops an oscillatory component as H is lowered through $H_\lambda^*(T)$. The wave vector $\vec{q}$ in the fan phase has a single component, parallel to a. Thus, if the PFF triple point is a LP then m=n=1.

The transitions on the ferro-fan line $H_1(T)$ have the features expected for first-order transitions.

B. Model for the Magnetic Interactions

Discussions of recent data near the PFF triple point are based on two complementary approaches: (1) scaling theory, without a specific model for the magnetic interactions, and (2) a specific model. Clearly, the first approach is more general. The second approach gives a greater physical insight, but has the risk of being oversimplified. In this section the basic features of the specific model are described. Scaling will be discussed later.

To interpret the low-temperature properties of MnP, Hiyamizu and Nagamiya[37] introduced a model which contains several implicit and explicit assumptions:

(i) The magnetic moments are treated as localized. This is in spite of the fact that MnP is a metal. (The resistivity at 300 K is two orders of magnitude higher than that of copper.[34]) Also, the saturation magnetization of MnP corresponds to a moment of $1.3\mu_B$ per Mn atom,[27,36] which is well below the value for manganese in insulators. Nevertheless, it has been stated that a localized-spin model should be applicable to MnP.[38]

(ii) The crystal structure of MnP is shown in Fig. 3(a). The Mn atoms are located on planes which are perpendicular to a. These planes are separated by 0.1a, 0.4a, 0.1a, 0.4a, etc. The magnetic structure of the screw phase is shown in Fig. 3(b). It consists of ferromagnetic planes with: (1) parallel spin directions for planes separated by 0.4a, and (2) a turn angle of about 20° between spins on planes separated by 0.1a. In the model of Hiyamizu and Nagamiya a pair of planes separated by 0.4a (with parallel spins) is replaced by a single plane. Thus, the model consists of a single set of equidistant parallel planes. In each of these planes the spins are ferromagnetically coupled.

(iii) The magnetic interactions (apart from the ferromagnetic coupling in each of the planes) consist of: (1) isotropic exchange coupling $J_{|n-m|}$ between the nth and mth planes; (2) anisotropic exchange coupling between planes; (3) a uniaxial anisotropy, with a as the hard axis; (4) second-order and fourth-order anisotropies in the bc plane; and (5) the Zeeman energy.

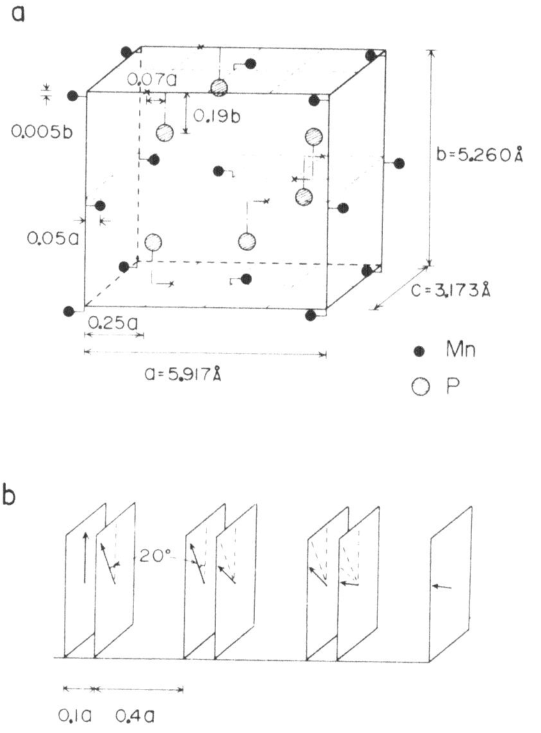

Fig. 3. (a) Crystal structure of MnP. (b) The magnetic structure of the screw phase at 4.2 K. Spins on each bc plane are parallel, and the propagation vector $\vec{q}$ is along the a axis. (Ref. 37).

The model of Hiyamizu and Nagamiya was simplified by Yokoi et al.[39] and by Tajima et al.[40] In both of these works the anisotropic exchange interaction is ignored. In addition, the anisotropy in the bc plane is restricted to be of second order. The helicoidal phases (screw and fan) arise from a competition between the effective isotropic exchange parameters $J_{|n-m|}$.

Using a mean-field approach, Yokoi et al. showed that the nature of the order-disorder transition on the λ line in Fig. 2 is governed by the Fourier transform $\tilde{J}(q)$ of $J_{|n-m|}$, for $\vec{q} \parallel \underline{a}$. When the maximum of $\tilde{J}(q)$ is at $q_0=0$, the transition is to a ferromagnetic phase. This corresponds to the segment $H_\lambda(T)$ of the λ line. When $\tilde{J}(q)$ is largest at $\pm q_0 \neq 0$, the transition is to a fan phase. This corresponds to the segment $H_\lambda^*(T)$. The fact that both types of transitions are observed implies that $\tilde{J}(q)$ changes along the λ line. Thus, the effective exchange constants $J_{|n-m|}$ depend on T, or on H, or on both.

The function $\tilde{J}(q)$ is even in q. Near q=0 it can be expanded as

$$\tilde{J}(q) = \tilde{J}(0) - \tilde{a}q^2 - \tilde{b}q^4 - \ldots \qquad (6)$$

Yokoi et al. assume that $\tilde{b}(T,H)$ is positive and that $\tilde{a}(T,H)$ changes its sign at the PFF triple point. If these assumptions are accepted, then the PFF triple point is a LP. However, it is essential to examine the alternative possibility that these assumptions are wrong. As the discussion in Sec. II indicates, the function $\tilde{J}(q)$ may have three or more peaks which change in such a way that q_0 jumps discontinuously from a finite value to zero. In that case the PFF triple point is not a LP. The evidence against this possibility is discussed in the next subsection and in Sec. V.

C. Dispersion Curves for Spin Waves at H=0

The dispersion curves $\omega(\vec{q})$ for spin waves were measured by Tajima et al. using neutron techniques.[40] The data were taken at H=0, between 38 and 150 K, and for $\vec{q}$ parallel to $\underline{a}$ and $\underline{b}$. Figure 4 shows their results. For $\vec{q} \parallel \underline{b}$ the dispersion curves do not show unusual features. More interesting are the dispersion curves for $\vec{q} \parallel \underline{a}$, which is the direction of the propagation vector in the helicoidal phases. These dispersion curves are unusually flat in the region of small q's, and they also show a marked dependence on T.

To interpret their data, Tajima et al. used the simplified version of the Hiyamizu-Nagamiya model, as discussed earlier. In the ferro phase, and at H=0, the expected dispersion curve is

$$\hbar\omega = 2S\{[\tilde{J}(0) - \tilde{J}(q) - 4D_2][\tilde{J}(0) - \tilde{J}(q) + D_1 - 2D_2]\}^{\frac{1}{2}}, \qquad (7)$$

where D_1 and D_2 are anisotropy constants. An expression for $\hbar\omega$ in the screw phase was also given.[40]

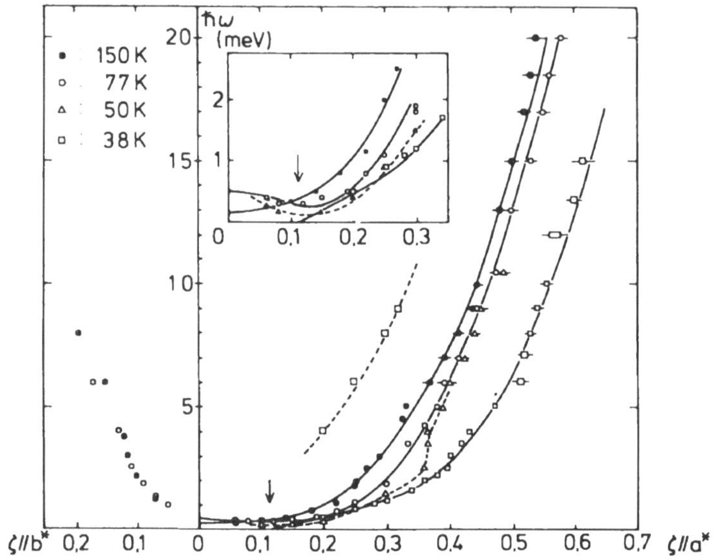

Fig. 4. Dispersion curves for spin waves along the a and b axes at various temperatures. Solid lines are fits to theoretical expressions. Dotted line with squares is the dispersion curve at 38 K (below T_α) arising from another Brillouin zone. ξ is the reduced wave vector. Arrow indicates the zone center for the screw phase. (Ref. 40).

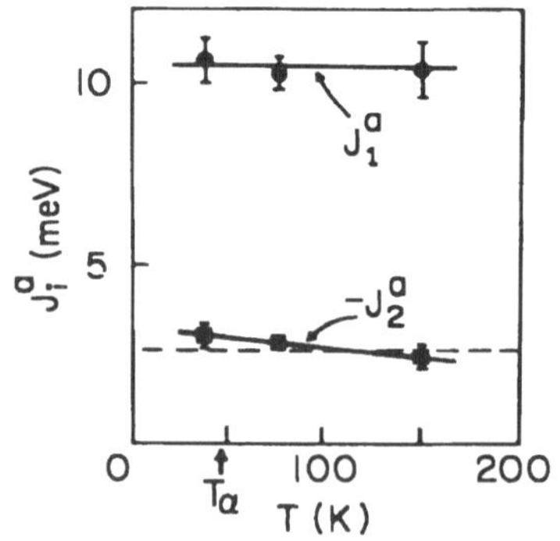

Fig. 5. T-dependence of the nearest-plane and next-nearest-plane exchange constants of MnP, J_1^a and J_2^a, respectively. The dashed line shows $\kappa = J_2^a/J_1^a = -0.25$. (Ref. 40).

The shapes of the dispersion curves for $\vec{q}\parallel \underline{a}$ are such that none of them can be accounted for by a single exchange constant J_1^a between nearest planes. To explain their data, Tajima et al. used two effective exchange constants: J_1^a for nearest planes, and J_2^a for next-nearest planes. With this additional simplification the model has many similarities to the R-S model. In particular, if $\kappa = J_2^a/J_1^a$ is equal to -0.25, then the parameter $\tilde{a}$ in Eq. (6) vanishes, and $\tilde{b}>0$. These are just the properties required for a LP in the mean-field treatment of Yokoi et al. The values of J_1^a and J_2^a obtained by Tajima et al. from fits of dispersion curves are shown in Fig. 5. Note that κ is close to -0.25, and is T-dependent. This T-dependence arises primarily from the variation of J_2^a. The temperature where κ is equal to -0.25 is close to the temperature of the PFF triple point in Fig. 2.

In a more general analysis of the dispersion curves, one may consider more than two effective interplanar exchange constants. The unusual flatness of the dispersion curves, in the ferro phase and for the region of small q's, implies that $\tilde{J}(q)$ has an unusually small curvature near q=0 [see Eq. (7)]. Thus, the parameter $\tilde{a}$ is very small. Moreover, the data in the inset of Fig. 4 strongly suggest that $\tilde{a}$ changes its sign between 77 and 150 K. Finally note that each of the dispersion curves in Fig. 4 seems to have only a single minimum in the range $q\geqslant 0$. This corresponds to a single maximum of $\tilde{J}(q)$ in this range (no more than two maxima if negative q's are also considered).

In summary, the effective interplanar exchange constants of MnP are favorable for the occurrence of a LP, and are T-dependent. A theoretical treatment of the temperature variation of the effective exchange constants of MnP was given by Takase and Kasuya,[38] although neither they nor Tajima et al. discussed the possibility of a LP.

IV. PHASE DIAGRAM NEAR THE PFF TRIPLE POINT

A. Experimental Results

Detailed measurements of the phase diagram near the PFF triple point were performed at the University of São Paulo and at MIT.[8,9] Three samples, cut from the same boule, were used. The results were in good agreement with each other. Figure 6 shows the data for one of the samples. The PFF triple point is at T_t=121±1 K.

The results in Fig. 6 have the qualitative features expected near a LP: All boundaries are tangent to each other at the PFF triple point, and the λ line has an inflection point at this triple point. It was also confirmed that the transitions on the λ line were of second order. The transitions on the ferro-fan line

$H_1(T)$ appeared to be of first order, in agreement with earlier data and with predictions based on a Landau-type treatment.[25]

B. Scaling Axes

To analyze the phase diagram, it is assumed that the PFF triple point is a multicritical point which obeys generalized scaling. Two linear scaling axes are introduced, as shown in Fig. 7(a). As usual, one axis ($\bar{p}$) is tangent to the phase boundaries at the triple point, and the other ($\bar{t}$) is not. The $\bar{t}$ axis is chosen to be parallel to the magnetic-field axis H_b, as discussed in Ref. 9.

In order to relate the phase diagram of MnP to the usual theoretical discussions of the LP, a schematic of the latter is shown in Fig. 7(b). The coordinates are P and T, and the scaling axes are μ_p and μ_t. Clearly, the scaling axes $\bar{p}$ and $\bar{t}$ for MnP correspond to μ_p and μ_t in the usual theory. Also, the axes T and H_b for MnP roughly correspond to the P and T axes in the usual theory, respectively.

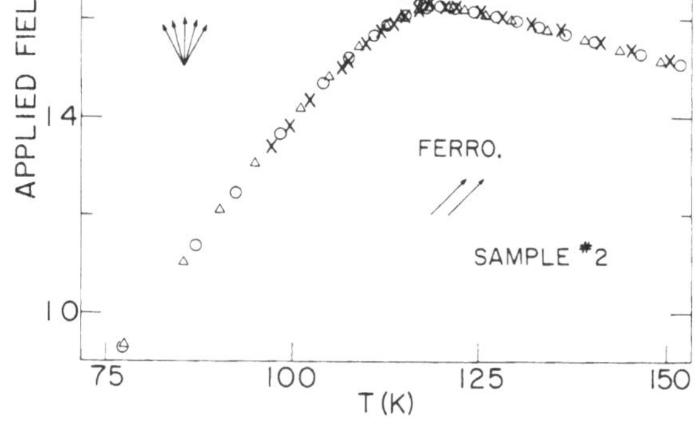

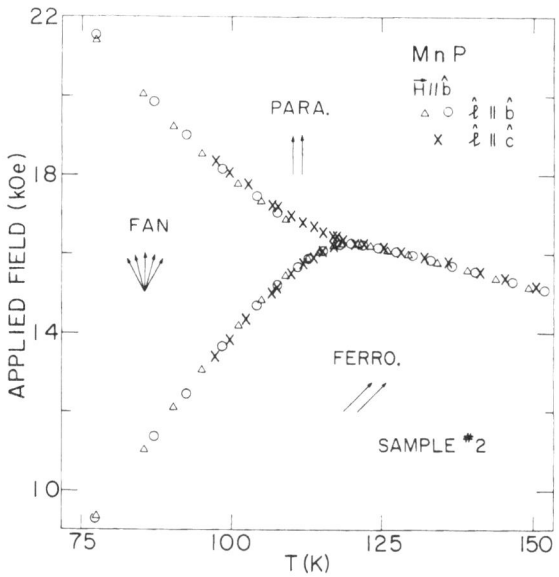

Fig. 6. The phase diagram near the para-ferro-fan (PFF) triple point for $\vec{H}\|\underline{b}$. (Refs. 8 and 9).

The correspondence between Figs 7(a) and 7(b) deserves some explanation. Consider first the parameter P in Fig. 7(b). This parameter is the analog of $\kappa=J_2/J_1$ in the R-S and ANNNI models. In the simple model for MnP (Sec. III C) the analogous parameter is J_2^a/J_1^a, which varies with T. Thus, a variation of T in the case of MnP corresponds to a variation of P in the standard treatments for a LP. A possible objection to this argument is that J_2^a/J_1^a in MnP may also depend on H. While this may be the case, some data which will be presented later suggest that J_2^a/J_1^a depends primarily on T and is not strongly dependent on H.

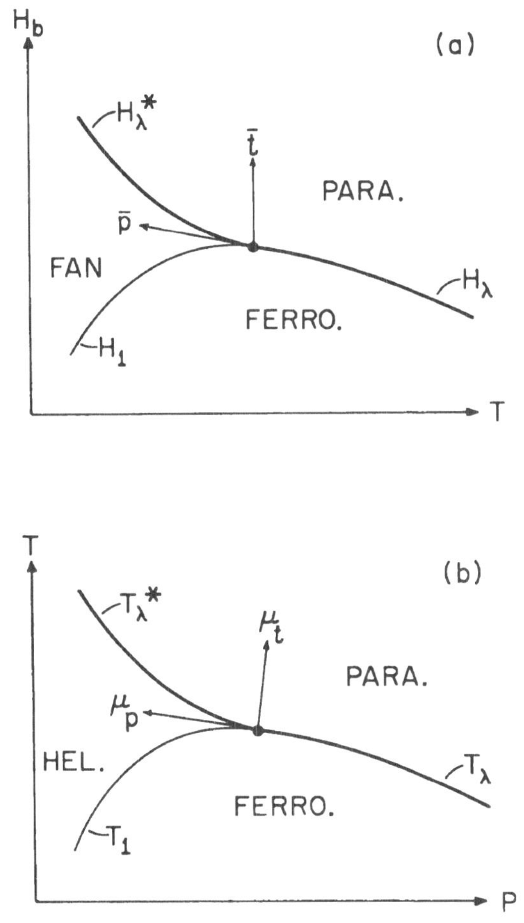

Fig. 7. (a) Scaling axes $\bar{p}$ and $\bar{t}$ at the PFF triple point of MnP. (b) Scaling axes μ_p and μ_t in the usual theory for a LP. (Ref. 9).

The correspondence between T in Fig. 7(b) and H_b in Fig. 7(a) can be explained in several ways. One explanation is based on the scaling theory of Riedel and Wegner for anisotropic ferromagnets.[41] This is discussed in Ref. 9. A more physical argument is the following. In the ordered phases of MnP there is a nonzero local magnetization component along c (uniform for the ferro phase, and oscillatory for the fan phase). These ordered phases disappear at high H_b because a high magnetic field along b destroys the c component of the local magnetization. The analogous process in the standard models for the LP is the destruction of the ordered phased at high T. Thus, H_b is analogous to T. Further support for this analogy can be given by comparing the results of Yokoi et al. for MnP (Ref. 39) with the corresponding mean-field results for the ANNNI model.

C. Crossover Exponent

Generalized scaling implies that near the PFF triple point the shape of either of the boundaries $H_\lambda(T)$ and $H_\lambda^*(T)$ is given by

$$\bar{t} = \text{const.}\ |\bar{p}|^{1/\phi}\ , \qquad (8)$$

where ϕ is the crossover exponent. Attempts to obtain ϕ from the shapes of the boundaries $H_\lambda(T)$ and $H_\lambda^*(T)$ were hindered by the fact that the direction of the $\bar{p}$ axis (tangent to the boundaries at the triple point) was not determined with sufficient precision. To get around this difficulty, generalized scaling was also applied to the ferro-fan line $H_1(T)$, which led to Eq. (8) for this line also. It then follows that the difference $H_\lambda^*-H_1$, for the same T, should be proportional to $(T_t-T)^{1/\phi}$, where T_t is the temperature of the triple point. Least squares fits of the data to this relation gave $\phi=0.63\pm0.04$.[8,9] This is higher than the mean-field value $\phi=0.5$, but is in good agreement with $\phi=0.625$ from first-order ε expansion.[26]

V. NEUTRON SCATTERING RESULTS FOR $\vec{q}$

Measurements of the variation of q in the fan phase of MnP, performed at Oak Ridge National Laboratory, showed that q decreases continuously as the PFF triple point is approached.[10] For technical reasons it was not possible to measure q when it became very small (very close to the PFF triple point). The smallest q which was actually observed corresponded to a period of 32 lattice spacings. The qualitative features of the data for q were consistent with a Lifshitz behavior.

Assuming that the PFF triple point is a LP, the analog of Eq. (5) is

$$q_0 = \text{const.} \; (T_t - T)^{\beta_k} , \tag{9}$$

where q_0 is the value of q on the line $H_\lambda^*(T)$. Fits of the data in the range 89<T<117 K to this equation gave β_k=0.44±0.05 and 0.49±0.03.[10] These two values correspond to slightly different choices of the boundary $H_\lambda^*(T)$ at which q_0 was evaluated; in one case the boundary was estimated from the neutron data themselves, while in the other it was taken from magnetostriction measurements. The above values for β_k are regarded as approximate. They are in reasonable agreement with the expected value for a LP (0.5 from mean-field theory, and 0.54 to order ε^2).

In a later work, R.M. Moon extended the measurements of q_0 to lower temperatures, down to 4 K.[42] His results are shown in Fig. 8. Here, $\delta=(a/2\pi)q$, with q evaluated at the para-fan boundary as determined from magnetostriction data. The value of δ at 4 K agrees with Ref. 36 and corresponds to a period of 7.2 lattice spacings. It is surprising that all the data in Fig. 8 can be described fairly well by Eq. (9). The fit to this equation (solid line in Fig. 8) gives β_k=0.480±0.013 and T_t=122.3±0.7 K. However, because Eq. (9) is for T close to T_t, the values obtained from this fit may not be significant.

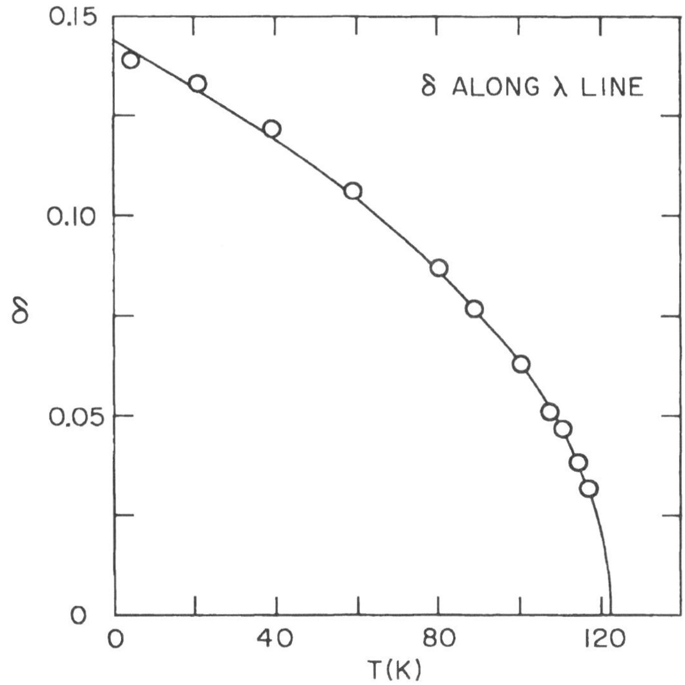

Fig. 8. Temperature variation of q_0 on the para-fan boundary $H_\lambda^*(T)$. Here, $\delta=(a/2\pi)q_0$. (R.M. Moon, private communication.)

A possible explanation as to why Eq. (9) holds reasonably well over a wide temperature range is the following. Equation (9) is the analog of Eq. (3), which is valid for κ near κ_L. In the case of MnP, the data in Fig. 5 suggest that: (1) $\kappa = J_2^a/J_1^a$ remains close to κ_L for all temperatures below T_t, and (2) the difference $(\kappa-\kappa_L)$ may be approximately proportional to (T_t-T).

VI. MEASUREMENTS WITH NONZERO H_c

In MnP the c component of $\vec{H}$ is the ordering field for the ferro phase. Thus, H_c is the analog of H in the usual discussions for a LP. Two types of experiments with $H_c \neq 0$ were performed:

(i) Measurements of the "transverse susceptibility" $\chi_c=\partial M_c/\partial H_c$, in which a small ac modulation field along c was superimposed on the large dc magnetic field along b. The susceptibility χ_c is the analog of $\chi(q=0)$ in Sec. IIA.

(ii) Measurements of the phase diagram when the dc magnetic field has both b and c components. This phase diagram, in the T-H_b-H_c space, is the analog of the theoretical phase diagram (near a LP) in the P-T-H space. That is, adding a coordinate H_c to Fig. 7(a) is analogous to adding a coordinate H to Fig. 7(b). Mean-field (or Landau-type) treatments of the latter situation were given in Refs. 43 and 44.

The analysis of the data obtained in these two experiments was complicated by the large demagnetization corrections for H_c.[9] Thus, only the qualitative features of χ_c, and some features of the phase diagram with $H_c \neq 0$, were obtained. These qualitative results were consistent with the interpretation that the PFF triple point is a LP.

VII. PHASE DIAGRAM FOR $\vec{H}$ PARALLEL TO THE HARD AXIS

The behavior of MnP when $\vec{H}$ is parallel to the a axis (hard axis) was studied only recently.[12] Figure 9 shows the phase diagram. Note the existence of a triple point at 123±3 K and at an applied magnetic field H_0=53 kOe. The corresponding internal field (after correcting for the demagnetization) is H=51 kOe.

Among the three ordered phases in Fig. 9, only the ferro phase can be identified with certainty. The magnetic structures of the other two ordered phases have not been determined as yet with neutron techniques. Nevertheless, in Ref. 9 these two phases were tentatively identified as cone and fan. It was also conjectured that the new fan phase is analogous to the fan phase for $\vec{H}\parallel$b. That is: (1) the propagation vector $\vec{q}$ is parallel to a, and (2) the magnetic moment oscillates about the direction of $\vec{H}$, remaining always in the plane containing $\vec{H}$ and the easy axis. If one accepts this interpretation then the triple point at 123±3 K in Fig. 9 is a PFF triple point.

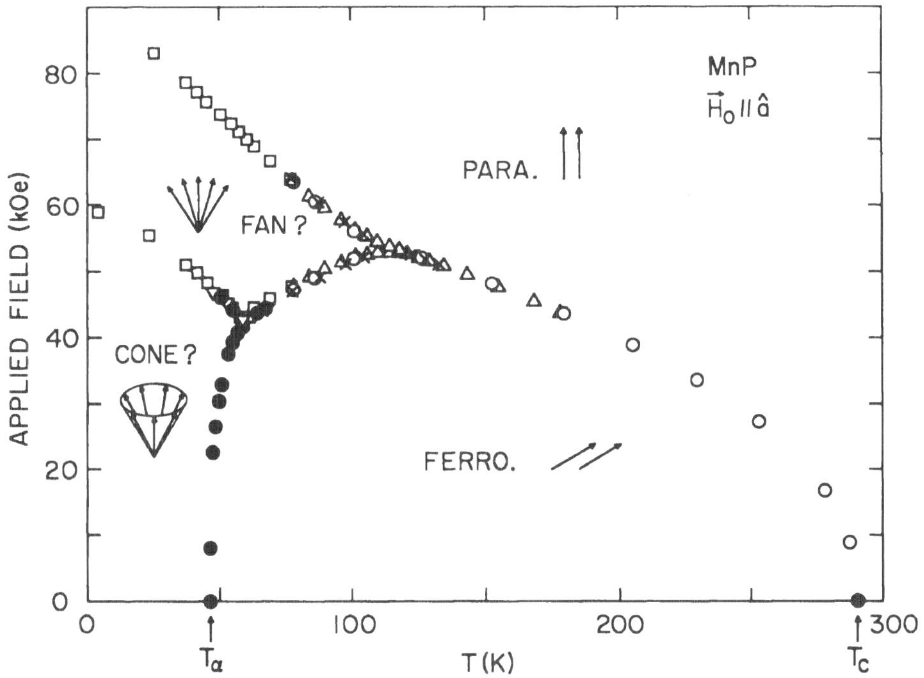

Fig. 9. Global phase diagram of MnP for applied magnetic field parallel to the a axis. (Ref. 12).

Figures 2 and 9 show that the phase diagrams for $\vec{H}\parallel b$ and $\vec{H}\parallel a$ have similar qualitative features near their respective PFF triple points. Moreover, these two triple points occur at practically the same temperature. A plausible interpretation is that both points are LP's. The difference in the values of H at these two points is explained by the difference in the anisotropies. The fact that both points occur at practically the same temperature, but at very different fields, suggests that the ratios of the exchange constants of MnP (J_2^a/J_1^a in the simple model) depend mainly on T and are practically independent of H. All these conjectures should be tested in future studies. One obstacle is the usual geometry of neutron spectrometers which are equipped with superconducting magnets. This geometry prevents measurements of $\vec{q}$'s which are parallel to $\vec{H}$.

VIII. CONCLUSION

The known properties of the PFF triple point, for $\vec{H} \parallel \underline{b}$, strongly suggest that it is a LP. However, additional experimental confirmations will be helpful. Among the studies which may be of interest are the following:

i) Measurements of $\chi(q)$ by neutron techniques. A non-Lorentzian line shape is expected near a LP.

ii) Measurements of critical exponents other than ϕ and β_k.

iii) Repeating some of the earlier measurements, but on samples which are ellipsoidal.

iv) Additional studies of the PFF triple point for $\vec{H} \parallel \underline{a}$.

ACKNOWLEDGEMENTS

Substantial contributions to the work reviewed here were made by C.C. Becerra, N.F. Oliveira, Jr., R.M. Moon, J.W. Cable, and T.S. Chang. The Francis Bitter National Magnet Laboratory is supported by the National Science Foundation.

REFERENCES

1. R.M. Hornreich, M. Luban, and S. Shtrikman, Phys. Rev. Lett. 35, 1678 (1975).
2. S. Legvold, P. Burgardt, and B.J. Beaudry, Phys. Rev. B 22, 2573 (1980).
3. S.K. Sinha, G.H. Lander, S.M. Shapiro, and O. Vogt, Phys. Rev. Lett. 45, 1028 (1980).
4. J.Y. Buzaré, J.C. Fayet, W. Berlinger, and K.A. Müller, Phys. Rev. Lett. 42, 465 (1979); A. Aharony and A.D. Bruce, Phys. Rev. Lett. 42, 462 (1979).
5. G. Sigaud, F. Hardouin, and M.F. Achard, Solid State Commun. 23, 35 (1977).
6. C.R. Safinya, R.J. Birgeneau, J.D. Litster, and M.E. Neubert, Phys. Rev. Lett. 47, 668 (1981).
7. I. Musevic, B. Zeks, R. Blinc, Th. Rasing, and P. Wyder, Phys. Rev. Lett. 48, 192 (1982).
8. C.C. Becerra, Y. Shapira, N.F. Oliveira, Jr., and T.S. Chang, Phys. Rev. Lett. 44, 1692 (1980).
9. Y. Shapira, C.C. Becerra, N.F. Oliveira, Jr., and T.S. Chang, Phys. Rev. B 24, 2780 (1981).
10. R.M. Moon, J.W. Cable, and Y. Shapira, J. Appl. Phys. 52, 2025 (1981).
11. Y. Shapira, J. Appl. Phys. 53, 1914 (1982).
12. Y. Shapira and N.F. Oliveira, Jr., Phys. Lett. 89A, 205 (1982).
13. R.M. Hornreich, J. Magn. Magn. Mater. 15-18, 387 (1980).
14. R.J. Elliot, Phys. Rev. 124, 346 (1961).
15. J. Villain, J. Phys. Chem. Solids 11, 303 (1959).

16. J.S. Smart, Effective Field Theories of Magnetism (Saunders, Philadelphia, 1966), Chap. 11; D.H. Martin, Magnetism in Solids (MIT Press, 1967), p. 266 ff.
17. S. Redner and H.E. Stanley, Phys. Rev. B 16, 4901 (1977), and J. Phys. C 10, 4765 (1977).
18. For simplicity I have put restrictions on the signs of J_0,J_1 and J_2. More general cases are discussed in Refs. 17 and 19.
19. W. Selke, Z. Phys. B 29, 133 (1978); W. Selke and M.E. Fisher, Phys. Rev. B 20, 257 (1979), and J. Magn. Magn. Mater. 15-18, 403 (1980); M.E. Fisher and W. Selke, Phys. Rev. Lett. 44, 1502 (1980); M.E. Fisher, J. Appl. Phys. 52, 2014 (1981).
20. Our definition of κ differs by a minus sign from that in Ref. 19.
21. W. Selke, Phys. Lett. 61A, 443 (1977), and J. Magn. Magn. Mater. 9, 7 (1978).
22. J.F. Nicoll, G.F. Tuthill, T.S. Chang, and H.E. Stanley, Phys. Lett. 58A, 1 (1976), and Physica 86-88B, 618 (1977). J.F. Nicoll, T.S. Chang, and H.E. Stanley, Phys. Rev. A 13, 1251 (1976).
23. R.M. Hornreich, Phys. Rev. B 19, 5914 (1979).
24. D. Mukamel, J. Phys. A L249 (1977).
25. A. Michelson, Phys. Rev. B 16, 577 (1977). See also Ref. 26.
26. D. Mukamel and M. Luban, Phys. Rev. B 18, 3631 (1978).
27. E.E. Huber, Jr. and D.H. Ridgley, Phys. Rev. 135, A1033 (1964).
28. T. Komatsubara, T. Suzuki, and E. Hirahara, J. Phys. Soc. Jpn. 28, 317 (1970).
29. A. Ishizaki, T. Komatsubara, and E. Hirahara, Prog. Theor. Phys. Suppl. 46, 256 (1970).
30. A list of references is given in Ref. 9.
31. G.P. Felcher, J. Appl. Phys. 37, 1056 (1966).
32. J.B. Forsyth, S.J. Pickart, and P.J. Brown, Proc. Phys. Soc. London 88, 333 (1966).
33. Y. Ishikawa, T. Komatsubara, and E. Hirahara, Phys. Rev. Lett. 23, 532 (1969).
34. A. Takase and T. Kasuya, J. Phys. Soc. Jpn. 48, 430 (1980).
35. R.M. Moon, J. Appl. Phys. 53, 1956 (1982).
36. H. Obara, Y. Endoh, Y. Ishikawa, and T. Komatsubara, J. Phys. Soc. Jpn. 49, 928 (1980).
37. S. Hiyamizu and T. Nagamiya, Intern. J. Magn. 2, 33 (1972).
38. A. Takase and T. Kasuya, J. Phys. Soc. Jpn. 47, 491 (1979); A. Yanase and A. Hasegawa, J. Phys. C 13, 1989 (1980).
39. C.S.O. Yokoi, M.D. Coutinho-Filho, and S.R. Salinas, Phys. Rev. B 24, 5430 (1981).
40. K. Tajima, Y. Ishikawa, and H. Obara, J. Magn. Magn. Mater. 15-18, 373 (1980).
41. E. Riedel and F. Wegner, Z. Phys. 225, 195 (1969).
42. R.M. Moon, private communication.
43. C.S.O. Yokoi, M.D. Coutinho-Filho, and S.R. Salinas, Phys. Rev. B 24, 4047 (1981).
44. M.D. Coutinho-Filho and M.A. de Moura, J. Magn. Magn. Mater. 15-18, 433 (1980).

NATURE OF THE SMECTIC-A-SMECTIC-C TRANSITION NEAR A NEMATIC-SMECTIC-A - SMECTIC-C MULTICRITICAL POINT

C. C. Huang and S. C. Lien

School of Physics and Astronomy
University of Minnesota
Minneapolis, MN 55455, USA

A mean-field Landau model with a sixth-order term is employed to analyze the x-ray and heat-capacity data along the smectic-A - smectic-C (AC) transition line near one nematic-smectic-A - smectic-C (NAC) multicritical point. Even though this model gives a satisfactory fitting to all data along the AC transition line, the coefficients obtained for this model suggest that the fluctuations near the NAC point may be important.

I. INTRODUCTION

Liquid crystals are organic molecules, anisotropic in shape, which have one or more mesophases between the crystalline state and the isotropic liquid. Unlike most materials which have a strong first order melting transition, the phase transitions between different mesophases, in many cases, can be continuous or weakly first order. Liquid crystals therefore provide a rich variety of orderings and are excellent systems for studying the three dimensional (d=3) aspects of phase transitions or multicritical points.

Among the various mesophases, three of them are relevant to our discussion here, namely, nematic (N), smectic-A (A) and smectic-C (C) phases. In the nematic phase, on average the long axis of each molecule (molecular director) is oriented along a preferred direction. The center of mass of each molecule is, however, free to diffuse throughout the system so that translational invariance is preserved. The smectic-A and the smectic-C phases are characterized by a one-dimensional density wave whose vector is along (A) or tilted with respect to (C) the molecular director. The molecules maintain translational invariance within the smectic planes. Between these three phases three phase transitions are possible, i.e., NA, AC and NC transitions. In 1973 de Gennes[1] proposed three theoretical models for these three transitions, respectively. Since then, the NA transition has been the most studied. So far the nature of the NA transition is still controversial. On the other hand, before the x-ray study by Safinya et al.[2] and our heat capacity study[3] it was generally believed that the continuous AC transition had helium-like exponents (n=2,d=3).[4] Our high-resolution heat capacity measurement demonstrated that the data could be uniquely described by a mean-field Landau model with an anomalously large sixth-order term.[3] Subsequently, x-ray, light scattering and heat capacity studies on the AC transition of another liquid crystal compound[5] confirmed our proposed model. Among these three transitions, the one least studied is the NC transition which has been found to be first order. In appropriate binary liquid-crystal mixtures it is possible to have lines of second order NA and AC transitions crossing over to a line of first order NC transitions. The triple point where these three transition lines intersect in the temperature-composition phase diagram is the NAC multicritical point.[6] Figure 1 shows the phase diagram for mixtures of octyl- and heptyloxy-p-pentylphenyl thiolbenzoate ($\bar{8}S5_{1-x}$ - $\bar{7}S5_x$).[7] Extensive experiments have been carried out on this mixture system in order to understand the nature of fluctuations near the NAC point.[8-11]

Recently there has been considerable interest in phase transitions associated with the Lifshitz point (LP),[12] i.e., a multicritical point connecting three distinct phases. The phases are, respectively, characterized by an order parameter, $\vec{M}(\vec{r})$, which is zero (disordered phase), finite but spatially uniform

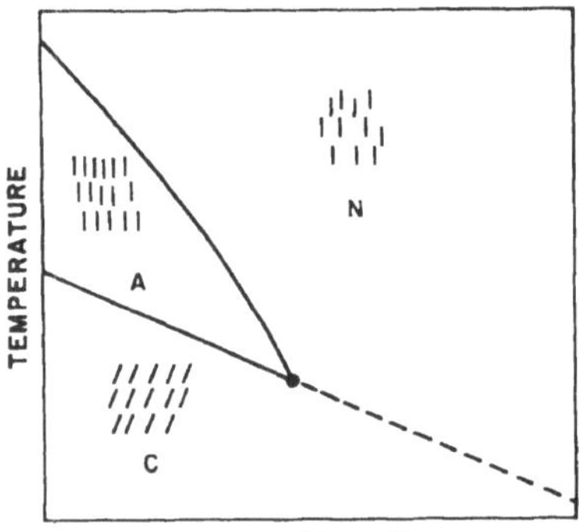

Fig. 1. Phase diagram near the NAC point. Two solid lines indicate continuous NA and AC transitions. The dotted line is the first-order NC transition line.

(uniformly ordered phase), and spatially varying (modulated phase). m denotes the number of spatial dimensions in which $\vec{M}$ can vary in the modulated phase. The realization of m=2 LP at a NAC point[13] has stimulated both experimental and theoretical interest. Theoretically it has been proven that d=3 is the lower critical dimensionality for m=2 LP.[14] Consequently, the NAC point is very different from m=1 LP which has been studied extensively on MnP.[15] Here we would like to report results obtained along the AC transition line near one NAC point.

II. DATA ANALYSIS AND DISCUSSION

Since there can be no long-range order on the uniformly ordered-modulated phase boundary in the n=m=d-1=2 LP problem, there can be no long-range (layered) positional order on the boundary between the A and C phases.[14] Here n is the dimensionality of the order parameter. However, in both pure[3,5] and mixture[16] systems, it has been found that the AC transition can be unambiguously described by the following mean-field Landau free energy with Ψ being the tilt-angle in the C phase:

$$F = F_0 + at\Psi^2 + b\Psi^4 + c\Psi^6. \tag{1}$$

Here F_0 is the nonsingular part of free energy, $t = (T-T_c)/T_c$ and T_c is the AC transition temperature. a, b and c are positive constants for a continuous transition. After minimizing the free energy with respect to Ψ, Ψ and anomalous part of the heat capacity ΔC are zero for $T > T_c$. In the case of $T < T_c$, one obtains

$$\Psi^2 = \frac{b}{3c}\left((1 + 3t'/t_0)^{1/2} - 1\right) \tag{2}$$

and

$$\Delta C = \frac{a^{3/2}\, T}{2(3c)^{1/2}\, T_c^{3/2}} (T_m - T)^{-1/2} \tag{3}$$

where $t'=-t$, $t_0=b^2/ac$ and $T_m=T_c(1+t_0/3)$. Actually, the full-width at half-height of the $(\Delta C/T)$ vs. T curve is t_0 in the reduced temperature scale. Furthermore, from Eq. (2), if $t' \ll t_0$, then Ψ is proportional to $(t')^{1/2}$ which represents the critical behavior for a second order mean-field transition. But if $t' \gg t_0$, then Ψ is proportional to $(t')^{1/4}$ which is a characteristic of a tricritical point. Therefore, the locus of t_0's will separate the second order transition region from the tricritical region in the phase diagram.

Using high-resolution x-ray scattering, Safinya *et al.*[8,9]

have measured both tilt-angles and layer-spacings in the C phase just below the AC transition line of ($\overline{8}$S5 - $\overline{7}$S5) mixtures. Here we calculated the ratio (R) of the measured tilt-angle to the quantity $((1+3t'/t_0)^{1/2} - 1)^{1/2}$ for each data point. By varying both T_c and t_0, the optimum values of T_c and t_0 are obtained by minimizing the relative deviation σ=D/M. Here D is the root-mean-squares deviation of the data from the calculated average ratio (M) for a given set of data. This was done with the experimental weighting factor for each data point. The same procedure was used to obtain T_c and t_0 for the 2-theta shift data (corresponding to layer spacings) which are proportional to Ψ^2. This kind of fitting scheme was found to be more satisfactory than a simple power law fitting in which, in general, one has to discard a good fraction of data with large tilt-angles.[16] The optimum values of T_c and t_0 for $\overline{8}$S5 and the four mixtures are listed in Table I along with other relevant physical parameters. The T_{AC}'s obtained by this method are consistent with the reported T_{AC}'s from a power law fitting.[9]

TABLE I. The parameters related to the NA and AC transition lines

Mole % of $\overline{7}$S5 in $\overline{8}$S5	(a)	T_{NA}(K)	T_{AC}(±5mK)$^{(b)}$	$T_{AC}$$^{(c)}$	$t_0 \times 10^3$(c)	T_{AC}/T_{NA}	$\sigma$$^{(d)}$
0	A	336.085	328.170	328.171	6.4	.9765	.044
	B			328.183	7.3		.105
15	A	331.990	326.620	326.618	4.5	.9838	.040
31	A	326.172	322.600	322.603	1.8	.9891	.049
	B			322.612	4.7		.066
41	A	324.163	323.258	323.259	0.62	.9972	.016
	B			323.271	1.6		.049
42	A	323.030	322.726	322.727	0.32	.9991	.020
	B			322.737	0.58		.069

a) Type of measurements. A: tilt angle; B: 2-theta shift.

b) These values obtained from the power law fitting (Ref. 9).

c) Optimum values from our fitting.

d) The minimum σ for the given set of data.

The fact that heat-capacity data show abrupt jumps at the T_{AC}'s and no discernible fluctuations above the T_{AC}'s also suggests mean-field-like AC transitions. Moreover, from the full -widths at half-height of the heat-capacity anomalies, one obtains t_0's which are shown in Fig. 2 as open circles. A reanalysis of x-ray data also gives t_0 for each mixture as shown with solid dots. These results clearly demonstrate that as T_{AC}/T_{NA} increases toward 1, t_0 monotonically decreases toward zero and the tricritical region (the region with $t' >> t_0$) monotonically increases.

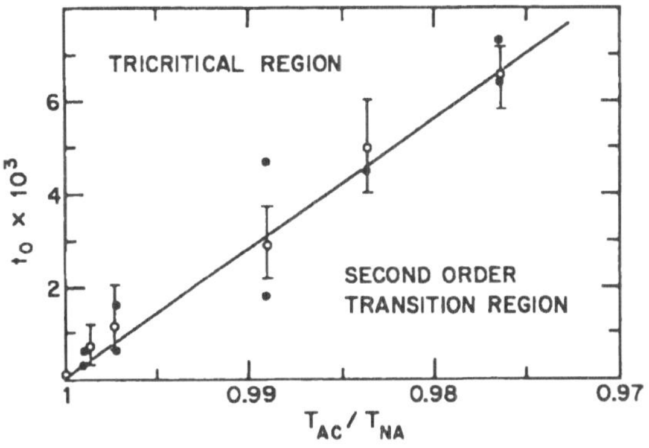

Fig. 2. t_0 vs (T_{AC}/T_{NA}). The solid line is a straight line through the data.

Figure 2 appears to suggest the existence of a mean-field tricritical point in the vicinity of the NAC point. A combination of x-ray[9] and heat-capacity[7] data allow one to gain further insight into the nature of the NAC point along the AC transition line. From $t_0 = b^2/ac$, $R^2 = b/3c$ and $\Delta C_J = a^2/2bT_c$ (the value of the heat-capacity jump at T_c), one can calculate the coefficients a, b and c associated with the Landau free energy. The relevant parameters and the results for a, b and c are shown in Table 2. Except for the 42% mixture, along the AC transition line the coefficient a and c do not vary too much and b decreases by about a factor of 3 in the 2-theta shift measurement. In the 42% mixture a, b and c all show increases above the values for the 41% mixture, with the 41% mixture having the minimum value of b. Consequently, the large increase in a and c and not the decrease in b in the 42% mixture causes the further decrease in t_0 near the NAC point. This result cannot be easily explained by a Landau free energy with coupled order parameters.[16] The fluctuations[8-11] near the NAC point may be responsible for the large increase in a and the moderate increase in b and c. Surprisingly, the x-ray data for the 42% mixture are also well described by the mean-field expression (Eq. 2) (see σ's in Table 1). On the other hand, the monotonic decrease in b from pure $\overline{8S5}$ to 41% also suggest that along the AC line the system has the tendency to approach a mean-field tricritical point with b=0. Finally

TABLE 2. Coefficients of Landau free energy

Mole % of $\overline{7}S5$ in $\overline{8}S5$	ΔC_J (b)	(a)	R^2	$t_0 \times 10^3$	a (c)	b (c)	c (c)
0	5.61	A	.0461	6.4	1410	66	470
		B	.0363	7.3	2050	137	1260
15	5.97	B	.0289	4.5	1680	87	1010
31	6.61	A	.0106	1.8	2000	114	3580
		B	.0151	4.7	1410	56	1240
41	10.75	A	.0101	0.62	1180	24	800
		B	.0198	1.6	1560	42	710
42	36.5	A	.0073	0.32	2890	42	1950
		B	.0109	0.58	3470	62	1880

(a) Type of measurements: A: tilt angle; B: 2-theta shift .

(b) Units in R_0 (the gas constant) (Ref. 7).

(c) Units in Joul/mole.

the large discrepancy in t_0 and c from tilt-angle and 2-theta shift measurements in the 31% mixture may be due to the presence of impurities because T_{AC} for this mixture is lower than that of both the 15 and 41% mixtures.[9] Simultaneous measurements on the tilt-angle and heat-capacity would be invaluable to explore the nature of AC transitions near the NAC point.

ACKNOWLEDGEMENTS

We are grateful to C. Dasgupta, Y. Shapira and J. Viner for useful discussions and C.R. Safinya and R.J. Birgeneau for providing us unpublished x-ray data. This work was supported in part by MEIS University of Minnesota, National Science Foundation-Solid State Chemistry- Grant DMR8204219 and the Department of Energy under contract DE-AC02-79-ER-10461 (SCL).

NOTE: In the near future the entire x-ray data near this NAC point will be published by authors in Ref. 10.

REFERENCES

1. P.G. de Gennes, Mol. Cryst. Liq. Cryst. 21, 49 (1973).
2. C.R. Safinya, M. Kaplan, J. Als-Nielsen, R.J. Birgeneau, D. Davidov, J.D. Litster, D.L. Johnson and M.E. Neubert, Phys. Rev. B21, 4149 (1980).
3. C.C. Huang and J.M. Viner, Phys. Rev. A25, 3385 (1982).
4. C.C. Huang and J.M. Viner, "Liquid Crystal and Ordered Fluids," Vol. 4, edited by A.C. Griffin and J.F. Johnson.
5. R.J. Birgeneau, C.W. Garland, A.R. Kortan, J.D. Litster, M. Meichle, B.M. Ocko, C. Rosenblatt, J.J. Yu and J. Goodby, Phys. Rev. A27, 1251 (1983).
6. D. Brisbin, D.L. Johnson, H. Fellner and M.E. Neubert, Phys. Rev. Lett. 50, 178 (1983) and references found therein.
7. R. DeHoff, R. Biggers, D. Brisbin and D.L. Johnson, Phys. Rev. A25, 472 (1982) and references found therein.
8. C.R. Safinya, R.J. Birgeneau, J.D. Litster and M.E. Neubert, Phys. Rev. Lett. 47, 668 (1981).
9. C.R. Safinya, MIT Thesis (1981) unpublished.
10. C.R. Safinya, L.J. Martinez-Miranda, M. Kaplan, J.D. Litster and R.J. Birgeneau, Phys. Rev. Lett. 50, 56 (1983).
11. S. Witanachchi, J. Huang and J.T. Ho, Phys. Rev. Lett. 50, 594 (1983).
12. R.M. Horneich, M. Luban and S. Shtrikman, Phys. Rev. Lett. 35, 1678 (1975).
13. J.H. Chen and T.C. Lubensky, Phys. Rev. A14, 1202 (1976).
14. G. Grinstein, Phys. Rev. B23, 4615 (1981) and references found therein.
15. Y. Shapira, invited talk in this conference.
16. C.C. Huang and S.C. Lien, Phys. Rev. Lett. 47, 1917 (1981).

DYNAMICS NEAR MULTICRITICAL POINTS

Volker Dohm

Institut für Festkörperforschung
Kernforschungsanlage Jülich
5170 Jülich, Germany

I. INTRODUCTION

Up to now the theory of critical dynamics has remained almost exclusively the domain of analytic mode-coupling and renormalization-group theories[1,2]. The mode-coupling approach is a selfconsistent method which is capable of a low-order treatment of reversible (non-dissipative) couplings, e.g. the strength of precession terms in magnetic systems. The renormalization-group theory goes beyond the mode-coupling approach not only because it provides a satisfactory foundation for the concepts of universality and dynamic scaling but also because it permits a systematic higher-order treatment of both reversible and dissipative couplings.

The dissipative couplings are of purely static origin. While near ordinary critical points they cause dynamic effects which are usually small compared to the strong effects induced by the reversible couplings, the static couplings play a crucial role upon approaching a multicritical point. Here it is just the change of static parameters which brings about the crossover from critical to multicritical behavior, both in statics and in dynamics, and may lead to qualitatively new dynamic properties. Therefore a simultaneous and proper treatment of both dissipative and reversible couplings is of primary importance in the theory of multicritical dynamics. The only approach which is expected to meet this requirement is the renormalization-group theory.

Particularly advantageous is the field-theoretical version of the renormalization-group approach[3-6]. Not only higher-order calculations of asymptotic critical properties and corrections thereof are facilitated by the field-theoretic techniques but also a convenient description of experimentally observable dynamic crossover phenomena well away from asymptotic criticality becomes possible[7].

The theoretical and experimental knowledge in the area of multicritical dynamics is quite incomplete. The present status of the theory is primarily based on studies of tricritical dynamics by Kawasaki[8] and by Siggia and Nelson[9], of bicritical and tetracritical dynamics by Janssen and the present author[10], and of Lifshitz-point dynamics by Huber[11] and by Folk and Selke[12]. For further contributions in the past decade back to the pioneering work by Kawasaki and Gunton [13] see Refs. 13-21. As will be discussed in Secs. V and VI, additional information is provided by recent work on the critical dynamics in ^4He and in ^3He-^4He mixtures[7,22]. On the experimental side, several dynamic measurements have been carried out near the tricritical point of ^3He-^4He mixtures as reviewed by Meyer et al.[23] and by Leiderer[24]. (See also Refs. 25-30.) No experimental information is available, however, on the dynamics near any other multicritical point. Also, nothing is known about multicritical dynamics in two dimensions or in random systems either theoretically or experimentally.

The general concept for describing dynamic critical phenomena [1,31] is applicable without principle modification to multicritical dynamics as well. The starting point are semiphenomenological Langevin-type equations of motion for all relevant "slow" variables. Near multicriticality the set of such variables may include, like in statics, the components φ_α of more than one order parameter, but also, unlike in statics, several hydrodynamic (conserved) variables m_β coupled to φ_α. Which variables φ_α and m_β should be included in the equations of motion depends on the symmetry properties and conservation laws of the multicritical system. Furthermore, like in ordinary critical dynamics[32], quantitative properties like the magnitude of coupling constants may determine whether the neglect or the inclusion of a hydrodynamic variable is a good approximation for a specific system. Several examples for dynamic models displaying multicritical behavior are presented in Sec. II.

The main task of the theory is to predict the critical behavior of dynamic correlation functions such as

$$C_{\varphi_\alpha \varphi_\alpha}(k,\omega) = \int d^3x \, dt \, e^{i(kx-\omega t)} < \varphi_\alpha(x,t)\varphi_\alpha(0,0) > \tag{1.1}$$

and similarly $C_{mm}(k,\omega)$. This can be compared with neutron- and light-scattering experiments, and also with macroscopic measurements of transport coefficients (e.g. thermal conductivity) in the hydrodynamic region $k\xi \ll 1$ (ξ is the correlation length).

So far the theory[8-21] has primarily investigated the traditional questions as to the asymptotic scaling behavior of these correlation functions in the vicinity of the λ-line and of the multicritical point. Let us recall the content of the dynamic-scaling hypothesis [33,34] applied to the k- and ω-dependence right at a multicritical point. First consider the simple case of a one-component order parameter φ. Then the validity of dynamic scaling implies that, for sufficiently small k and ω, $C_{\varphi\varphi}(k,\omega)$ can be written as

$$C_{\varphi\varphi}(k,\omega) = \frac{C_{\varphi\varphi}(k)}{A_\varphi k^z} F_\varphi\left(\frac{\omega}{A_\varphi k^z}\right) . \tag{1.2}$$

Here the frequency scale of the shape function F_φ is set by a characteristic frequency

$$\Omega_\varphi = A_\varphi k^z \tag{1.3}$$

which vanishes for $k \to 0$ (critical slowing down) according to a power law, with a dynamic exponent z characteristic for the multicritical point. With a suitable normalization of F_φ, $C_{\varphi\varphi}(k)$ represents the static (equal-time) correlation function according to the sum-rule

$$C_{\varphi\varphi}(k) = \int_{-\infty}^{\infty} \frac{d\omega}{2\pi} C_{\varphi\varphi}(k,\omega) . \tag{1.4}$$

At multicriticality we have $C_{\varphi\varphi}(k) \sim k^{2-\eta}$ with a static multicritical exponent η. The renormalization-group theory may prove or disprove the asymptotic validity of (1.2). The theory also provides (approximate) results for the deviations of z from the conventional value z=2 (nonconserved order parameter) or z=4 (conserved order parameter) and for the deviations of the universal shape function F_φ from the conventional Lorentzian shape. "Extended dynamic scaling" holds if a conserved variable m coupled to φ has a correlation function of the scaling form

$$C_{mm}(k,\omega) = \frac{C_{mm}(k)}{A_m k^z} F_m\left(\frac{\omega}{A_m k^z}\right) \tag{1.5}$$

with the same exponent z as in (1.2). In this case the fluctuations of φ and m relax on the same time scales (provided that A_m/A_φ is of order 1). This case will be shown to hold at the tricritical point of uniaxial antiferromagnets (Sec.IV). It is conceivable, however, both at ordinary critical and at multicritical points, that m and φ have different dynamic exponents $z_m \neq z_\varphi$ and that the shape function F_m depends on two characteristic frequencies Ω_m and Ω_φ even asymptotically (breakdown of extended dynamic scaling). This seems to happen at the bricritical point in MnF_2 (Sec.V) and possibly at the tricritical point of ^3He-^4He mixtures (Sec.VI). In case of a multicomponent order parameter φ_α the dynamic-scaling form at multicriticality reads

$$C_{\varphi_\alpha\varphi_\alpha}(k,\omega) = \frac{C_{\varphi_\alpha\varphi_\alpha}(k)}{A_\alpha k^z} f_\alpha\left(\frac{\omega}{A_\alpha k^z}\right) \tag{1.6}$$

with only one exponent z for all α (and with finite ratios A_α/A_β). This turns out to be valid for the dynamics at bicritical and tetracritical points (Secs.IVC, VA). In the presence of a conserved variable, however, some doubt remains concerning a nonanalytic dependence of $A_\alpha/A_\beta \sim \exp(-c/\varepsilon)$ on $\varepsilon = 4-d$ (Sec.VA).

More relevant than the asymptotic validity of critical and multicritical dynamic scaling is a quantitative description of the observable features of the dynamic correlation functions including the crossover from λ-line to multicritical behavior at finite distance from criticality. The observable dynamics may be quite different from the dynamics in the asymptotic scaling region due to the presence of large correction terms and also due to a strong influence of noncritical ("background") effects even close to criticality, like those found recently in ^4He [7,35]. We shall discuss this possibility and therefore formulate part of these lectures in the spirit of a non-asymptotic description similar to that employed recently[7,22]. This description still includes the asymptotic critical and multicritical behavior as special limiting cases and thus will also account for the traditional questions concerning the asymptotic power laws.

II. DYNAMIC MODELS *

A. Relaxational models

The simplest model for multicritical dynamics is the time-dependent Ginzburg-Landau model for a nonconserved order parameter $\varphi(x,t)$ (model A[32])

$$\dot{\varphi}(x,t) = -\Gamma\delta H/\delta\varphi(x,t) + \theta_\varphi(x,t) \ . \tag{2.1}$$

*All quantities of this Section are unrenormalized quantities which in Secs.III-VI will be indexed by a subscript o.

The stochastic force θ_φ simulates the effect of nonhydrodynamic degrees of freedom and is assumed to have a Gaussian distribution with

$$<\theta_\varphi> = 0, \quad <\theta_\varphi(x,t)\theta_\varphi(0,0)> = 2\Gamma\delta(x)\delta(t) \quad . \tag{2.2}$$

This model applies to the tricritical dynamics of anisotropic antiferromagnets with the Hamiltonian[36]

$$H = \int d^d x[\tfrac{1}{2}r\varphi^2+\tfrac{1}{2}(\nabla\varphi)^2+u\varphi^4+v\varphi^6] \tag{2.3}$$

provided that the coupling to the conserved energy density can be neglected. Another application is the relaxational dynamics[12] near a Lifshitz point in which case[37]

$$H = \int d^d x\,[\tfrac{1}{2}r\varphi^2+\tfrac{1}{2}(\nabla_\alpha\varphi)^2+\tfrac{c}{2}(\nabla_\beta\varphi)^2+(\nabla_\beta^2\varphi)^2+u\varphi^4] \tag{2.4}$$

with ∇_α and ∇_β being gradient operators of dimensionality m and d-m respectively. Also the relaxational dynamics with Hamiltonians describing higher-order multicritical points have been studied[19] on the basis of (2.1). Similarly, bicritical and tetracritical dynamics of strongly anisotropic systems can be described by the relaxational model[10]

$$\frac{\partial}{\partial t}\vec{S}(x,t) = -\Gamma_s \frac{\delta H}{\delta\vec{S}} + \vec{\theta}_s(x,t) \quad , \tag{2.5a}$$

$$\frac{\partial}{\partial t}\vec{\sigma}(x,t) = -\Gamma_\sigma \frac{\delta H}{\delta\vec{\sigma}} + \vec{\theta}_\sigma(x,t) \quad , \tag{2.5b}$$

with Gaussian-Markovian random forces $\vec{\theta}_i$ and with the Hamiltonian[38]

$$H = \int d^d x\{\tfrac{1}{2}[r_\sigma\vec{\sigma}^2+(\nabla\vec{\sigma})^2+r_s\vec{S}^2+(\nabla\vec{S})^2]+u\vec{\sigma}^4+2w\vec{\sigma}^2\vec{S}^2+v(\vec{S}^2)^2\} \quad . \tag{2.6}$$

Here $\vec{S}$ and $\vec{\sigma}$ denote n_s- and n_σ-component variables.

In uniaxial antiferromagnets with rotational symmetry around the easy axis (z-direction) the z-component of the magnetization is a conserved quantity and must be treated explicitly in dynamics. Then in (2.1) one should use the modified tricritical Hamiltonian

$$H = \int d^d x[\tfrac{1}{2}\tau\varphi^2+\tfrac{1}{2}(\nabla\varphi)^2+U\varphi^4+v\varphi^6+\tfrac{1}{2}m^2+\gamma m\varphi^2-hm] \quad . \tag{2.7}$$

The equation of motion for m reads (model C[32])

$$\frac{\partial}{\partial t}m(x,t) = \lambda\nabla^2\frac{\delta H}{\delta m(x,t)} + \theta_m(x,t) \quad , \tag{2.8}$$

$$< \theta_m >=0, \quad < \theta_m(x,t)\theta_m(0,0) >=-2\lambda\nabla^2\delta(x)\delta(t) \quad . \tag{2.9}$$

In (2.7) h represents a magnetic field parallel to the z-direction. Eqs.(2.8), (2.9) are consistent with the conservation property

$$\frac{d}{dt}\int d^d x \; m(x,t) = 0 \quad . \tag{2.10}$$

Huber[15] has also discussed the more general case of tricritical dynamics in uniaxial antiferromagnets where both the z component of the magnetization and the energy density are treated as two separate conserved variables. The asymptotic critical and tricritical dynamics in this case are believed[1,9] to be identical with those of (2.1), (2.7), (2.8), but non-negligible differences may arise in the experimentally accessible nonasymptotic region.

B. Bicritical model with rotational symmetry

We consider a uniaxial antiferromagnet with rotational symmetry about the easy axis (z-direction) which has a spin-flop bicritical point in the presence of a magnetic field h applied in the z-direction. The transitions along the two λ-lines that meet at the bicritical point are assumed to be Ising-like, with a single component order parameter σ, or XY-like, with a two-component order parameter $\vec{S}=(S_x,S_y)$, respectively. This situation applies to MnF_2. Due to the rotational symmetry, the z-component of the magnetization is a conserved quantity whose density m(x,t) must be included in the set of dynamic variables[18]. Because of analogies with the dynamics of ^4He to be discussed below it will be convenient to replace $\vec{S}$ by the complex order parameter

$$\psi = S_x+iS_y, \quad \psi^* = S_x-iS_y \quad . \tag{2.11}$$

The main difference with the relaxational models is a reversible coupling between m(x,t) and ψ(x,t) through precessional terms ~g. The appropriate equations of motion read[10]

$$\frac{\partial}{\partial t}\psi = -2\Gamma\frac{\delta H}{\delta\psi^*} + ig\psi\frac{\delta H}{\delta m} + \theta_\psi \quad , \tag{2.12}$$

$$\frac{\partial}{\partial t} m = \lambda\nabla^2\frac{\delta H}{\delta m} - 2g\,\mathrm{Im}(\psi^*\frac{\delta H}{\delta\psi^*}) + \theta_m \quad , \tag{2.13}$$

$$\frac{\partial}{\partial t}\sigma = -\Gamma_\sigma\frac{\delta H}{\delta\sigma} + \theta_\sigma \quad , \tag{2.14}$$

with the bicritical Hamiltonian

$$H = \int d^dx \{ \frac{1}{2}[\tau_\sigma\sigma^2+(\nabla\sigma)^2+\tau_\psi|\psi|^2+|\nabla\psi|^2+m^2] \\ +U\sigma^4+2W\sigma^2|\psi|^2+V|\psi|^4+\gamma_\sigma m\sigma^2+\gamma m|\psi|^2-hm\} \quad , \tag{2.15}$$

and with Gaussian-Markovian random forces θ_i. Note that σ is coupled to m and ψ only through the static couplings γ_σ and W. If $\gamma_\sigma \equiv 0$ and $W \equiv 0$, ψ and m satisfy equations of motions which are formally idenitcal with those for the superfluid order parameter and the entropy density in ^{4}He (model F[39]). This means that the XY-like λ-line dynamics of MnF_2 belongs to the same universality class as the dynamics of the superfluid transition in ^{4}He.

C. Tricritical model for ^{3}He-^{4}He mixtures

In a mixture of liquid ^{3}He and ^{4}He we have two hydrodynamic variables to be treated explicitly in addition to the superfluid order parameter $\psi(x,t)$, namely the entropy and the concentration of ^{3}He atoms[40]. It is convenient to choose a linear combination q(x,t) of the entropy and of the concentration fluctuations c(x,t) such that there is no coupling $\sim$ qc in the bilinear part of the tricritical Hamiltonian

$$H = \int d^dx \{ \frac{1}{2}[\tau|\psi|^2+|\nabla\psi|^2+\chi_q^{-1}q^2+\chi_c^{-1}c^2] \\ +U|\psi|^4+v|\psi|^6+\gamma_1 q|\psi|^2+\gamma_2 c|\psi|^2-h_1 q-h_2 c \} \quad . \tag{2.16}$$

The appropriate equations of motions describing the dynamics near the superfluid transition are even more complex than for bicritical dynamics. They read[9]

$$\frac{\partial}{\partial t}\psi = -2\Gamma\frac{\delta H}{\delta\psi^*} + ig_1\psi\frac{\delta H}{\delta q} + ig_2\psi\frac{\delta H}{\delta c} + \theta_\psi \quad , \tag{2.17}$$

$$\frac{\partial}{\partial t}q = K\nabla^2\frac{\delta H}{\delta q} + L\nabla^2\frac{\delta H}{\delta c} - 2g_1\,\mathrm{Im}(\psi^*\frac{\delta H}{\delta\psi^*}) + \theta_q \quad , \tag{2.18}$$

$$\frac{\partial}{\partial t}c = L\nabla^2\frac{\delta H}{\delta q} + \lambda\nabla^2\frac{\delta H}{\delta c} - 2g_2\,\mathrm{Im}(\psi^*\frac{\delta H}{\delta\psi^*}) + \theta_c \quad , \tag{2.19}$$

with appropriate correlations between the Langevin forces θ_i. The off-diagonal kinetic coefficient L implies a dynamic coupling between q and c in lowest order even in the absence of all static and dynamic nonlinear couplings. This describes the phenomenon of thermodiffusion characteristic for binary mixtures[41]. In the absence of nonlinear couplings, the kinetic coefficients are related to the ^{3}He mass diffusion coefficient D, to the thermal diffusion ratio k_T and to the thermal conductivity κ according to

$$D = \lambda/\chi_c, \; \kappa = K-L^2/\lambda, \; k_T = L/D \quad . \tag{2.20}$$

In Sec.VI we shall discuss the critical and tricritical temperature dependence of D, κ, k_T arising from the nonlinear couplings.

D. Dynamic functional and response fields

All of the above models have the structure

$$\frac{\partial}{\partial t} \varphi_\alpha(x,t) = -L_{\alpha\beta} \frac{\delta H}{\delta\varphi_\beta} + V_\alpha\{\varphi\} + \theta_\alpha(x,t) \tag{2.21}$$

where V_α represents the reversible terms, $L_{\alpha\beta}$ is a symmetric matrix of kinetic coefficients and θ_α are random forces with a Gaussian probability distribution

$$w\{\theta_\alpha\} \sim \exp -\frac{1}{4} \int d^dx\, dt\, \vec{\theta}(xt) L^{-1} \vec{\theta}(xt) \quad . \tag{2.22}$$

An equivalent description can be given in terms of a probability distribution of φ_α rather than θ_α by eliminating $\vec{\theta}$ in favor of $\dot{\varphi}_\alpha$ and φ_α by means of (2.21)[42]. Particularly advantageous is the introduction of auxiliary "response fields" $\tilde{\varphi}_\alpha(x,t)$[43] in addition to $\varphi_\alpha(x,t)$ by means of a Gaussian transformation[44,45]. This leads to the probability distribution

$$w\{\varphi_\alpha,\tilde{\varphi}_\alpha\} \sim \exp -J\{\varphi_\alpha,\tilde{\varphi}_\alpha\} \tag{2.23}$$

with the dynamic functional[44,45] (apart from a Jacobian term)

$$J\{\varphi_\alpha,\tilde{\varphi}_\alpha\} = \int d^dx\, dt[-\tilde{\varphi}_\alpha L_{\alpha\beta}\tilde{\varphi}_\beta + \tilde{\varphi}_\alpha(\dot{\varphi}_\alpha + L_{\alpha\beta} \frac{\delta H}{\delta\varphi_\beta} - V_\alpha)] \ . \tag{2.24}$$

The physical meaning of $\tilde{\varphi}_\beta(xt)$ is elucidated by the relation

$$< \varphi_\alpha(x_1t_1)\tilde{\varphi}_\beta(x_2t_2) > = \delta < \varphi_\alpha(x_1t_1) > / \delta\tilde{F}_\beta(x_2t_2) \tag{2.25}$$

where $\tilde{F}_\beta(x,t)$ is an external force added to the r.h.s. of (2.21). Thus $<\varphi\tilde{\varphi}>$ is to be interpreted as a response function. Dynamic correlation and response functions can now be calculated in close analogy with static correlation functions, with the static statistical weight e^{-H} being replaced by e^{-J}. Owing to this fundamental analogy, also the advanced field-theoretic methods[3] of statics can be extended to dynamics[4-6].

III. STATIC PROPERTIES

In the following we summarize some results concerning multicritical statics as far as this is needed for the discussion of dynamics in Secs.IV-VI. We shall always employ the field-theoretic formulation[3]. The quantities of Sec.II will be indexed by the subscript o in order to distinguish them from renormalized quantities.

A. Field-theoretic approach

Consider the standard Ginzburg-Landau Hamiltonian (2.3) with $u_0>0$, $v_0=0$. The deviations of the correlation functions from mean-field critical behavior can be conveniently calculated at infinite cut-off by means of renormalized perturbation theory in $4-\varepsilon$ dimensions. In order to obtain finite expressions for correlation functions for $d\leq 4$ it is necessary to introduce renormalized quantities

$$\varphi=Z_\varphi^{-1/2}\varphi_0,\quad r=Z_r^{-1}Z_\varphi r_0,\quad u=\mu^{-\varepsilon}K_d Z_u^{-1}Z_\varphi^2 u_0 \quad . \tag{3.1}$$

Here μ is a nonuniversal reference wave number and $K_d=2^{1-d}\pi^{-d/2}\Gamma(d/2)^{-1}$ is a conventional geometric factor. The Z factors serve to absorb the (wave-vector and temperature independent) poles $\sim\varepsilon^{-m}$ of the bare correlation functions for $\varepsilon\to 0$. Thus (within the minimal renormalization procedure[46]) the Z-factors are constants of the form

$$Z_i(u,\varepsilon) = 1 + \sum_{m=1}^{\infty} f_{i,m}(u)\varepsilon^{-m} \quad . \tag{3.2}$$

The coefficients $f_{i,m}$ are determined (order by order in u) such that renormalized correlation functions such as $C_{\varphi\varphi}\equiv\langle\varphi(x)\varphi(0)\rangle = Z_\varphi^{-1}\langle\varphi_0(x)\varphi_0(0)\rangle$ remain finite for $\varepsilon\to 0$ if they are expressed in terms of renormalized quantities. The Z-factors enter the theory via the functions $\zeta_\varphi(u)=(\mu\partial_\mu \ln Z_\varphi^{-1})_0$, $\zeta_r(u)=(\mu\partial_\mu \ln Z_r^{-1}Z_\varphi)_0$ and $\beta_u(u,\varepsilon)=(\mu\partial_\mu u)_0$ where the derivatives are taken at fixed bare parameters. From the fact that $\langle\varphi_0\varphi_0\rangle$ is independent of μ one can derive the renormalization group-equation

$$(\mu\partial_\mu+\beta_u\partial_u+r\zeta_r\partial_r-\zeta_\varphi)C_{\varphi\varphi} = 0 \quad . \tag{3.3}$$

This means that a change of the reference length μ^{-1} is equivalent to a change of u and r, apart from the term ζ_φ. More explicitly, in terms of the (Fourier transformed) dimensionless part

$$G_{\varphi\varphi}(\frac{k}{\mu},\frac{r}{\mu^2},u) = \mu^2 C_{\varphi\varphi}(k,r,u,\mu) \quad , \tag{3.4}$$

the integration of (3.3) yields the relation

$$G_{\varphi\varphi}(\frac{k}{\mu},\frac{r}{\mu^2},u) = G_{\varphi\varphi}(\frac{k}{\bar\mu},\frac{\bar r}{\bar\mu^2},\bar u)\exp\int_\ell^1[2+\zeta_\varphi(\bar u)]\frac{d\ell'}{\ell'} \tag{3.5}$$

with $\bar\mu\equiv\mu\ell$ and effective parameters $\bar r\equiv r(\ell)$, $\bar u\equiv u(\ell)$ determined by

$$\ell\frac{dr(\ell)}{d\ell} = r(\ell)\zeta_r(\bar u),\quad \ell\frac{du(\ell)}{d\ell} = \beta_u(\bar u) \quad . \tag{3.6}$$

The important content of (3.5) is that it relates the correlation function at some k,r with the correlation function at some other $k'=k/\ell$, $r'=\bar r/\ell^2$ (with a different effective coupling $\bar u$). The "flow parameter" ℓ is as yet unspecified, except that at $\ell=1$, $r(1)=r$ and

$u(1)=u$. The arbitrariness of ℓ can be exploited by choosing ℓ (as a function of k and $\bar{r}$) such that $G_{\varphi\varphi}$ on the r.h.s. of (3.5) remains noncritical even if $G_{\varphi\varphi}$ on the l.h.s. of (3.5) becomes critical ($k\to 0$, $r\to 0$). This enables one to employ on the r.h.s. of (3.5) a simple approximate form for $G_{\varphi\varphi}$ as calculated by ordinary perturbation theory in the noncritical region.

B. λ-line and tricritical behavior

The dimensionless renormalized coupling u can be regarded as a smooth function of the physical "fields", e.g. of the temperature and (in antiferromagnets) the magnetic field. A line of ordinary critical points ("λ-line") $T_c(u)$ is generated by varying u. Criticality is determined by $r=0$, hence we may make the identification

$$r=at, \quad t=[T-T_c(u)]/T_c(u), \quad a>0 \tag{3.7}$$

The asymptotic λ-line behavior is obtained in the limit $\ell\to 0$ where $u(\ell)$ tends to the fixed point value $u^* = u(0)$. Then (3.5) becomes a scaling relation with universal critical exponents

$$\zeta_\varphi(u^*) = -\eta, \quad \zeta_r(u^*) = 2-\nu^{-1} \quad , \tag{3.8}$$

independent of the nonuniversal initial value $u(1)=u > 0$.

The lambda line terminates in $u=0$ which defines the tricritical temperature $T_t=T_c(0)$. For $u<0$ a first-order transition occurs. In this region it is of course necessary to keep the sixth-order term $v_0\varphi_0^6$ in the Hamiltonian (2.3). For a discussion of this term see the review by Lawrie and Sarbach[47,48]. Here we confine ourselves to $u\geq 0$ and $r\geq 0$. Even at $u=0$, $r>0$ the term $v_0\varphi_0^6$ is in principle needed in order to obtain the correct (d-independent) classical tricritical exponents[36] for $d\geq 3$ (and logarithmic corrections in $d=3$) rather than Gaussian exponents. On the other hand, since both sets of exponents coincide for $d=3$, the approximation[9,49] $v_0=0$ appears to be acceptable for the present purpose. For $u=0$ we have $u(\ell)=u(1)=0$, $\zeta_r(\bar{u})=\zeta_\varphi(\bar{u})=0$, $r(\ell)=r(1)$ which implies the expected classical tricritical behavior $G_{\varphi\varphi}=(k^2+r)^{-1}$.

The crossover between λ-line and tricritical behavior is controlled by varying u and r in (3.6). As an explicit example we use the Z-factors (3.2) in first order in u which yields $\zeta_r= -f_{r,1} = -4(n+2)u$ and $\beta_u=u(-\varepsilon+f_{u,1})$, $f_{u,1} = 4(n+8)u$ with $u^* = \varepsilon/4(n+8)$. Then (3.6) can be integrated analytically,

$$u(\ell) = u^*(1+\ell^{\varepsilon}/g)^{-1}, \quad g = u/(u^*-u) \quad , \tag{3.9}$$

$$r(\ell) = r[(1+g)/(1+g\ell^{-\varepsilon})]^{(n+2)/(n+8)} \quad . \tag{3.10}$$

Furthermore we approximate $G_{\varphi\varphi}$ on the r.h.s. of (3.5) by its zeroth-order form $G_{\varphi\varphi}(x,y,0)=(x^2+y)^{-1}$. A reasonable choice for ℓ is in this case

$$[k^2+r(\ell)]/\mu^2\ell^2 = 1 \tag{3.11}$$

which, together with (3.10) and (3.7), identifies ℓ as a function of k,t and u. This completes, in lowest order in u, the task of determing the l.h.s. of (3.5) as a function of experimentally controllable parameters both in the tricritical region $g \ll \ell^{\varepsilon}$ and in the λ-line critical region $\ell^{\varepsilon} \ll g$. Note, however, that there are still the nonuniversal parameters μ,a and u which need a quantitative identification. This can be achieved by means of comparison with experiments for a specific system. All of these considerations also pertain to dynamics. There, in principle, ℓ may become a function of the frequency ω as well.

C. Bicritical and tetracritical behavior

We extend our discussion to the Hamiltonian (2.6). Now we have three β-functions β_u, β_v, β_w which have a symmetric fixed point[38] $u^*=v^*=w^*$ corresponding to bicritical behavior. This fixed point is expected to be stable for $n=n_\sigma+n_s \lesssim 3$ in d=3. It implies static bicritical exponents which are identical with those of an n-component Ginzburg-Landau or isotropic Heisenberg model. For larger n a "biconical" fixed point $u^* \neq v^* \neq w^*$ becomes stable corresponding to tetracritical behavior[38].

The deviations from the two λ-lines which meet at the bicritical or tetracritical point are described in terms of the renormalized counterparts r_1 and r_2 of $r_{\sigma o}$ and r_{so}. They are introduced as[10]

$$\begin{pmatrix} r_{\sigma o} \\ r_{so} \end{pmatrix} = \begin{pmatrix} Z_1 & A_\sigma \\ A_s & Z_2 \end{pmatrix} \begin{pmatrix} r_1 \\ r_2 \end{pmatrix} \tag{3.12}$$

with off-diagonal renormalization constants $A_i = -4n_i w/\varepsilon+O(w^2)$. The generalization of (3.6) is

$$\ell\frac{dr_i(\ell)}{d\ell} = \Sigma_j r_j(\ell)\zeta_{ji}(\bar{p}), \quad \ell\frac{dp(\ell)}{d\ell} = \beta_p(\bar{p}) \tag{3.13}$$

where $\bar{p}=p(\ell)$, p=u,v,w, and $\zeta_{j,i}(p)$ is defined by $(\mu\partial_\mu r_i)_o=\Sigma_j r_j\zeta_{ji}$. The appropriate scaling variables[50] r and g are certain linear combinations of r_1 and r_2 such that the corresponding transformed matrix ζ'_{ij} becomes diagonal at the fixed point. This property implies that the eigenvalues ζ_r^* and ζ_g^* of $\zeta_{ij}(p^*)$ can be identified as

$$\zeta_r^* = 2-\nu^{-1}, \quad \zeta_g^* = 2-\phi/\nu \tag{3.14}$$

where ν is the exponent related to the correlation length and ϕ is crossover exponent[38]. At the symmetric fixed point $u^* = v^* = w^*$, the appropriate bicritical scaling variables are[50,51]

$$r=(n_\sigma r_1+n_s r_2)/n, \quad g=r_1-r_2 \quad . \tag{3.15}$$

For a study of the crossover from λ-line to bicritical ($g=0$) behavior see e.g. Ref.51.

D. Conserved variable : λ-line and tricritical behavior

As long as one is interested only in static properties of the order parameter one may integrate out the variable m_0 of the Hamiltonian (2.7). This yields (2.3) with $u_0=U_0-\gamma_0^2/2$ and $r_0=\tau_0+2\gamma_0 h_0$. In dynamics, however, we need m_0 explicitly. An important simplification comes from the fact that (2.7) can be decomposed according to

$$H = H_\varphi + \frac{1}{2}\int d^dx \; M_0(x)^2 \tag{3.16}$$

where H_φ is identical with (2.3) (apart from a constant term $\sim h_0^2$) and where M_0 is given by

$$M_0(x) = m_0(x) - h_0 + \gamma_0 \varphi_0(x)^2 \quad . \tag{3.17}$$

Thus $M_0(x)$ has a Gaussian probability distribution which is not correlated with the Ginzburg-Landau distribution for φ_0. This implies exact properties such as $\langle M_o \rangle=0$, $\langle \varphi_o^2 M_0 \rangle=0$, $\int d^dx \langle M_o(x)M_o(0)\rangle=1$ etc. In particular

$$\int d^dx(\langle m_o(x)m_o(0)\rangle - \langle m_o\rangle^2)=1+\gamma_o^2 \int d^dx(\langle \varphi_o^2(x)\varphi_o(0)^2\rangle - \langle \varphi_o^2\rangle^2). \tag{3.18}$$

From these relations it follows that all Z-factors which are necessary to renormalize (2.7) are expressible in terms of those renormalizing the standard Ginzburg-Landau Hamiltonian (2.3). For example from (3.18) one obtains[52] for Z_m ($m = Z_m^{-1/2}m_0$)

$$Z_m^{-1} = 1+\gamma^2 A \tag{3.19}$$

where A is the known additive constant[3] which renormalizes the correlation function on the r.h.s. of (3.18), and where the renormalized coupling γ is defined as $\gamma=Z_\gamma^{-1}Z_\varphi Z_m^{1/2}\mu^{-\varepsilon/2}K_d^{1/2}\gamma_0$. Other relations are[10,52,53]

$$\begin{aligned} Z_\gamma Z_m^{-1} &= Z_r, \quad Z_U U = Z_u u+\tfrac{1}{2}Z_\gamma^2 Z_m^{-1}\gamma^2 \quad , \\ Z_\tau &= Z_\gamma, \quad \tau = r, \quad U = u+\tfrac{1}{2}\gamma^2 \quad , \end{aligned} \tag{3.20}$$

where Z_U and Z_τ renormalize U_o and τ_o according to $U=Z_U^{-1}Z_\varphi^2\mu^{-\varepsilon}K_d U_o$ and $\tau=Z_\tau^{-1}Z_\varphi\tau_o$. An important consequence is the simple structure of the flow equation[52]

$$\ell\frac{d\gamma(\ell)}{d\ell} = \frac{\gamma(\ell)}{2}(-\varepsilon+2\zeta_r+\zeta_m),\ \zeta_m=(\mu\partial_\mu \ln Z_m^{-1})_o = \gamma^2 B \quad , \tag{3.21}$$

with $B(u)=2n+O(u^2)$ being determined by the poles $\sim\varepsilon^{-1}$ of A. Depending on the sign of $2\zeta_r^*-\varepsilon = d-2\nu^{-1} = -\alpha/\nu$ the stable fixed point of (3.21) is

$$\zeta_m^*=\gamma^{*2}B(u^*) = \begin{cases}\alpha/\nu & \text{if } \alpha>0\\ 0 & \text{if } \alpha\le 0\ .\end{cases} \tag{3.22}$$

At the tricritical point this yields the large value $\zeta_m^* = \alpha_t/\nu_t = 1$ which has a strong effect on the susceptibility $\chi_m=C_{mm}(k=0)$ and on the specific heat. For χ_m the relation corresponding to (3.5) reads

$$\chi_m(\frac{r}{\mu^2},u,\gamma) = \chi_m(\frac{\bar{r}}{\bar{\mu}^2},\bar{u},\bar{\gamma},)\exp\int_\ell^1 \zeta_m \frac{d\ell'}{\ell'} \tag{3.23}$$

which implies asymptotically $\chi_m \sim \ell^{-\alpha/\nu}$ if $\alpha>0$. Thus a weak divergence along the λ-line $(0<\alpha<1)$ or a cusp $(\alpha<0)$ turns into a strong divergence $\sim\ell^{-\alpha_t/\nu_t}\sim\ell^{-1}$ in the tricritical region. The change of γ^{*2} from a vanishing or small value along the λ-line to a much larger value at the tricritical point is the origin for the large departures of tricritical dynamics from both conventional dynamics and λ-line dynamics (Secs.IVB and VI).

E. Conserved variable: bricritical behavior

On the basis of the results of subsections A-D above we are now in the position to discuss the fairly complicated Hamiltonian (2.15). Since m_o enters only in linear and quadratic order, there exists a variable

$$M_o(x) = m_o + \gamma_{\sigma o}\sigma_o^2 + \gamma_o|\psi_o|^2 - h_o \tag{3.24}$$

which has only Gaussian fluctuations not correlated with ψ_o or σ_o. This is seen explicitly by decomposing (2.15) according to

$$H = H_{\psi\sigma} + \frac{1}{2}\int d^dx M_o(x)^2 \quad , \tag{3.25}$$

where $H_{\psi\sigma}$ is identical with (2.6) for the case $n_s=2$ and $n_\sigma=1$ (apart from a constant term $\sim h_o^2$), with $r_{\sigma o}=\tau_o+2\gamma_{\sigma o}h_o$, $u_o=U_o-\gamma_{\sigma o}^2/2$, $w_o=W_o-\gamma_{\sigma o}\gamma_{so}/2$, and similarly for r_{so} and v_o. As in Sec.D above this implies exact properties such as $\langle M_o\rangle=0$, $\langle|\psi_o|^2M_o\rangle=0$, $\langle\sigma_o^2M_o\rangle=0$ etc. which again lead to exact relations between the Z-factors and renormalized parameters of the Hamiltonians (2.15) and (2.6). In particular one obtains[10]

$$u = U-\gamma_\sigma^2/2, \quad w = W-\gamma_\sigma\gamma_s/2, \quad v = V-\gamma_s^2/2 \tag{3.26}$$

and $r_i=\tau_i$. Most important is the generalization of the flow equation (3.21). The corresponding β-functions have the structure[10]

$$\begin{pmatrix} \beta_{\gamma_\sigma} \\ \beta_{\gamma_s} \end{pmatrix} = \begin{pmatrix} \bar{\zeta}_{11} & \zeta_{21} \\ \zeta_{12} & \bar{\zeta}_{22} \end{pmatrix} \begin{pmatrix} \gamma_\sigma \\ \gamma_s \end{pmatrix} \tag{3.27}$$

with $\bar{\zeta}_{ii}=\zeta_{ii}+(\zeta_m-\varepsilon)/2$ where ζ_{ij} has been defined in Sec.C above. For $\beta_{\gamma_i}=0$, (3.27) yields two possible fixed points $\zeta_m^*=\varepsilon-2\zeta_r^*=\alpha/\nu$ and $\zeta_m^*=\varepsilon-2\zeta_g^*=2\phi\nu^{-1}-d$ in terms of the eigenvalues (3.14). For the symmetric fixed point $u^*=v^*=w^*$ the latter turns out to be stable,

$$\zeta_m^* = 2\phi\nu^{-1}-d \quad , \tag{3.28}$$

which leads to the bicritical fixed point values[10]

$$\gamma_\sigma^*/\gamma_s^* = -2, \quad \gamma_\sigma^{*2} = \tfrac{4}{3}(2\phi\nu^{-1}-d) + O(\varepsilon^3) \quad . \tag{3.29}$$

Substituting this into (3.26) determines also $U^*\neq V^*\neq W^*$. A physical interpretation of the opposite signs of γ_σ^* and γ_s^* is the following. Since m is coupled to $\vec{S}^2$ and σ^2, a fluctuation of m corresponds to a local fluctuation of the variables τ_i which measure the distance from the λ-lines. Since $\gamma_\sigma^*/\gamma_s^*<0$ a fluctuation of m near bicriticality corresponds to τ_i fluctuations in opposite directions; depending on the sign of m the system is (locally) driven further up towards the spin-flop phase or down towards the Ising phase, similarly as if the magnetic field had been changed.

The result (3.28) has an important effect on the asymptotic behavior of the susceptibility [compare (3.23)]

$$\chi_m = C_{mm}(k=0) \sim \ell^{-\zeta_m^*} \quad . \tag{3.30}$$

While it diverges weakly along the Ising λ-line ($n_\sigma=1$, $\zeta_m^*=\alpha/\nu \ll 1$) and has a finite cusp along the spin-flop λ-line ($n_s=2$, $\alpha<0$, $\zeta_m^*=0$) it diverges relatively strongly in the bicritical region. Using[54] $\phi \approx 1.25$, $\nu\approx 0.70$ for n=3, d=3 yields $\zeta_m^* \approx 0.56$.

IV. RELAXATIONAL MULTICRITICAL DYNAMICS

The close analogy between statics and dynamics discussed in Sec.II D permits to proceed similarly as in Sec.III. Although we now consider the variables $\varphi_0, m_0, \sigma_0, \vec{S}_0$ as time dependent quantities, no new renormalizations of static quantities are necessary[5] (within the minimal renormalization procedure[46]), only the genuine dynamic quantities (response fields and kinetic coefficients) require additional renormalizations. In the absence of reversible couplings, a basic simplification comes from dissipation-fluctuation relations[44,55] between response and correlation functions, for example for the case of (2.1)

$$<\varphi_o\varphi_o>(k,\omega) = \frac{2\Gamma_o}{\omega} \text{ Im } <\varphi_o\tilde{\varphi}_o> (k,\omega) \quad . \tag{4.1}$$

Upon rewriting (4.1) in terms of renormalized quantities $\tilde{\varphi} = \tilde{Z}_\varphi^{-1/2}\tilde{\varphi}_o$ and $\Gamma = Z_\Gamma^{-1}\Gamma_o$ it follows that[4]

$$Z_\Gamma \tilde{Z}_\varphi^{1/2} = Z_\varphi^{1/2}, \tag{4.2}$$

i.e. $Z_\Gamma \tilde{Z}_\varphi^{1/2}$ is already determined by statics. Similar relations hold for the other relaxational models. In all cases the effective kinetic coefficient does not diverge at criticality or multicriticality in contrast to systems with reversible couplings.

A. λ-line and tricritical dynamics without conserved variables

Consider the tricritical model (2.1)-(2.3). The basis for the following discussion is provided by the extension of the important static relation (3.5) to dynamics. For example for the dimensionless part $G_{\varphi\tilde{\varphi}} = \Gamma\mu^2 < \varphi\tilde{\varphi}>$ of the response function one obtains a similar relation (see Ref.4)

$$G_{\varphi\tilde{\varphi}}\left(\frac{k}{\mu},\frac{\omega}{\Gamma\mu^2},\frac{r}{\mu^2},u\right) = G_{\varphi\tilde{\varphi}}\left(\frac{k}{\bar{\mu}},\frac{\omega}{\bar{\Gamma}\bar{\mu}^2},\frac{\bar{r}}{\bar{\mu}^2},\bar{u}\right)\exp\int_\ell^1 [2+\zeta_\varphi(u)]\frac{d\ell'}{\ell'} \tag{4.3}$$

with the effective kinetic coefficient

$$\bar{\Gamma} \equiv \Gamma(\ell) = \Gamma\exp\int_1^\ell \zeta_\Gamma(\bar{u})\frac{d\ell'}{\ell'} \tag{4.4}$$

where $\zeta_\Gamma(u) = (\mu\partial_\mu \ln Z_\Gamma^{-1})_o$. The role of a <u>characteristic frequency</u> is played by

$$\Omega_\varphi(\ell) = \mu^2\Gamma\exp\int_1^\ell (2+\zeta_\Gamma)\frac{d\ell'}{\ell'} \tag{4.5}$$

which behaves asymptotically as $\sim \ell^z$ with the <u>dynamic critical exponent</u>

$$z = 2+\zeta_\Gamma^*, \quad \zeta_\Gamma^* = \zeta_\Gamma(u^*) \quad . \tag{4.6}$$

For $u=u^*=0$ we have $\zeta_\Gamma^*=0$, hence the tricritical dynamic exponent z_t has the classical value $z_t=2$ like the static tricritical exponents. Since the λ-line exponent $z=2+c\eta$ is expected to be only slightly larger than 2 (with c of the order of 1)[32,56] in three dimensions, there is no pronounced difference between λ-line and tricritical dynamics in this case.

While the effective kinetic coefficient $\Gamma(\ell)$ vanishes as $\sim \ell^{z-2}$ along the λ-line it remains finite at the tricritical point. Therefore the conventional theory of critical slowing down (see e.g. Ref. 31) applies to the tricritical point in the present case. The results of a computer simulation on tricritical relaxation of a three-dimensional kinetic Ising model by Müller-Krumbhaar and Landau[17] are consistent with the conventional theory ($z_t=2$).

As an illustration analoguous to the static case (Sec.III B) we approximate $G_{\varphi\tilde{\varphi}}$ on the r.h.s. of (4.3) by its zeroth-order form $G_{\varphi\tilde{\varphi}}(x,p,y,0)=[-ip+\Gamma(x^2+y)]^{-1}$ and choose ℓ such that $|G_{\varphi\tilde{\varphi}}|=1$ on the r.h.s. of (4.3), i.e.

$$\mu^2\ell^2 = \left\{\frac{\omega^2}{\Gamma(\ell)^2} + [k^2+r(\ell)]^2\right\}^{1/2} \tag{4.7}$$

Together with (3.6), (4.1), (4.3) and (4.4) this determines the dynamic crossover of $\langle\varphi\tilde{\varphi}\rangle$ and of $\langle\varphi\varphi\rangle$ from λ-line to tricritical behavior (by letting $u\to 0$). In particular the asymptotic scaling form (1.2) is confirmed at the tricritical point, with $z=z_t=2$ and a Lorentzian shape function F_φ.

B. λ-line and tricritical dynamics with a conserved variable

Consider now the model (2.1), (2.7)-(2.9). As pointed out by Siggia and Nelson[9], it exhibits interesting tricritical dynamic behavior. The basic dynamic quantity of the model is the ratio of effective kinetic coefficients

$$w(\ell) = \Gamma(\ell)/\lambda(\ell) \quad . \tag{4.8}$$

Here $\lambda(\ell)$ is defined by an equation analogous to (4.4), with $\zeta_\lambda=(\mu\partial_\mu \ln Z_\lambda^{-1})_0$ and $\lambda=Z_\lambda^{-1}\lambda_0$. In order to obtain the fixed point value $w^*=w(0)$ one must calculate the zero of the β-function

$$\beta_w=w(\zeta_\Gamma-\zeta_\lambda) \quad . \tag{4.9}$$

In one-loop order the result is (for n=1)

$$\zeta_\Gamma=4\gamma^2 w/(1+w), \qquad \zeta_\lambda=2\gamma^2 \quad . \tag{4.10}$$

Since $\gamma^{*2}>0$ ($\alpha>0$) in three dimensions, this implies $w^*=1$ for the critical and tricritical case, apart from higher-order corrections. A finite $w^*>0$ means that φ and m relax asymptotically on the same time scale, therefore this leads to the validity of extended dynamic scaling. For $w^*>0$ one obtains from (4.9) $\zeta_\Gamma^*=\zeta_\lambda^*$, thus the dynamic-scaling exponent is

$$z=2+\zeta_\Gamma^*=2+\zeta_\lambda^* \tag{4.11}$$

In analogy to (4.2) we have $Z_\lambda \tilde{Z}_m^{1/2}=Z_m^{1/2}$, furthermore the conservation property of m_0 implies $\tilde{Z}_m=Z_m^{-1}$. This yields the exact relation

$$\zeta_\lambda=\zeta_m \quad . \tag{4.12}$$

According to (3.22) we therefore arrive at[32]

$$z=2+\alpha/\nu \tag{4.13}$$

with $0<\alpha \ll 1$ along the λ-line. The tricritical value becomes[9]

$$z_t=2+\alpha_t/\nu_t=3 \tag{4.14}$$

which differs strongly from conventional theory. As suggested by Siggia and Nelson[9], the large difference between λ-line and tricritical dynamics may be experimentally observable in magnetic systems.

The description of the dynamic crossover behavior of $\langle\tilde{\varphi}\varphi\rangle$ and $\langle\varphi\varphi\rangle$ is now parallel to Sec.A above. The main difference is that $\Gamma(\ell)$ becomes dependent on $\gamma(\ell)$ and $w(\ell)$ through $\zeta_\Gamma(\gamma,w,u)$ where $\gamma(\ell)$ and $w(\ell)$ are determined by (3.21) and by the flow equation $\ell dw(\ell)/d\ell=\beta_w$. The identification of ℓ according to (4.7) remains applicable also in the present case. Siggia and Nelson[9] have calculated a crossover scaling function for the temperature and u dependence of $\Gamma(\ell)$ at $k=\omega=0$. In this case ℓ is determined implicitly[57] by $r(\ell)=\mu^2\ell^2$.

In addition one may also study the response and correlation functions related to m which have an effective characteristic frequency

$$\Omega_m(\ell) = \mu^2\ell^2\lambda(\ell) = \mu^2\lambda\exp\int_1^\ell(2+\zeta_m)\frac{d\ell'}{\ell'} \quad . \tag{4.15}$$

The ratio (4.8) may be written as [see (4.5)]

$$w(\ell) = \Omega_\varphi(\ell)/\Omega_m(\ell) \tag{4.16}$$

i.e. as a ratio of relaxation rates between the order parameter and the conserved variable, therefore the value of $w^*=w(0)$ is crucial for the validity or breakdown of extended dynamic scaling.

C. Bicritical and tetracritical dynamics

The main questions concerning bicritical and tetracritical dynamics within the simple model (2.5), (2.6) are (i) whether the two order parameters $\vec{S}$ and $\vec{\sigma}$ have a common multicritical dynamic exponent $z_S=z_\sigma=z$; (ii) whether z differs from a model-A type exponent. The relevant dynamic quantity is the ratio

$$\rho(\ell) = \frac{\Gamma_\sigma(\ell)}{\Gamma_s(\ell)} = \frac{\Omega_\sigma(\ell)}{\Omega_s(\ell)} \quad . \tag{4.17}$$

The effective renormalized kinetic coefficients $\Gamma_i(\ell)$ and the characteristic frequencies $\Omega_i(\ell)$ are completely analogous to $\Gamma(\ell)$, (4.4), and $\Omega_\varphi(\ell)$, (4.5). Again we look for the fixed point value $\rho^*=\rho(0)$ of the flow equation

$$\ell\frac{d\rho(\ell)}{d\ell} = \beta_\rho[\rho(\ell),u(\ell),v(\ell),w(\ell)] = \rho(\ell)[\zeta_{\Gamma_\sigma}-\zeta_{\Gamma_s}] \quad . \tag{4.18}$$

A calculation of β_ρ to second order in u,v,w shows that the stable dynamic fixed point value ρ^* is finite in all cases,[10] hence $z=z_s=z_\sigma$.

Specifically, at the static Heisenberg fixed point[38] $u^*=v^*=w^*$ it turns out that not only the exponents but even the amplitudes of the characteristic frequencies of the two order parameters become equal, $\rho^*=1$, independent of n_s and n_σ, at least up to second order in ε. Thus we have a maximally symmetric dynamic behavior of the $n_s+n_\sigma=n$ component order parameter at the bicritical point. Not surprising that the value for z also agrees with that of an n-component isotropic Ginzburg-Landau model (model A). This implies only small differences between the bicritical and λ-line dynamic exponents.

At the biconical fixed point[38] $(u^*\neq v^*\neq w^*)$, ρ^* becomes slightly larger than 1, and ρ^* and z depend on n_s and n_σ separately[10]. Also in this case 2-z is of order η and is hardly distinguishable experimentally from the λ-line exponents.

The relaxational bicritical dynamics may be extended to the presumably more interesting case where the energy density is taken into account as an additional slow variable. This corresponds to the Hamiltonian (2.15) discussed in Sec.III.E and to the dynamic equations (2.12)-(2.14) with g=0. Due to the strong divergence of χ_m, (3.30), we expect significant differences in this case from the λ-line dynamics*, similar as in tricritical dynamics with a conserved variable. An analysis of this case has not yet been performed. Furthermore it would be interesting to see whether fixed point values of the model-C type $\sim \exp(-c/\varepsilon)$ arise in two-loop order (see Sec.V A).

While one believes that a relaxational model provides an appropriate description for anisotropic systems like $GdAlO_3$ it is difficult to say in which case the coupling to the energy density is im-

---

*We expect $\lambda\sim\ell^{\zeta_m^*}$ with ζ_m^* given by (3.28).

portant in the experimentally accessible region. Like in the tricritical case dynamic experiments would be highly desirable.

D. Lifshitz point dynamics

The relaxational dynamics near a Lifshitz point have been studied by Folk and Selke[12] using (2.1) with the Hamiltonian (2.4) for a one-component order parameter. Because of the anisotropy in k-space the characteristic frequency Ω_c depends on two wave numbers p and q corresponding to m and d-m dimensional components $\vec{p}$ and $\vec{q}$ of $\vec{k}$. At the Lifshitz point it has the form[11,12] $\Omega_c(p,q)=\ell^z f(p/\ell, q\ell^{-x})$, i.e. $\Omega_c \sim p^z$ for q=0 and $\sim q^{z/x}$ for p=0, with the dynamic critical exponents z and z/x where

$$z=(4-\eta_{L4})/(1-\lambda), \quad x=(4-\eta_{L4})/(2-\eta_{L2}) \quad . \tag{4.19}$$

Here η_{L4} and η_{L2} denote the pair of static critical exponents introduced by Hornreich et al.[37], and λ is determined by a dynamic two-loop diagram $\sim u^{*2} \sim \varepsilon^2$ where $\varepsilon=d_c-d$ with $d_c=4+m/2$. The results for λ are found as $\lambda=0.9\cdot 10^{-3}\varepsilon^2$ for m=1 and $1.4\cdot 10^{-2}\varepsilon^2$ for m=2 which determine the small deviations from conventional behavior ($\lambda=0$). Huber[11] and Folk and Selke[12] have also studied the reversible dynamics of three-component spin systems near a Lifshitz point. The results for the dynamic exponents are

$$z/x = \frac{1}{2}(d-m+2-\eta_{L2}+m/x) \quad , \quad z/x = \frac{1}{2}(d-m+m/x) \quad , \tag{4.20}$$

in case of ferromagnets and antiferromagnets, respectively. For m=0 this agrees with the ordinary critical exponents[1].

V. BICRITICAL DYNAMICS WITH REVERSIBLE COUPLINGS

The dominant role played by the reversible couplings is clearly shown by the critical divergence of transport coefficients which does not occur in systems with purely dissipative (relaxational) dynamics. An interesting situation arises when at a multicritical point a λ-line of transitions with reversible dynamics meets another λ-line of transitions with purely dissipative dynamics. This leads to a competition in the coupled motion of the two order parameters whose outcome is far from obvious even in a qualitative sense. As a non-trivial example for this situation we discuss in the following the dynamics near a spin-flop bicritical point of magnetic systems like MnF_2.

A. Asymptotic bicritical dynamics

We first consider the asymptotic vicinity of bicriticality as characterized by sufficiently small r, k, ω at g=0 [see (3.15)]. In this region the dynamics of ψ, σ and m (Sec.II B) are governed by power laws with exponents

$$z_\psi = 2+\zeta_\Gamma^* , \quad z_\sigma = 2+\zeta_{\Gamma_\sigma}^* , \quad z_m = 2+\zeta_\lambda^* \tag{5.1}$$

where $\zeta_i=(\mu\partial_\mu \ln Z_i^{-1})_0$, $i=\Gamma,\Gamma_\sigma,\lambda$ (compare ζ_Γ in Sec.IV A). These exponents were first calculated by Huber and Raghavan[18] by means of mode-coupling theory. The interesting result was $z_\psi=z_m=\phi/\nu$ (≈ 1.78) $\neq z_\sigma$ (≈ 2) which implies a breakdown of dynamic scaling of the order parameter ($z_\psi \neq z_\sigma$) as well as a divergence of the kinetic coefficients $\Gamma(\ell)\sim\ell^{z_\psi-2}$ and $\lambda(\ell)\sim\ell^{z_m-2}$, $\ell\sim\xi^{-1}$. This problem was subsequently reexamined by means of a renormalization-group analysis[10] which indicated that the breakdown of scaling is only a spurious result appearing in lowest nontrivial order (one-loop order). It was found that in two-loop order bicritical dynamic scaling is restored with a common exponent z slightly larger than 2,

$$z_\psi = z_\sigma = z = 2+c\eta+O(\rho^*) \quad , \tag{5.2}$$

which implies that $\Gamma(\ell)$ and $\Gamma_\sigma(\ell)$ do not diverge at the bicritical point. In (5.2) $\rho^*\ll 1$ is the fixed point value of the ratio $\rho=\Gamma_\sigma/\Gamma$ and $\eta\ll 1$ is the static (i.e. n=3 Heisenberg) exponent. [The constant $c \ll 1$, however, has a value different from that for the n=3 Ginzburg-Landau model.] In the following we briefly discuss the origin for the nontrivial qualitative difference between one- and two-loop order.

In addition to ρ, the relevant dynamic parameters are the ratio $w=\Gamma/\lambda$ and the dimensionless coupling $f=g^2/\Gamma\lambda$ where Γ,λ are renormalized kinetic coefficients and $g=Z_g^{-1}\mu^{-\varepsilon/2}K_d^{1/2}g_o$ is the renormalized dynamic coupling. The fixed point values ρ^*, w^* and f^* are determined by the zeros of

$$\beta_\rho=\rho(\zeta_{\Gamma_\sigma}-\zeta_\Gamma),\ \beta_w=w(\zeta_\Gamma-\zeta_\lambda),\ \beta_f=f(-\varepsilon-\zeta_\Gamma-\zeta_\lambda+\zeta_m) \quad . \tag{5.3}$$

(The term ζ_m in β_f is due to $Z_g=Z_m^{1/2}$ derived from a Ward identity [5,39]). From $\beta_f=0$ one obtains $f^*=18\varepsilon/11+O(\varepsilon^2)$. Calculating ζ_i in two-loop order (second order in the couplings $U,V,W,f,\gamma^2,\gamma_\sigma^2$) and taking the couplings (but not the ratio ρ and w) at their fixed point values $\sim\varepsilon$ leads to the structure[10]

$$\beta_\rho= \rho[\varepsilon F_1 + \varepsilon^2 F_2 \ln\rho + \varepsilon^2 F_3 + O(\varepsilon^3)] \tag{5.4}$$

where the functions $F_i(\rho,w)$ are well behaved. In $O(\varepsilon)$, one obtains $\rho^*=0$ because of $F_1>0$ which leads to $z_\psi \neq z_\sigma$. In $O(\varepsilon^2)$, however , the term $\sim \ln\rho$ with $F_2>0$ dominates even for $\varepsilon\ll 1$ and prevents $\rho(\ell)$ from vanishing for $\ell\to 0$. Instead, ρ^* becomes finite which, according to (5.3), implies $\zeta_{\Gamma_\sigma}^*=\zeta_\Gamma^*$ and therefore leads to (5.2). The value[59] for ρ^* is determined by the implicit equation $\beta_\rho=0$ which can be solved numerically for $\varepsilon=1$ and need not be further iterated with respect to ε. If one insists, however, on a strict ε-expansion for ρ^* itself one may, for $\varepsilon\to 0$, evaluate F_i at $\rho=0$ 9in the sense of an iteration) and neglect the F_3 term; this leads to the nonanalytic ε-dependence[10] $\rho^* = \exp(-F_1/\varepsilon F_2)$ with $\rho^* \to 0$ for $\varepsilon\to 0$. This type of ε-dependenc is known to occur in model C[1,32,52,58] and in the SSS-model [60,61] as well. The unusual property of the two-loop term $\sim \varepsilon^2 \ln \rho$ is that it is nonnegligible as compared to the one-loop term even for arbitrarily small $\varepsilon>0$. This does not, however, indicate a general inconsistency of the loop expansion for the β-functions since it still constitutes a systematic expansion in powers of the couplings (not necessarily in powers of ε) which may well be suited to pick up also a nonanalytic ε-dependence of the ratio ρ^*. Nevertheless this point requires further investigation.

The remaining fixed point value w^* has been found to vanish in two-loop order[10] which implies a breakdown of extended dynamic scaling ($z_m \neq z$). The value for z_m follows from $f^*>0$,

$$z_m=2+\zeta_\lambda^*=2-\varepsilon-\zeta_\Gamma^*+\zeta_m^*=2\phi/\nu-z_\psi \quad . \tag{5.5}$$

According to (5.2) this differs from the mode-coupling result[18] $z_m=\phi/\nu$. [Eq.(5.5) also corrects the value $z_m=\phi/\nu$ given in Ref.10]

B. Comparison with asymptotic λ-line dynamics

In order to identify specific bicritical effects a comparison with λ line properties is necessary. Along the Ising-like λ-line the dynamics are described by (2.13), (2.14) with g=0, γ=0, W=0 which is equivalent to model C[32]. Therefore the discussion given in Sec.IV B is immediately applicable. In particular, $(\Gamma_\sigma/\lambda)^* \equiv w_\sigma^*>0$ and $z_\sigma=z_m= =2+\alpha/\nu$ which implies that $\Gamma_\sigma(\ell)$ and $\lambda(\ell)$ vanish asymptotically as $\sim \ell^{\alpha/\nu}$.

The dynamics of the transition along the XY-like λ-line belong to the same universality class as the dynamics of the superfluid transition in ^{4}He, as noted in Sec.II B, and therefore the exponents z_ψ and z_m are identical* with those for ^{4}He. They are determined via the zeros of the β-functions (5.3). While there is no doubt that f^* is finite and of order 1 in three dimensions, the finiteness of $w^*=1-O(\varepsilon)$ near four dimensions[39] may not hold in d=3. Instead the weak-scaling fixed point[5] $w^*=0$ may become stable in d=3. The borderline dimension d^* at which this happens appears to be close to 3. If $d^*< 3$, then $w^*>0$ which according to (5.3) yields the dynamic scaling exponent[33,34] $z_\psi=z_m=3/2$. If $d^*>3$, then $w^*=0$ and $z =(3+x)/2 \neq z_m= =(3-x)/2$ with $x \sim O(d^*-3)$. Two loop calculations[5,62] indicate $d^* \lesssim 3$ with $0 \lesssim w^* \ll 1$. In any case, both $\Gamma(\ell)$ and $\lambda(\ell)$ will diverge asympto-

* There is, however, an important physical difference; the order-parameter correlation function and the characteristic frequency $\Omega_\psi \sim \ell^{z_\psi}$ are directly measureable in magnetic systems like MnF_2 but not in ^{4}He.

tically, with exponents $z_\psi-2\simeq z_m-2\simeq-1/2$. The present results for the asymptotic λ-line and bicritical dynamics for d=3 are summarized in the following Table (with w_σ^* and f^* evaluated in $O(\varepsilon)$ at $\varepsilon=1$).

	Ising-line	bicritical	XY-line
z_ψ	-	$\gtrsim 2$	$\approx 3/2$
z_σ	$2+\alpha/\nu \approx 2.1$	$\gtrsim 2$	-
z_m	$2+\alpha/\nu \approx 2.1$	≈ 1.6	$\approx 3/2$
w^*	-	0	$\ll 1$
w_σ^*	≈ 1	0	-
f^*	-	$\approx 18/11$	≈ 1

One may also study the asymptotic shape of $\langle mm \rangle$ as a function of frequency. At the bicritical point, for sufficiently small k, the low frequency portion of the spectrum will be dominated by the smallest characteristic frequency $\Omega_\sigma \ll \Omega_\psi \ll \Omega_m$ and will be determined essentially by the relaxational dynamics of the σ variable. A quantitative study [63] shows that $\langle mm \rangle$ has an asymptotic shape not significantly different from a Lorentzian centered around ω=0, in contrast to the (asymptotic) shape along the XY λ-line[62,64].

C. Nonasymptotic dynamics

From the study of ^{4}He it is known that the small w* is accompanied by a small dynamic transient exponent[5] $\omega_w \sim w^* \ll 1$ in d=3 due to the existence of the borderline dimension $d^* \lesssim 3$. This is a universal feature which will also occur at the XY-like transition in MnF_2 and will therefore affect the crossover to bicriticality. (Whether this transient remains small in the bicritical region cannot be answered at present). An important consequence of $\omega_w \ll 1$ is that the asymptotic dynamic-scaling region of the XY λ-line becomes experimentally inaccessible. Even the inclusion of the usual additive corrections to scaling ($\sim \ell^{\omega_w}$) corresponding to a linearization of β_w and β_f around w* and f* does not provide an adequate description of observable properties. Instead, a complete integration of the nonlinear flow equations

$$\ell\frac{dw(\ell)}{d\ell} = \beta_w\ , \qquad \ell\frac{df(\ell)}{d\ell} = \beta_f \tag{5.6}$$

becomes necessary[7,65]. Within the loop-expansion (up to second-order [5,62] in f and to fourth order[66] in γ) the nonlinear dependence of β_w and β_f on the nonperturbative parameter w is known exactly, therefore the integration of β_w and β_f properly describes corrections $\sim (\ell^{\omega_w})^n$ to all orders n.

In addition, a nonuniversal feature enters via the initial conditions $w(\ell_o)$ and $f(\ell_o)$ (at some ℓ_o in the experimental range) which depend on the specific system. Because of the possible relevance also for systems like MnF_2 we show in Fig. 1 the result of a quantitative analysis in case of ^4He [65] (with the approximation $\gamma(\ell) = 0$; a similar result is obtained if the static coupling $\gamma(\ell) \ll 1$ is taken into account [67,68]). Since in the experimental region ($t \sim \ell^{1/\nu} > 10^{-6}$) the effective ratio $w(t)$ is far from $w^* \simeq 0$, $w(t)$ remains dependent on nonuniversal parameters (pressure in case of ^4He, magnetic field in case of MnF_2) even close to criticality. Furthermore, $f(t)$ exhibits a crossover in the experimental range from a quasicritical value $f(t) \approx f^*$ to a small noncritical background value $f(1) \ll 1$, with a leading ℓ-dependence $f(\ell) \sim \ell^{-1}$ in the "precritical" region[35,67,69]. The latter ℓ-dependence is a general (universal) consequence of the structure of β_f in d=3 and may therefore play a role not only along the XY λ-line of MnF_2 but even in the bicritical region. The actual size of this precritical region, however, depends on the (nonuniversal) magnitude of $f(\ell_o)$ in the noncritical background [67]. As shown for various dynamic quantities above and below T_λ[7,65] the ℓ-dependence of f and w shown in Fig. 1 is crucial for explaining the experiments. Therefore also for systems like MnF_2 an identification of $w(\ell_o)$, $f(\ell_o)$ and presumably $w_\sigma(\ell_o)$, at least in order of magnitude, seems to be necessary before one may attempt to predict quantitatively the observable λ-line and bicritical dynamics. Once these parameters have been identified in the disordered phase one is also able to make predictions on the dynamics in the ordered phases without adjustable parameters[7,65].

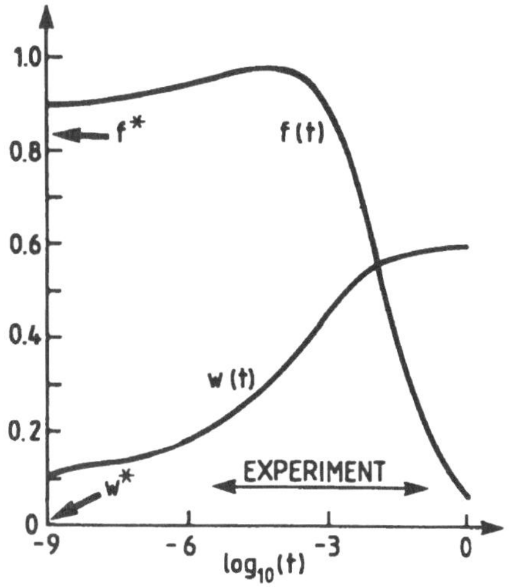

Fig. 1: Effective dynamic parameters for ^4He

VI. TRICRITICAL DYNAMICS OF ^3HE-^4HE MIXTURES

^3He-^4He mixtures represent the most favorable system for an experimental test of the theory of multicritical dynamics. So far the main objective of the theories[9,13,14,16,20,21] and of the analysis of the experimental data[23-30] was a study of asymptotic properties such as tricritical dynamic exponents and universal scaling functions. In principle there exist four dynamic exponents describing the asymptotic behavior of the effective kinetic coefficients (see Sec. II C)

$$\Gamma(\ell)\sim\ell^{z_\psi-2}, \quad \lambda(\ell)\sim\ell^{z_\lambda-2}, \quad K(\ell)\sim\ell^{z_K-2}, \quad L(\ell)\sim\ell^{z_L-2}. \tag{6.1}$$

A renormalization-group calculation in one-loop order by Siggia and Nelson[9] (SN) yielded the predictions

$$z_\psi = 2-\varepsilon/3, \; z_\lambda = z_K = z_L = 2-2\varepsilon/3 . \tag{6.2}$$

It was pointed out by SN themselves, however, that this result may be subject to qualitative changes due to possible two-loop contributions of the model-C type, similar as for bicritical dynamics discussed in Sec.V above. In the present case these contributions are as yet unknown.

Apart from this uncertainty concerning the asymptotic behavior, the previous theories of ^3He-^4He mixtures do not describe two important nonasymptotic features which have been very recently shown[22] to occur along the λ-line: (i) the existence of the same small dynamic transient exponent $\omega_w \ll 1$ as for pure ^4He, (ii) the considerable departures from the asymptotic dynamic behavior in a precritical region $t>10^{-3}$, $t\equiv(T-T_\lambda)/T_\lambda$, similar as for pure ^4He (see Fig.1 and Fig.2 below). It is natural to expect that also in the corresponding tricritical temperature range there may exist effects similar to those of (ii). Furthermore, even if the smallness of ω_w does not extend into the tricritical region, it may affect the crossover to tricriticality. In order to deal with these problems a nonasymptotic theory of tricritical dynamics is necessary.

A major step in this direction has been performed very recently in the context of the λ-line dynamics[22]. But the theory does not yet contain all contributions which are potentially important in the tricritical region. In the following we describe some essential elements of the present status of the theory[9,22,66].

A. Transport coefficients and light-scattering spectrum

The physical quantities of interest are the mass diffusion coefficient D, the thermal conductivity κ, and the thermal diffusion ratio k_T. Within the model (2.16)-(2.19) the leading contributions to their (tri)critical temperature dependence are determined by the effective transport coefficients $\lambda(\ell)$, $K(\ell)$, and $L(\ell)$ according to

$$D=\lambda(\ell)/\chi_c(\ell), \quad \kappa=K(\ell)-L(\ell)^2/\lambda(\ell), \quad k_T=L(\ell)\chi_c(\ell)/\lambda(\ell) \tag{6.3}$$

where $\chi_c(\ell)$ is the concentration susceptibility. The flow parameter ℓ is to be identified as $\mu^2\ell^2=r(\ell)$ with $r(\ell)$ given by (3.10). Apart from static corrections to scaling this implies $\ell \sim t^\nu$, $\nu \approx 2/3$, in the λ-line critical region and $\ell \sim t_t^{1/2}$ in the tricritical region with $t_t=(T-T_t)/T_t$. Here t_t means the relative temperature at constant chemical potential difference $\mu_3-\mu_4$ rather than at constant tricritical concentration c_t. In the latter case the effect of the Fisher renormalization[70] is to change the exponent by a factor $1/(1-\alpha_t)=2$ which yields $\ell \sim t_t$.

Another quantity of interest is the dynamic structure factor whose dominant contribution comes from the concentration correlation function $C_{cc}=\langle cc\rangle$. In the hydrodynamic region it has the form[22]

$$C_{cc}(k,\omega) = 2k^2\chi_c \frac{D\omega^2+D_s\hat{\Gamma}_o\hat{\Gamma}_s}{(\omega^2+\hat{\Gamma}_o^2)(\omega^2+\hat{\Gamma}_2^2)} \tag{6.4}$$

with the linewidths

$$\hat{\Gamma}_{o,2} = \frac{k^2}{2}\{D+D_s \pm [(D+D_s)^2 - 4DD_T]^{1/2}\} \tag{6.5}$$

(not to be confused with the kinetic coefficient Γ_o of the order parameter). In (6.4), (6.5), D_s and D_T are thermal diffusivities

$$D_T = \kappa/\chi_q, \quad D_s = D_T + k_T^2 D/\chi_c\chi_q \tag{6.6}$$

with χ_q being proportional to the specific heat $Cp_{,c}$. The linewidth $\hat{\Gamma}_o$ can be measured by light-scattering experiments[24] while D,κ and k_T can be determined by macroscopic measurements[23].

In principle both static quantities χ_q and χ_c have a critical behavior arising from the couplings γ_1 and γ_2 in (2.16). Following SN[9] we drop the less important coupling γ_1 (because of the weak tricritical temperature dependence of $Cp_{,c}$) and consider only $\gamma_2 \equiv \gamma \neq 0$[71]. According to Sec.III this accounts for the important temperature dependence of the concentration susceptibility which diverges as $\chi_c(\ell) \sim \ell^{-1}$ in the tricritical region.

It is convenient to work with dimensionless (renormalized) parameters (analogous to w and f in Secs.IV and V) as defined by[9]

$$w_1 = \frac{\Gamma\chi_q}{K}, \quad w_2 = \frac{\Gamma\chi_c}{\lambda}, \quad w_3 = \frac{L}{(K\lambda)^{1/2}}, \quad f_1 = \frac{g_1^2}{\Gamma K}, \quad f_2 = \frac{g_2^2}{\Gamma\lambda} \tag{6.7}$$

where $g_i^2=\mu^{-\varepsilon}K_d\, g_{io}^2 Z_i$ with $Z_1=1$ and $Z_2=Z_c$. (The coupling g_{10} is not renormalized because of the approximation $\gamma_1=0$). The corresponding effective parameters $w_i(\ell)$ and $f_i(\ell)$ satisfy the flow equations

$$\ell dw_i(\ell)/d\ell = \beta_{w_i}, \quad \ell df_i(\ell)/d\ell = \beta_{f_i}, \tag{6.8}$$

with β-functions similar in structure to (5.3). For the r.h.s. of

(6.3) this leads to the following expressions

$$D = g_2[\ell^{\varepsilon}\chi_c(\ell)w_2(\ell)f_2(\ell)]^{-1/2} \quad , \tag{6.9}$$

$$\kappa = g_1\chi_q^{1/2}[\ell^{\varepsilon}w_1(\ell)f_1(\ell)]^{-1/2}[1-w_3(\ell)^2] \quad , \tag{6.10}$$

$$k_T = \chi_c(\ell)w_3(\ell)\frac{g_1}{g_2}\left(\frac{f_2(\ell)}{f_1(\ell)}\right)^{1/2} \quad . \tag{6.11}$$

In deriving (6.11) we have used the interesting exact relation between the dynamic parameters[22]

$$\frac{f_1(\ell)w_2(\ell)}{f_2(\ell)w_1(\ell)} = \frac{g_1^2\chi_c(\ell)}{g_2^2\chi_q} \equiv B(\ell) \tag{6.12}$$

with $B(\ell)$ being a purely static parameter. While $B(0)$ is finite along the λ-line, it diverges as $\sim \chi_c(\ell)$ upon approaching the tricritical point.

B. Asymptotic dynamics

One can show in all order of perturbation theory that[22]

$$f_1^* = f_2^* \equiv f^*, \; w_3^* = 1 \tag{6.13}$$

along the λ-line and presumably also in the tricritical case (in the latter case at least in one-loop order[9]). According to (6.11) this implies the asymptotic behavior ($\ell \to 0$)

$$k_T = \chi(\ell)g_1/g_2 \quad , \tag{6.14}$$

thus k_T remains finite along the λ-line but diverges as ℓ^{-1} in the tricritical region (g_1/g_2 remains finite), in agreement with experiment[28]. The relation (6.14) was first proposed by Papoular[16] for the superfluid phase. The proportionality $k_T \approx \chi_c$ is already contained in the predictions of mode-coupling theory[13,14].

The vanishing of $1-w_3(\ell)^2$ as $\ell \to 0$ leads to a finite value for κ along the λ-line[9,13,14], with a slow approach to its critical value $\kappa(T_\lambda)$ because of the small transient exponent $\omega_w \ll 1$ [22]. Also at tricriticality κ is predicted[9,13,14] to remain finite, in agreement with experiment[28], with a fast transient exponent[9] $2\varepsilon/3$ which needs confirmation by a two-loop calculation.

As a general result for the λ-line behavior it has been shown[22] that f^*, (6.13), and

$$w^* = (w_1^{*-1} + w_2^{*-1})^{-1} \tag{6.15}$$

are identical with f* and w* of pure ^{4}He (compare Sec.V B above). This property, together with

$$w_2^*/w_1^* = B(0) \ , \tag{6.16}$$

completely determines the asymptotic λ-line behavior and confirms to all orders the mode-coupling predictions[13,14] $z_\psi=z_\lambda=z_K=z_L=d/2$ provided that $w^*>0$. In the tricritical case, (6.16) implies $w_2^*/w_1^* = \infty$. A one-loop calculation of SN[9] yielded $w_1^* = 0$, $w_2^* = \infty$, $f^* = 4\varepsilon/3$, $\gamma^{*2} = \varepsilon/4$ which leads to the tricritical dynamic exponents (6.2). It is possible, however, that w_1^* or w_2^* become finite in two-loop order (without violation of $w_2^*/w_1^* = \infty$).

Among the four exponents (6.2) there is only z_λ which enters the experiments sensitively (via the mass diffusion D). The present experimental information on $z_\lambda-2$ comes from Refs.26 and 29, with the results $z_\lambda-2 \approx -0.68$ (Ref.26) and $z_\lambda-2 \approx -0.6$ (Ref.29). This appears to be consistent with (6.2). It should be noted, however, that the data of Ref.26 (which are presumably the more accurate ones) lie in a temperature range of only $2\cdot10^{-2} \leq t_t \lesssim 10^{-1}$. This raises some doubt about the asymptotic nature of the value of the measured exponent $z_\lambda-2$.

The above results imply $D_S \gg D_T \gg D$ near tricriticality. Substituting this into (6.5) leads to the following asymptotic behavior for the linewidths

$$\hat{\Gamma}_0/k^2 \sim \chi_c(\ell)^{-1}\kappa \sim \ell,\ \hat{\Gamma}_2/k^2 \sim D_s \sim \ell^{z_K-2} \ . \tag{6.17}$$

The result for $\hat{\Gamma}_0$ is in agreement with light-scattering experiments[25] along the superfluid branch of the coexistence curve. Note that this yields confirmation on the finiteness of $\kappa(T_t)$ but no information on the values of the exponents (6.2).

C. Nonasymptotic dynamics

Near the λ-line, the effective parameters $w_3(\ell)$ and $f_2(\ell)/f_1(\ell)$ have a fast approach to their fixed point values whereas $w_1(\ell)$ and $w_2(\ell)$ come close to w_1^* and w_2^* only for inaccessibly small ℓ[22]. In addition, the background values $w_i(10^{-1})$ and $f_i(10^{-1})$ were found to be very far from the fixed point values. We illustrate this in Fig.2 for the example of $w_2(t)$ and $w_3(t)$, $t=(T-T_\lambda)/T_\lambda$, at the ^{3}He concentration c=0.04 as obtained from a fit of two-loop β-functions[22] to recent data[72].

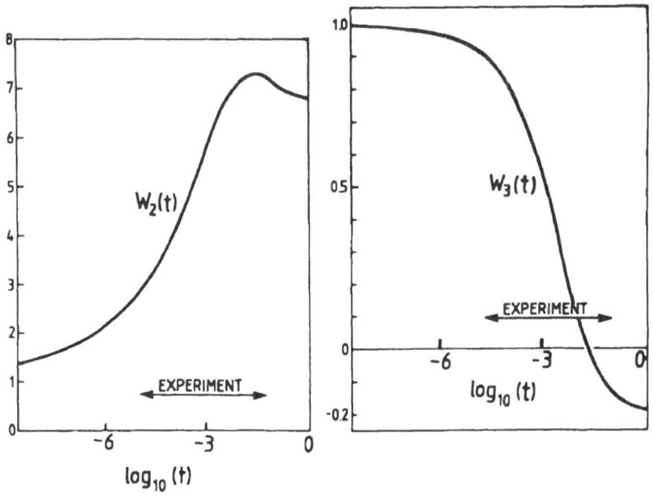

Fig.2: Effective dynamic parameters for $^3He-^4He$ mixtures at the ^3He concentration c=0.04 (from Ref.22)

We see that $w_3(t) \approx w_3^* = 1$ for $t < 10^{-4}$ but $w_2(t) > 1 \gg w_2^*$ even for $t \approx 10^{-8}$. Note also the large value of $w_2(10^{-1})$ and the smallness of $|w_3(10^{-1})|$. Furthermore $w_1(t)$ as well as $f_1(t)$ and $f_2(t)$ exhibit a strong t-dependence[22] (similar to w(t) and f(t) in Fig.1). By contrast the usual asymptotic theories completely neglect this t-dependence. Figs.1 and 2 demonstrate that this concept is remarkable unrealistic in the experimentally accessible region, at least for dilute $^3He-^4He$ mixtures. At present there is no reason to expect that the nonasymptotic ℓ-dependence becomes negligable for tricritical $^3He-^4He$ mixtures. A definite answer can be given only on the basis of (approximate) knowledge on 1) the five β-functions β_{w_i} and β_{f_i}, 2) the initial values $w_i(\ell_o)$ and $f_i(\ell_o)$ at $\ell_o \sim 10^{-1}$.

1) The existing knowledge on the β-functions relevant for tricritical dynamics comes form three sources: (i) A one-loop calculation by SN[9], (ii) a two-loop calculation by the present author[66] for $w_3=f_1=0$ but $\gamma \neq 0$ and $f_2 \neq 0$ corresponding to model F[39], (iii) a two-loop calculation[22] for $\gamma=0$ but $w_3 \neq 0$, $f_1 \neq 0$, $f_2 \neq 0$. Unfortunately, even if one combines these pieces of information there are potentially important two-loop terms missing such as $\sim \gamma^2 f_1$ and

$\sim \gamma^2 (f_1 f_2)^{1/2}$. While the approximation $\gamma=0$ seems to be reasonable along the λ-line where $\gamma(\ell) \ll 1$ and $\gamma^*=0(\alpha<0)$, this is definitely not applicable in the tricritical case were $\gamma^*=(\alpha_t/4\nu_t)^{1/2}=1/2$ in three dimensions (according to Sec.III D). Therefore we suspect that significant progress on this problem can be achieved only by means of a complete two-loop calculation. While at first sight this appears to be a task which one would suggest to postpone to the next century it looks better at second sight. If one combines the recently introduced "diagonal representation" of the equations of motion[22] (which circumvents the off-diagonality of the coupling L) with a model-F like calculation in two-loop order[66] the problem may become feasible.

2) Concerning the initial conditions $w_i(\ell_o)$, $f_i(\ell_o)$ nothing is known at present, but sufficient information can probably be extracted from existing experimental data. We note that these data need not be very close to the tricritical concentration since $w_i(\ell_o)$, $f_i(\ell_o)$ in the noncritical background region $\ell_o \gtrsim 10^{-1}$ should depend on the concentration only smoothly. An estimate of these parameters, at least in order of magnitude, as well as of their (smooth) variation along the λ-line would be of considerable interest and would also allow to make predictions on the critical behavior in the superfluid phase without adjustable parameters.

VII. CONCLUDING REMARKS

This review is necessarily as imcomplete as the present status of the research on multicritical dynamics. For reasons of space limitations we have not discussed the theoretical work on tricritical dynamics of multicomponent fluids[8], on sound attenuation[21] and on nucleation and spinodal decomposition phenomena[73,74] near the tricritical point of ^3He-^4He mixtures, and on Lifshitz point dynamics of liquid crystals[75]. We do not attempt here to present a complete list of the many other systems which are interesting candidates for future studies of multicritical dynamics (for example random systems). We only mention the dynamics of polymer solutions near the θ-point[76,77] and the dynamics of surfaces and interfaces near a multicritical point (for multicritical statics see Ref.78).

Finally we wish to emphasize that there are basic problems unsolved concerning the multicritical behavior, even for the presumably best accessible multicritical point - the tricritical point of ^3He-^4He mixtures. So far neither the theory nor the experiments provide convincing answers even to qualitative questions, for example which of the leading dynamic tricritical exponents (6.2) are equal or not equal to one another. Considerable additional effort both theoretically and experimentally is necessary to improve the present unsatisfactory situation.

ACKNOWLEDGEMENTS

I wish to thank R. Folk and H.K. Janssen for fruitful collaboration on the critical dynamics of helium and on bicritical dynamics.

REFERENCES

1. P.C. Hohenberg and B.I. Halperin, Rev. Mod. Phys. 49, 435 (1977).
2. (a) K. Kawasaki, in Phase Transitions and Critical Phenomena, Vol. Vb, edited by C. Domb and M.S. Green (Academic Press, New York, 1976);
(b) J.D. Gunton, in Proceedings of the International Conference on Dynamic Critical Phenomena, p.1, edited by C.P. Enz (Springer, Berlin, Heidelberg, New York, 1979).
3. E. Brézin, J.C. Le Guillou, and J. Zinn-Justin, in Ref.2a, Vol. VI; D.J. Amit, Field Theory, The Renormalization Group and Critical Phenomena (Mc Graw Hill, New York, 1978).
4. R. Bausch, H.K. Janssen, and H. Wagner, Z. Physik B24, 113 (1976).
5. C. De Dominicis and L. Peliti, Phys. Rev. B18, 353 (1978).
6. H.K. Janssen, in Ref, 2b. p.25.
7. V. Dohm and R. Folk, Phys. Rev. Lett. 46, 349 (1981); and in Festkörperprobleme (Advances in Solid State Physics), Vol. XXII, p.1, edited by P. Grosse (Vieweg, Braunschweig, 1982).
8. K. Kawasaki, J. Phys. A8, 262 (1975).
9. E.D. Siggia and D.R. Nelson, Phys. Rev. B15, 1427 (1977).
10. V. Dohm and H.K. Janssen, Phys. Rev. Lett. 39, 946 (1977); J. Appl. Phys. 49, 1347 (1978); V. Dohm, Report of the Kernforschungsanlage Jülich Nr. 1578 (1979).
11. D.L. Huber, Phys. Lett. 55A, 359 (1976); 70A, 500 (1979).
12. R. Folk and W. Selke, Phys. Lett. 69A, 255 (1978).
13. K. Kawasaki and J.D. Gunton, Phys. Rev. Lett. 29, 1661 (1972).
14. M.K. Grover and J. Swift, J. Low Temp. Phys. 11, 751 (1973).
15. D.L. Huber, Phys. Rev. B10, 3992 (1974).
16. M. Papoular, J. Low Temp. Phys. 24, 105 (1976).
17. H. Müller-Krumbhaar and D.P. Landau, Phys. Rev. B14, 2014 (1976).
18. D.L. Huber and R. Raghavan, Phys. Rev. B14, 4048 (1976); D.L. Huber, Phys. Lett. 49A, 345 (1974); R. Raghavan, thesis (University of Wisconsin, 1976).
19. V.V. Prodnikov and G.B. Teitelbaum, JETP Lett. 23, 296 (1976).
20. L. Peliti, in Ref. 2b, p.189.
21. M. Braun, Diplomarbeit (Universität München, 1982).
22. V. Dohm, R. Folk and J.K. Bhattacharjee, Phys. Rev. B (1983).
23. H. Meyer, G. Ruppeiner, and M. Ryschkewitsch, in Ref.2b, p.171.
24. P. Leiderer, in Quantum Fluids and Solids, edited by S.B. Trickey, E.D. Adams, and J.W. Dufty (Plenum, New York, 1977).
25. P. Leiderer, D.R. Nelson, D.R. Watts, and W.W. Webb, Phys. Rev. Lett. 34, 1080 (1975).
26. D.B. Roe, G. Ruppeiner, and H. Meyer, J. Low Temp. Phys. 27, 747 (1977).
27. D. Roe and H. Meyer, J. Low Temp. Phys. 28, 349 (1977).

28. G. Ruppeiner, H. Ryschkewitsch, and H. Meyer, J. Low Temp. Phys. 41, 179 (1980).
29. R.P. Behringer and H. Meyer, J. Low Temp. Phys. 46, 407 (1982).
30. D.N. Sinha and J.K. Hoffer, Phys. Rev. Lett. 50, 515 (1983).
31. S.K. Ma and G.F. Mazenko, Phys. Rev. B11, 4077 (1975).
32. B.I. Halperin, P.C. Hohenberg, and S.K. Ma, Phys. Rev. B10, 139 (1974); B13, 4119 (1976).
33. R.A. Ferrell, N. Menyhárd, H. Schmidt, F. Schwabl, and P. Szépfalusy, Ann. Phys. (N.Y.) 47, 565 (1968).
34. B.I. Halperin and P.C. Hohenberg, Phys. Rev. 177, 952 (1969).
35. R.A. Ferrell and J.K. Bhattacharjee, Phys. Rev. Lett. 42, 1638 (1979).
36. E.K. Riedel and F.J. Wegner, Phys. Rev. Lett. 29, 349 (1972).
37. R.M. Hornreich, M. Luban, and S. Shtrikman, Phys. Rev. Lett. 35, 1678 (1975).
38. J.M. Kosterlitz, D.R. Nelson and M.E. Fisher, Phys. Rev. B13, 412 (1976).
39. B.I. Halperin, P.C. Hohenberg, and E.D. Siggia, Phys. Rev. B13, 1299 (1976).
40. I.M. Khalatnikov, An Introduction to Theory of Superfluidity (Benjamin, New York, 1965).
41. L.D. Landau and I.M. Lifshitz, Fluid Mechanics (Pergamon, London, 1959), Chap. VI.
42. R. Graham, in Springer Tracts in Modern Physics, 66 (Springer Berlin, Heidelberg, New York, 1973).
43. P.C. Martin, E.D. Siggia, and H.A. Rose, Phys. Rev. A8, 423 (1973).
44. H.K. Janssen, Z. Physik B23, 377 (1976).
45. C. De Dominicis, J. Phys. (Paris), Colloq. 37, C1-247 (1976).
46. G. t'Hooft, Nucl. Phys. B61, 455 (1973).
47. I.D. Lawrie and S. Sarbach, in Ref. 2a, Vol. VIII, to be published.
48. See also D. Blankschtein and A. Aharony, Phys. Rev. Lett. 47, 439 (1981).
49. J. Rudnick and D.R. Nelson, Phys. Rev. B13, 2208 (1976).
50. M.E. Fisher, Phys. Rev. Lett. 34, 1634 (1975).
51. D.R. Nelson and E. Domany, Phys. Rev. B13, 236 (1976).
52. E. Brézin and C. de Dominicis, Phys. Rev. B12, 4954 (1975).
53. E. Eisenriegler and B. Schaub, Z. Phys. B38, 65 (1980).
54. M.E. Fisher, AIP Conf. Proc. 24, 273 (1975).
55. U. Deker and F. Haake, Phys. Rev. A11, 2043 (1975).
56. R. Bausch, V. Dohm, H.K. Janssen and R.K.P. Zia, Phys. Rev. Lett. 47, 1837 (1981).
57. The explicit identification $\ell \sim t^{1/2}$ in Ref.9 is justified within an ε-expansion [$\nu=1/2+O(\varepsilon)$] in the asymptotic scaling region. In this spirit also the w-dependence of $\Gamma(\ell)$ has been dropped by setting $w(\ell)=w^*=1+O(\varepsilon)$ in (5.1)-(5.4) in Ref.9.
58. K.K. Murata, Phys. Rev. B13, 2028 (1976).
59. The precise two-loop value for ρ^* is not known since F_3 has not yet been calculated completely.

60. L. Sásvari, F. Schwabl, and P. Szépfalusy, Physica 81A, 108 (1975).
61. V. Dohm, Z. Physik B31, 327 (1978).
62. V. Dohm, Z. Physik B33, 79 (1979).
63. V. Dohm and M. Lücke, unpublished.
64. R.A. Ferrell, V. Dohm and J.K. Bhattacharjee, Phys. Rev. Lett. 41, 1818 (1978).
65. V. Dohm and R. Folk, Z. Physik B40, 79 (1980); B41, 251 (1981).
66. V. Dohm, manuscript in preparation
67. G. Ahlers, P.C. Hohenberg, and A. Kornblit, Phys. Rev. B25, 3136 (1982).
68. V. Dohm and R. Folk, Z. Physik B45, 129 (1981).
69. R.A. Ferrell and J.K. Bhattacharjee, in Ref. 2b, p. 152.
70. M.E. Fisher, Phys. Rev. 176, 257 (1968).
71. Very recently both couplings γ_1 and γ_2 have been treated by A. Onuki, J. Low Temp. Phys. (to be published). This is relevant for intermediate concentrations well away from the tricritical point.
72. D. Gestrich, R. Walsworth, and H. Meyer, J. Low Temp. Phys. (1983).
73. P.C. Hohenberg and D.R. Nelson, Phys. Rev. B20, 2665 (1979).
74. J.D. Gunton, M. San Miguel, and P.S. Sahni, in Ref. 2a, Vol.X, to be published
75. K.A. Hossein, J. Swift, J.H. Chen, and T.C. Lubensky, Phys. Rev. B19, 432 (1979).
76. P.G. de Gennes, Scaling Concepts in Polymer Physics, Cornell University Press, London (1979).
77. K. Binder, J. Chem. Phys. to be published
78. K. Binder, in Ref. 2a, Vol.X, to be published

CRITICAL DYNAMICS OF SOUND

Kristian Fossheim and Jon Otto Fossum

Department of Physics
University of Trondheim
The Norwegian Institute of Technology
7034-NTH Trondheim

INTRODUCTION

Measurements of acoustic velocity and attenuation is one of the best methods available for studying the dynamics of solids, liquids and gases near phase transitions. Acoustic waves act as a very sensitive probe of order parameter fluctuations, and hence often display strong "anomalies" near the phase transition temperature T_c. In principle, many experimental methods used in investigations of phase transitions are capable of giving some information about critical dynamics, but in different regions of the space of variables q, ω, ξ^{-1}, τ^{-1} illustrated in Fig.1. Here q is the wavevector, ω is frequency, ξ is the correlation length and τ is the characteristic time given by $\tau = \xi^z f(q\xi)$ in the scaling region. In Fig.1 the vertical axis is essentially a temperature axis, but it assumes a double meaning: both as an inverse length axis ξ^{-1}, and as an inverse time axis τ^{-1}. At the origin q, ω, ξ^{-1}, $\tau^{-1} = 0$ true divergence of the order parameter susceptibility occurs. The inner "core" of Fig.1, where q, ω, ξ^{-1}, τ^{-1} are small but finite, is the critical, or asymptotic scaling region. Outside this region corrections to scaling are important.

In ultrasound experiments the mechanical susceptibility (the elastic constant) is measured, but we show below that it is simply related to the order parameter susceptibility. The applied sound wave may be considered as spatially homogeneous, i.e. $\xi^{-1} \gg q \approx 0$ in the experimentally accessible region. Recent work has shown also that it is possible to perform experiments inside the asymptotic region and to observe dynamical scaling.

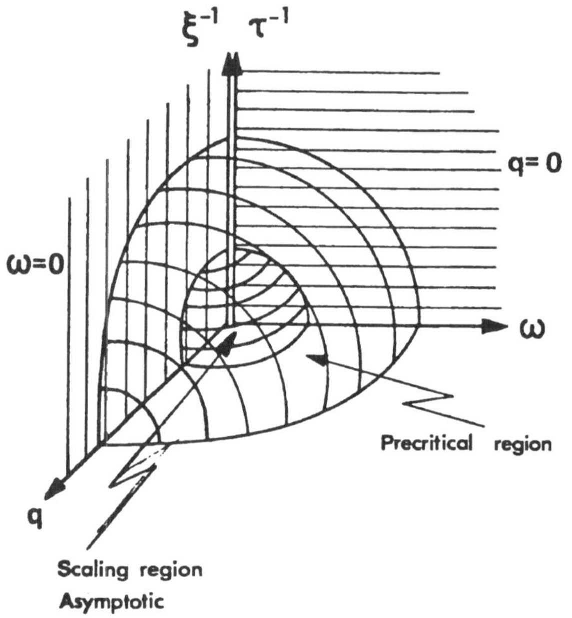

Fig.1. q, ω, ξ^{-1}, τ^{-1} space of order parameter - fluctuations with $\tau = \xi^z f(q\xi)$ + corrections to scaling.

Depending on the symmetry, different types of coupling between strain ε_{ij} and order parameter Q_1 will exist in different cases, resulting in characteristically different types of anomalies. A good example, with two successive structural transitions[1], is shown in Fig.2. In this lecture we will try to outline (i) how a systematic treatment may be given of the dynamics of sound by use of the fluctuation-dissipation theorem, (ii) give examples of experimental results which substantiate the theory, and (iii) point out that dynamical scaling has been observed in connection with phase transitions in all the main classes of condensed matter systems. Since the problems we treat have not yet been investigated at multicritical points we shall discuss the behaviour at ordinary continuous transitions. We note however that many of the systems investigated so far do indeed possess multicritical points. Eventually therefore even the dynamics of multicritical points are accessible to experimental investigation.

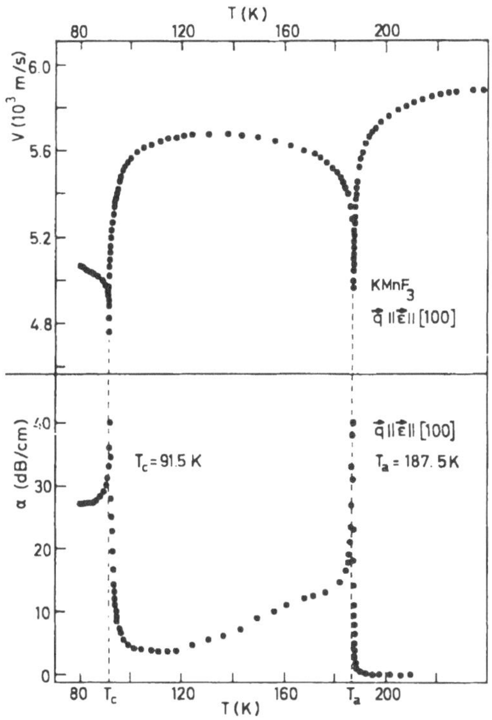

Fig.2. Attenuation, α , and velocity , v , at 11,7 MHz with longitudinal sound along [100] in $KMnF_3$. Fossheim et. al. (1974) (Ref.1).

THEORY

The response function, or susceptibility, which may be directly studied by ultrasonic methods is the complex elastic stiffness tensor defined through the (classical) fluctuation-dissipation theorem[2] as the following integral over the correlation of fluctuating internal stress $\delta\sigma_j(0,0)$ and its time derivative $\delta\dot{\sigma}_i(\vec{r},t)$:

$$C_{ij}(\vec{q},\omega) = \frac{1}{kT}\int d^d\vec{r}\int_0^\infty dt\; e^{-i\vec{q}\cdot\vec{r}}\, e^{i\omega t}\, \langle\delta\dot{\sigma}_i(\vec{r},t)\,\delta\sigma_j(0,0)\rangle \qquad (1)$$

where the spatial integration is over the volume of the system. $\vec{q}$ and ω are the wave vector and the frequency of the acoustic wave, k is Boltzmann's constant, and T is temperature. The best way to proceed is by Fourier transformation of the spatial dependence of $\delta\sigma_i$. In addition, we choose to extract the statics by performing a partial integration over the time variable. After these changes Eq.(1) becomes:

$$C_{ij}(\vec{q},\omega) = -\frac{V}{kT}\langle\delta\sigma_i(\vec{q},0)\,\delta\sigma_j(-\vec{q},0)\rangle$$

$$- i\omega\,\frac{V}{kT}\int_0^\infty dt\; e^{i\omega t}\langle\delta\sigma_i(\vec{q},t)\,\delta\sigma_j(-\vec{q},0)\rangle \qquad (2)$$

Here the first term is purely static, and hence, as we shall see, contains all the information about the sound velocity $v(\vec{q},\omega,T)$ in the low frequency limit. As will become apparent the velocity may in addition contain a dynamic component coming from the second term. From this term we also obtain the damping, or attenuation $\alpha(\vec{q},\omega,T)$. The relations between these quantities are (dropping tensor indices for a moment)

$$\frac{\Delta v}{v_o} = \frac{Re\Delta C}{2\rho v_o^2} \quad ; \quad \frac{v_o\Delta\alpha}{\omega} = -\frac{Im\Delta C}{2\rho v_o^2} \qquad (3)$$

Here v_o is the "background" velocity measured far away from T_c ; Δv , ΔC and $\Delta\alpha$ are the respective changes from "background" values well away from T_c , and ρ is the density of the material.

Coupling, and dynamical models

We now have to specify the coupling between the acoustic wave and the order parameter, i.e. give the coupling term H_c in the Hamiltonian. Depending on symmetry three main types of coupling may dominate:

(i) Linear : $H_c = a_{ij}\varepsilon_i Q_j \Rightarrow \sigma_i = \partial H_c/\partial\varepsilon_i = a_{ij}Q_j$ (4a)

(ii) Quadratic : $H_c = b_{ijl}\varepsilon_i Q_j Q_l \Rightarrow \sigma_i = \partial H_c/\partial\varepsilon_i = b_{ijl}Q_j Q_l$ (4b)

(iii) Quadratic : $H_c = d_{ijl}\varepsilon_i\varepsilon_j Q_l \Rightarrow \sigma_i = \partial H_c/\partial\varepsilon_i = d_{ijl}\varepsilon_j Q_l$ (4c)

Of these the first and last ones can only exist for certain symmetries, while the second one, which is the strictive type, will exist in all matter. It may therefore be said to be the most important one. For this reason and due to space limitations we choose to discuss this case in some detail, and give only the results for the first case. We note that in the case of normal-to-incommensurate transitions b_{ijl} may have to be considered a wavevector dependent complex quantity[3]. A more complete account of calculations and data will be given in a forthcoming paper[4]. We refer also to reviews[5]. However, the calculations cannot be made without some assumptions being made about the decay behaviour of the order parameter fluctuations δQ_l . The simplest case will of course be to assume purely relaxational behaviour $\delta Q_1(\vec{q},t) = \delta Q_1(\vec{q},0)\exp(-t/\tau_q)$, corresponding to an overdamped phonon description in structural systems when τ_q is real (giving essentially mean field like results). To generalize the treatment, however, one may allow τ_q to be a time-dependent quantity. Further generalization requires the use of scaling forms for the dynamical susceptibility.

Such modification can easily be done at the end of the calculation, since the results will be expressed in terms of the total relaxation time τ_q . Even more sophisticated models, including the central mode in structural systems may be introduced in the same way. For discussions of the central mode, see Refs. 6,7,8 . We return briefly to this question below.

In the simplest case the relaxation time is expressed as[9] $\tau_q = \xi^z f(q\xi)$ where ξ is the correlation length, z is the dynamical exponent which is near 2 for a non conserved order parameter and near 4 for a conserved one.

εQ^2-coupling

To perform the calculation of ΔC_{ij} we take $\delta\sigma_i$ from Eq.(4b) :

$$\delta\sigma_i(\vec{r},t) = b_{ijl}[Q_j(\vec{r},t)Q_l(\vec{r},t) - \langle Q_j(\vec{r},t)Q_l(\vec{r},t)\rangle]$$

or

$$\delta\sigma_i(\vec{q},t) = \frac{Vb_{ijl}}{(2\pi)^d}\int d^d\vec{k}[Q_j(\vec{k},t)Q_l(-\vec{k}+\vec{q},t) - \langle Q_j(\vec{k},t)Q_l(-\vec{k}+\vec{q},t)\rangle] \tag{5}$$

Writing $Q_j = Q_{0j} + \delta Q_j$, where Q_{0j} is the average of Q_j we get two terms in $\delta\sigma_i$:

$$\delta\sigma_i(\vec{q},t) = 2b_{ijl}Q_{0j}\;\delta Q_l(\vec{q},t) + \frac{V}{(2\pi)^d}\,b_{ijl}\int d^d\vec{k}[\delta Q_j(\vec{k},t)\,\delta Q_l(-\vec{k}+\vec{q},t) - \langle\delta Q_j(\vec{k},t)\,\delta Q_l(-\vec{k}+\vec{q},t)\rangle] \tag{6}$$

Inserting this in Eq.(2) we obtain two types of non-vanishing correlation functions, a 2-point correlation which gives the so-called Landau-Khalatnikov[10] term (LK) and a 4-point correlation which we may call the critical scattering term. We discuss first the LK-term:

$$\Delta C_{ij}^{LK}(\vec{q},\omega) = -\frac{4V}{kT} b_{imn} b_{jpr} Q_{0m} Q_{0p} \times$$

$$[<\delta Q_n(\vec{q},0)\delta Q_r(-\vec{q},0)>+i\omega\int_0^\infty dt e^{i\omega t}<\delta Q_n(\vec{q},t)\delta Q_r(-\vec{q},0)>]$$

$$= 4b_{imn} b_{jpr} Q_{0m} Q_{0p} \chi_{nr}(\vec{q},\omega) \quad (7)$$

Using the simple relaxation form for δQ_n mentioned before would now lead to the well-known mean-field results for the LK contributions[5]. To obtain more general forms we have to use scaling arguments. We write the order parameter susceptibility as

$$\chi_{nr}(\vec{q},\omega) \equiv \chi = \chi(\vec{q},0) f(i\omega\tau)$$

where $f(i\omega\tau)$ is some scaling function. By expansion of $f(i\omega\tau)$ we see that

$$\text{Re}\,\chi \sim t^{-\gamma} \quad ; \quad \text{Im}\,\chi \sim t^{-\gamma}\,\omega\tau \quad \text{for} \quad \omega\tau << 1$$

Also we must require that χ does not diverge at T_c when $\omega \neq 0$, i.e. $\chi \sim t^{0}$ for $\omega\tau >> 1$.

$$\text{Re}\,\chi \sim t^{-\gamma}(\omega\tau)^{-\frac{\gamma}{\nu z}} \quad ; \quad \text{Im}\,\chi \sim t^{-\gamma}\omega\tau(\omega\tau)^{-1-\frac{\gamma}{\nu z}} \quad \text{for } \omega\tau >> 1 .$$

The resulting forms, correct in both limits, for α and v are given in table 1. We note that these may be combined in a scaling form which satisfies both limits, for instance

$$\frac{\Delta v}{v_0} = -A_1' \, t^{2\beta-\gamma}\frac{1}{[1+(\omega\tau)^2]^{\frac{\gamma}{2\nu z}}} \quad (8a)$$

$$\frac{v_0 \Delta\alpha}{\omega} = A_1 \, t^{2\beta-\gamma} \frac{\omega\tau}{[1+(\omega\tau)^2]^{\frac{1}{2}+\frac{\gamma}{2\nu z}}} \quad (8b)$$

Our results above are different from those of Suzuki[11] who demanded nondivergence for the entire mechanical susceptibility at T_c, while such a restriction should only be placed on the order parameter susceptibility when the order parameter is not a strain.

Next we treat the critical part which contributes both above and below T_c. Combining Eqs. (2) and (6) :

$$\Delta C_{ij}^{Crit}(\vec{q},\omega) = \frac{V^3}{(2\pi)^{2d}\,kT}\, b_{imn}\, b_{jpr} \int d^d\vec{k}\; d^d\vec{k}' \times$$

$$\{<\delta Q_m(\vec{k},0)\,\delta Q_n(-\vec{k}+\vec{q},0)\,\delta Q_p(-\vec{k}',0)\,\delta Q_r(\vec{k}'-\vec{q},0)> \tag{9}$$

$$- <\delta Q_m(\vec{k},0)\,\delta Q_n(-\vec{k}+\vec{q},0)><\delta Q_p(-\vec{k}',0)\,\delta Q_r(\vec{k}'-\vec{q},0)>$$

$$+ i\omega \int_o^\infty dt\; e^{i\omega t}[<\delta Q_m(\vec{k},t)\,\delta Q_n(-\vec{k}+\vec{q},t)\,\delta Q_p(-\vec{k}',0)\delta Q_r(\vec{k}'-\vec{q},0)>$$

$$- <\delta Q_m(\vec{k},t)\,\delta Q_n(-\vec{k}+\vec{q},t)><\delta Q_p(-\vec{k}',t)\,\delta Q_r(\vec{k}'-\vec{q},t)>]\}$$

Performing the integration using $\delta Q(\vec{k},t) = \delta Q(\vec{k},0)\exp(-t/\tau_k)$, one obtains for $\vec{q} = 0$

$$\Delta C_{ij}^{Crit}(0,\omega) = \frac{V^3}{kT\cdot(2\pi)^{2d}}\, b_{imn}\, b_{jpr} \int d^d\vec{k}\; d^d\vec{k}'\; \frac{1}{1-i\frac{1}{2}\omega\tau_k} \times$$

$$[<\delta Q_m(\vec{k},0)\,\delta Q_n(-\vec{k},0)\,\delta Q_p(-\vec{k}',0)\,\delta Q_r(\vec{k}',0)>$$

$$- <\delta Q_m(\vec{k},0)\,\delta Q_n(-\vec{k},0)><\delta Q_p(-\vec{k}',0)\,\delta Q_r(\vec{k}',0)>] \tag{10}$$

Below we shall see what approximate <u>general</u> forms this may take. Here we proceed to develop the mean field expressions. For this purpose we use a Gaussian factorization approximation for the four-point correlation function, equivalent to neglecting interactions between fluctuations. In this approximation the correlation function is a sum of three terms one of which cancels upon substitution in eq.(10) against the product of two-point correlation functions, while the other two, because of symmetry, give exactly the same contribution. Eq.(10) then becomes

$$\Delta C_{ij}^{Crit}(0,\omega) = \frac{2V^3}{(2\pi)^{2d}kT}\, b_{imn}\, b_{jpr} \int d^d\vec{k}\; d^d\vec{k}'$$

$$\frac{1}{1-\frac{1}{2}i\omega\tau_k} <\delta Q_m(\vec{k},0)\,\delta Q_p(-\vec{k}',0)><\delta Q_n(-\vec{k},0)\,\delta Q_r(\vec{k}',0)> . \tag{11}$$

Using $<\delta Q_m(\vec{k},0)\,\delta Q_p(-\vec{k}',0)> = \frac{kT}{V}\chi_{mm}(\vec{k},0)\,\delta_{mp}\,\delta(\vec{k}-\vec{k}')$ one obtains

$$\Delta C_{ij}^{Crit}(0,\omega) \approx \frac{2kT}{(2\pi)^d} b_{imn} b_{jmn} \int \frac{d^d\vec{k}}{1-i\frac{1}{2}\omega\tau_k} \chi_{mm}(\vec{k},0)\chi_{nn}(-\vec{k},0) \quad (12)$$

Making use of the mean field relations, in case of a real τ_k we can write

$$\tau_k \approx \xi^2/[1+(k\xi)^2] \quad ; \quad \chi_{jj}(\vec{k},0) \sim \xi^2/[1+(k\xi)^2] \ .$$

Performing an isotropic k-integration we find for $d=3$

$$\Delta C_{ij}^{Crit}(0,\omega) = - iB^2\pi \frac{\xi}{\omega\tau} \left(\sqrt{1-i\frac{1}{2}\omega\tau} - 1\right) \quad (13)$$

where B^2 is a constant and $\tau \sim \xi^2$. A similar result was obtained by Bhattacharjee[12], by Pytte[13], and by Levanyuk[14]. In the limit $\omega\tau \ll 1$ this gives

$$\Delta v \sim t^{-\frac{1}{2}} \quad ; \quad \frac{\Delta\alpha}{\omega} \sim \omega t^{-\frac{3}{2}} \quad (14)$$

for $\omega \gg 1$:

$$\Delta v \sim \omega^{\frac{1}{2}} t^{0} \quad ; \quad \frac{\Delta\alpha}{\omega} \sim \omega^{-\frac{1}{2}} t^{0} \quad (15)$$

The total result which reproduces Eq.(13) in the asymptotic limits $\omega\tau \ll 1$ and $\omega\tau \gg 1$ is :

$$\frac{\Delta v}{v_o} = - A_2' \, t^{-\frac{1}{2}} \frac{1}{1+(\omega\tau)^{1/2}} \quad (16a)$$

$$\frac{v_o \Delta\alpha}{\omega} = A_2 \, t^{-\frac{1}{2}} \frac{\omega\tau}{1+(\omega\tau)^{3/2}} \quad (16b)$$

By a straightforward generalization based on the same scaling arguments used for the two-point correlation function above, we find the following approximate dynamical scaling forms:

$$\frac{\Delta v}{v_o} = - A_3' \, t^{-\mu} \frac{1}{[1+(\omega\tau)^2]^{\mu/2\nu z}} \quad (17a)$$

$$\frac{v_o \Delta\alpha}{\omega} = A_3 \, t^{-\mu} \frac{\omega\tau}{[1+(\omega\tau)^2]^{\frac{1}{2}+\frac{\mu}{2\nu z}}} \quad (17b)$$

where $\mu = \alpha + 2(\phi-1)$ is the velocity exponent[15] in the static limit, α is the specific heat exponent, and ϕ is the crossover exponent. $\phi \neq 1$ for symmetry-breaking modes. For non-symmetry-breaking strains i.e. the sound wave couples to energy density fluctuations in the order parameter, $\phi = 1$ and $\mu = \alpha$. The form (17a,b) is similar to that proposed by Kawasaki[16], and consistent with the analysis given by Fossheim and Holt[17] . It also agrees with the limiting form given by Schwabl[18] for $t \to 0$.

Further results

The results of the above analysis together with those of similar treatment for εQ-coupling to lowest order in the fluctuations are given in Table 1 . Since the results have been expressed here in terms of the relaxation parameter τ various cases may be discussed by simple insertion for τ in table 1 , for example in structural systems:

$$\tau = \frac{2\gamma}{\omega_o^2} \quad \text{purely relaxational (overdamped) mode}$$

$$\tau = \frac{2\gamma}{\omega_o^2} - i\frac{\omega}{\omega_o^2} \quad \text{propagating damped mode}$$

$$\tau = \frac{2\gamma}{\omega_o^2} - i\frac{\omega}{\omega_o^2} + \frac{\delta^2\tau_o\omega_o^{-2}}{1-i\omega\tau_o} \quad \text{soft mode + central peak.}$$

The case of systems with an acoustic soft mode has not been explicitly discussed here. We refer to papers by Schwabl[19] and coworkers, and Cowley[20] on that subject. We expect that this case will be similar to εQ-coupling. The case of logarithmic corrections to mean field results for marginal dimensions has been discussed for a dipolar system by Meissner and Pirc[21].

EXPERIMENTAL RESULTS

A great number of substances have been studied by ultrasonic methods near their phase transitions. A compilation of experimental results at structural transitions was recently given by Lüthi and Rehwald[22]. However, no general review of dynamics is available. Here we would like to stress the wide range of systems recently studied for their dynamical behaviour, lending evidence

Table 1. Results obtained by use of the fluctuation-dissipation theorem and dynamical scaling in calculating velocity and attenuation change to lowest order near T_c .

Temperature range	Quantity calculated	εQ-coupling	εQ^2-coupling
$T>T_c$	$\frac{\Delta v}{v_o}$	$\frac{t^{-\gamma}}{[1+(\omega\tau)^2]^{\gamma/2\nu z}}$	$\frac{t^{-\mu}}{[1+(\omega\tau)^2]^{\mu/2\nu z}}$
	$\frac{v_o\Delta\alpha}{\omega}$	$t^{-\gamma}\frac{\omega\tau}{[1+(\omega\tau)^2]^{\frac{1}{2}+\frac{\gamma}{2\nu z}}}$	$t^{-\mu}\frac{\omega\tau}{[1+(\omega\tau)^2]^{\frac{1}{2}+\frac{\mu}{2\nu z}}}$
$T<T_c$	$\frac{\Delta v}{v_o}$	$\frac{t^{-\gamma}}{[1+(\omega\tau)^2]^{\gamma/2\nu z}}$	$t^{2\beta-\gamma}\frac{1}{[1+(\omega\tau)^2]^{\gamma/2\nu z}} + t^{-\mu}\frac{1}{[1+(\omega\tau)^2]^{\mu/2\nu z}}$
	$\frac{v_o\Delta\alpha}{\omega}$	$t^{-\gamma}\frac{\omega\tau}{[1+(\omega\tau)^2]^{\frac{1}{2}+\frac{\gamma}{2\nu z}}}$	$t^{2\beta-\gamma}\frac{\omega\tau}{[1+(\omega\tau)^2]^{\frac{1}{2}+\frac{\gamma}{2\nu z}}} + t^{-\mu}\frac{\omega\tau}{[1+(\omega\tau)^2]^{\frac{1}{2}+\frac{\mu}{2\nu z}}}$

to the claim that ultrasound is indeed an excellent method for such work.

The superfluid XY-transition in ^{4}He is one of the most thoroughly studied examples[23,24] . It belongs to the εQ^2 case discussed above. The theory of the dynamical scaling behaviour has been discussed by Ferrell and Battacharjee[25] and dynamical scaling is found both above and below T_c . It has been pointed out[25] that a separation into a critical part and an LK-part is not possible, due to the existence of a mixing term. The data[24] as analyzed[25] are shown in Fig.3.

Critical dynamics and dynamic scaling has been studied in the magnetic compounds[11,26] MnP , and in[27] Rb_2CoF_4 (Fig.4a,b) both of which belong to the εQ^2-coupling systems. More recently EuO_2 has been studied[28] . In MnP apparent discrepancies exist between the static behaviour and that observed dynamically. MnP is considered to be a d=3 uniaxial Heisenberg ferromagnet from statics[29], but appears to behave in mean field like fashion with respect to dynamics, with a dynamic exponent close to z=2 expected for a non-conserved order parameter. The discrepancy remains to be explained. In Rb_2CoF_4 from statics[30] the system is d=2 Ising like, while dynamics[27] indicate a d=2 isotropic antiferromagnet, $z \approx 1$.

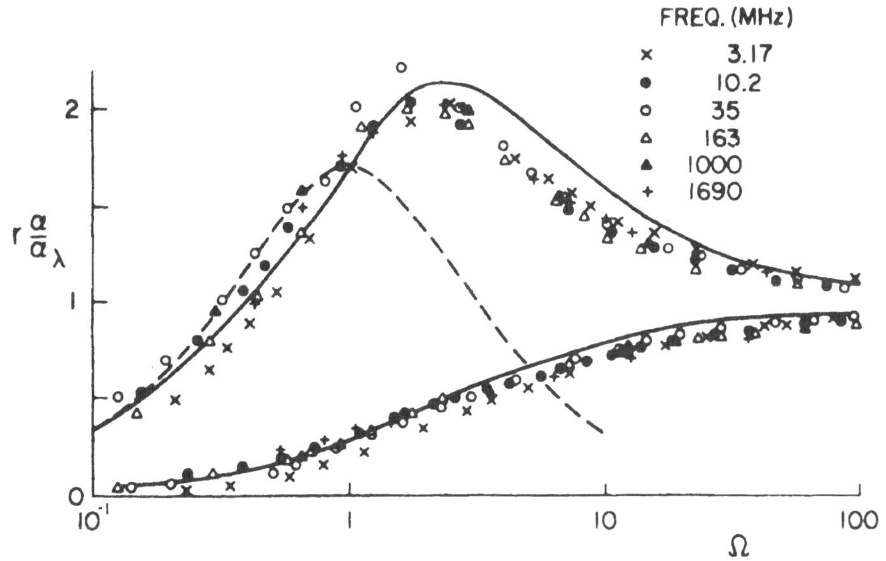

Fig.3. Scaling functions of experimental attenuation data near the λ-transition ^{4}He . The upper and lower curves show scaling below and above the lambda point respectively. The dashed curve shows the Landau-Khalatnikov contribution. Ferrell and Bhattacharjee (1981). (Ref.25).

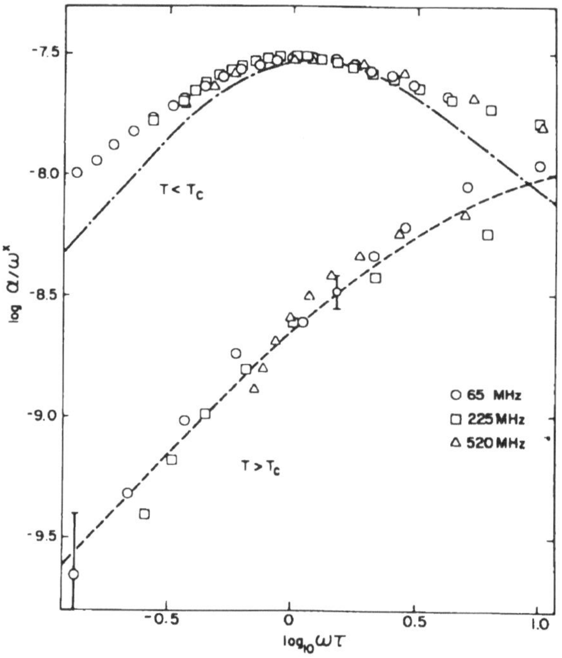

Fig.4a. Reduced attenuation α/ω^{x} (x=0,82) versus $\omega\tau$ near the magnetic transition in MnP. Golding (1975) (Ref.26).

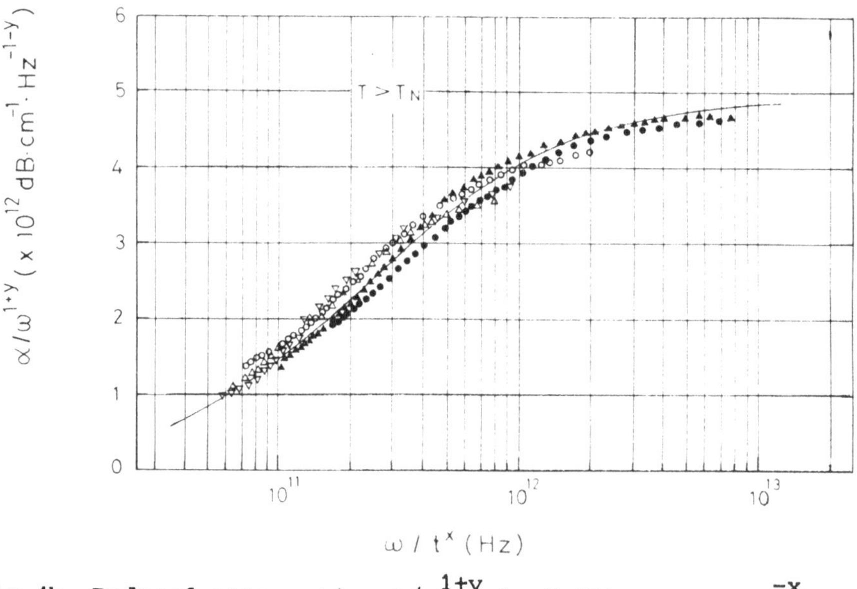

Fig.4b. Reduced attenuation α/ω^{1+y} (y =0.31) versus ωt^{-x} (x=1.20) for four values of ω above the magnetic transition in Rb_2CoF_4 . The full line is a scaling function similar to Eq.17(b), giving $\tau=\tau_0 t^{-x}$ where $\tau_0=4{,}2\cdot10^{-12}$s . Suzuki et.al (1982) (Ref.27).

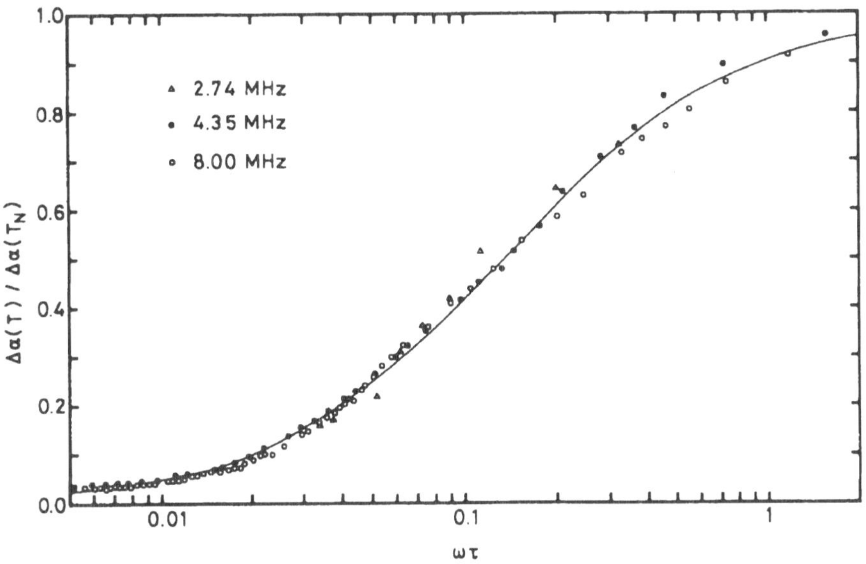

Fig.5a. Reduced attenuation $\Delta\alpha(T)/\Delta\alpha(T_N)$ versus $\omega\tau$ for three frequencies above the commensurate to incommensurate transition temperature T_N in $NaNO_2$. $\tau=\tau_0\ t^{-1}$ where $\tau_0=2{,}3\cdot10^{-9}$s . Hatta (1980) (Ref.33).

The first observation of critical behaviour in structural transitions were in fact made by ultrasound by Berre, Fossheim and Müller[31]. Critical behaviour under external stress was also studied[32]. At structural transitions dynamical scaling has been reported at the commensurate-to-incommensurate transition[33] in $NaNO_2$, and at the structural transition[17,34] in $KMnF_3$, Fig.5a,b. Again the coupling in these cases is εQ^2, with the interesting modification of a complex and k-dependent coupling constant in $NaNO_2$. In $NaNO_2$ the interpretation of data does not seem to be quite clear. The exponent νz takes a mean field like value of 1 , while[35] $\alpha=0.38$ in apparent contradiction to scaling. Further work seems to be needed here. In $KMnF_3$ the results[17] for $T>T_c$ appear to be in agreement with theoretical expectations discussed above. Critical dynamics in an XY structural transition was recently reported[36]. We mention a couple of other perhaps not so well known cases, namely critical velocity and attenuation of surface waves[37] in $SrTiO_3$ and $KMnF_3$, and the observation of an LK-like attenuation peak below T_c in superconductors[38].

Cases of linear coupling εQ like in[39] KH_2PO_4 and[40] rare earth vanadates are so well known in the literature that we merely refer to them here.

The case of $\varepsilon^2 Q$-coupling is quite rare, and may have been observed only in very few cases[41].

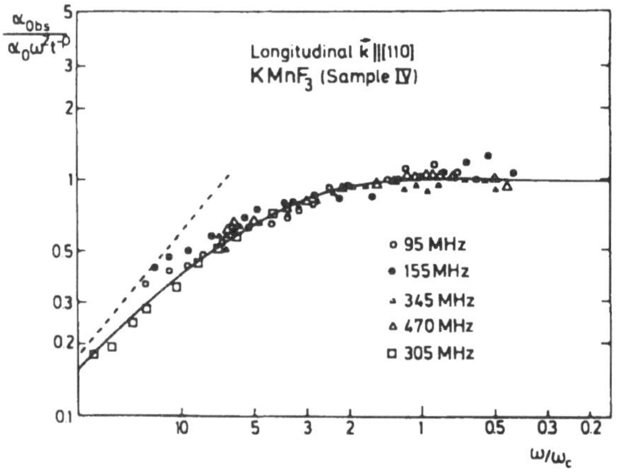

Fig.5b. Reduced attenuation $\frac{\alpha_{obs}}{\alpha_o \omega^2 t^{-\rho}} = \frac{\alpha(\omega\tau)}{\alpha(\omega\tau<<1)}$ versus $\omega\tau$ above the 186 K structural transition in $KMnF_3$. Data points for five frequencies for longitudinal sound propagating in the [110]-direction are shown. The fully drawn curve is the theoretical scaling function of Iro and Schwabl[18], while the dashed line is their theoretical assymptotic behaviour. $\tau=\tau_o t^{-1.41}$ where $\tau_o \sim 1,4\cdot10^{-9}$s . Iro and Schwabl (1983) (Ref.18). Measurements by Fossheim and Holt (1980,1981) (Ref.17).

Finally we mention that critical dynamics of sound has been observed in liquid crystals for the smectic-A to smectic-C phase transition[42], in binary fluids[43] and in liquid gas transitions[44] where dynamic scaling is again obtained.

In future work we expect also that the role of defects will be more systematically investigated, as already indicated by theoretical work by several groups[45,46,47,48,49] . As regards critical dynamics at multicritical points we refer to the paper given by Dohm[50] in the present volume.

The above review has necessarily been sketchy and incomplete due to the limited space available. Therefore a lot of interesting cases could not be discussed at all, or only mentioned without further discussion. We plan to remedy this situation in a forthcoming paper[4].

The authors wish to acknowledge valuable comments from E. Pytte and D. Bedeaux.

REFERENCES

1. K. Fossheim, O. Martinsen and A. Linz, in Anharmonic Lattices, Structural Transitions and Melting, T. Riste, ed., p.141. Noordhoff, Groningen (1974).
2. R. Kubo, Reports on Progress in Physics 29:255 (1966).
3. A.D. Bruce and R.A. Cowley, J.Phys. C 11:3609 (1978).
4. K. Fossheim and J.O. Fossum, to be published.
5. K. Fossheim, Physica Scripta 25:665 (1982).
   W. Rehwald, Advances in Physics 22:721 (1973).
6. J. Feder, in Local Properties at Phase Transitions, K.A. Müller and A. Rigamonti, ed., p.113, North Holland, Amsterdam (1976).
7. S.M. Shapiro, J.D. Axe, G. Shirane and T. Riste, Phys. Rev. B6:4332 (1972).
8. F. Schwabl, in Anharmonic Lattices, Structural Transitions and Melting, T. Riste ed., p.87, Noordhoff, Groningen (1974).
9. P.C. Hohenberg and B.I. Halperin, Rev.Mod.Phys. 49:435 (1977).
10. L.D. Landau and I.M. Khalatnikov, Sov.Phys.Dokhlady 96:469 (1951).
11. M. Suzuki and T. Komatsubara, J.Phys. C 15:4559 (1982).
12. J.K. Bhattacharjee, Phys.Rev. B 25:3404 (1982).
13. E. Pytte, Phys.Rev.B 1:924 (1971).
14. A.P. Levanyuk, Sov.Phys. JETP 22:901 (1966).
15. K.K. Murata, Phys.Rev.B 13:4015 (1976).
16. K. Kawasaki, ICIFUAS-6, in International Friction and Ultrasonic Attenuation in Solids R.R. Hasiguti and N. Mikoshiba, ed., University of Tokyo Press, Tokyo (1977).
17. K. Fossheim and R.M. Holt, Phys.Rev.Lett. 45:730 (1980).
    R.M. Holt and K. Fossheim, Phys.Rev. B 24:2680 (1981).
18. F. Schwabl and H. Iro, Ferroelectrics 35:215 (1981).
    H. Iro and F. Schwabl Sol.State Comm. 46:205 (1983).
19. F. Schwabl, Ferroelectrics 24:171 (1980).
20. R.A. Cowley, Phys.Rev. B 13:4877 (1976).
21. G. Meissner and R. Pirc, Sol.State Comm. 33:253 (1980).
22. B. Lühti and W. Rehwald, in Topics in Current Physics vol. 23, K.A. Müller and H. Thomas ed., Springer, Berlin, p.131 (1981).
23. R.D. Williams and I. Rudnick, Phys.Rev.Lett. 25:276 (1970).

24. K. Tozaki and A. Ikushima, J.Low Temp.Phys. 32:379 (1978).
25. R.A. Ferrell and J.K. Bhattacharjee, Phys.Rev.Lett. 44:403 (1980).
R.A. Ferrell and J.K. Bhattacharjee, Phys.Rev.B 23:2434 (1981).
26. B. Golding, Phys.Rev.Lett. 34:1102 (1975).
27. M. Suzuki, K. Kato and H. Ikeda, J.Phys.Soc. Japan 49:2 (1980).
28. R.L. Melcher and T. Baumann, Private communication.
29. H. Terui, T. Komatsubara and E. Hirahara, J.Phys.Soc.Japan 38:383 (1975).
30. E.J. Samuelsen, Phys.Rev.Lett. 31:936 (1973).
31. B. Berre, K. Fossheim and K.A. Müller, Phys.Rev.Lett. 23:589 (1969).
32. K. Fossheim and B. Berre, Phys.Rev.B 5:3292 (1972).
33. I. Hatta, J.Phys.Soc.Japan 49, Suppl. B: 163 (1980).
34. M. Suzuki, J.Phys. C 13:549 (1980).
35. I. Hatta and A. Ikushima, J.Phys.Chem.Solids 34:57 (1973).
36. M. Yoshizawa, T. Goto, T. Fujimura, Phys.Rev. B 26:1499, (1982).
37. L. Bjerkan and K. Fossheim, in Phonon Scattering in Solids L.J. Challis, V.W. Rampton and A.F.G. Wyatt, ed., Plenum, New York, p.265, (1976).
L. Bjerkan and K. Fossheim, Sol.State Comm. 21:1147 (1977).
38. J.R. Leibowitz and K. Fossheim, Phys.Rev.Lett. 21:1246 (1968).
N.T. Opheim and K. Fossheim, in Proceedings of International Conference on Phonon Scattering in Solids, H.J. Albany ed., Service de Documentation du CEN Saclay, Paris, p.193 (1972).
39. E. Litov and C.W. Garland, Phys.Rev.B 2:4597 (1970).
E.M. Brody and H.Z. Cummins, Phys.Rev.B 9:179 (1974).
40. R.W. Cohen, G.D. Cody and J.J. Halloran, Phys.Rev.Lett. 19:840 (1967).
J.R. Sandercock, S.B. Palmer, R.J. Elliot, W. Hayes, S.R.P. Smith and A.P. Young, J.Phys.C 5:3126 (1972).
R.L. Melcher and B.A. Scott, Phys.Rev.Lett. 28:607 (1972).
41. U.T. Höchli and J.F. Scott, Phys.Rev.Lett. 26:1627 (1971).
42. S. Bhattacharya, B.Y. Cheng, B.K. Sarma and J.B. Ketterson, Phys.Rev.Lett. 49:1012 (1982).
43. J.K. Bhattacharjee and R.A. Ferrell, Phys.Rev.A 24:1643 (1981).
D.B. Fenner, Phys.Rev.A 23:1931 (1981).
44. D.M. Kroll and J.M. Ruhland, Phys.Lett. 80 A:45 (1980).
D. Roe, B. Wallace and H. Meyer, J.Low Temp.Phys. 16:51 (1974).
D. Roe and H. Meyer, J.Low Temp.Phys. 30:91 (1978).
45. A.P. Levanyuk, V.V. Osipov, A.S. Sigov and A.A. Sobyanin, Sov.Phys.JETP 49:176 (1979).
46. K.-H. Höck and H. Thomas, Z.Physik B 27:267 (1977).
47. L. Sasvári and F. Schwabl, to be published.
48. B.I. Halperin and C.M. Varma, Phys.Rev.B 14:4030 (1976).
49. G. Grinstein, S. Ma and G.F. Mazenko, Phys.Rev.B 15:258 (1977).
50. V. Dohm, in Proceedings of this conference.

MULTICRITICAL PHENOMENA AT STRUCTURAL PHASE TRANSITIONS

Kristian Fossheim

Department of Physics, University of Trondheim
The Norwegian Institute of Technology
7034-NTH Trondheim

INTRODUCTION

Multicritical phase diagrams in structural systems under the influence of external fields have been the subject of considerable theoretical interest during recent years[1-5]. Hamiltonians containing cubic anisotropy have been discussed extensively[1-11], usually in a socalled quadratic field, i.e. a field which couples quadratically to the order parameter. In the present paper we focus particularly on such cases since they represent a very important class of multicritical systems, and since they constitute the bulk of systems which have so far been subject to systematic experimental investigations.

Since the theory of these systems is dealt with in a separate paper by Aharony and Blankschtein in these proceedings I shall here merely refer aspects of the theory which are necessary in order to elucidate the particular experimental results to be discussed. Furthermore, since the specific heat measurements are the most recent ones, and since EPR results are discussed separately, the present review will emphasize the former, while referring more briefly to EPR and ultrasonic data.

THE CUBIC HAMILTONIAN WITH SYMMETRY BREAKING FIELD

The Hamiltonian employed in studying the systems under discussions here is of the Landau-Ginzburg-Wilson (LGW)-type[1,9]

$$H = \int d^3x\{\frac{1}{2}r|\vec{Q}|^2 + \frac{1}{2}|\vec{\nabla}\vec{Q}|^2 + u_o|\vec{Q}|^4$$
$$+ v_o \sum_{\alpha=1}^{3} Q_\alpha^4 - \frac{1}{2} f \sum_{\alpha=1}^{3} (\partial Q_\alpha/\partial x_\alpha)^2 \tag{1}$$
$$- \sum_{\alpha=1}^{3} T_\alpha[(L_1-L_2)Q_\alpha^2+L_2|\vec{Q}|^2]-L_3(T_4Q_2Q_3+T_5Q_1Q_3+T_5Q_1Q_3+T_6Q_1Q_2)\}$$

where the terms in the first line are isotropic, those in the second line represent cubic anisotropy and anisotropic dispersion respectively[5,9,10], while the last line are terms generated by coupling to an external field, here stress T_α, resulting in a socalled quadratic field. The latter terms clearly are of a form which, besides being able to break the symmetry and hence change the effective number n of order parameter components Q_α, will also affect the temperature variable $r = r'(T-T_o)$. An alternative and convenient way of introducing the symmetry breaking effects of an external field (see for instance Refs. 1 and 6) is by

$$H_g = \int d^3x \frac{g}{n}[(n-m) \sum_{\alpha=1}^{m} Q_\alpha^2 - m \sum_{\alpha=m+1}^{n} Q_\alpha^2] \tag{2}$$

This is the equivalent of the third line of Eq.(1). The advantage is that the external field here is specified by one quantity g, the disadvantage is that the coupling constants L_i of Eq.(1) are not shown explicitly here, and that no shear stresses are included.

The main importance of the cubic term lies in its ability to control in what direction the "spins" $\vec{Q}$ will align below T_c. Since the function $\sum_{\alpha=1}^{3} Q_\alpha^4$ has minima along <111> and maxima along <100> a positive v_o will create alignment along <111>. The case $v_o>0$ corresponds to $LaAlO_3$, while negative values are found in $SrTiO_3$, $KMnF_3$, $RbCaF_3$. We shall restrict ourselves to $v_o < 0$, i.e. alignment along <100>. In addition, v_o is important in deciding whether the transition is first order $(u_o + v_o < 0)$ or second order $(u_o + v_o > 0)$. We have here referred to a classical picture.

For convenience of treatment it is sometimes better to group the terms of the cubic Hamiltonian in a slightly different manner, namely

$$H = \int d^3 \{\frac{1}{2}r|\vec{Q}|^2 + \frac{1}{2}|\vec{\nabla}\vec{Q}|^2 + u \sum_{i=1}^{3} Q_i^4 + v\sum_{i<j=1}^{3} Q_i^2Q_j^2\} \tag{3}$$

where now $u = u_o + v_o$ and $v = 2u_o$. In $d = 3$ dimensions this system possesses an isotropic fixed point[11] which is common to both the <100> and the <111> oriented sectors of the u,v

parameter plane. Therefore the cubic term may not be essential in determining the critical exponents. Although ultrasonic experiments[12] on $KMnF_3$ may be interpreted in accord with this view, the presicion is not good enough to decide the issue which is further discussed by Aharony and Blankschtein in these proceedings.

The effects of the anisotropy in soft-mode dispersion, controlled by the parameter f, is discussed below. We mention here only that the term may be generalized further[5] to allow the possibility of Lifshitz behaviour[13].

In terms of the effective Hamiltonian introduced in Eq.(3) the parameter space of u and v is divided into a sector (Fig.1) of attraction of the stable isotropic fixed point, and sectors (hatched areas) outside this domain, where a runaway is encountered. The experiments to be discussed below are all relevant to systems where one is above the dividing line $u + 3v = 0$ such that the stable fixed point is inaccessible at stress $T_\alpha = 0$. Note that $SrTiO_3$ is an especially intriguing case here in view of the vast literature claiming that $SrTiO_3$ is second order. We return to this question below.

The phase diagrams to be expected in these systems under stress have been discussed in great detail in theoretical papers referred to above. We now look at the experimental situation.

OBSERVATIONS OF MULTICRITICAL PHENOMENA

Since the hatched regions of Fig.1 are first order in RG-theory, but not in Landau theory one refers to systems belonging to these sectors as systems with fluctuation driven first order transitions[6]. In other words: Fluctuations tend to reduce the available u,v space as far as second order transitions are concerned. However, the very interesting possibility arises that by application of an external field the boundary may be moved[4,6,7] by breaking the symmetry, thus reducing the number of components n of the order parameter $\vec{Q}$. Generally this will have the effect of expanding the second order region of parameter space so that many first order transitions may become second order. This implies the existence of tricritical points as a function of applied field. Another possibility which is equally interesting is the existence of a bicritical point[1] in systems which are already second order. We discuss this case first.

$\vec{p}\,||$ [100]: Bicritical point and crossover exponent ϕ

Although evidence now exists[14] that well annealed $SrTiO_3$ may be first order at zero stress (see below) this material has been a natural one in which to look for a bicritical point under pressure. Already many years ago both EPR work probing the

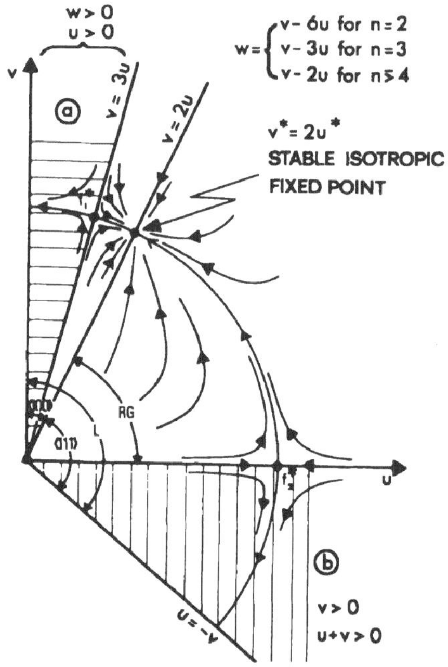

Fig.1. Stable and unstable sectors in the u , v plane with Hamiltonian flow and fixed points for d=3 , n=3 . Sectors L and RG indicate which parts of the u , v plane are second order according to Landau theory (L) and renormalization group (RG) respectively. Hatched areas lie outside of the attraction of the stable isotropic fixed point according to RG analysis and hence correspond to first order transitions. f_1^* and f_2^* are unstable borderline fixed points. Adapted from Refs.4,6.

nature of fluctuations under stress[15], as well as specific heat measurements[16] clearly indicated the existence of the crossover from isotropic fixed point behaviour to XY and Ising behaviour under pressure and tension along [100] , respectively. This is presicely as expected from a simple analysis of the symmetry breaking term (2). However a direct measurement of the crossover exponent ϕ which governs this behaviour in terms of critical exponents was not performed until recently[17]. The analyzed data are shown in Fig.2. They give a best value $\phi = 1.27 \pm 0.06$ which agrees very nicely with the predicted value[18] $\phi = 1.25 \pm 0.015$. The function fitted to the data was

$$T_o(p) = T_o(0) + Wp^{1/\phi} + Ap \tag{4}$$

where the possibility of an analytic (classical) part is taken

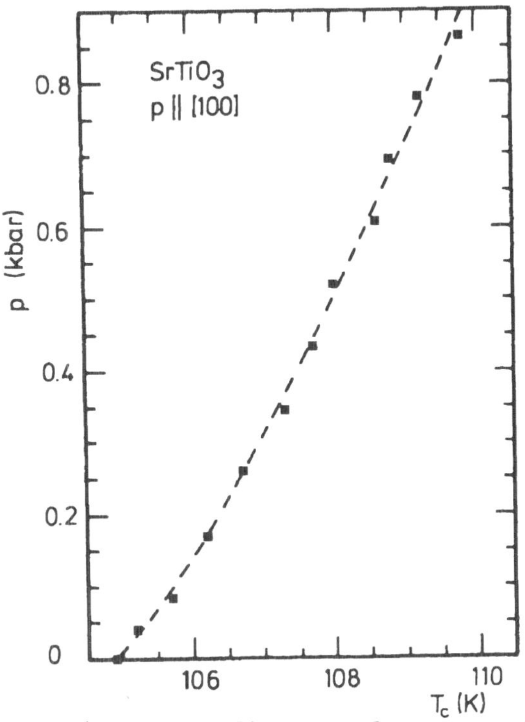

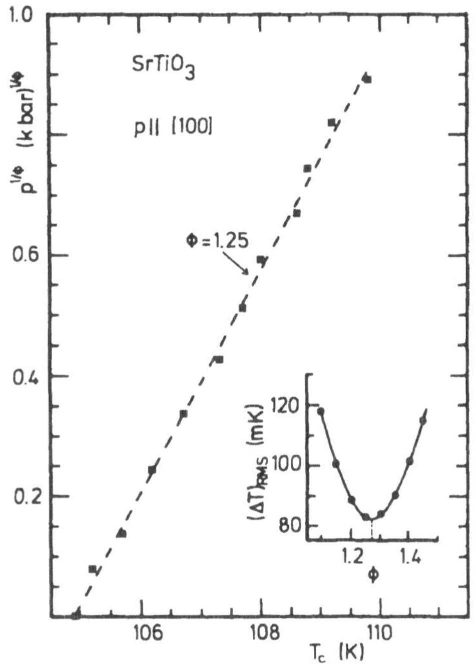

Fig.2a. Phase diagram of $SrTiO_3$ with uniaxial pressure p along [100] showing T_o vs p. The dashed line is $T_o(p) = T_o(0)+Wp^{1/\phi}$, with $T_o(0)=104.89$ K, W=5.40 K/(kbar)$^{0.8}$ and $\phi=1.25$.(From ref.17).

Fig.2b. The data given in Fig.2a are shown here with T_o plotted as a function of $p^{1/\phi}$ with =1.25. The straight dashed line is the same as the line shown in Fig.2a. In the inset is shown the root-mean-square deviation as a function of ϕ when fitting the data to $T_o(p)=T_o(0)+Wp^{1/\phi}$.(From ref.17).

into account. The data were carefully examined with respect to possible contents of the linear term, but it was found to be negligibly small. Pressures less than 1 kbar were applied. The possible[14] first order nature of the $SrTiO_3$ transition apparently did not influence the data, either because the transition at p=0 already was second order or because very small $\vec{p} \parallel [110]$ brings it into the second order sector of u,v space. We note that this is apparently still the only case where the crossover exponent has been directly measured in a structural system. Ultrasound data exist which are in either good agreement[12] or in fair agreement[19] with the value of ϕ found here.

Tricritical behaviour $\vec{p} \parallel [100]$, $\vec{p} \parallel [110]$

A very interesting possibility mentioned above is the observation of tricritical behaviour in systems which are first order at zero stress, since pressure may take them into the second order sector of u,v space. Experiments have been performed in[20] $RbCaF_3$ and in[21] $KMnF_3$, both of which are also candidates for Lifshitz behaviour. This latter fact of course complicates matters. Here we have a competition between two effects: On the one hand the strongly cubic fluctuations with $f \approx 1$ which drive the transition first order[10] , on the other hand the shift of T_c caused by the stress so that the renormalized quartic coefficient u_{eff} may become positive[5]. The theoretical situation has been carefully analyzed by Aharony and Bruce[5] along the lines sketched by Hornreich et. al.[13]

The EPR measurements by Buzaré et. al[20] demonstrated in a convincing way the existence of a tricritical point in $RbCaF_3$ at a stress of (1.9 ± 0.3) kg/mm^2 . Below this stress the transition is first order, above it is second order. The order parameter exponent β was also measured, and found to be in accord with that expected for a uniaxial Lifshitz system. However, this conclusion is not the only one possible since the result could well be interpreted as ordinary Ising tricritical behaviour. Further work is needed to clarify the situation, including more accurate measurements of the soft mode dispersion.

More recently extensive specific heat measurements were performed[21] on $KMnF_3$ under uniaxial stress along [100] and [110]. Indeed a tricritical point was located on the phase boundary for [110]-stress, above which the transition remained second order as shown in Fig.3. The tricritical pressure was 0.45 kbar, i.e. about twice that of $RbCaF_3$, not a great difference considering the fact that the frame in which the F_6-octehedra sit is quite different in the two cases. A change of slope of T_c vs.p was also noted, just like in[20] $RbCaF_3$ with pressure along [100] . In the second order region no hysteresis was detected beyond the 12 mK resolution of the apparatus.

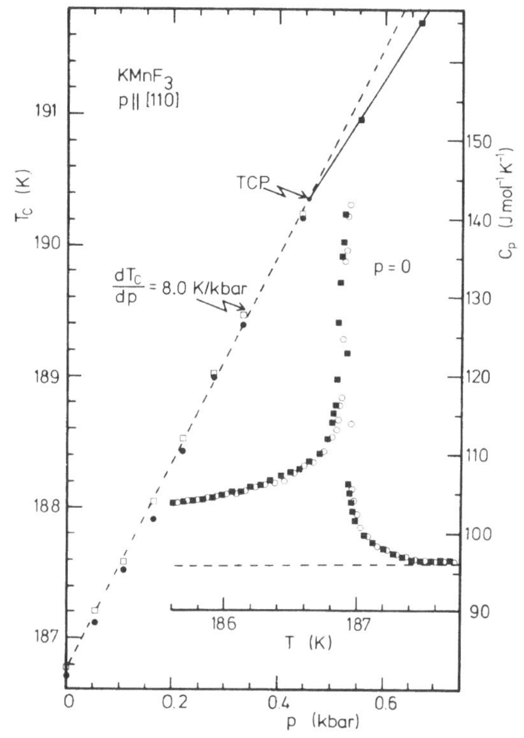

Fig.3. Phase diagram of $KMnF_3$ for p along [110] showing $T_c(p)$ both for cooling (solid circles) and heating (open squares) runs. The inset shows specific-heat curves at p=0, for both cooling (solid squares) and heating (open circles) runs. (From ref.21b).

It is quite possible, considering the similarities of dispersion etc. in two materials, that the tricritical point in $KMnF_3$ is a Lifshitz tricritical point. However, the fact that the specific heat data in $KMnF_3$ could be described by mean field expressions except in a region of about 0.3 K to each side of T_c made it impossible to analyze the limiting behaviour near T_c in terms of nonclassical exponents. Therefore other methods must be pursued also here.

With pressure along [100] a new picture was found: Instead of one tricritical point as in the previous case, two consecutive tricritical points were located along the phase diagram as shown in Fig.4. Such a behaviour was not previously observed and is not expected from the theory of fluctuation driven first order transitions. However, Blankschtein and Aharony[3,8] have found exactly this type of tricritical behaviour in the course of a

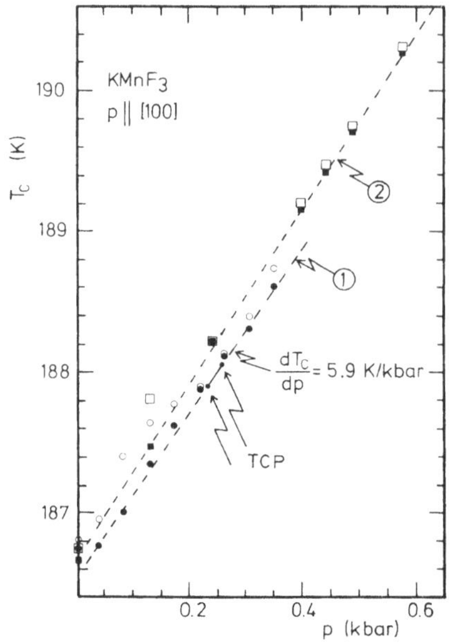

Fig.4. Phase diagram of $KMnF_3$ for p along [100] showing $T_c(p)$ for both cooling (solid circles, solid squares) and heating (open circles, open squares) runs. The difference between curve 1 (solid circles, open circles) and curve 2 (solid squares, open squares) is explained in Ref. 21. (From ref.21b).

study of systems driven second order by fluctuations. The new feature included in their Hamiltonian is to take careful account also of 6th order terms and so show that they do indeed contribute to the renormalization of the quartic coefficient. This leads to a picture where a certain range of negative values along the axis of the quartic coefficient u in fact provide for second order behaviour. Thus such transitions may be called fluctuation driven second order transitions, and they should occur near tricritical points. Furthermore, when this occurs then after a certain change of the symmetry breaking parameter the transition should again become first order[3,8]. This situation is so strikingly similar to that observed with pressure along [100] in $KMnF_3$ that the conclusion seem inescapable, i.e. that this mechanism is indeed responsible. We refer to a paper by Aharony in these proceeding for further details on this subject.

The mean field analysis of data in $KMnF_3$ gave, furthermore, evidence of a certain characteristic variation of parameters near the tricritical points. This is shown in Fig.5. The specific heat has been analyzed here using the expressions given in the Appendix.

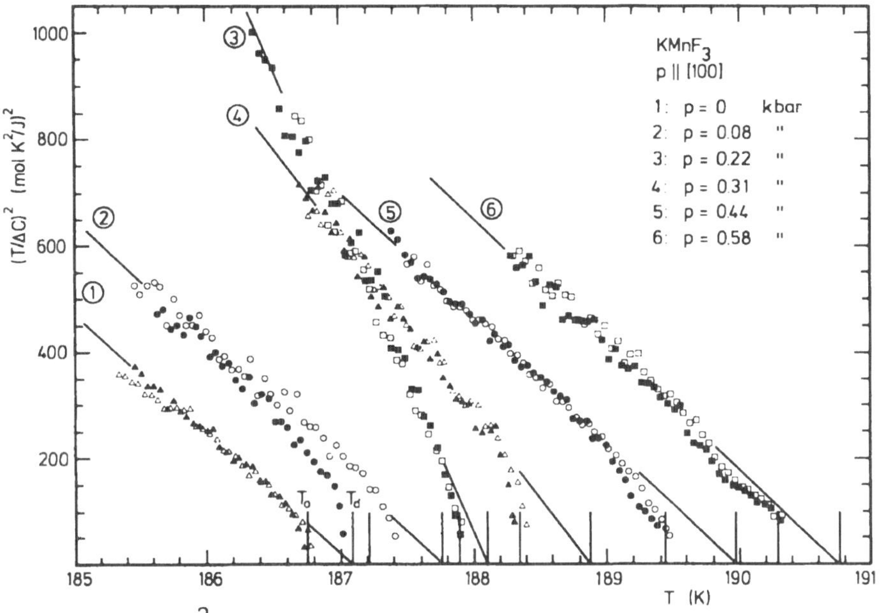

Fig.5. $(T/\Delta C)^2$ for $KMnF_3$ (sample MIT5) for six different pressures along [100] showing both cooling (▲,●,■) and heating (Δ,O,□) curves. ΔC is the Landau contribution to the specific heat and is calculated as described in the Appendix. The straight lines are the prediction. For each pressure T_o is defined as the mean value of T_c for the cooling and heating curve. Every fourth point measured is shown. A, p=0 kbar; B, p=0.08 kbar; C,p=0.22 kbar: p=0.31 kbar; E,p=0.44 kbar; F,p=0.58 kbar. (From ref.21a)

Note that there is an increase of the slope of the straight line near the tricritical point. This situation is similar for both directions of pressure. Regarding specific heat above T_c we note that the observations[21] give a very high effective value of α , namely $\alpha \approx 1$ or larger when determined at some distance from T_c . A possible explanation is in terms of defects which would give such exponents, even in the mean field range (see Appendix).

<u>Critical end point: $\vec{p}\,||$ [111]</u>

One of the most interesting features of recent theories on phase diagrams[4,7] is the prediction of a critical end point (CEP) under pressure along [111] . We have[21] tested this idea in $KMnF_3$, again by specific heat measurements. The result is shown in Fig.6. We find in fact that a new branch (α) in the phase diagram shoots off from the first order curve (γ) , at a pressure near 0.2 kbar. In the specific heat data this branch is

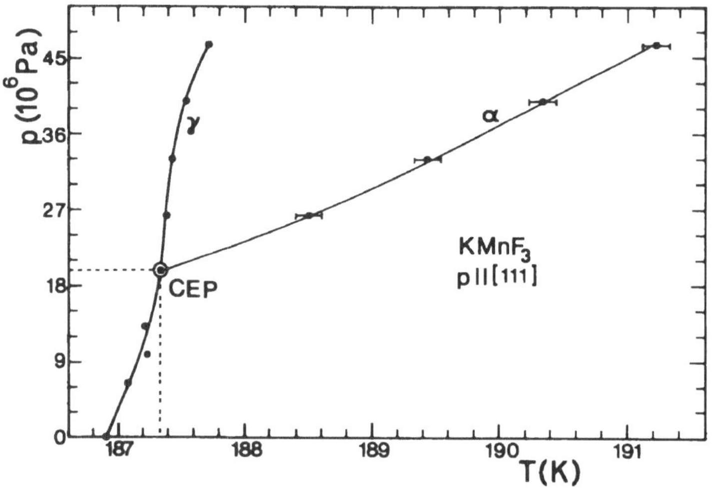

Fig.6. Phase diagram of $KMnF_3$ under pressure along [111] as observed by specific heat measurements. The α-line is tentatively interpreted as the Ising transition, and the γ-line as the Potts transition.

found as a new anomaly with the appearence shown in Fig.7. The new feature in the C_p vs. T curve appears at 0.2 kbar and moves out with increasing pressure. There can be little doubt that this is the second order Ising curve with ordering along [111] predicted[4]. The starting point of this branch is then the critical end point. We emphasize that these data are as yet preliminary in the sense that a careful analysis of the temperature dependences of C_p vs T has not yet been done. Therefore the critical temperature in the upper (Ising) transition may be more carefully established in our further work. Using a criterion previously employed[17] in $SiTiO_3$ for determination of T_c the branch (α) was established with fair accuracy. It is worth noting that although the new anomaly in Cp is not a sharp peak, it resembles very closely that found in $SrTiO_3$ under similar conditions[22].

What is further of interest here is the idea[23] that the lower transition at the first order line (γ) is a 3-state Potts transition due to ordering in the (111)-plane with a rotational symmetry of 120° . This sort of behaviour has previously been seen[23,14] in $SrTiO_3$. However, in that case the CEP is not clearly resolved, nor is the curvature of the (α)-branch established due to difficulties in assigning the transition temperatures. In specific heat measurements[22] with pressure along [111] in $SrTiO_3$ the anomaly along the (α)-branch was so extremely

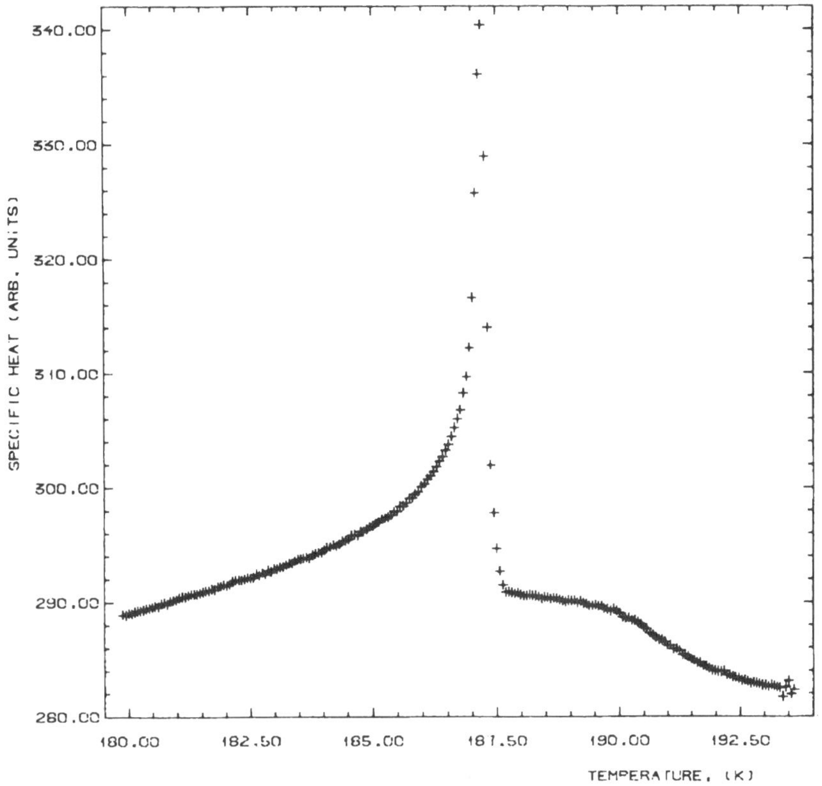

Fig.7. Specific heat in $KMnF_3$ under pressure along [111] . The shoulder-like feature is interpreted as the Ising transition with ordering along [111] , and the sharp peak as the Potts transition. The pressure in this experiment was $p=47\cdot10^6$ Pa .

weak (<0.1% of the total Cp) that the Ising transition could not be resolved.

In this review we have limited the discussion to perovskites. Other realizations of multicritical structural phase diagrams may be studied[24], but so far other systems have not been subject to similar systematic treatment either theoretically or experimentally.

For a review of the experimental situation in $LaAlO_3$ we refer to the accompanying paper by Müller.

I should like to express my sincere appreciation for cooperation and discussions with S. Stokka, K.A. Müller, J.O. Fossum, A. Aharony and D. Blankschtein on the subjects dealt with above.

APPENDIX

REMARKS ON SPECIFIC HEAT

Classical theory

In order to discuss specific heat observations on multicritical phenomena a few remarks on the nature of specific heat anomalies are in order. We note first that recent work[21,25] on the specific heat C_p of several structural systems has shown that the dominant part often is well described by mean field theory. An example is given in Fig.8, from measurements[25] on the ferroelectric 1. order transition in SbSI. It is seen that mean field theory describes the excess specific heat for all $T<T_C$. As a general rule one should therefore be cautious not to attempt a more "sophisticated" analysis of critical exponents until the possibility of a mean field description has been carefully tried out. It was found[21,25], furthermore, that in order to obtain a correct mean field description it is essential to include sixth order terms in the free energy F when expanded in powers of the ordering quantity Q. Hence we take the general

$$F = F_Q + \frac{1}{2}a(T-T_0(p))Q^2 + \frac{1}{4}b(p)Q^4 + \frac{1}{6}c(p)Q^6 \qquad \text{(A1)}$$

where the coefficients as well as T_0 may depend on pressure p. The excess specific heat may be expressed as follows[25]:

$$\Delta C = \begin{cases} AT(T_d - T)^{-\frac{1}{2}} & T < T_c(p) \\ 0 & T > T_c(p) \end{cases} \qquad \text{(A2)}$$

$$A = a^{3/2}/4c^{\frac{1}{2}} \quad ; \qquad \text{(A3)}$$

$$T_d = T_d(p) = T_0(p) + b^2/4ac \qquad \text{(A4)}$$

$$T_c = \begin{cases} T_0(p) & \text{for } b > 0 \\ T_0(p) + 3b^2/16ac & \text{for } b < 0 \end{cases}$$

The temperature T_d is the upper stability limit, while T_0 is the lower one. If b is positive the transition is continuous and occurs at $T_0(p)$. Note however, that T_d is the temperature at which C_p diverges even when the transition is continuous. Worth noting is also the fact that although $\alpha=0$ in Landay theory the specific heat may still be quite strongly temperature dependent. Such a possibility is caused by the existence of the temperature T_d even in continuous transitions. T_d will be equal to T_0 only where $b=0$, i.e. at a Landau tricritical point.

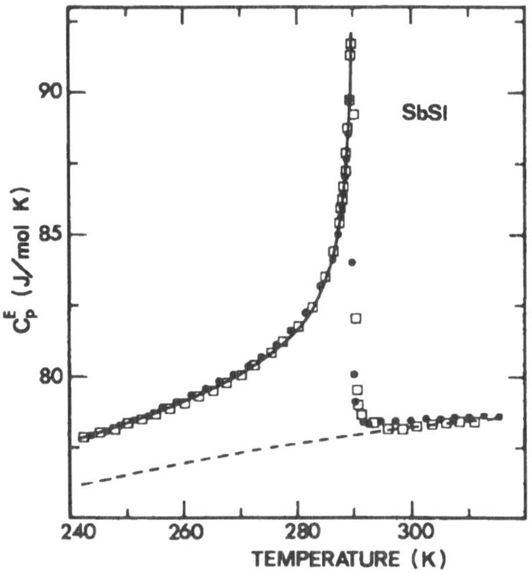

Fig.8. Specific heat of SbSI measured with decreasing (●) and increasing (□) temperature. (Ref.25). The fully drawn curve is the Landau expression.

Here $\alpha=\frac{1}{2}$. In all other cases Landau theory predicts that $T_d > T_0$ and $\alpha=0$. By this observation one can also explain the great difference in the size of the specific heat peaks in for example $SrTiO_3$, where it is very small, and in $KMnF_3$, where it is quite large, (we refer here to $p=0$ data).

We also remark that even in cases where Landau theory gives a good description below T_c and hence should be consistent with a completely nonsingular Cp above T_c in reality one always finds (at some distance from T_c) an increase of Cp on approaching T_c and with quite high exponents, often $\alpha \approx 1\text{-}1.5$. We ascribe most of these effects to defects[26].

Regarding recent development of specific heat measurement technique, we refer to a paper by Stokka and Fossheim[27].

REFERENCES

1. A.D. Bruce and A. Aharony, Phys.Rev. B 11:478 (1975).
A. Aharony and A.D. Bruce, Phys.Rev.Letters 33:427 (1974).
2. D. Blankschtein and A. Aharony, J.Phys.C 14:1919 (1981).
3. D. Blankschtein and A. Aharony, Phys.Rev.Letters 47:139 (1981).
4. D. Blankschtein and D. Mukamel, Phys.Rev.B 25:6939 (1982).
5. A. Aharony and A.D. Bruce, Phys.Rev. Letters 42:462 (1979).

6. E. Domany, D. Mukamel, and M.E. Fisher, Phys.Rev. B 15:5432 (1977).
7. M. Kerszberg and D. Mukamel, Phys.Rev. B 23:3943 (1981).
8. D. Blankschtein and A. Aharony, Phys.Rev. B (in press).
9. A.D. Bruce, J.Phys. C 7:2089 (1974).
10. T. Nattermann and S. Trimper, J.Phys. A8:2000 (1975). T. Nattermann, J.Phys.C9:3337 (1976).
11. E. Brézin, J.C. Le Guillou, and J. Zinn-Justin, Phys.Rev. B 10:892 (1974).
12. R. Holt and K. Fossheim, Phys.Rev. B 24:2680 (1981).
13. R.M. Hornreich, M. Luban, and S. Shtrikman, Phys.Rev. Letters 35:1678 (1975).
14. K.A. Müller and W. Berlinger, Z. Phys.B-Condensed Matter 46:81 (1982).
15. K.A. Müller and W. Berlinger, Phys.Rev. Letters 35:1547 (1975).
16. I. Hatta, Y. Shiroishi, K.A. Müller, and W. Berlinger, Phys.Rev.B 16:1138 (1977).
17. S. Stokka and K. Fossheim, Phys.Rev.B 25:4896 (1982).
18. See for example P. Pfeuty, D. Jasnow, and M.E. Fisher, Phys.Rev.B 2088 (1974).
19. W. Rehwald, Solid State Commun.21:667 (1977).
20. J.Y. Buzaré, J.C. Fayet, W. Berlinger, and K.A. Müller, Phys.Rev. Letters 45:465 (1979).
21a.S. Stokka and K. Fossheim, J.Phys.C 15:1161 (1982).
21b.S. Stokka, K. Fossheim, and V. Samulionis, Phys.Rev. Letters 47:1740 (1981).
22. S. Stokka, K.A. Müller, and K. Fossheim, (unpublished results).
23. A. Aharony, K.A. Müller, and W. Berlinger, Phys.Rev. Letters 38:33 (1977).
24. A. Aharony, Ferroelectrics 24:313 (1980).
25. S. Stokka and K. Fossheim, Phys.Rev. B 24:2807 (1981).
26. A.P. Levanyuk, V.V. Osipov, A.S. Sigov, and A.A. Sobyanin, Zh. Eksp. Teor. Fiz. 76:345 (1979); Sov.Phys, JETP 49:176 (1979).
27. S. Stokka and K. Fossheim, J.Phys.E 15:123 (1982).

BI- AND TETRA-CRITICAL BEHAVIOR OF UNIAXIALLY STRESSED $LaAlO_3$

K.A. Müller, W. Berlinger, J.E. Drumheller,* and J.G. Bednorz

IBM Zurich Research Laboratory
8803 Rüschlikon, Switzerland

INTRODUCTION

Multicritical points with different topologies result from a variety of reasons. Consider the isotropic Landau-Ginsburg-Wilson Hamiltonian[1]

$$\mathcal{H} = \int dx^d \left(\frac{1}{2}(r|\vec{s}(\vec{x})|^2 + c|\nabla\vec{s}(\vec{x})|^2) + u(s(\vec{x}))^4 + \text{h.o.} \right) \tag{1}$$

with $|s(\vec{x})|^2 = \sum_{i=1}^{h} s_i^2(\vec{x})$, where d is the lattice dimensionality, and n the dimensionality of the order parameter, $r = T - T_0$. If u vanishes, a tricritical and if $c \to 0$, a Lifshitz point will result. Nonvanishing higher-order terms, h.o., then stabilize $\mathcal{H}$.

Another class of multicritical points will occur if the symmetry of $\mathcal{H}$ is broken, for example, by a uniform external field in an antiferromagnetic system, as emphasized at an early stage by Fisher, Nelson, and Kosterlitz.[2]

In structural transitions, unaxial stress can break the symmetry as first investigated by Aharony and Bruce[3] for the second-order structural phase transitions (SPT) $SrTiO_3$ and $LaAlO_3$.[4]

*Permanent address: Physics Department, Montana State University, Montana, USA

These transform from cubic to tetragonal and trigonal phases, respectively. In these systems, the order parameter is the rotational angle of the octahedra $\vec{\varphi}$, with components $(\varphi_{100}, \varphi_{010}, \varphi_{001})$.[5] Therefore in Eq. (1), d = n = 3 and the s_i components of the continuous spin system have to be replaced by the rotational φ_i ones. The high symmetry phase being cubic rather than isotropic in (1), a cubic term $\mathscr{H}_c$ is allowed. Since at the transition, the octahedra alternate in rotation by $\pm\varphi_i$, i.e., one has a so-called antiferrodistortive rotation, the stress p can by symmetry only couple *quadratically* to the order parameter. The symmetry-breaking term thus has the form $p_{ij}\varphi_i\varphi_j$, and the total Hamiltonian is

$$\mathscr{H}^t = \mathscr{H} + \mathscr{H}_c + \mathscr{H}_{sb}$$

$$= \mathscr{H} + v \sum_{i=1}^{3} \varphi_i^4(\vec{x}) + \sum_{ij=1}^{3} p_{ij}\varphi_i\varphi_j. \quad (2)$$

For example, for a uniaxial stress p along [100], the last term becomes $g[\varphi_1^2 - 1/2(\varphi_2^2 + \varphi_3)^2]$ with 1, 2, 3 meaning [100], [010] and [001], respectively, and g = Cp.

For $v < 0$ and $g = 0$, the *tetragonal* phase is realized as in $SrTiO_3$. For p along the *tetragonal* axis, bicritical behavior is found as first shown qualitatively by EPR,[6] later quantitatively by ultrasound[7] and most recently by specific-heat measurements.[8]

For $v > 0$ and $g = 0$, the *trigonal* phase is realized as in $LaAlO_3$. For p along the *trigonal* axis, a bicritical point occurs for $g \neq 0$. This and its critical behavior are reported for the first time in the section on the bicritical shift exponent under [111] stress. It is analogous to the bicritical point in $SrTiO_3$.[7,8]

For p along a *tetragonal* [100] direction in $LaAlO_3$, a tetracritical point was predicted by Bruce and Aharony.[4] Figure 1 recapitulates these findings. On the left side, the mean-field, and on the right side, the predicted phase diagram due to critical fluctuations are drawn. For $g > 0$, a second-order boundary $T_1(g)$ separates the cubic from an orthorhombic phase, on cooling, where the AlO_6 octahedra rotate around [011] directions. At lower temperatures, $T_2(g,v)$ separates this phase from an intermediate phase with rotational components $[\varphi_{100}, \varphi_{010}, \varphi_{001}]$, $\varphi_{100} < \varphi_{010} = \varphi_{001}$ which for $g = 0$ are, of course, $\varphi_{100} = \varphi_{010} = \varphi_{001}$. For $g < 0$, the components are $\varphi_{100} > \varphi_{010} = \varphi_{001}$. On heating, $\varphi_{010} = \varphi_{001}$ vanish at T_2' and ordering is along [100], i.e., the usual $SrTiO_3$ I4/mcm phase. On heating more, the rotational [001] order ceases at T_1'.

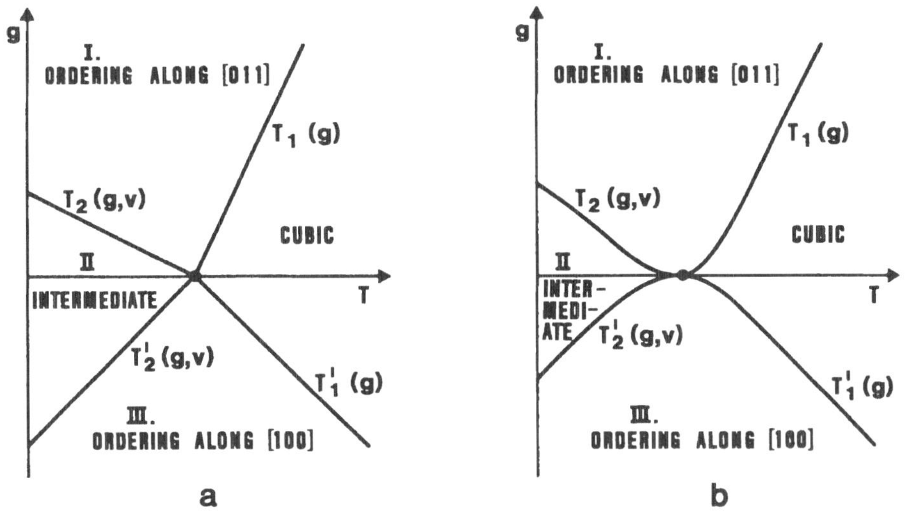

Fig. 1. Tetracritical phase boundaries of [100]-stressed $LaAlO_3$, after Ref. 4.

Under [100] stress, the four phase boundaries for $v > 0$ ($LaAlO_3$), are given by[4]

$$\overbrace{\begin{aligned} T_1(g) &= T_o - bg + g/2; \\ T_2(g,v) &= T_o - bg - \left(\frac{3u+v}{v}\right) g; \end{aligned}}^{g>0} \qquad \overbrace{\begin{aligned} T_1'(g) &= T_o - bg - g \\ T_2'(g,v) &= T_o - bg + \left(\frac{3u+v}{v}\right)\frac{g}{2} \end{aligned}}^{g<0} \tag{3}$$

From Eqs. (3), we see that there are three independent parameters which determine the four phase boundaries, with $g = Cp$, these are

$$B = Cb; \; C; \; D = C(1 + 3\,\frac{u}{v}). \tag{4}$$

In the next section, the first observation of the [110] phase between the T_1 and T_2 boundaries is summarized as well as a phase boundary for [110] stress reported.[9] In the last section, the interest in observing critical behavior as proposed in Fig. 1(b) will be commented on.

MEAN-FIELD PHASE BOUNDARIES OF TETRACRITICAL $LaAlO_3$

In order to carry out EPR experiments under uniaxial stress at such high temperatures as the $T_c \approx 800$ K of $LaAlO_3$, new apparatus was constructed. It is described in detail elsewhere[10] and basically consists of a K-band TE_{011} cylindrical cavity with a quartz heating tube axially through the cavity. Cylindrical samples of 4 mm long × 1.45 mm diameter were mounted inside the heater tube in such a way as to allow pressure to be applied along its longitudinal axis. To apply positive stress along [100] is straightforward, but negative stress, i.e., pulling on an oxide crystal at such high temperatures is much harder to realize. A negative pressure p was simulated by applying stress along a [011] direction of the crystal. Such a pressure has tensor components $T_{010} = T_{001} = T_{011} = -p/2$, and yields a phase diagram with the same topology as that resulting in pulling the crystal along [100][11] but with two otherwise inclined boundaries, $T_1^\dagger$ and $T_2^\dagger$. These boundaries have been derived to be[12]

$$T_1^\dagger = T_o - Bp + \frac{C}{2} p$$

$$T_2^\dagger = T_o - Bp - \frac{3u + v}{4v} \cdot \frac{C}{2} p - \frac{u + v}{2v} \cdot E\, p. \qquad (4')$$

From the above equations, we see that $T_1^\dagger$ is the same as T_1 and no new information is obtained, however, $T_2^\dagger$ contains the parameters u and v in a different combination than T_2, but a term with E resulting from the shear stress is now also present.

For the EPR experiments, Fe^{3+} ions replacing Al^{3+} ions were used as local symmetry probes. The Fe-doped $LaAlO_3$ single-crystal samples were from the same boule on which the first rotation of AlO_6 octahedra was detected in the absence of stress.[5] The spin Hamiltonian of the Fe^{3+}, S = 5/2 ion consists mainly of a cubic and axial crystal-field term as well as the magnetic one:[13]

$$\mathscr{H} = a(S_x^4 + S_y^4 + S_z^4) + D\, S_\zeta^2 + g\, \beta\, \vec{s}\, \vec{H} + \text{const.} \qquad (5)$$

The x,y,z directions of the cubic term point *locally* from the center of the Fe^{3+} to the corners of the oxygen octahedra. They rotate by $\vec{\varphi}$ with the octahedra. Varying the constant magnetic field in a crystallographic plane yields rotation patterns of the resonance fields which allow measuring of the octahedral rotation in this phase, i.e., the *component of* $\vec{\varphi}$ *perpendicular to this plane*. For example, for $\vec{H}$ *in* (001), φ_{001} is measured. The second term results from the distortion of the octahedron. There are two causes: (i) the strain due to the applied stress, and (ii), the strain due to the antiferrodistortive rotation of octahedra $\vec{\varphi}(t)$. As these two are linearly superposed, in the present experiments the ζ direction varied with stress and temperature. With the cubic term, the T_2 and $T_2^\dagger$ phase boundaries, and with the $D\, S_\zeta^2$ term, the T_1 and $T_1^\dagger$ boundaries can be determined.

In our high-temperature cavity, the uniaxial stress was always perpendicular to the plane in which the magnetic field was rotated.[10] This, because the pivoting axis of the EPR magnet was along the perpendicular stress axis. Thus, for the upper part of the phase diagram with stress along a [001] axis, $\vec{H}$ could be rotated in a (001) plane. Similarly, for the lower phase diagram with [110] stress, rotation was possible in a (110) plane. In the former case, the φ_{001} component, and in the latter, the φ_{100} and φ_{010} components can be observed due to the cubic term in the spin Hamiltonian. Now, φ_{001} vanishes at T_2 and both φ_{100} and φ_{010} at $T_2^{\dagger}$. Thus, these two phase boundaries can be determined by observing the disappearance of the respective rotation splittings. For T_2, this was carried out by varying either the temperature at constant stress or vice versa. For $T_2^{\dagger}$, only the former procedure worked because $T_2^{\dagger} = T_c$, i.e. was found to be independent of [110] stress. These results are shown in Fig. 2.

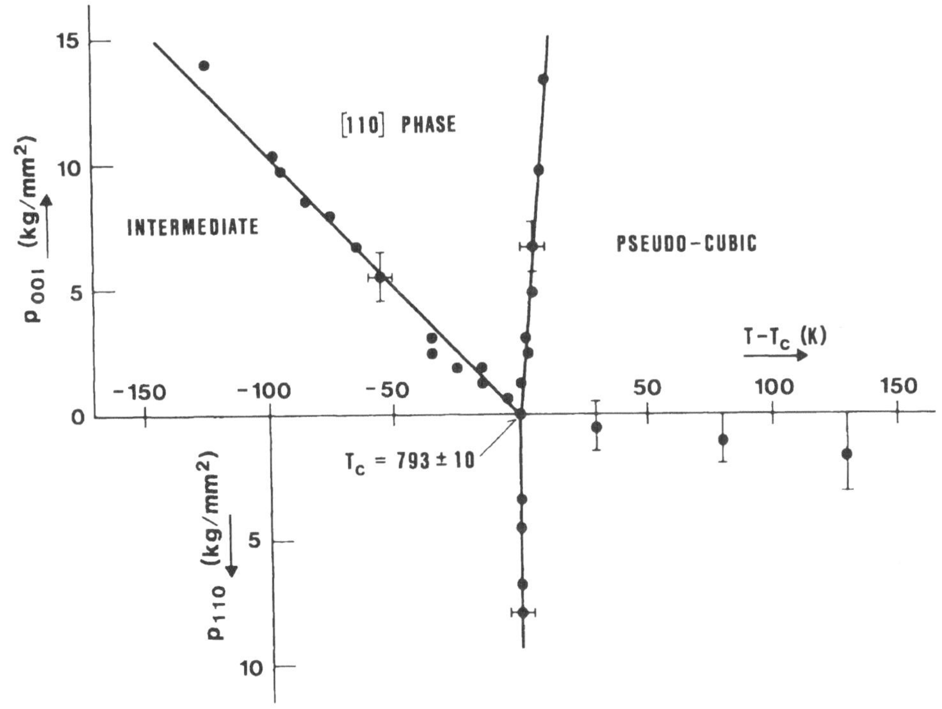

Fig. 2. Experimental phase boundaries of $LaAlO_3$ under [001] and [110] stress.

To determine the T_1 and $T_1^{\dagger}$ phase boundaries from the 110, or 001 phases to the pseudo-cubic phase, a different procedure was necessary. We have seen above that in either phase, the rotational component of the order parameter parallel to the stress axis has vanished at T_2 and $T_2^{\dagger}$. This means the vector of $\vec{\varphi}$ lies in the plane of the magnetic field $\vec{H}$ perpendicular to the stress axis. The octahedral rotations $\pm\vec{\varphi}$ are therefore degenerate in this plane and no splittings due to the cubic term in (5) are observable. The circumvention was the second term which results from the deformed octahedron. The part of the deformation induced by the order parameter is proportional to its absolute magnitude squared $|\varphi^2|$ and along $(\vec{\varphi})/\varphi$. This deformation is orthorhombic with the main axis along [110] in the 110 phase and along [001] for the 001 phase, i.e., in the plane of the magnetic field $\vec{H}$. Thus, $\vec{\zeta}$ in (5) had a component along [110] or [001] in these phases, respectively, which vanished passing T_1 or $T_1^{\dagger}$. In the high-temperature phase, ζ lay exactly along the main stress axis p. In this manner, T_1 was determined by scanning T at constant p or vice versa. This phase boundary is also shown in Fig. 2.

The slopes of T_1 and T_2 are, with 10 kg/mm^2 = 1 Kilobar,

$$\frac{dT_1}{dp} = -B - \frac{C}{2} = (7.5 \pm 10)°/\text{Kb}$$

$$\frac{dT_2}{dp} = -B - D = -(97 \pm 3)\ °/\text{Kb}. \tag{6}$$

Upon scanning the EPR lines on the lower diagram (σ_{110}), no further phase boundary was observed. A change in EPR splittings detected on the far right side of T_c could be due to stresses in the sample. Now $T_1^{\dagger}$ and T_1 should have the same slopes. Thus, the perpendicular boundary is close to the one shown. In the next section, quantitative arguments will be given that $T_1^{\dagger}$ and $T_2^{\dagger}$ lie close to one another, i.e., the existence range of the (001) phase between $T_1^{\dagger}$ and $T_2^{\dagger}$ may be very small. Therefore, we put

$$\frac{dT_1^{\dagger}}{dp} = -B + \frac{C}{2} = (0 \pm 3)\ °/\text{Kb}. \tag{6'}$$

Averaging between the T_1 and $T_1^{\dagger}$ data gives $C/2 = B + (3.8 \pm 1.8)°/\text{Kb}$.

Comparing Figs. 1(a) a and 2, one sees that the topology is the one predicted for $g > 0$. Thus, we can assume this also to be the case in the lower part of the diagrams which requires $dT_2^{\dagger}/dp \leq dT_1^{\dagger}/dp$. Knowing E will then allow arrival at an upper bound for u/v. Feder determined the quantity $A = u + v = 0.741 \times 10^{43}$ cgs for $LaAlO_3$ from various experiments.[14] Thus, an approximate location of the physical point of $LaAlO_3$ in the u-v plane is possible. To do so, a lower bound on u is also necessary. This and a combination of E and B parameters can be obtained from uniaxial [111] stress experiments.

THE BICRITICAL SHIFT EXPONENT UNDER [111] STRESS

Under R.G. flow, $LaAlO_3$ and $SrTiO_3$ will flow to the isotropic Heisenberg or the cubic fixed point if $u > 0$. Whether the Heisenberg or the cubic fixed points are stable for the cubic Hamiltonian of Eq. (2) in d = 3 dimensions, is still not entirely settled theoretically.[15] In either case, $LaAlO_3$ should show bicritical behavior under [111] stress, like $SrTiO_3$ under [100] stress,[4] with the same shift exponent ψ. The mean field boundaries are described by

$$T_1 = T_0 - Bp - Ep \quad (p > 0)$$

$$T_1' = T_0 - Bp + \frac{E}{2} p \quad (p < 0). \qquad (7)$$

In order to possibly reveal the critical character of the T_1 phase boundary, it was decided to improve considerably the accuracy of the EPR experiments as compared with those reported in the previous section. To do so at these high temperatures under uniaxial stress was quite a difficult task and one not previously accomplished.

Improvements were achieved in four different ways: (i) The temperature stability was enhanced to 10^{-5}. This is as good as in magnetic and structural studies at low temperatures, and resulted from a better voltage compensation and feedback plus choice of grounding point of the thermo-element employed. (ii) To determine T_1 and $T_1{}^{\dagger}$-type phase boundaries, the $(S_x^4 + S_y^4 + S_y^4)$ term in Eq. (5) is not necessary. The bicritical point has only such boundaries, whereas the T_2 and $T_2{}^{\dagger}$ are collapsed onto a first-order line.[3,4] Thus, the Fe^{3+}, S = 5/2, ion was replaced by the same size and valency Cr^{3+} with a spin s = 3/2. The cubic splitting yields nothing for such a spin and considerably simplifies the EPR analysis. (iii) New single crystals with a higher concentration of paramagnetic ions grown in a zone-leveling furnace by one of us (JGB)[16] have become available. These crystals contain one order of magnitude fewer dislocations than the Verneuil crystals previously used, thus, the EPR lines were narrower. (iv) The cylindrical sample with a [111] axis transformed in the absence of applied [111] stress into a monodomain crystal. Thus, the machining and mounting procedure had strained the sample along the cylinder axis. We annealed it in the cubic phase at 810 K for a day. After this annealing, all four equivalent ⟨111⟩ domains were present proving the disappearance of the built-in strain. The bicritical T_c had dropped by some 1.5 K, this being proof that previously, a "built-in stress p" of $\approx 1/2$ kg/mm^2 had induced the crystal to transform along the bicritical line.

We measured D(T) in Eq. (5) as a function of temperature at fixed stresses with many data points. Above the phase boundary $T_1(p)$ with $\varphi_{111} = 0$, a small splitting D_0 from the stress-distorted $LaAlO_3$ exists. Below T_1, for a given stress $\sigma_{111} = -p_{111} = -p$, $D^{(p)}(T) = D_0^{(p)} + \rho\langle\varphi^2\rangle$. Nearly straight lines resulted by plotting $[D^{(p)}(T) - D_0^{(p)}]^{3/2}$ as a function of temperature because $\langle\varphi^2\rangle \approx (T_c^{(p)} - T)^{2/3}$. The intersection of these empirical nearly

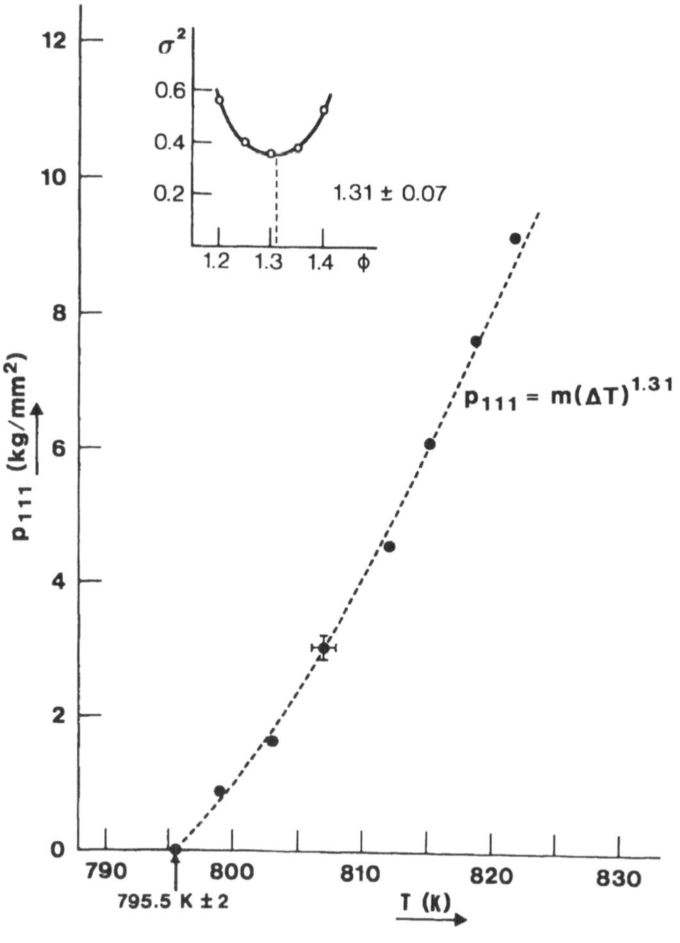

Fig. 3. Bicritical phase boundary T_1 of $LaAlO_3$ under [111] stress.

straight lines to zero yields $T_c(p)$. These are plotted in Fig. 3 as a function of T and p_{111}. Then a curve $p_{111} = m\,(T_c(p) - T_c(o))^{\phi}$ was fitted to them, with $T_c(o)$ not fixed. The insert in the upper left part of the figure shows the mean-squared deviations of the least-square fit by varying ϕ. It has a minimum for $\phi = 1.31$, with a standard deviation of 0.07. The dashed line is drawn with this exponent.

The shift exponent ϕ, within experimental accuracy, is that of a crossover between a Heisenberg and Ising system under symmetry breaking as Bruce and Aharony[4] predicted. For such a condition, $\psi_1 = \phi = 1.25$. Our value is higher, as also that observed by Stokka and Fossheim[8] for the bicritical point in $SrTiO_3$ $\phi = 1.27 \pm 0.06$. Possibly the high end of our experimental margin $\phi = 1.38$ is also compatible with a cubic fixed-point behavior.[15]

The mean field slope of $(23 \pm 2)^{\circ}$/Kb in Fig. 3 away from the bicritical point gives an additional relation between B and E employing Eqs. (7). With it and $dT_2^{\dagger}/dp < 0$ from the topology argument an upper bound of $u/v \leq 0.2$ was deduced[12] with the aid of Eqs. (6). On the other hand, the clear crossover behavior is proof of the second-order character of the bicritical point, i.e., of T_c in the absence of stress being continuous. Therefore $u \geq 0$ and the final result if $u/v = 0.1 \pm 0.1$, u is at least an order of magnitude smaller than v in $LaAlO_3$. The ratio of u/v allowed with Eqs. (4') bounding of the $T_2^{\dagger}$ boundary under [110] stress, $0 \pm 3\ ^{\circ}/\mathrm{Kb} \geq dT_2^{\dagger}/dp \geq -(7 \pm 1.5)\ ^{\circ}/\mathrm{Kb}$. Its negative inclination is of the same order as the positive one from $d\,T_1/dp = d\,T_1^{\dagger}/dp = +\,(7.5 \pm 1)\ ^{\circ}/\mathrm{Kb}$ [Eqs. (4')]. Experimentally, a resolution between $T_1^{\dagger}$ and $T_2^{\dagger}$ was apparently impossible, and a temperature-independent average of both was observed.

Knowing $A = 7.41 \times 10^{42}$ cgs[14] and u/v, the separate values of u and v were obtained. They are listed in Table 1 together with those of $SrTiO_3$. In the critical u-v plane, the physical point of $LaAlO_3$ lies very near the v-axis with $u \geq 0$, whereas the one of $SrTiO_3$ is close to, but below the u-axis. To our knowledge, it is the first time that the topology of a multicritical phase diagram has been used, as here for $LaAlO_3$, to locate the physical point of a system.

POSSIBLE TETRACRITICAL SHIFT EXPONENTS UNDER [100] STRESS

There are two tetracritical shift exponents:[4] $\psi_1 = \phi$ for T_1 and $T_1^{\dagger}$ and $\psi_2 = \phi - \Theta_2\phi_v$ with $\phi_v < 0 < \Theta_2$, i.e., $\psi_2^H > \psi_1$, for Heisenberg fixed-point behavior as shown in Fig. 1(b), and $\phi_v > 0$ for cubic fixed-point behavior, i.e., $\psi_2^c < \psi_1$. Thus, an attempt

Table 1. u and v parameters of Eqs. (1) and (2) in cgs units divided by 10^{43}. Parameters from Ref. 14 and this paper, see text.

	$SrTiO_3$	$LaAlO_3$
u	+ 1.91	+ 0.06 ± 0.06
v	− 0.068	+ 0.68 ± 0.06

was started to measure ψ_1 and ψ_2 under [100] stress (see upper part of Fig. 2). These experiments were still under way at the time of writing this manuscript.

ACKNOWLEDGMENT

The authors would like to thank A. Aharony for the close collaboration in understanding the phase diagram under [011] stress, and communication to us of his theoretical results and numerical estimates.[12]

REFERENCES

1. K. G. Wilson, Phys. Rev. Lett., 28:548 (1972); K. G. Wilson and M. E. Fisher, Phys. Rev. Lett. 28:240 (1972).
2. M. E. Fisher and D. R. Nelson, Phys. Rev. Lett. 32:1350 (1974); D. R. Nelson, J. M. Kosterlitz, and M. E. Fisher, Phys. Rev. Lett. 33:813 (1974).
3. A. Aharony and A. D. Bruce, Phys. Rev. Lett. 33:427 (1974).
4. A. D. Bruce and A. Aharony, Phys. Rev. B 11:478 (1975).
5. K. A. Müller, W. Berlinger and F. Waldner, Phys. Rev. Lett. 21:814 (1968).
6. K. A. Müller and W. Berlinger, Phys. Rev. Lett. 35:1547 (1975).

7. W. Rehwald, Solid State Commun. 21:667 (1977); B. Lüthi and W. Rehwald, Topics in Current Physics 23, Springer, Berlin, p. 169 (1981).
8. S. Stokka and K. Fossheim, Phys. Rev. 25:4896 (1982).
9. J. E. Drumheller, W. Berlinger, and K. A. Müller, Europhysics Conference Abstracts (Manchester, England Conference) Abstract GA:354 (1982).
10. W. Berlinger, Rev. Sci. Instr. 53:338 (1982).
11. D. Blankschtein and A. Aharony, J. Phys. C 14:1919 (1981).
12. A. Aharony, private communication and to be published.
13. G. E. Pake and T. L. Estle, "The Physical Principles of Electron Paramagnetic Resonance," Benjamin Inc., New York (1973).
14. J. Feder, in: "Structural Phase Transitions and Soft Modes," and references therein, p. 171, E.J. Samuelsen, E. Andersen and J. Feder, eds., Universitetsforlaget, Oslo (1971).
15. A. Aharony, Proc. 13th IUPAP Conf. on Stat. Physics, Ann. Isr. Phys. Soc. 2:13 (1978).
16. J. G. Bednorz, Ph.D. thesis, ETH Zurich 1981, unpublished.

FLUCTUATION DRIVEN MULTICRITICAL POINTS

Amnon Aharony

Department of Physics and Astronomy
Tel Aviv University
Tel Aviv 69978, Israel

and

Daniel Blankschtein

Department of Physics
Massachusetts Institute of Technology
Cambridge, Massachusetts 02139, U.S.A.

INTRODUCTION

The purpose of the present lecture is to emphasize the important effects of critical fluctuations on the order of phase transitions. There are many cases in which the simple Landau theory,[1] which ignores fluctuations, does not predict the correct order of the transition. We shall review several such cases, and suggest experimental situations in which (due to the fluctuations) the order of the transition changes via tricritical points.

Modern critical phenomena theory has given a general explanation of the role played by fluctuations, associating their effects with fixed points of the renormalization group (RG) transformation.[2,3] A continuous (second order) transition occurs only if the Hamiltonian describing the system is in a region from which it can "flow" under these transformations to a stable fixed point. Otherwise, the transition is usually first order.[3-6] The borderlines between these two regions represent tricritical surfaces. The location of these surfaces in the parameter space, determined by the RG recursion relations and therefore by the fluctuations, is not the same as that predicted by the Landau theory. It is for this reason that some systems exhibit fluctuation-driven continuous transitions even when Landau's theory predicts first order ones,[7,8] and vice versa.[4-6]

A convenient tool for changing parameters and moving Hamiltonians from one region to another is that of symmetry breaking fields.[3] In particular, breaking the symmetry between the n components of the order parameter yields bicritical phase diagrams.[3] As we shall see below, the order of the transitions in these diagrams may change due to fluctuations.

ISOTROPIC TRICRITICAL POINTS

We start with the simple n-component isotropic tricritical point. An appropriate Ginzburg-Landau Hamiltonian may be written as[9]

$$H = \int d^d x \, \{\tfrac{1}{2} r|\vec{S}|^2 + \tfrac{1}{2}(\vec{\nabla}\vec{S})^2 + u_4|\vec{S}|^4 + u_6|\vec{S}|^6 + \ldots\} \, , \quad (1)$$

where $\vec{S}(\vec{x})$ is a continuous n-component order parameter. If the coefficient of the gradient term $(\vec{\nabla}\vec{S})^2$ is much larger than all the other terms then one can ignore the spatial variations of $\vec{S}$, and minimize the Landau free energy

$$F = \tfrac{1}{2} r|\vec{S}|^2 + u_4|\vec{S}|^4 + u_6|\vec{S}|^6 \, . \quad (2)$$

The resulting phase diagram is shown, for $u_6>0$, in Fig.1. For $u_4>0$, a second order transition occurs at $r=0$ [hence one writes $r=a(T-T_0)$]. For $u_4<0$, a first order transition occurs at $r=u_4^2/2u_6$. The point $u_4=r=0$, which separates the first - and second - order lines, is tricritical. Adding the u_6-axis to the figure, we end up

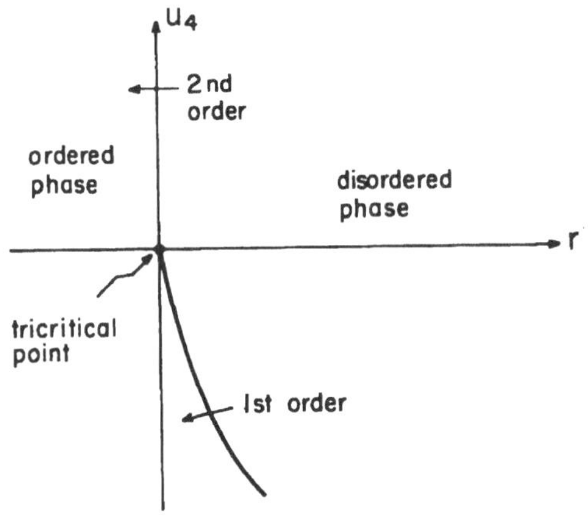

Fig. 1. Landau theory phase diagram for $u_6>0$.

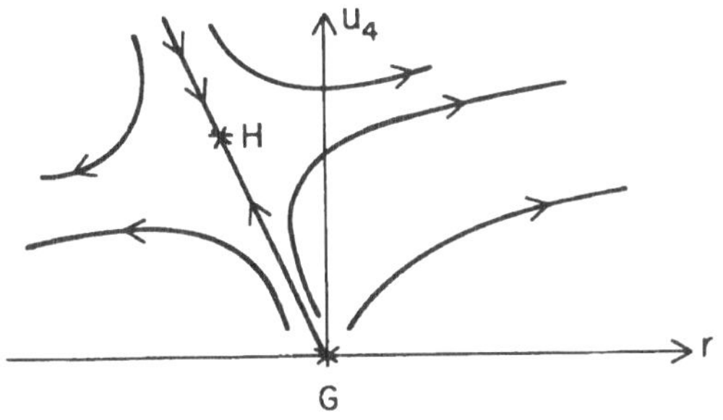

Fig. 2. Renormalization group flow lines for $u_6=0$.

with a three dimensional phase diagram, which exhibits a critical surface, a first order surface and a tricritical line (at $u_4=r=0$, $u_6>0$).

When the gradient term is not too large then one must take care of the fluctuations. This is conveniently done using the renormalization group technique, by which short wave length fluctuations are gradually integrated over in the partition function.[2,3] After one eliminates all the fluctuations with wave numbers larger than Λ/e^{ℓ} (Λ is the original cutoff in momentum space) one ends up with an effective Hamiltonian $\mathcal{H}(\ell)$, with parameters $r(\ell)$, $u_4(\ell)$, $u_6(\ell)$, etc. Typical flows, for $u_6=0$, are shown in Fig.2. Since the correlation length ξ varies under iteration as $\xi\to\xi/e^{\ell}$, it follows that $\xi=\infty$ for all the points which flow to the isotropic "Heisenberg" fixed point H. The critical line, which was predicted by the Landau theory to be at r=0, is thus shifted by fluctuations to the line GH, given to leading order in u_4 by $r=-2(n+2)K_d u_4$, with $K_d^{-1}=2^{d-1}\pi^{d/2}\Gamma(d/2)$ (d is the dimensionality).

Adding u_6, the flows must be considered in the three dimensional space r-u_4-u_6. Fixing r at its critical value, the flows on the critical surface are shown, for dimensionalities $d\gtrsim 3$, in Fig. 3. All the points which lie to the right of the line

$$u_4 = -C(n)u_6 \quad , \quad C(n) = 3(n+4)K_d/(d-2) \; , \tag{3}$$

flow to the Heisenberg fixed point H, and thus correspond to a second order transition ($\xi=\infty$). However, points with $u_4<-C(n)u_6$ flow towards more and more negative values of u_4. By iterating the full recursion relations (including r) until $r(\ell^*)=1$, and by eliminating the remaining fluctuations, one can now show that the resulting effective Landau free energy exhibits a first order transition.

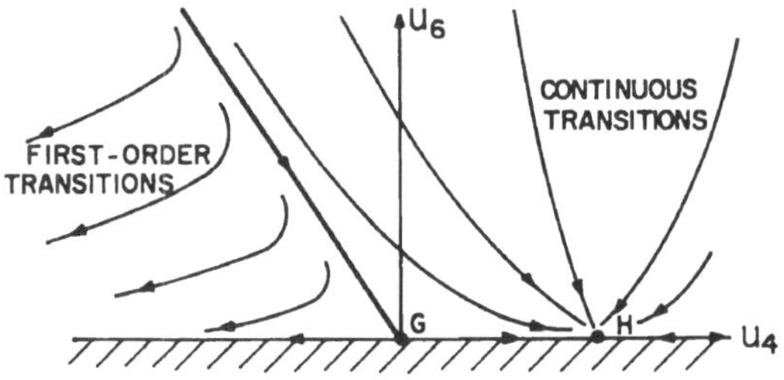

Fig. 3. Renormalization group flows on the critical surface. The heavy line, given by eq.(3), represents tricritical points.

The points on the line $u_4= -C(n)u_6$, which flow to the "Gaussian" fixed point $r=u_4=u_6=0$, separate the two regions and are thus identified as tricritical. Again, fluctuations shift the tricritical points from $u_4=0$ to $u_4= -C(n)u_6$. The transitions in the range $-C(n)u_6<u_4<0$ are thus continuous only due to the fluctuations.

BICRITICAL POINT

We now break the rotational invariance of Eq.(1) by adding a (quadratic) symmetry breaking term,[3]

$$\mathcal{H}_g = \tfrac{1}{2}g \int d^dx \; [m\,|\vec{S}_{n-m}|^2 - (n-m)\,|\vec{S}_m|^2] \,/\, n \quad , \tag{4}$$

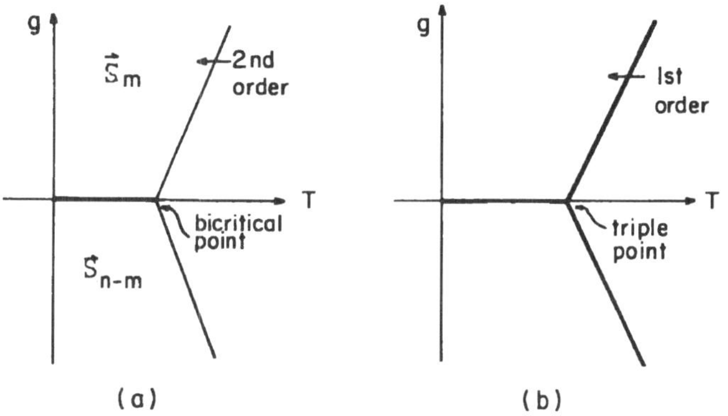

Fig. 4. Landau theory phase diagrams for quadratic symmetry breaking: (a) $u_4>0$, (b) $u_4<0$.

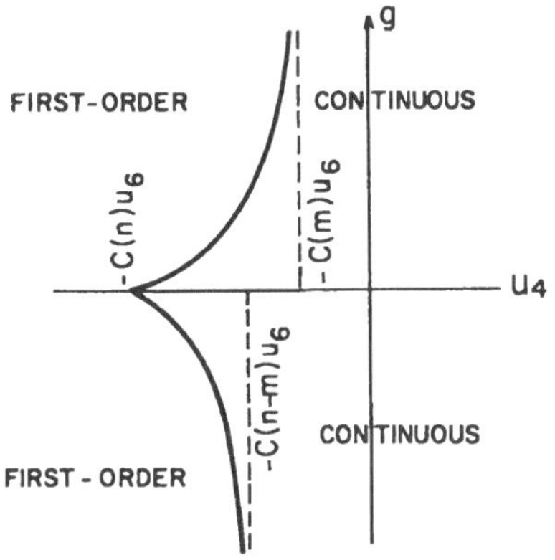

Fig. 5. Phase diagram in g-u_4 plane for m<(n-m). The heavy lines represent tricritical points.

where $\vec{S}_m$ ($\vec{S}_{n-m}$) contains m (n-m) components of the vector $\vec{S}$. Landau's theory now predicts the phase diagrams shown in Fig. 4a (for $u_4>0$) and Fig. 4b (for $u_4<0$). As indicated above, fluctuations shift the critical lines to lower temperatures. However, in the fluctuation driven region $-C(n)u_6<u_4<0$ we find additional new effects.

When g is large, one can simply integrate the fluctuations in $\vec{S}_{n-m}$ out of the partition function. The result is an effective

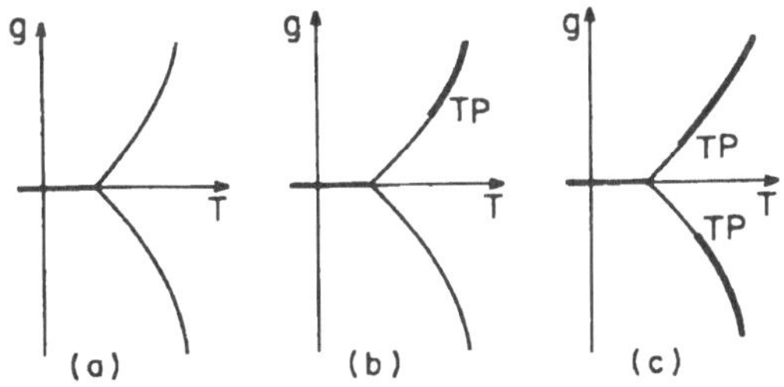

Fig. 6. Fluctuation driven g-T phase diagrams. Thin (heavy) lines represent continuous (first order) transitions. TP is a tricritical point. (a),(b) and (c) correspond, respectively, to $u_4>-C(m)u_6$, $-C(m)u_6>u_4>-C(n-m)u_6$ and $-C(n-m)u_6>u_4>-C(n)u_6$.

m-component Hamiltonian, with[7,8], e.g.

$$u_4^{eff} = u_4+3(n-m)u_6 \int \frac{d^dq}{[r_{n-m}+q^2]} - 4(n-m)u_4^2 \int \frac{d^dq}{[r_{n-m}+q^2]^2} + \ldots, \tag{5}$$

where $r_{n-m}=r+mg/n$ (>>1). Comparing u_4^{eff} to $-C(m)u_6^{eff}$ then yields the tricritical lines shown in Fig.5. If the original n-component system had a fluctuation driven continuous transition then the transition may turn first order for large g. The resulting new phase diagrams are now shown in Fig.6. We emphasize again that all of these diagrams are predicted by the Landau theory to have the shape shown in Fig.4b!

When g is not very large, one must first iterate the recursion relations for the full n-component Hamiltonian. Since g is "relevant" near the isotropic "Heisenberg" fixed point, it grows under these iterations. After ℓ^* iterations, when $g(\ell^*) \gtrsim 1$, one can integrate $\vec{S}_{n-m}$ out and reproduce the phase diagrams described above.[8]

CUBIC SYMMETRY

The situation becomes more complicated when quartic symmetry breaking terms are introduced. A "classical" example involves cubic symmetry, with[3]

$$H_c = \int d^dx \{v \sum_{\alpha=1}^{n} (S^\alpha)^4 + x \sum_{\alpha=1}^{n} (S^\alpha)^6 + y \sum_{\alpha\neq\beta} S_\alpha^4 S_\beta^2 + \ldots\}. \tag{6}$$

Ignoring fluctuations, Landau's theory predicts a continuous transition if the quartic terms are positive, i.e. if

$$u_4 + v > 0 \quad , \quad u_4 + v/n > 0 \ . \tag{7}$$

These Landau tricritical lines are shown in Fig.7. However, fluctuations again modify the picture. Fig.7 also exhibits the flows of u_4 and v under the renormalization group iterations, for $n \lesssim 3$ [3] (keeping r at its critical value and $u_6=x=y=0$). The region of attraction of the Heisenberg fixed point is smaller than that given by Eq.(7), and also depends on u_6, x, y, etc. Points outside of this region flow across the Landau tricritical borderlines, and can be shown (by iteration until $r(\ell^*)=1$ and matching to Landau's theory[10]) to represent fluctuation driven first order transitions.

One can now add the quadratic anisotropy, Eq.(4), and end up with the tricritical lines shown in Fig.8 (drawn for the special

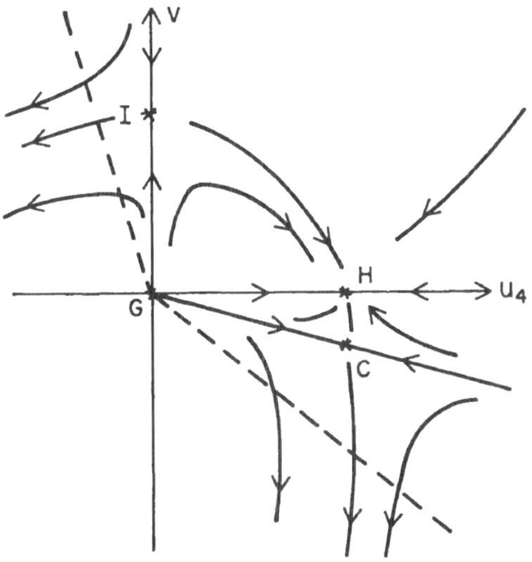

Fig. 7. Renormalization group flows on the critical surface, for the cubic problem ($u_6=0$, $n \lesssim 3$). The broken lines represent the Landau theory stability border lines.

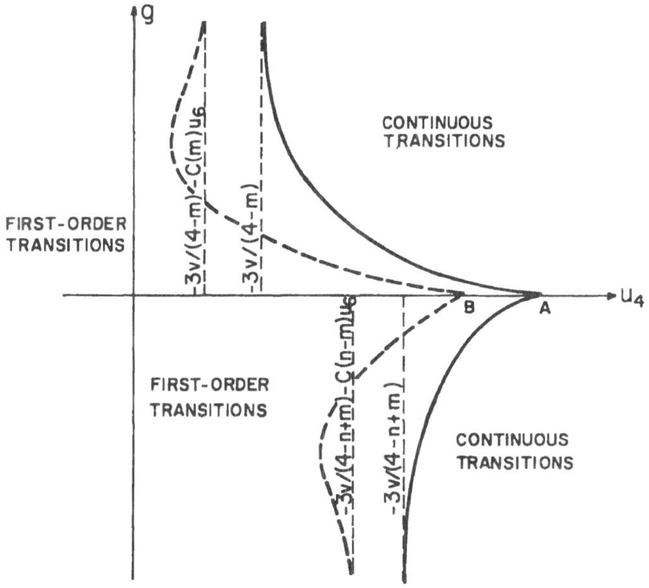

Fig. 8. Phase diagrams in g–u_4 plane for $m<(n-m)$. The (full) tricritical line is for $u_6=0$, and the broken one is for $u_6>0$.

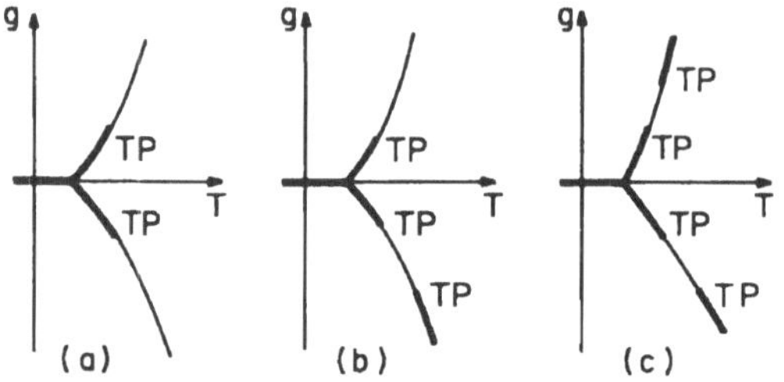

Fig. 9. Fluctuation driven phase diagrams for the cubic case, $v<0$. The value of u_4 decreases from (a) to (c), with u_6 and v fixed.

case $x=y=0$). The competition between the cubic effect (yielding fluctuation driven first order transitions) and the effect of the sixth order terms (yielding fluctuation driven continuous transitions) now results in regions in which one may expect two consecutive tricritical points as function of g. The corresponding phase diagrams are shown in Fig.9. For $u_6=x=y=0$, the diagram shown in Fig.9a was first predicted in Ref.10.

REALIZATIONS IN STRUCTURAL PHASE TRANSITIONS

The Hamiltonians (1) and (6) are appropriate for the description of displacive structural phase transitions, e.g. from cubic to trigonal or to tetragonal symmetry. In the last ten years there has been much interest in such transitions, which occur in perovskites like $SrTiO_3$, $RbCaF_3$, $KMnF_3$, etc.[11] In these three systems, $v<0$ and the transitions are into the tetragonal phase (the order parameter $\vec{S}$ roughly measures the amount of rotation of the ionic octahedra about [100] axes). A uniaxial anisotropy of the form (4) (with $n=3$, $m=1$) can be generated by uniaxial stress along one of the cubic axes[11]. Indeed, detailed predictions concerning the phase diagram shown in Fig.6a were experimentally confirmed for $SrTiO_3$.[11,12]

The situation for $RbCaF_3$ and for $KMnF_3$ is complicated by the strongly cubic character of the order-parameter fluctuations[13,14]. This tends to enhance the effect of fluctuations, and to turn continuous transitions into first order ones. It is for this reason that both $RbCaF_3$ and $KMnF_3$ exhibit fluctuation driven first order transitions at zero stress. The application of uniaxial stress has indeed confirmed all the theoretical predictions. Fig.9a has been observed in $RbCaF_3$ [14], and the tricritical point exhibits Lifshitz tricritical exponents, as predicted[13]. Fig.9b has recently been observed in $KMnF_3$ [15].

CONCLUSIONS

In the above examples, we hope we have demonstrated that the interplay between fluctuations and symmetry breaking yields many new kinds of tricritical points.

The short space has not allowed us to discuss tetracritical points, which are expected for $v>0$[11], or additional directions of the uniaxial stress[16]. In these cases, fluctuations also induce critical end points[17].

It should be emphasized that any mechanism which weakens critical fluctuations may have the same effect as described above, i.e. to change the transition from second to first order. Another example occurs when the coupling between layers of a quasi-two-dimensional system is strengthened, e.g. by uniaxial pressure[18].

In conclusion, we emphasize again the main message of this paper: The simple result of the Landau theory should be considered with care in the vicinity of tricritical points.

This paper was supported by the U.S.-Israel Binational Science Foundation.

REFERENCES

1. L. D. Landau and E. M. Lifshitz, "Statistical Physics", Pergamon, New York, Chap. XIV, (1968).
2. K. G. Wilson and J. Kogut, Phys. Repts. 12C, 75 (1974).
3. A. Aharony, in "Phase Transitions and Critical Phenomena", C. Domb and M. S. Green, editors, Academic, New York, Vol.6, p.357, (1976).
4. P. Bak, S. Krinsky and D. Mukamel, Phys. Rev. Lett. 36, 52 (1976).
5. D. Mukamel and S. Krinsky, Phys. Rev. B13, 5065, 5078 (1976); P. Bak and D. Mukamel, Phys. Rev. B13, 5086 (1976).
6. S. A. Brazovskii and I. E. Dzyaloshinskii, Pis'ma Zh. Teor. Fiz. 21, 360 (1975)[JETP Lett. 21, 164 (1975)].
7. D. Blankschtein and A. Aharony, Phys. Rev. Lett. 47, 439 (1981).
8. D. Blankschtein and A. Aharony, Phys. Rev. B, in press (1983).
9. E. K. Riedel and F. J. Wegner, Phys. Rev. Lett. 39, 629 (1977).
10. E. Domany, D. Mukamel and M. E. Fisher, Phys. Rev. B15, 5432 (1977).
11. A. Aharony, Ferroelectrics 24, 313 (1980).
12. S. Stokka and K. Fossheim, J. Phys. C15, 1161 (1981).

13. A. Aharony and A. D. Bruce, Phys. Rev. Lett. 42, 462 (1979).
14. J. Y. Buzaré, J. C. Fayet, W. Berlinger and K. A. Müller, Phys. Rev. Lett. 42, 465 (1979).
15. S. Stokka, K. Fossheim and V. Samulionis, Phys. Rev. Lett. 47, 1740 (1981).
16. D. Blankschtein and A. Aharony, J. Phys. C14, 1919 (1981).
17. D. Blankschtein and D. Mukamel, Phys. Rev. B25, 6939 (1982).
18. D. Blankschtein and A. Aharony, Phys. Rev. B26, 415 (1982).

A NEW MULTICRITICAL POINT IN A SEVEN DIMENSIONAL PARAMETER SPACE

Serge Galam

Department of Physics
New York University
New York, N.Y. 10003

1. INTRODUCTION

Anisotropic n-component d-dimensional ferromagnets with both random (non-ordering) and uniform (ordering) magnetic fields exhibit a rather large variety of multicritical behavior.[1,2,3] In the associated seven dimensional parameter space there exist a New Multicritical Point which is of the highest degree found up to now. It occurs at the intersection of various multicritical hypersurfaces which involve critical, tricritical, fourth-order, critical end point and bicritical behavior. Although there is still controversy regarding the value of the lower critical dimensionality for the Ising random field model[4], the richness of the model presented here motivated further theoretical study and search for experimental realization.

2. THE MODEL

The simplest Hamiltonian describing the model presented above is written in the form

$$H = -\frac{1}{2} J^1 \sum_{(i\neq j)} (S_i^1 S_j^1 + a\, S_i^2 S_j^2 + b \sum_{\alpha=3}^{n} S_i^\alpha S_j^\alpha) \tag{1}$$
$$- \vec{H} \sum_i \vec{S}_i - \sum_i \vec{H}_i \vec{S}_i ,$$

where $\vec{S}_i \equiv \{S_i^1, \ldots , S_i^n\}$ is a classical n-component (unit) vector located at the site i of a d-dimensional lattice. The sum $(i \neq j)$ is over nearest-neighbor pairs, J^1 is the (positive) exchange coupling in the 1-direction, a is the uniaxial anisotropy which makes the 1-direction the easy one $(0 \leq a \leq 1)$ and b makes the

2-direction the medium axis of magnetisation ($0 \leq b \leq a$). In the 1-2 plane, $\vec{H}$ is a uniform field while $\vec{H}_i$ is a local field with its own independent symmetrical distribution $p(\vec{H}_i) = p(-\vec{H}_i)$.

Focusing on the Heisenberg case n = 3, we present the mean field analysis of (1). Taking for the random field distribution

$$p(\vec{H}_i) = \frac{1}{2}\{\delta(H_i^1 + H_0)\,\delta(H_i^2 + H_\perp)\,\delta(H_i^3) + \delta(H_i^1 - H_0)\,\delta(H_i^2 - H_\perp)\,\delta(H_i^3)\}, \tag{2}$$

and assuming one single average magnetisation $\vec{M}$ per spin, the average free-energy per spin is[1,2,3]

$$\bar{F} = \frac{1}{2} c\, J^1 \{(M^1)^2 + a(M^2)^2 + b(M^3)^2\} - \frac{1}{2\beta} \mathrm{Ln}\,[g(\beta H_+)\, g(\beta H_-)], \tag{3}$$

where c is the coordination number, $\beta = 1/k_\beta T$, k_β is the Boltzmann constant, T is the temperature, $g(\kappa) = 4\pi \sinh(\kappa)/\kappa$ and

$$H\pm = \{(c\, J^1 M^1 + H^1 \pm H_0)^2 + (a\, c\, J^1 M^2 + H^2 \pm H_\perp)^2 + (b\, c\, J^1 M^3)^2\}^{1/2}. \tag{4}$$

The space of parameters has thus seven dimensions which are respectively: the temperature; the two anisotropies; the two components of the uniform field, and the two components of the random field. In the following we exhibit some portions of the global phase diagram associated with (3).

3. CASE: $H^1 = H^2 = H_\perp = 0$ and $a = b$

The three dimensional ($a - T - H_0$) phase diagram[1], is shown in Fig.1. For strong anisotropy ($0 \leq a \leq 1/2$) the system behaves as Ising ferromagnet. There exists one longitudinal phase with a tricritical line. For weak anisotropy ($0.8 < a \leq 1$) the system shows a bicritical line. A spin flop surface separates the longitudinal (low field) and the transverse (high field) phases. At higher anisotropy the bicritical line splits into a tricritical line and a critical end point line. This split first occurs at a New Multicritical Point whose coordinates are $a \simeq 0.797$, $\beta H_0 \simeq 1.372$ and $\beta c J^1 \simeq 4.213$.

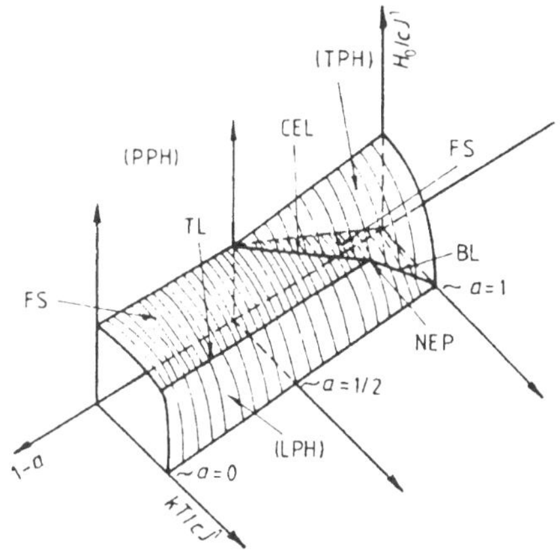

Fig. 1. Phase diagram as function of longitudinal random field component, temperature, anisotropy. PH=phase, L=longitudinal, P=paramagnetic, T=transverse. BL=bicritical line CEL=critical end line, TL=tricritical line, NEP=new multicritical point, FS=first-order surface.

4. CASE: $H^1 = H^2 = 0$ and $a = b$

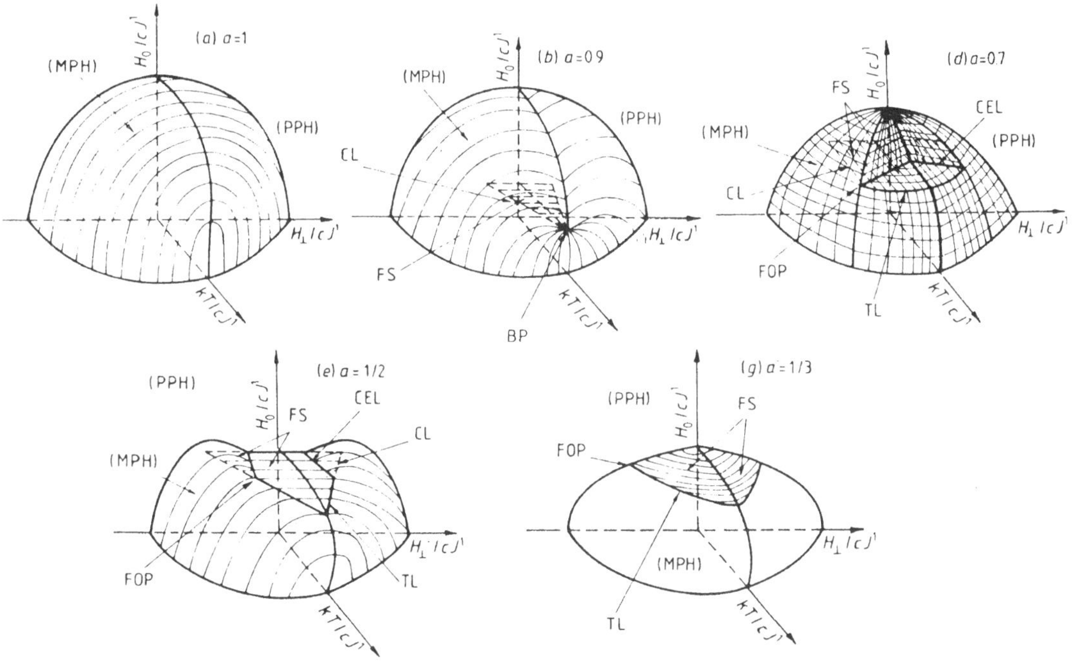

Fig. 2. Temperature, longitudinal transverse random field components for various values of the anisotropy. PG=phase, M=mixed, P=paramagnetic, CL=critical line, BP=bicritical point, CEL=critical end line, TL=tricritical line, FOP=fourth-order point, NEP=new multicritical point.

The introduction of a transverse component ($H^2_i = \pm H_\perp$) to the random field gives more insight into the nature of the new multicritical point in the four-dimensional space $(H_\perp - H_0 - T - a)^2$. The phase diagram for various fixed values of the anisotropy is shown in Fig.2. There exists a single ordered phase for $H_\perp \neq 0$. For large anisotropy (0.5 > a > 0.5), it is bounded by three-dimensional hypersurfaces of first- and second-order transitions, separated by a tricritical surface. At intermediate anisotropy (0.8 > a > 0.5) the tricritical surface splits into a critical surface (inside the ordered phase) and a critical end surface. This occurs at a line of fourth-order critical points. For weak anisotropy (1 > a > 0.8) the transition into the disordered phase is always second order. There exists a spin-flop 'shelf' inside the ordered phase ending at a bicritical line. This bicritical line meets the fourth-order line exactly at the new multicritical point in the $H_0 - T$ plane for one value of the anisotropy.

5. CASE: $H_\perp = H^2 = 0$ and a = b

The introduction of staggered fields has provided a deeper understanding of the usual tricritical point[5]. The effect of a non-zero ordering field[3] (uniform) on the phase diagram of Fig.1 is shown for various fixed values of the anisotropy in Fig.3.

It turns out that the main features of the four-dimensional phase diagrams ($H_\perp$-H_0-T-a) and (H-H_0-T-a) are similar. However, in place of the fourth-order points of the earlier case there exist special bicritical points. The particularity of these bicritical points is the presence of cubic terms for the longitudinal order. The 'classical wings' of the tricritical point are recovered. We find that the bicritical point for fixed anisotropies H=0 is associated with two symmetrical 'horns' (H > 0 and H < 0), containing two tricritical lines. This picture agrees with partial Monte Carlo results on antiferromagnets[6] and contrasts with the 'balloon' form predicted from renormalisation-group calculations for small staggered fields[7]. The new multicritical point occurs when the overlap of the 'horns' with the 'wings' is complete. It is located at the intersection of three bicritical lines of which two are special.

6. CONCLUSION

We have found in the seven dimensional parameter space a New Multicritical Point whose coordinates are

$$a = b \simeq 0.797;\ \beta H_0 \simeq 1.372;\ \beta c J^1 \simeq 4.123;\ H^1 = H^2 = H_\perp = 0. \tag{5}$$

At this stage it is worth noticing that there exists a mathematical mapping between the model presented here and the anisotropic antiferromagnet with both uniform and staggered fields[8]. The global phase diagrams of Figs.1,2,3 are thus relevant for both dilute antiferromagnets and pure antiferromagnets in a field.

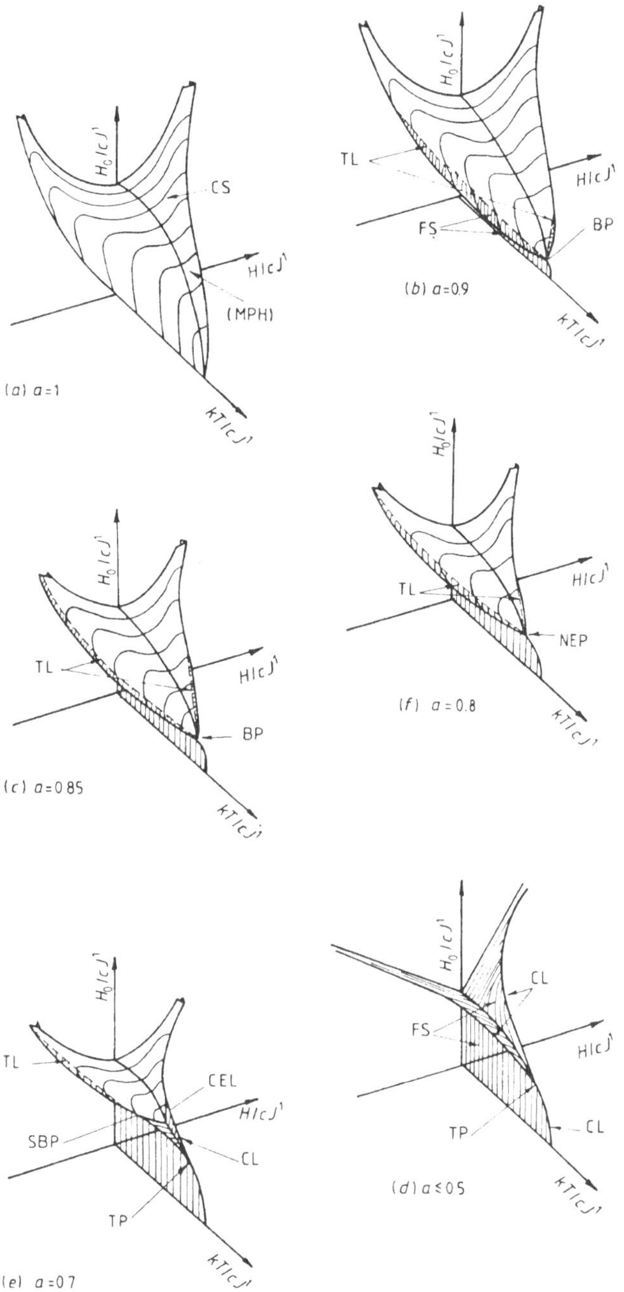

Fig. 3. Temperature, random, uniform longitudinal components for various values of the anisotropy. CS=critical surface, FS= first-order surface, CEL=critical end line., PH=phase, M= mixed, BP= bicritical point, SBP=special bicritical point, TP=tricritical point, TL=tricritical line, NEP=new multi-critical point.

ACKNOWLEDGEMENT

The author is particularly grateful to Amnon Aharony for having introduced him to the field of multicritical phenomena. This work was supported in part by National Science Foundation grant #DMR-12399 and PSC-BHE Faculty Research grant # RF 13958.

REFERENCES

1. S. Galam and A. Aharony, J. Phys. C: Sol. St. Phys. 13, 1065 (1980).
2. S. Galam and A. Aharony, J. Phys. C: Sol. St. Phys. 14, 3603 (1981).
3. S. Galam, J. Phys. C: Sol. St. Phys. 15, 529 (1982).
4. A. Aharony, "Theory of Critical Phenomena in Random Systems", in this proceeding; R.A. Cowley, R.J. Birgeneau, G. Shirane and H. Yoshizawa, "Disorder, Random Fields and Competing Interactions in Antiferromagnets", in this proceeding.
5. R.B. Griffiths, Phys. Rev. Lett. 35, 1399 (1970).
6. D.R. Landau and K. Binder, Phys. Rev. B17, 2328 (1978).
7. J.M. Kosterlitz, D.R. Nelson and M.E. Fisher, Phys. Rev. B13, 412 (1976).
8. S. Galam, Phys. Lett.A, in Press (1984).

MULTICRITICAL POINTS IN INCOMMENSURATE SYSTEMS

Pierre Toledano

Groupe de Physique théorique
Faculté des Sciences d'Amiens
33, rue Saint-Leu, 80039 Amiens Cedex - France

INTRODUCTION

Since Hornreich et al[1] introduced the concept of a Lifshitz point - i.e. a triple point in the Pressure-Temperature phase diagram, separating two lines of continuous transitions to ferromagnetic and helicoïdal phases - a large number of theoretical studies have discussed the symmetry and critical properties of this multicritical point[2-5]. Experimental evidence of a Lifshitz point was also claimed for a large variety of compounds undergoing magnetic[6], structural[7], liquid crystal[8], as well as surface phase transitions[9]. However, as noted by Aslanyan and Levanyuk[10] the Lifshitz point considered in Ref. 1 corresponds to a physical situation which is not the more likely to occur. Accordingly, these authors defined three other types of Lifshitz points[10].

In this paper, we generalize the procedure used in Refs. 1 and 10 and present a systematic investigation of multicritical points which may be found in incommensurate systems.

MULTICRITICAL POINTS IN THE LANDAU THEORY OF INCOMMENSURATE PHASE TRANSITIONS

In the ordinary Landau theory of phase transitions[11], the coefficient α of the quadratic order-parameter (OP) invariant depends on the sole external variable, say the temperature T, and not on the wave-vector $\vec{k}$, as $\vec{k}$ possesses a fixed value which corresponds either to the center, or to a high-symmetry boundary point of the relevant Brillouin-zone (BZ). Among the coefficients associated to the different irreducible representations (IR's) figuring in the transition free-energy F, α is singled out as it vanishes first at the Curie-temperature T_c, while the coefficients pertaining to the inactive IR's remain strictly positive.

In the study of incommensurate phase transitions, as $\vec{k}$ varies with temperature in the modulated phase, one has to consider its influence on the variation of α. The value of $\vec{k}$ which corresponds to the onset of the incommensurability is the one, say $\vec{k}_1$, minimizing $\alpha(\vec{k})$ at T_c. However $\vec{k}_1$ determines a minimum of F only in the vicinity of T_c. When the temperature is lowered, other values, namely $\vec{k}_2$, $\vec{k}_3$,... will be associated to the successive minima of F, i.e. to the successive equilibrium values of the OP in the incommensurate state. So, to each of the preceding $\vec{k}_j$, one should relate a set of OP components η_i^j whose symmetry properties would determine the successive forms F_j of the free-energy. Actually, the phenomenological Landau-type theory developed by Dzialoshinskii[12] and later by Levanyuk and Sannikov[13], consists in choosing among the $\vec{k}_j$ one particular vector $\vec{k}_o$ which involves the more symmetric free-energy F_o. Generally $\vec{k}_o$ is the lock-in vector arising at the low-temperature lock-in transition, which is located at a high-symmetry point of the BZ. But as it does not correspond to the minimum of $\alpha(\vec{k})$ in the whole incommensurate phase, the actual OP components η_i^j appear as a slowly modulated function of the η_i^o which are the values of the η_i^j at $\vec{k} = \vec{k}_o$.

The advantage of the preceding description is that for any temperature and value of $\vec{k}$, only one free-energy F_o is used, which describes the complete set of OP's. However one has to introduce in F_o, invariants containing derivatives of the η_i^o with respect to space coordinates. Such invariants express the k-dispersion of $\alpha(\vec{k})$ in the vicinity of $\vec{k}_o$, which can be written under the form of an expansion of the successive degrees of $\vec{k} - \vec{k}_o$:

$$\alpha(\vec{k}) = \alpha_o\ (T,P) + \sum_n \alpha_n\ (T,P)(\vec{k} - \vec{k}_o)^n \qquad (1)$$

where P is an external parameter (e.g. pressure, or composition for mixed compounds, etc...)

Various multicritical points may then be defined, depending on the actual form of expansion (1), which is determined by the symmetry of the system.

1. We first examine the case where only even degrees of $\vec{k}-\vec{k}_o$ are allowed in (1). Fig 1.a) shows the variation of $\alpha(\vec{k})$ for $n_{max} = 4$, i.e. for $\alpha(\vec{k}) = \alpha_o(T,P) + \alpha_2(\vec{k}-\vec{k}_o)^2 + \alpha_4(\vec{k}-\vec{k}_o)^4$. It corresponds to the situation considered in Ref. 1, with a Lifshitz point defined by :

$$\alpha\ (T\ ,P\ ) = \alpha_2(T\ ,P\ ) = 0$$

This point divides two lines of continuous transitions ($\alpha\ (T_c,P)=0$) from a high-symmetry normal (N) phase to an incommensurate (INC) phase (where $k_i = k_o \pm (\frac{-\alpha_2}{2\alpha_4})^{1/2}$ and $\alpha_2 = \alpha_{2o}(P-P_c) < 0$),

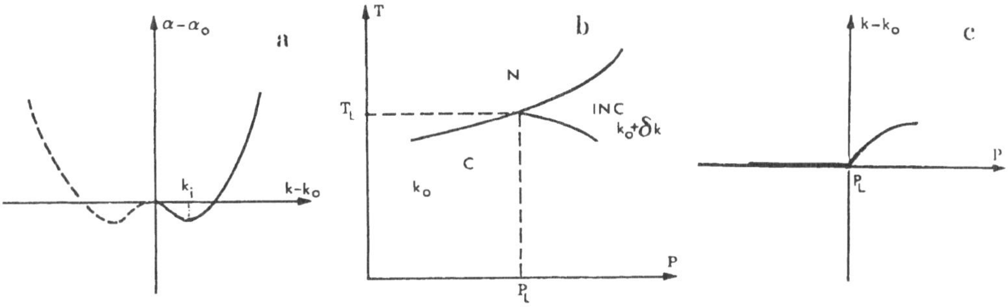

Fig. 1. Lifshitz point for n_{max}=4. a) Dispersion curve, b) Phase diagram T-P, c) P-variation of the wave-vector $\vec{k} - \vec{k}_o$

and to a commensurate (C) phase ($k = k_o$, $\alpha_2 > 0$). The two low-temperature phases ($T < T_c$) are separated by a line of second or first-order transitions (Fig.1,b), depending on the OP symmetries[3].

A characteristic feature of this multicritical point is that $\vec{k}_i$ increases continuously from k_o (Fig. 1,c), the stability of the INC phase being insured by the positiveness of α_4. As remarked by Aslanyan and Levanyuk[10], the condition $\alpha_4 > 0$ is unlikely as it contradicts the calculated value obtained in the nearest neighbor interaction :

$$\alpha(\vec{k}) \sim \omega^2(\vec{k}) \simeq 1 - \cos ka \simeq \frac{k^2a^2}{2} - \frac{k^2a^4}{4}$$

Accordingly these authors have considered the case with n_{max}=6 and $\alpha_4 < 0$, $\alpha_6 > 0$ (Fig. 2,a). Here a Lifshitz point can be defined by the condition :

$$\alpha_4^2 - 4\,\alpha_2\alpha_6 = 0 \qquad (\alpha_2 > 0)$$

with a first-order transition line between the INC and C phases (Fig. 2,b), and a discontinuous jump at (T_c, P_L) (Fig.2,c).

For $n_{max} > 6$, the minimization of (1) must be performed numerically. When n_{max} = 8,10 and 12, we obtain the phase diagrams represented on Fig. 3. For n_{max} = 8 or 10, two multicritical points can be found on the line $\alpha(T_c, P) = 0$ at which two distinct

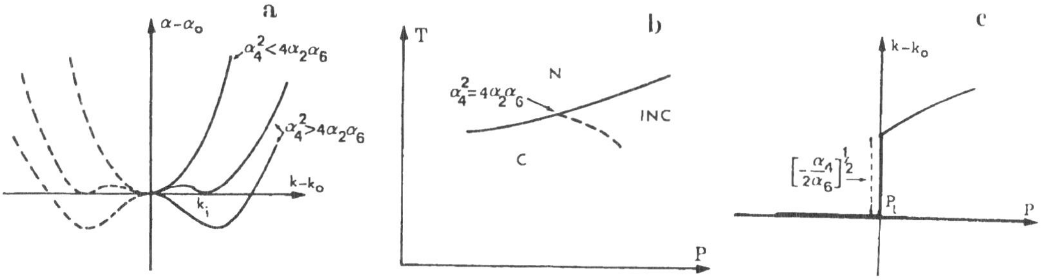

Fig. 2. Lifshitz point for n_{max} = 6 a) Dispersion curves for various values of $\alpha_4^2 - 4\alpha_2\alpha_6$ b) Phase diagram T-P (dots=first-order transition line c) Variation of the wave-vector $\vec{k} - \vec{k}_o$

incommensurate regimes appear associated respectively to wave-vectors k_{i1} and k_{i2}. For n_{max} = 8, k_{i1} increases discontinously from (T_c, P_{L1}) then jumps discontinuously at (T_c, P_{L2}), while for n_{max} = 10, k_i displays two discontinuous jumps. For n_{max} = 12, three multicritical points can be obtained by minimization of (1), with a succession of three INC phases separated by two lines of first-order transitions.

Taking into account terms of degree n in the expansion of $\alpha(k)$ means that in the free-energy density of the transition, one must retain invariants which are of the second degree in the η_i^o and of the nth order in the derivatives with respect to space coördinates. For n_{max} = 4, Hornreich et al[1] considered invariants of the type $(\frac{\partial \eta_i^o}{\partial x_j})^2$ and $(\frac{\partial^2 \eta_i^o}{\partial x_j^2})^2$. Here, the condition n even indicates that no odd order derivatives should be allowed by the symmetry of the system. It implies that the invariance group of $\alpha(k)$ should contain the inversion or a central point. In particular, the fact that no linear invariant is allowed in (1) expresses the fact that the Lifshitz condition is satisfied by the IR inducing the transition. We now consider the case where this condition is violated.

2.When odd degrees are allowed in (1), the simplest case n_{max}=2 is represented on Fig.4 a).The stability of the INC phase is insured by the positiveness of α_2,$\alpha(k)$ having an oblique tangent at k=k₀. The

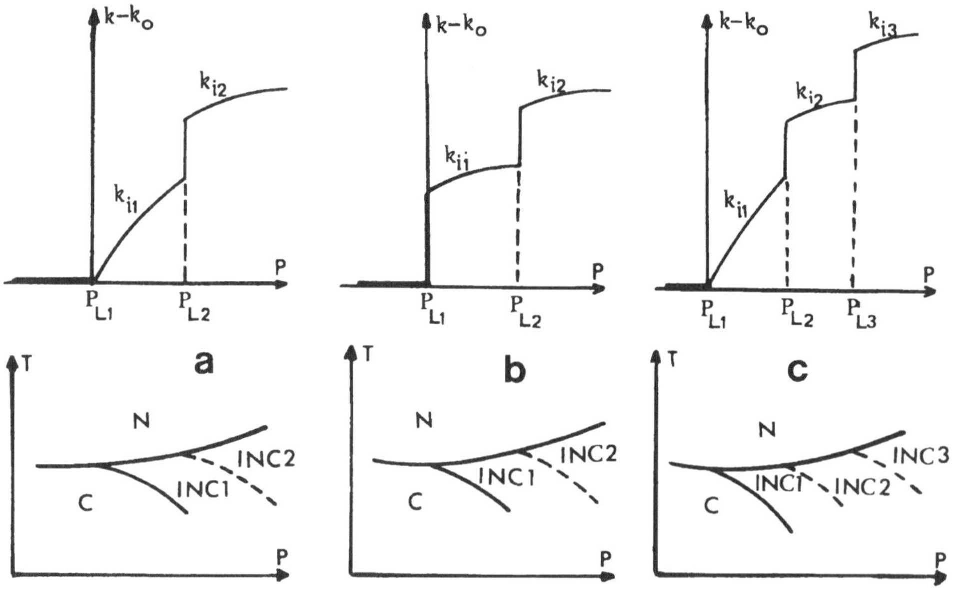

Fig. 3. Lifshitz point for n_{max} = 8(a), 10(b) and 12(c): Phase diagrams (bottom) and P-variation of the wave vectors (top).

wave-vector $k_i = k_\circ - \alpha_1/\alpha_2$ decreases linearly as α_1 increases. The second -order N-INC line is tangent to the first-order INC-C transition line at a multicritical point, defined by $\alpha = \alpha_1 = 0$, at which one goes across continuously from the N to the C-phase. When $\alpha_2 < 0$, the preceding value of k_i is unstable and one has to take into account the third degree term in (1), with $\alpha_3 > 0$ (Fig.4 b). Here the $\alpha(k)$ curve has again only one minimum corresponding to:

$$k_i = k_\circ - \alpha_2/2\alpha_3 + (\alpha_2^2/4\alpha_3^2 - \alpha_1/\alpha_3)^{1/2}$$

The curve $k(\alpha_1)$ decreases with increasing α_1 and falls discontinuously at $\alpha_1 = \alpha_2^2/4\alpha_3$. This last condition defines on the T-P diagram a tricritical point which separates a line of continuous N-INC transitions from a N-C first-order line. On Fig.4 c we show the case where $n_{max} = 4$ $(\alpha_2 < 0, \alpha_3 < 0, \alpha_4 > 0)$. The $k(\alpha_1)$ curve possesses two minima, associated to two INC phases with wave-vectors k_{i1} and k_{i2}. These phases are separated by a first-order transition line ending at a tricritical point lying on the α (T)=0 line.

The P-variation of α_1 -the coefficient of the linear term in (1)- appears to be essential for the existence of the preceding multicritical

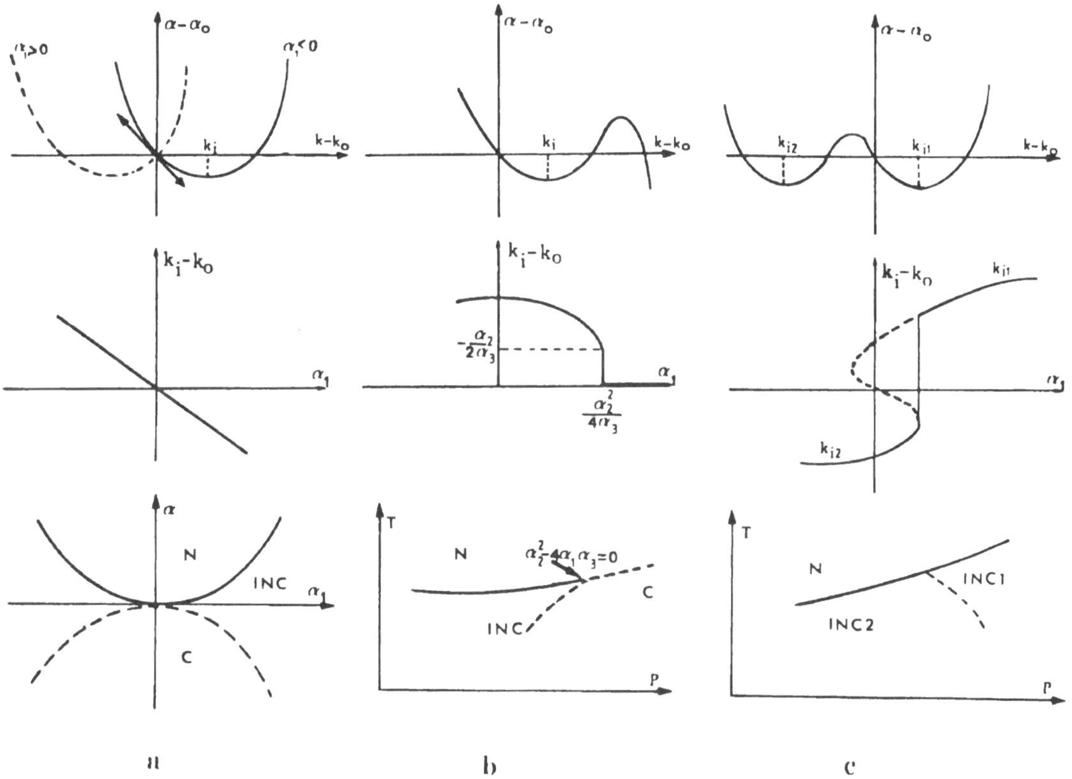

Fig.4. $\alpha(k)$ curves (top), variation of $k(\alpha_1)$ (middle) and phase diagrams (bottom), for $n_{max} = 2$ (a), 3 (b) and 4 (c).

points. A linear term in (1) corresponds, in the transition free-energy F_0, to antisymmetric terms of the type: $\eta_i \frac{\partial \eta_j}{\partial x} - \eta_j \frac{\partial \eta_i}{\partial x}$ (Lifshitz invariant) which always exist when the Lifshitz condition is not satisfied. More generally in all the above considered cases, the stability of the INC phases and their onset at multicritical points, have been determined by considering exclusively those terms in F_0 which are quadratic with respect to the OP components η_i^0.

Some authors[10,14,15] have shown that third order invariants of the type $\eta_i \eta_j \frac{\partial \eta_k}{\partial x}$ also lead to stability of INC phases. Such terms are shown to induce, for particular relationships between the coefficients in F_0, new minima for $\alpha(k)$ and thus allow to define other types of multicritical points, dividing lines of transitions to INC phases that will be generally first-order[10]. A more complex set of possibilities can still arise if $\alpha(k)$ is expressed as a function of the various components k_i of $\vec{k}$, the coefficients α_i in (1) then becoming tensorial quantities. The analysis of such situations, together with the discussion of the critical behaviour of physical quantities in the vicinity of Lifshitz points, will be published elsewhere[16].

REFERENCES

1. R.M. Hornreich, M. Luban and S. Shtrikman, Phys. Rev. Letters 35, 1678 (1975).
2. R.M. Hornreich, M. Luban and S. Shtrikman, Phys. Rev. B19, 3799 (1979).
3. A. Michelson, Phys. Rev. B16, 577, 585 and 5121 (1977).
4. D. Mukamel and M. Luban, Phys. Rev. B18, 3631 (1978).
5. R.M. Hornreich, Phys. Rev. B19, 5914 (1979).
6. S.K. Sinha, G.H. Lander, S.M. Shapiro and O. Vogt, Phys. Rev. Letters 45, 1038, (1980).
7. E. Abrahams and I.E. Dzyaloshinskii, Solid State Commun. 23, 883 (1977).
8. G. Sigaud, F. Hardouin and M.F. Achard, Solid State Commun. 23, 35 (1977).
9. I.F. Lyuksyutov, JETP Letters, 32, 580 (1980).
10. Y.S. Aslanyan and A.P. Levanyuk, Sov. Phys. Solid State 20, 1925 (1978).
11. L.D. Landau and E.M. Lifshitz, Statistical Physics, (Pergamon, 1959).
12. I.E. Dzialoshinskii, Sov. Phys. JETP, 19, 960 (1964).
13. A.P. Levanyuk and D.G. Sannikov, Sov. Phys. Solid State 18, 245 (1976).
14. E.B. Loginov, Sov. Phys. Crystallogr. 24, 637 (1980).
15. A.L. Korzhenevskii, Sov. Phys. JETP 54, 568 (1981).
16. P. Toledano, unpublished.

NEUTRON INVESTIGATION OF MODULATED STRUCTURES

R. Currat

Institut Laue-Langevin

156X, 38042 Grenoble Cedex, France

I. INTRODUCTION

I.1. Modulated Structures

Incommensurably-modulated solids are often described[1] as long-range ordered structures which lack translational symmetry due to the superposition within the same system of several mutually incompatible periodicities. As such IC solids provide an intermediate step between the ordinary crystalline state and the disordered non-periodic amorphous state.

In fact, part of the current work on IC structures is aimed at testing the validity of the above definition. As shown by de Wolff[2] and Janner and Janssen[3], spatial periodicity is hidden rather than absent in the ideal IC solid and may be recovered via a suitable generalization of the concept of symmetry operation. On the other hand, experimental and theoretical evidence indicate that real IC solids close to the C limit[4] or in the presence of frozen defects[5,6] do not qualify as fully long-range ordered structures.

A wide range of physical systems are known to present compositional or displacive modulations and the corresponding physical situations are necessarily very diverse. For instance 2-dimensional systems, such as adsorbed monolayers or intercalates exhibit qualitatively unique features (see Nielsen[7] for an experimental review). In this lecture the emphasis will be on displacive modulations i.e. modulations of the equilibrium atomic positions, in ordinary 3-dimensional solids.

Perhaps the single feature common to all IC systems is the occurrence of irrational satellite reflections, in addition to the Bragg reflections associated with the unmodulated reference system. The observation of a continuous shift in the satellite diffraction pattern as a function of temperature or external field has been taken, perhaps too readily, as acceptable evidence for the true IC nature of the modulation. As suggested by theoretical models (see Bak[8] for a recent review), an apparent smooth change in modulation wavevector may correspond to an infinite sequence of high-order C phases separated or not by true IC phases (incomplete or complete devil's staircase). In such models the concept of discommensuration (DC) or domain wall separating locally C regions is of central importance[9]. In particular the value of the modulation wavevector in the IC state is determined by the average DC spacing, the C state corresponding to zero DC density.

Another characteristic feature of IC systems is the presence of global hysteresis affecting the temperature and pressure dependence of the modulation wavevector and of those macroscopic variables (e.g. dielectric constants, optical birefringence coefficients, etc.) which couple to the IC order parameter. As pointed out by Aubry[10], global hysteresis is not compatible with a smooth analytic behaviour . Thus its observation lends support to the (complete) devil's staircase picture where irreversible effects arise as a result of pinning of the DC's by the underlying lattice. Recently, experimental evidence have shown that defects play an important rôle in the observed irreversibilities, giving rise to specimen-dependent behaviour and memory effects[11-14]. Thus, in general, both types of pinning mechanisms, intrinsic and defect-induced, are expected to contribute to the observed irreversible behaviour and incomplete ordering.

I.2 Phasons

IC solids are characterized by new types of excitations[15,16], corresponding to fluctuations in the phase and amplitude of the static modulation wave. In the long-wavelength limit the phase mode consists in an overall phase shift of the modulation with respect to the atomic lattice (sliding mode). For a well-behaved IC modulation, for which the atomic displacements can be described in terms of analytic functions of space, such a shift costs no energy and the corresponding phase branch is gapless. For a quasi-C modulation the phase branch becomes identical to the vibrational "phonon" branch of the DC lattice. Its gapless character is preserved as long as the DC's are not pinned.

Recently, it has been shown that phase modes are overdamped, at least at long wavelengths[17]. The damping arises from the fact that the atomic motions involved are inhomogeneous on a microscopic scale and are thus affected by ordinary anharmonic (i.e. dissipative) interactions.

Hence, unlike acoustic modes, phasons do not qualify as true Goldstone modes. In terms of conservation laws, the difference lies in the fact that the total particle-momentum operator commutes with the total crystal Hamiltonian, while the analogous phase-momentum operator commutes with the harmonic part only. The overall phase shift is thus associated with friction.

Practically, the phase mode response is expected to be of the damped harmonic oscillator type :

$$S(q,\omega) = \frac{k_T}{\Pi} \frac{\Gamma_o}{(\omega^2 - v^2q^2)^2 + \omega^2\Gamma_o^2}$$

where the velocity v and damping coefficient Γ_o are slowly-varying with q. The response is diffusive at low q :

$$\omega_{ph}(q) = i\,\frac{v^2q^2}{\Gamma_o} \quad (q < \frac{\Gamma_o}{v})$$

becoming progressively underdamped at large q :

$$\omega_{ph}(q) = vq + i\Gamma_o \quad (q > \frac{\Gamma_o}{v}).$$

Rough estimates for v and Γ_0 indicate that the two regimes may be probed by Brillouin and neutron scattering, respectively[18]. Underdamped phase modes have been observed by neutron scattering in phase III of biphenyl[19] and in $ThBr_4$[20] (see Fig. 1).

I.3 Experimental Techniques

Most of the experimental information on IC solids is based on X-ray, neutron and electron diffraction data. The best q-resolutions ($\simeq 10^{-4}$ Å^{-1}) are obtained from X-ray sources. In 2H-$TaSe_2$, minute deviations from commensurability have thus been revealed. In structurally-modulated $BaMnF_4$, satellite reflections of the type $(h+\xi, k+1/2, 1+1/2)$, with ξ ranging from 0.39 to 0.40 depending upon temperature (and specimen origin), have been shown to present a finite q-width along ξ, corresponding to a correlation length of a few hundred cells[22]. In view of the absence of any lock-in transition at low temperatures, this result suggests a defect-induced pinning of the modulation.

Real space visualization of DC arrays in 2H-$TaSe_2$ has been achieved by dark-field electron microscopy[23,24]. In particular, dynamic studies as a function of temperature show that the basic

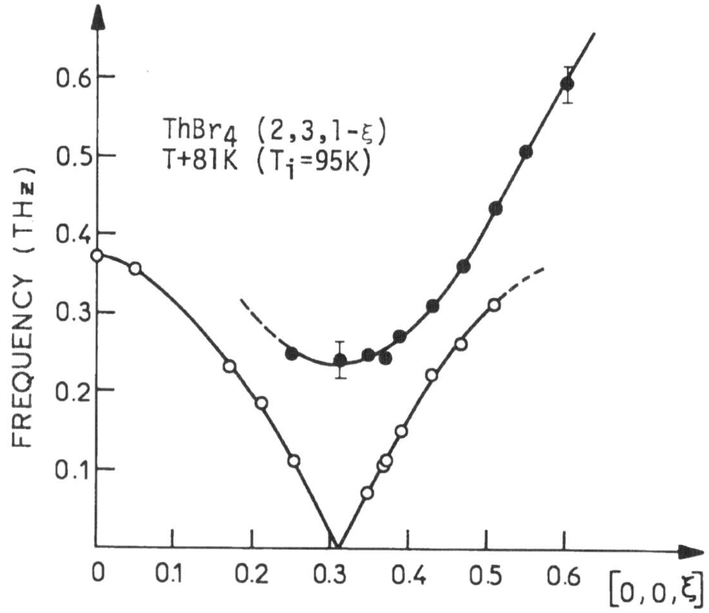

Figure 1 : Phase mode (open circles) and amplitude mode (closed circles) in $ThBr_4$ (from ref. 20).

mechanism through which the DC array readjusts its average spacing is "stripple" nucleation (or evaporation) rather than homogeneous expansion.

Local information may be obtained from resonance techniques (see Blinc[25] for a review), or at low temperatures, from optical crystal-field spectroscopy[26]. Resonance lines are broadened inhomogeneously due to the distribution of local environments introduced by the modulation. The lineshape of the resonance spectrum mirrors the distribution of local fields, provided the latter fluctuates slowly on the characteristic timescale of the measurement (inverse linewidth). This condition should be (and appears to be) generally fulfilled except in the immediate vicinity of the IC ordering temperature where motional narrowing should in principle occur due to phase fluctuations. In Rb_2ZnBr_4 a partial crossover to the fast fluctuation regime has been reported[27] at about 10K below the ordering temperature (T_i = 200K). The ^{87}Rb NMR linewidth at that temperature was $\sim$ 20 kHz.

II. NOWOTNY PHASES : AN EXAMPLE OF UNIAXIAL INTERGROWTH COMPOUND

II.1 Structure

The structure of Nowotny phases[28] is based on two interpenetrating tetragonal lattices with identical cell dimensions in the (a,b) plane, but different c parameters. Model systems of this type have been considered by Theodorou and Rice[29] and Axe and Bak[30]. The chemical formula is TX_x where T is a transition metal element and X is an element of group III or IV. The observed variation of the atomic ratio x with the chemical nature of T and X suggests that the stability of the structure is largely determined by the value of the electron concentration per T-atom[31].

For each of the two sublattices one can define an idealized subcell (cf. Fig. 2), of dimensions (a, a, c_T) and (a, a, c_X), with 4 atoms/cell in each case (i.e. $c_T = xc_X$). The atomic positions on sublattice X are modulated in such a way as to avoid overlap with the surrounding T-atoms. The modulation consists in a rotation of the X-X dumbells in the (a,b) plane, together with a shift along c. The X-atomic positions along c are expressed as

$$z_X(n) = nc_X + u_X + g_T \quad (nc_X + u_X - u_T) \tag{1}$$

where $u_{X,T}$ are appropriate origins and $g_T(z)$ is a periodic modulation function with period c_T :

$$g_T(z + c_T) = g_T(z) \tag{2}$$

An analogous expression can be written down for the T positions, although the observed distortions are much smaller.

In order to calculate the diffracted intensity from such a system let us write the density function of each sublattice as :

$$\tilde{\rho}_{X,T}(\vec{r}) = \rho_{X,T}(\vec{r} - \vec{g}_{T,X}(z)) \tag{3}$$

where $\vec{g}_{T,X}$ is a displacement vector satisfying (2) and $\rho_{X,T}(\vec{r})$ is the unmodulated sublattice density function :

$$\rho_{X,T}(\vec{r}) = \sum_{G_\perp,\lambda} \rho_{X,T}(G_\perp, \lambda) \exp\left\{ i(\vec{G}_\perp \cdot \vec{r}_\perp + \lambda c^*_{X,T} z)\right\} \tag{4}$$

where $\vec{G}_\perp + \lambda \vec{c}^*_{X,T}$ is a reciprocal lattice vector of sublattice X,T.

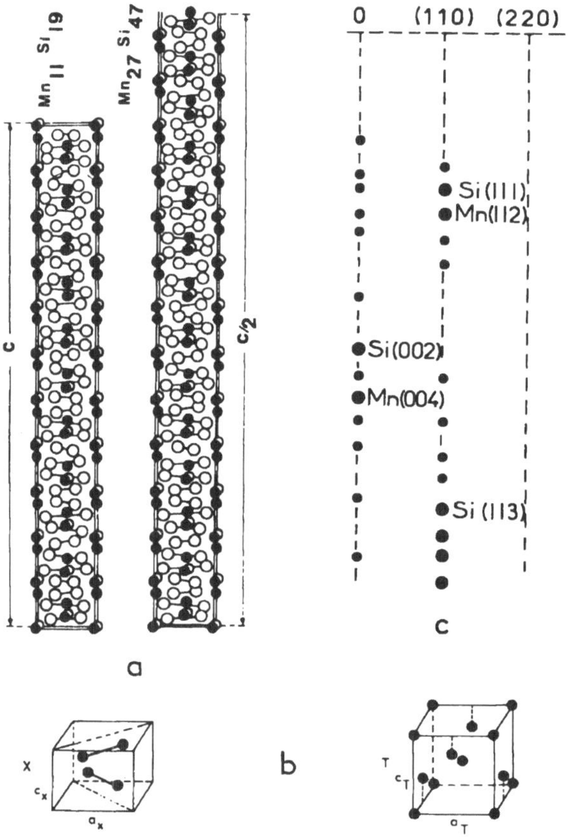

Figure 2 : Structure of Nowotny phases TX_x (T = Ti, Zr, V, Mo, Cr, Mn, Rh; X = Si, Ge, Sn). (a) sketch of unit cell for two stoichiometric compounds in the $MnSi_x$ series (x = 1.73±0.02) (from ref. 35); (b) idealized subcells (a_T = a_X; C_T = x C_X); (c) typical diffraction pattern. All spots can be indexed according to eq. (6). Only the main reflections (λ or μ = 0) are labelled in the figure.

Inserting (4) into (3) yields :

$$\tilde{\rho}_X(\vec{r}) = \sum_{G_\perp,\lambda} \rho_X(G_\perp,\lambda) \exp\left\{ i(\vec{G}_\perp \cdot \vec{r}_\perp + \lambda c_X^\star z)\right\} \times \exp\left\{ -i\,[\vec{G}_\perp \cdot \vec{g}_T^\perp(z) + \lambda c_X^\star g_T^{\parallel}(z)]\right\} \quad (5)$$

The last exponential factor in (5) is periodic with period c_T, whence :

$$e^{-i[\ ]} = \sum_\mu w_T(\mu;G_\perp,\lambda)\, e^{i\mu c_T^\star z}$$

and :

$$\tilde{\rho}_x(\vec{r}) = \sum_{G_\perp,\lambda,\mu} \rho_x(G_\perp,\lambda)\, w_T(\mu;G_\perp,\lambda) \times \exp\left\{ i\,[\vec{G}_\perp \cdot \vec{r}_\perp + (\lambda c_X^\star + \mu c_T^\star)\, z]\right\}$$

Bragg peaks will be observed at :

$$\vec{Q} = \vec{G}_\perp + \lambda \vec{c}_x^{\,\star} + \mu \vec{c}_T^{\,\star} \quad (\lambda,\mu = \text{integers}) \quad (6)$$

with structure factors :

$$F_{G_\perp,\lambda,\mu} = \rho_x(G_\perp,\lambda)\, w_T(\mu;G_\perp,\lambda) + \rho_T(G_\perp,\mu)\, w_x(\lambda;G_\perp,\mu) \quad (7)$$

Thus in general both sublattices contribute to each reflection. Because the T-sublattice is relatively rigid ($w_X(\lambda;G_\perp,\mu) \simeq 0$ if $\lambda \neq 0$) the reflections corresponding to $\lambda \neq 0$ will essentially be pure X-reflections. The only pure T-reflections are obtained for $\lambda = G_\perp = 0$ ($w_T(\mu;0,0) = \delta\mu$).

For rational values of $x (x=m/n)$, the supercell dimension along c is $\bar{c} = nc_T = mc_X$. The space groups of the corresponding compounds are D_{2d}^6, D_{2d}^{12} or D_{2d}^8, depending upon the parities of m and n.

In the $MnSi_x$ system, the following compounds have been reported by X-ray diffraction[32,33]:

Mn_nSi_m	x	a(Å)	($\bar{c}$/n)(Å)
$Mn_{11}Si_{19}$	1.727	5.518	4.376
$Mn_{15}Si_{26}$	1.733	5.525	4.370
$Mn_{27}Si_{47}$	1.741	5.530	4.368

Commensurate structures have also been observed by de Ridder et al.[34] by electron diffraction. In addition, these authors observe compositional inhomogeneities across the volume of their specimen, giving rise to an alternance of C and IC diffraction patterns[35].

Since the observed concentration fluctuations cannot be eliminated by annealing, they indicate that the equilibrium Si concentration is inhomogeneous even at high temperature. This feature can be tentatively interpreted as resulting from a competition between the band structure energy, which is minimized for some IC value of x, and short-range lock-in terms, which favour simple rational fractions.

II.2 Dynamics

To the extent that the displacement functions $g_{T,X}$ (cf. eq.(1)) are continuous, analytic functions with mutually IC periodicities, excitations polarized along c will propagate independently on each sublattice. Indeed inelastic neutron scattering data taken near a pair of strong Mn(Si) reflections reveal two longitudinal acoustic branches with different slopes[36].

In the long-wavelength limit, however, the microscopic incommensurability of the two sublattices become irrelevant and several coupling mechanisms arise[30,37]. As a result, a single combined acoustic mode should emerge. In addition, one expects to observe quasielastic scattering corresponding to the (diffusive) phase mode response. For unscreened ionic crystals, however, the longitudinal phase branch should present a gap, corresponding to the two-sublattice plasmon frequency. All these predictions remain to be tested experimentally.

III. CHARGE DENSITY WAVES

Fermi surface instabilities in low-dimensional metals provide a specific mechanism leading to modulated structures. The modulation in that case, results from the screening of the electronic charge density wave by the nuclei (the "$2k_f$" instability). It is essentially through the lattice modulation that the instability is observed in a diffraction experiment.

The IC character of the modulation is connected to that of the Fermi wavevector, the latter arising from non-stoichiometry, as in KCP, charge transfer as in TTF-TCNQ, or just energy band structure as in the transition-metal dichalcogenides. In the latter case the lock-in potential from the lattice is sufficient to eventually drive the CDW periodicity to a commensurate value[38]. This is not so in one-dimensional metals where commensurability can only be achieved by altering the value of the Fermi wavevector, i.e. by changing the conduction electron concentration.

The existence within the IC-state of a transition between the analytic, Peierls-conducting regime and the pinned insulating state has been predicted[39,40] and possibly observed in doped polyacetylene[41].

IV. INSULATORS

IV.1 General

In insulators the IC-state results either from a soft mode-type of instability (displacive case) of from a collective ordering mode (order-disorder case), both of general wavevector.

Within the soft mode picture, the IC ordering transition appears as a mere extension of the conventional ferro- and antiferrodistorsive transitions. While the occurrence of a local minimum around an arbitrary wavevector on a phonon dispersion curve suggests a competition between interactions of different characteristic ranges, it does not in general provide any insight into the nature of these interactions.

It is likely that in ionic crystals the competition involves Coulomb vs short-range forces but this applies equally well to most phase transitions in ionic crystals[42]. Furthermore, microscopic calculations are generally too complex to yield a simple physical picture of the IC instability[43].

In a non-polar molecular crystal such as biphenyl[44] the IC modulation is believed to arise from the competition between the intramolecular potential favouring a finite torsional angle between the two molecular phenyl rings (as in the free molecule) and intermolecular interactions which favour a co-planar conformation.

On a phenomenological level, incommensurability may sometimes be associated with the non-fulfillment of Lifshitz criterion[45]. For example, in non-symmorphic structures, most of the irreducible space-group representations corresponding to special zone-boundary wavevectors are pairwise degenerate and give rise to Lifshitz invariants. In that case a zone boundary phonon softening leads first to a modulated structure with a wavevector close to the high-symmetry zone-boundary point[46].

This type of argument, however, does not apply to IC phases with wavevectors close to the zone-center (and which eventually lock-in at $q = 0$, as encountered in $NaNO_2$ [47], quartz[48] and Thiourea[49], and alternative models based on two-mode interactions have been proposed[50-52].

IV.2 Thiourea

Thiourea crystallizes in the orthorhombic space group Pnma with four molecules per unit cell (see Fig. 3). Each molecule is planar and possesses a permanent dipole moment oriented in the (a,c) plane. The net dipole moment per cell vanishes due to the (n) and (a) glide planes.

Below room temperature, the softening of a polar optic mode (B_{3u}) is observed by infra-red spectroscopy[53] and neutron inelastic scattering[54]. The soft-mode dispersion minimum lies in the $\vec{b}^*$-direction where mixing with a TA branch of some symmetry ($\tau 4$) occurs. For $T < 220K$, a heavily damped coupled mode response is observed.

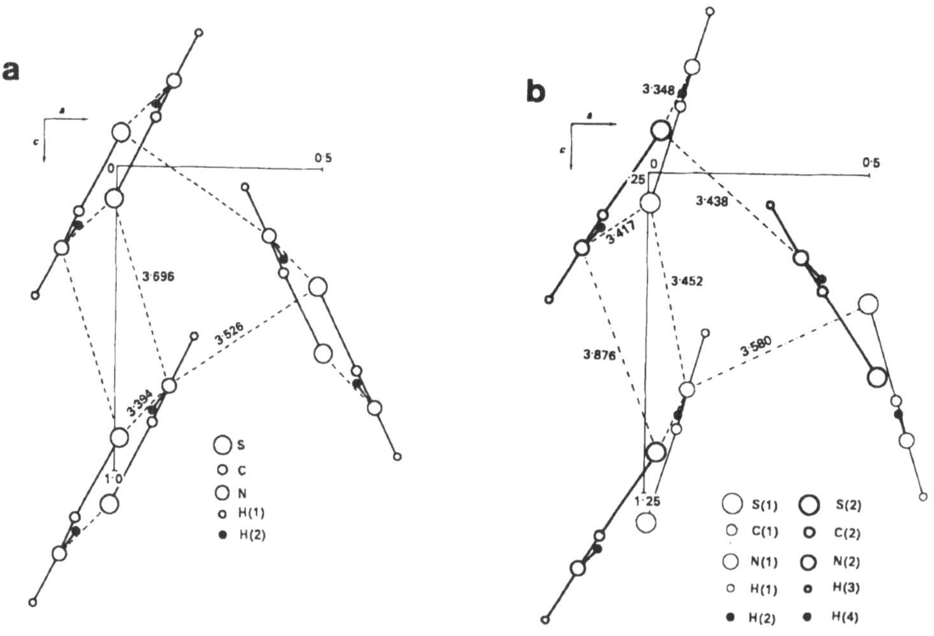

Fig. 3 : Structure of thiourea in the paraelectric (Pnma) phase (a) and ferroelectric ($P2_1ma$) phase (b). Molecules labelled (3) and (4) are displaced by b/2 with respect to (1) and (2). (after ref. 57).

Below T_i = 218K (all transition temperatures refer to the deuterated compound), satellite reflections appear, located at :

$$\vec{q}_\delta(T) = \delta(T)\vec{b}^* \qquad (T < T_i)$$

$\delta(T)$ decreases continuously from $\delta(T_i) = 0.141$ to $\delta(T_9) = 0.115$. Between T_9 = 193K and T_c = 191K, δ locks to the commensurate value 1/9. Below T_c, δ vanishes giving rise to a ferroelectric phase with spontaneous polarization along $\vec{a}(P2_1ma)$. Both lock-in transitions are distinctly first-order. A moderate (< 1K) wavevector hysteresis is observed in the IC phase.

On cooling from T_i, first, second and third order satellites appear successively. Only small intensity discontinuities are observed at T_9. Critical exponents for the first and second order satellite intensities are found to be in good agreement[55] with the critical behaviour of the d = 3 XY-model, including a uniaxial perturbation[56].

Fig. 4 shows the isothermal variation of δ with pressure for a few selected temperatures. Pressure hysteresis is within experimental error (± 20 bars). Note the additional step at δ = 1/7 for isotherms below 207K, and the characteristic shape of the $\delta(P)$ curve just above and below the 1/7 plateau. Continuous lock-in transitions

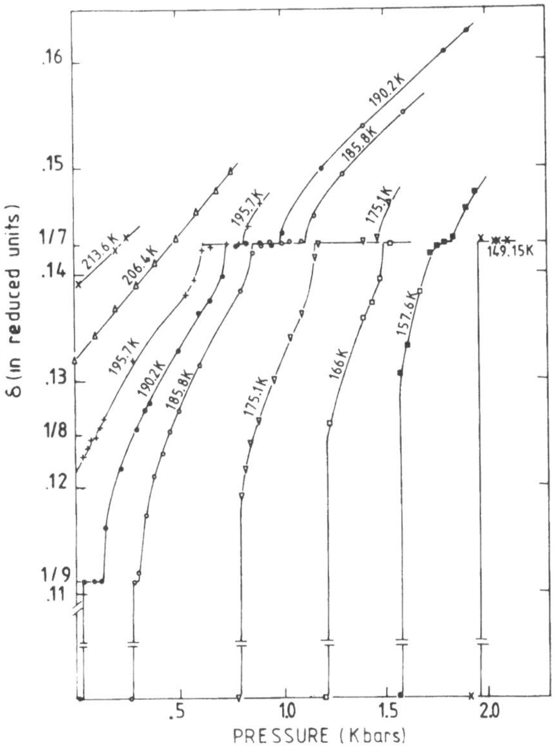

Figure 4 : Thiourea : isothermal variation of modulation wavevector δ with pressure (after ref. 49).

with logarithmic singularities have been predicted theoretically[10,8]. Note also the absence of a well-defined plateau on the 157.6K isotherm.

Results are summarized in the (P,T) phase diagram shown in Fig. 5. The values of δ along the $T_i(P)$ and $T_c(P)$ lines are shown as δ_i and δ_c, respectively. The negative sign of dT_i/dP and dT_c/dP is typical of a ferroelectric-type instability[42].

Single crystal diffraction patterns[54] indicate that the high-pressure $\delta = 1/3$ phase belongs to the same class of lock-in phases as the $\delta = 1/7$ and $1/9$ phases. Its persistence to temperatures much above the $T_i(P)$ line is, however, not understood.

As far as the lower part of the diagram is concerned ($P < 2$ kbars), the results may be formulated as follows :

i) a steady decrease of $\delta(T,P)$ with decreasing T and P is observed across the IC phase.

ii) when the "natural" value of $\delta(T,P)$ approaches a simple rational value such as 1/7 or 1/9, small lock-in free energy terms stabilize the corresponding C-phase over a narrow T or P range. Note the absence of a lock-in phase at 1/8 and the presence of two distinct 1/7 phases.

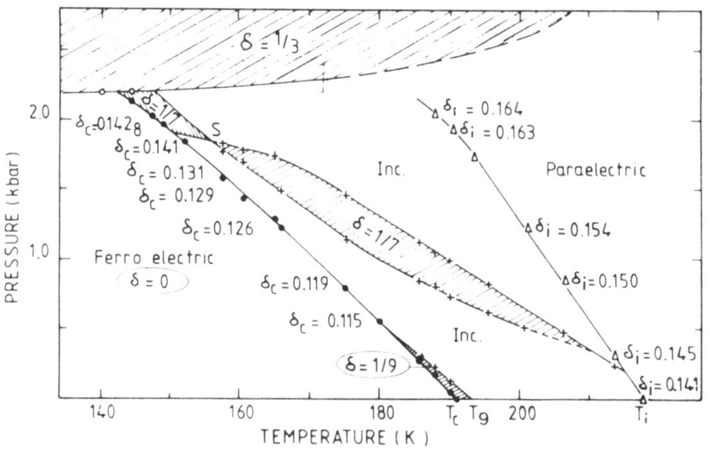

Figure 5 : Thiourea : (P,T) phase diagram (after ref. 49).

Point i) above may be accounted for in terms of a phenomenological Landau-Ginzburg free-energy of the type[58,59] :

$$F = \int \left\{ \frac{1}{2} A_o P_x^2 + \frac{1}{4} BP_x^4 + \frac{1}{2}\alpha\left(\frac{dP_x}{dy}\right)^2 + \frac{1}{4}\beta\left(\frac{d^2P_x}{dy^2}\right)^2 + \gamma\left(P_x \frac{dP_x}{dy}\right)^2 \right\} dy \tag{8}$$

with : $A_o = a(T-T_o)$; $B > 0$; $\alpha < 0$; $\beta > 0$; $\gamma > 0$

Approximating the polarization wave $P_x(y)$ by a sine-wave :

$$P_x(y) = P_o + 2P_\delta \cos(2\pi\ \delta y) \tag{9}$$

expression (8) yields :

$$F = \frac{1}{2} A_o P_o^2 + \frac{1}{4} BP_o^4 + A_\delta P_\delta^2 + \frac{3}{2} B_\delta P_\delta^4 + \frac{3}{2}(B+B_\delta)P_o^2 P_\delta^2 \tag{10}$$

where : $A_\delta = A_o + \alpha\delta^2 + \frac{1}{2}\beta\delta^4$

and : $B_\delta = B + \frac{4}{3}\gamma\delta^2$.

Model (10) gives a second-order phase transitions (at $T = T_i > T_o$), from the paraelectric state ($P_\delta = P_o = 0$) to the IC-state ($P_o = 0$; $P_\delta \neq 0$) and a first-order transition (at $T = T_c < T_o$) from the IC-state to the ferroelectric state ($P_o \neq 0$; $P_\delta = 0$). Minimization of (10) with respect to δ, leads to the required variation of δ across the IC-phase. Presumably the pressure dependence of the various parameters in the model can be adjusted in such a way as to reproduce the observed phase diagram. The inclusion of higher-order harmonics in (9) will simply enhance the T-dependence of δ.

In order to discuss point ii) it is necessary to introduce microscopic order parameters and make use of their symmetry properties. Let $Q(q_\delta,\tau_4)$ denote the complex order parameter corresponding to the first Fourier component of the modulation. A commensurate phase with $\delta = 1/n$ is stabilized by an Umklapp free energy term of the form :

$$\Delta F_n = V^{(n)}(\vec{q}_\delta,\tau_4)\ Q^n(\vec{q}_\delta,\tau_4) + \text{c.c.} \tag{11}$$

$V^{(n)}$ is an appropriate anharmonic coefficient whose symmetry properties are determined by the space group of the paraelectric phase. In particular $V^{(n)}$ vanishes unless:

$$n\vec{q}_\delta = m\vec{b}^* \quad \text{i.e.} \quad \delta = \frac{m}{n} \quad (m,n = \text{integers})$$

Furthermore point group selection rules require m and n to be of same parity. Hence the 1/7 and 1/9 phases correspond to lock-in terms of order 7 and 9, while the 1/8 phase would require a lock-in term of order 16, which is too weak to be effective. It has been predicted[49], however, and confirmed experimentally[60,61] that the 1/8-phase could be stabilized by a uniform electric field applied along $\vec{a}$. The corresponding lock-in term is of the form :

$$\propto Q^8(\vec{q}_\delta,\tau_4)\; Q(0,\tau_4)$$

where $Q(0,\tau_4)$ is induced by the applied field.

The stability range of the 1/7-phase appears to vanish in the vicinity of the point labelled S in fig. 5. In fact, satellite intensity patterns recorded on either side of point S suggest two distinct 1/7-phases. This observation can be accounted for if one assumes that the pressure and temperature dependence of the $V^{(7)}$ Umklapp coefficient is such that it vanishes on a curve in the (P,T) plane. Point S would then be at the intersection of the two curves defined by :

$$V^{(7)}(P,T) = 0$$

$$\tilde{\delta}(P,T) = 1/7$$

where $\tilde{\delta}(P,T)$ is the natural value of δ, i.e. the value of δ which minimizes the free energy in the absence of Umklapp lock-in terms. Introducing real variables :

$$V^{(7)}(\vec{q}_{1/7},\tau_4) = iv(T,P) \quad ; \quad Q(\vec{q}_{1/7},\tau_4) = \eta e^{i\phi}$$

(11) becomes :

$$\Delta F_7 = v\eta^7 \sin 7\phi$$

Thus the two 1/7 phases are characterized by opposite values of the phase

$$\phi = \pm \frac{\pi}{14} \; (\text{mod. } \frac{2\pi}{7})$$

or, equivalently, by opposite signs of the atomic displacements $\vec{u}_{\ell k}$:

$$\vec{u}_{\ell k} = (m_k)^{-1/2}\, \eta\, \vec{w}_k(\vec{q}_{1/7},\tau_4)\, e^{i(\vec{q}_{1/7}\vec{r}_{\ell k}+\phi)} + \text{c.c.}$$

The two phases together with the IC phase would then coexist at S.

In fact, in a very narrow T and P range around S the contribution from the next Umklapp term :

$$V^{(14)}(\vec{q}_{1/7},\tau_4)\ Q^{14}(\vec{q}_{1/7},\tau_4) \equiv w\,\eta^{14}\ \cos 14\phi \qquad (12)$$

should be included into ΔF_7. Depending upon the sign of w, two different situations may arise, but in either case the isolated character of point S is removed.

Finally we wish to stress that although expressions such as (11) are written in term of the primary Fourier component of the modulation alone, this does not preclude the existence of higher order distortion harmonics, which do contribute additional lock-in free-energy terms. Formally one assumes that the minimization of the free-energy with respect to higher-order harmonics has already been performed and that the $V^{(n)}$'s have been renormalized accordingly.

This formulation, however, is not adequate for a quantitative discussion of the stability of the IC phase with respect to the higher-order C phases since it does not allow for the formation of phase distortions in the IC wave, corresponding to discommensurations with respect to the high-order C phase. These DC's, however, will presumably be broad on an atomic scale due to the small amplitude of the lock-in term and apart from making the lock-in transitions continuous, no qualitatively new feature will be introduced.

In the same context, it is of interest to determine the harmonics content of the modulation in the C phases. Hence a detailed structural study of the 1/9 phase has been initiated and is currently in progress[62]. Preliminary results indicate that the simple ferroelectric-domain model (i.e. a square-wave with 9 planes polarized up and 9 planes polarized down) is not realistic. The harmonic content of the modulation appears to be considerably lower than predicted by the above model. If fomain walls are present some degree of disorder must also be present, in such a way as to smooth out the average wave profile. This result is not unexpected since the DC picture is not applicable to the case $\delta = 0$[63].

IV.3 TMATC-Zn

Tetramethylammonium tetrachlorozincate (TMATC-Zn) belongs to a class of compounds of formula $(N(CH_3)_4)_2\ XC\ell_4$ (X = Zn Co, Mn, Fe, etc.), isomorphous to $Rb_2ZnC\ell_4$ and K_2SeO_4 in their undistorted phase (Pnma). Below room temperature these compounds undergo various sequences of phase transitions involving IC, C, and quasi-C modulated structures, with modulation wavevector :

$$\vec{q}_{\delta} = \delta(T)\vec{a}^*$$

Fig. 6 shows the $\delta(T)$ curve measured on deuterated TMATC-Zn, using neutron diffraction[64]. In addition to pronounced thermal hysteresis, the experimental curve exhibits a quasi-C plateau prior to the onset of the 3/7-phase. A similar effect has been observed in $Ba_2NaNb_5O_{15}$, where the C phase is in fact never reached[65]. It has been ascribed to the stabilization by impurities of phase defects in the C wave[66].

In a later (cooling) run, using the same sample the 1/2 plateau was found to be much reduced (points marked ⋆ in fig. 6). In other samples the same plateau appears only under stress[67].

Different behaviours are observed on deuterated and non-deuterated samples[68]. In particular non-deuterated samples exhibit a plateau at 2/5, which is also observed on the deuterated compound under pressure (a few hundred bars).

Clearly the balance of interactions which determines the value of the modulation wavevector, is affected by differences in crystal

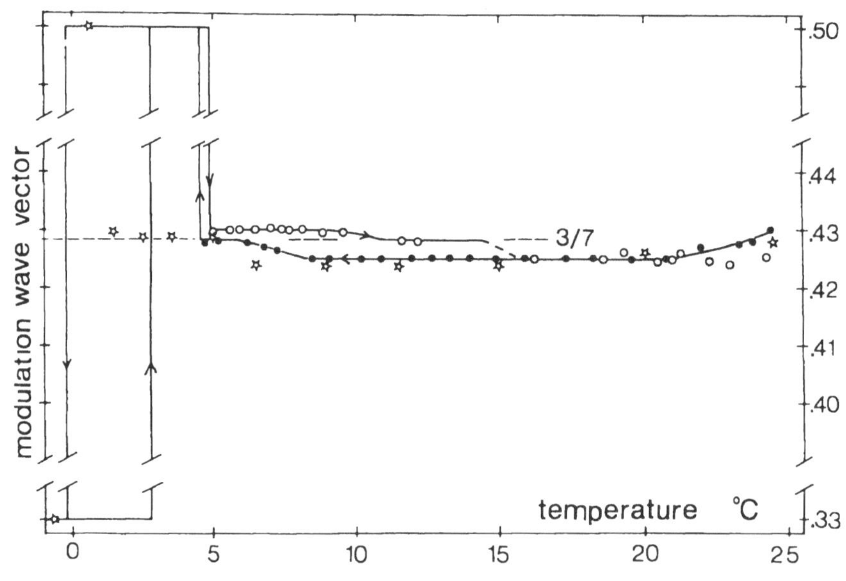

Figure 6 : TMATC-Zn : variation of modulation wavevector with decreasing (closed circles) and increasing temperature (open circles). The data points marked as ⋆ correspond to a later cooling run. (deuterated compound; from ref. 64).

quality (residual strains, impurities) and by thermal history. The corresponding effects in thiourea are either absent or much attenuated.

Another qualitative difference between the two compounds is the non-monotonic variation of $\delta(T)$ as shown in fig. 6. Aubry[1] has shown in the context of the discrete Frenkel-Kontorova model, that the average lattice spacing misfit ℓ of the adsorbed chain (here the modulation periodicity) needs not vary monotonically with the substrate potential λ (here the temperature). He also points out that very small changes in the chain tension μ, the other parameter in the model, can produce large changes in the $\ell(\lambda)$ curve. Fig. 7 shows two such curves calculated for slightly different values of μ : one shows a monotonous decrease of $\ell/2a$ from 2/5 to zero while the other exhibits a plateau at a maximum value of 1/2. The absence of an IC phase in the low-λ part of the calculated curves is an artefact of the model, related to the particular form of substrate potential used.

In summary, it seems that the complex behaviour shown by the present compound and by others in the same class, cannot be under-

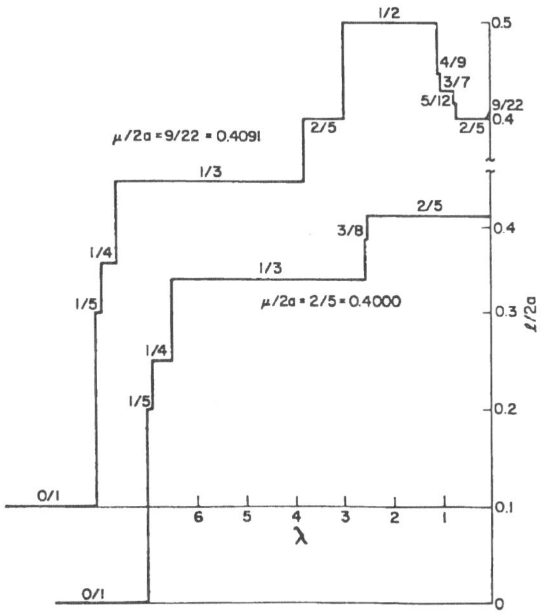

Figure 7 : Discrete Frenkel-Kontorowa model : Average lattice-spacing misfit versus substrate potential for two values of the chain tension (from ref. 74).

stood, even qualitatively, without taking explicit account of the discrete nature of the modulated lattice.

V. A MAGNETIC EXAMPLE : CERIUM ANTIMONIDE

Below the magnetic ordering temperature (T_N = 16.2K) CeSb displays a particularly complex behaviour[69] as evidenced by the many specific heat anomalies observed[70] (cf. fig. 8). The complete (H,T) phase diagram is shown in fig. 9 for H along <001> . The magnetic structures of the various phases observed are all based on a periodic stacking sequence of non-magnetic and ferromagnetic (001) planes.

The three end-structures are :

- the paramagnetic (P) phase at high T
- the ferromagnetic (F) phase at high H and low T
- the antiferromagnetic (AF) phase at low H and low T.

The other phases can be regarded as intermediate between two of the above three.

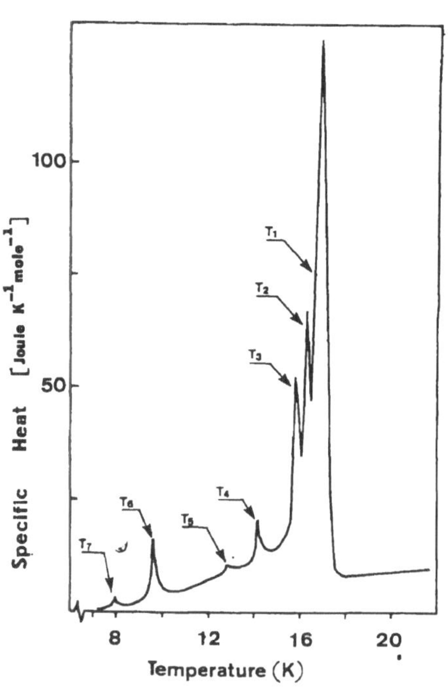

Figure 8 : Specific heat of CeSb in zero applied magnetic field (from ref. 70).

The structure of the AF phase consists in a sequence of two pairs of Ce^{3+} layers with opposite magnetization, i.e. (++--). This sequence can be described by a square-wave of wavevector $k = (0, 0, 1/2)\frac{2\pi}{a}$ in the f.c.c. Brillouin zone (interlayer spacing = a/2).The structures labelled AFF2 and AFF1 are obtained from AF by removing one down layer (-) every 7th and 3rd layer, respectively, i.e. (++--++-) and (++-). The corresponding square-wave wavevectors are k = 4/7 and 2/3.

Similarly the AFP phases are generated from the AF phase by introducing one non-magnetic plane every 2n-1 layer. The resulting wavevector is of the form k = n/2n-1 with n = 2,4,5,6,...∞ (an exception to the rule is the AFP2 phase of wavevector k = 8/13). In all cases no net magnetization is introduced.

Finally between the F and P phases the substitution involves non-magnetic and ferromagnetic up (+) planes. The corresponding structures can be visualized as resulting from the superposition of a ferromagnetic component and of a square-wave of wavevector n/2n+1 (FP3, FP4) or n/2n-1 (FP1).

The overall shape of the phase diagram is well accounted for in terms of a simple thermodynamic model in which each Ce^{3+} ion is assumed to contribute an entropy of $k\ell n2$ in the non-magnetic

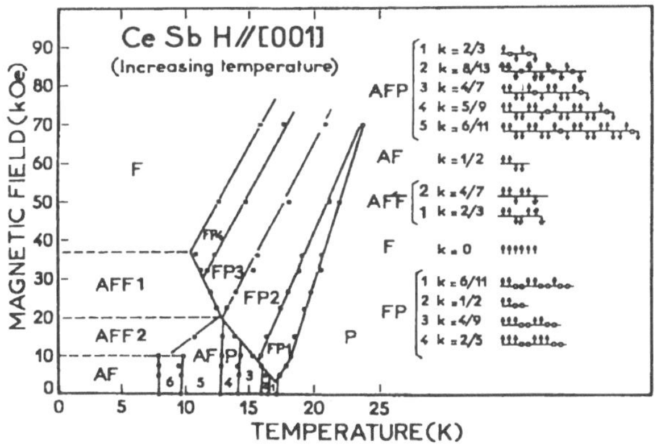

Figure 9 : (H,T) phase diagram of CeSb as deduced from specific heat anomalies (from ref. 70).

state and a moment of 2.I μ_B in the magnetic state. For example the transitions AF → AFF and AFF → F involve a change in magnetization and no change in entropy ($\Delta M \neq 0$; $\Delta S = 0$). Hence from Clapeyron's relation :

$$\frac{\partial H}{\partial T} = - \frac{\Delta S}{\Delta M} = 0$$

in agreement with fig. 9. Similarly the AF → AFP → P transitions correspond to $\Delta S \neq 0$ and $\Delta M = 0$ and thus should be field-independent, while the P → FP → F ones involve changes in both S and M and thus should have a finite slope in the (H,T) diagram.

The physical origin of the non-magnetic planes has not been fully clarified, so far. The possibility of two different electronic groundstates for the magnetic and non-magnetic Ce^{3+} ions has been considered. A related problem is the large anisotropy of the exchange interactions between in-plane (J_0) and out-of-plane (J_1,J_2) Ce^{3+} neighbours. The stability of the AF (++--) structure at T = H = 0, suggests $J_1 < 0$,$J_2 < 0$ and $J_0 >> |J_1|,|J_2|$ which is somewhat surprising in view of the cubic symmetry of the structure for $T > T_N$.

On the other hand, the large exchange anisotropy makes CeSb an attractive model system and several attempts have been made at explaining the main features of the experimental phase diagram, particularly the H = 0 AF → P sequence, within the ANNNI model[71,72,73]. Noting that this sequence corresponds to introducing an increasing proportion of non-magnetic defects, Villain and Gordon[73] argue that the true devil's staircase behaviour will only be realized if the effective interaction between defects is repulsive at all distances. This condition imposes constraints on the relative values of J_1 and J_2 which are not met in the case of CeSb. Alternatively[8] one may argue that the missing intermediate phases are suppressed by the metastability of some of the adjacent ones. It has been noted that some samples miss more steps than others which implies that at least in some cases the number of observed phases is limited by (extrinsic) pinning mechanism.

In any case, it is clear that the experimental limit has not yet been reached, in the sense that finer steps could still be resolved, if present. Until then, the search for the ideal devil's staircase system is likely to continue.

ACKNOWLEDGMENTS

I am grateful to R. Almairac, G. Marion, J.D. Axe, R.A. Cowley and S. Aubry for cumminication of results prior to publication. Useful discussions with F. Denoyer, L. Bernard, C. Vettier, J. Rossat-Mignod and J. Villain are acknowledged.

REFERENCES

1. S. Aubry, J. Physique 44:147 (1983).
2. P.M. de Wolff, Acta Cryst. A30:777 (1974).
3. A. Janner and T. Janssen, Phys. Rev. B15:643 (1977), Acta Cryst. A36:399 (1980); Acta Cryst. A36:408 (1980).
4. P. Bak, Phys. Rev. Lett. 46:791 (1981) and this volume.
5. Y. Imry and S.K. Ma, Phys. Rev. Lett. 35:1399 (1975).
6. L.J. Sham and B.R. Patton, Phys. Rev. B13:3151 (1976).
7. M. Nielsen, J. Bohr, K. Kjaer and J.P. McTague, Inst. Phys. Conf. Ser. 64 p. 289 (Schofield P. Editor) (1982).
8. P. Bak, Rep. Prog. Phys. 45:587 (1982).
9. W.L. McMillan, Phys. Rev. B14:1496 (1976).
10. S. Aubry, in "Solitons in Condensed Matter Physics", ed. by A.R. Bishop and T. Schneider (Springer, Berlin)(1978).
11. H. Mashiyama, S. Tanisaki and K. Hamano. J. Phys. Soc. Jpn.(1982) 51:2538; K. Hamano, Y. Ikeda, T. Fujimoto, K. Ema and S. Hirotsu, J. Phys. Soc. Jpn 49:2278 (1980).
12. H.G. Unruh. To appear in J. Phys. C. (1983).
13. J.P. Jamet and P. Lederer, to appear in J. Phys. (Paris)(1983).
14. G. Errandonea, this volume.
15. A.W. Overhauser, Phys. Rev. B3:3173 (1971).
16. A.D. Bruce and R.A. Cowley, J. Phys. C 11:3609 (1978).
17. R. Zeyher and W. Finger, Phys. Rev. Lett. 49:1833 (1982).
18. V.A. Golovko and A.P. Levanyuk, Sov. Phys. JETP 54:1217 (1981).
19. H. Cailleau, F. Moussa, C.M.E. Zeyen and J. Bouillot, Solid State Comm. 33:407 (1980).
20. L. Bernard, R. Currat, P. Delamoye, C.M.E. Zeyen, S. Hubert and R. de Kouchkovsky, J. Phys. C 16:433 (1983); L. Bernard, R. Currat and P. Delamoye, to be published.
21. R.M. Fleming, D.E. Moncton, D.B. McWhan and F.J. DiSalvo, Phys. Rev. Lett. 45:576 (1980).
22. D.E. Cox, S.M. Shapiro, R.J. Nelmes, T.W. Ryan, H.J. Bleif, R.A. Cowley, M. Eibschütz and H.J.Guggenheim, preprint (1983).
23. K.K. Fung, S. McKernan, J.W. Steeds and J.A. Wilson, J. Phys. C 14:5417 (1981).
24. C.H. Chen, J.M. Gibson and R.M. Fleming, Phys. Rve. Lett. 47:723 (1981).
25. R. Blinc, Phys. Rep. 79:331 (1981).
26. P.D. Delamoye and R. Currat, J. Physique Lett. 43:L-655 (1982).
27. R. Blinc, D.C. Ailion, P. Prelovsek and V. Rutar, Phys. Rev. Lett. 50:67 (1983).
28. H. Nowotny. The Chemistry of Extended Defects in Non-Metallic Solids, ed. Le Roy, Eyring and M.O'Keefe, North-Holland Pub. Co., Amsterdam (1970).
29. G. Theodorou and T.M. Rice, Phys. Rev. B18:2840 (1978).
30. J.D. Axe and P. Bak, Phys. Rev.B 26:4963 (1982).
31. W. Jeitschko and E. Parthé, Acta Cryst. 22:417 (1967).
32. G.H. Flicher, H. Völlenkle and H. Nowotny, Mh.Chem.98:2173 (1967).
33. H.W. Knott, M.H. Mueller and L. Heaton, Acta Cryst. 23:549 (1967).
34. R. de Ridder and S. Amelinckx, Mat. Res. Bull. 6:1223 (1971).

35. R. de Ridder, G. Van Tendeloo and S. Amelinckx, Phys. St. Sol. (a) 33:383 (1976).
36. J.D. Axe (private communication).
37. W. Finger and T.M. Rice, Phys. Rev. Lett. 49:468 (1982).
38. For a review on 2H-$TaSe_2$ see J.D. Axe (1982).
39. P. Bak and V.L. Pokrovsky, Phys. Rev. Lett. 47:958 (1981).
40. P.Y. Le Daeron and S. Aubry, Metal Insulator Transition in the Peierls Chain, submitted to J. Phys. C. (1982).
41. C.K. Chiang, C.R. Fincher, Y.W. Park, A.J. Heeger, H. Shirikawa, E.J. Louis, S.C. Gau and A.G. MacDiarmid, Phys. Rev. Lett. 39:1098 (1977).
42. G.A. Samara, Comments Sol. St. Phys. 8:13 (1977).
43. M.S. Haque and J.R. Hardy, Phys. Rev. B 21:245 (1980).
44. H. Cailleau, F. Moussa and J. Mons, Sol. St. Comm. 31:521 (1979).
45. See A. Michelson, Phys. Rev. B 18:459 (1978) for a discussion of the applicability of Lifshitz criterion to IC phase transitions.
46. M. Iizumi and K. Gesi, Sol. St. Comm. 22:37 (1977).
47. D. Durand, F. Denoyer, M. Lambert, L. Bernard and R. Currat, J. Physique 43:149 (1982); and ref. therein.
48. G. Dolino, J.P. Bachheimer and C.M.E. Zeyen, Sol. St. Comm. 45:295 (1983).
49. F. Denoyer, A.H. Moudden, R. Currat, C. Vettier, A. Bellamy and M. Lambert, Phys. Rev. B 25:1697 (1982).
50. A.P. Levanyuk and D.G. Sannikov, Sov. Phys. Sol. St. 18:1122 (1976).
51. T.A. Aslanian and A.P. Levanyuk, Sol. St. Comm. 31:547 (1979).
52. V. Heine and J.D.C. McConnell, Phys. Rev. Lett. 46:1092 (1981).
53. F. Brehat, J. Claudel, P. Strimer and A. Hadni, J. Physique Lettres. 37:L229 (1976).
54. A.H. Moudden, Thesis, Université Paris-Sud (1980) unpublished.
55. A.H. Moudden and J. Als-Nielsen (unpublished).
56. R.A. Cowley and A.D. Bruce, J. Phys. C 11:3577 (1978).
57. M. Elcombe and J.C. Taylor, Acta Cryst. A24:410 (1968).
58. Y. Ishibashi and H. Shiba, J. Phys. Soc. Jap. 45:409 (1978).
59. P. Lederer and C.M. Chaves, J. Physique Lett. 42:L127 (1981).
60. J.P. Jamet, P. Lederer and A.H. Moudden, Phys. Rev. Lett. 48:442 (1982).
61. A.H. Moudden, E.C. Svensson and G. Shirane, Phys. Rev. Lett. 49:557 (1982).
62. Simonson, F. Denoyer, C. Vettier and R. Currat (to be published).
63. A.D. Bruce, R.A. Cowley and A.F. Murray, J. Phys. C 11:3591 (1978).
64. G. Marion, R. Almairac, J. Lefebvre and M. Ribet, J. Phys. C. 14:3177 (1981).
65. J. Schneck, J.C. Toledano, C. Joffrin, J. Aubree, B. Joukoff and A. Gabelotaud, Phys. Rev. B 25:1766 (1982).
66. K. Nakanishi, J. Phys. Soc. Jpn. 46:1434 (1979).
67. H. Mashiyama, S. Tanisaki and K. Gesi, J. Phys. Soc. Jpn. 50:1415 (1981).
68. G. Marion (private communication).

69. J. Rossat-Mignod, P. Burlet, J. Villain, H. Bartholin, Wang Tcheng-Si, D. Florence and O. Vogt, Phys. Rev. B 16:440 (1977).
70. J. Rossat-Mignod, P. Burlet, H. Bartholin, O. Vogt and R. Lagnier, J. Phys. C 13:6381 (1980).
71. P. Bak and J. von Boehm, Phys. Rev. B 21:5297 (1980).
72. W. Selke and M.E. Fisher, Phys. Rev. B 20:257 (1979).
73. J. Villain and M. Gordon, J. Phys. C. 13:3117 (1980).
74. S. Aubry, to appear in J. Phys. C. (1983).

NATURE OF THE INCOMMENSURATE PHASE OF $Rb_2ZnC\ell_4$

S.R. Andrews

Department of Physics
University of Edinburgh
Mayfield Road
Edinburgh, EH9 3JZ
Scotland, UK

H. Mashiyama

Department of Physics
Yamaguchi University
Yamaguchi 753
Japan

$Rb_2ZnC\ell_4$ is one of the many numbers of isomorphous compounds of the form A_2BX_4 (A = K,Rb etc., B = Se, Zn etc., X = O, Cℓ, Br etc.) which undergo a continuous transition from a normal (N) phase (space group Pmcn) at a temperature T_c to a one dimensionally modulated incommensurate (IC) phase (see e.g. Iizumi et al 1977, de Pater and van Dijk 1978) characterised by a distortion with a periodicity that is an irrational fraction of the periodicity of the underlying lattice. The modulation wavevector of the incommensurate distortion, referred to the unit cell of the N phase, is given by $\underline{q}_s = (\frac{1}{3} - \delta)\underline{C}^*$ where c is the pseudo-hexagonal axis. In $Rb_2ZnC\ell_4$ $\delta \sim .029$ just below T_c ~303 K (Mashiyama et al 1982) decreasing monotonically to zero at $T_L \sim 190$ K where the IC phase locks into a commensurate (C) ferroelectric phase ($P2_1cn$).

The theory of the N-IC transition, the evolution of the IC phase with decreasing temperature and of the IC-C transition reveals a number of unusual features which have attracted much recent interest (Cowley 1980, Currat: this volume). At T_c the $\pm\underline{q}_s$ soft modes in the N phase become coupled and the excitations of the IC phase include two branches, one corresponding to oscillations in the amplitude and the other to oscillations in the phase of the distortion. The amplitude mode has a frequency spectrum with a gap which increases as $(T_C - T)^{1/2}$ in mean field theory whilst the phason is gapless, if unpinned by impurities etc., reflecting the fact that a uniform shift in ϕ throughout the crystal does not change the energy of the system and the phason has a linear, only weakly temperature dependent dispersion relation. In a refined Landau theory (McMillan 1976, Bak and Emery 1976), in which ϕ is allowed to vary with position, a simple plane wave modulation will break up into almost commensurate regions separated by equally spaced domain walls (phase solitons) where, in $Rb_2ZnC\ell_4$, the phase changes through an angle $2\pi/3$. The lock-in transition can be viewed as a soft mode transition in which the order parameter is the number of domain walls.

X-ray scattering experiments have been performed in both the N and IC phases of $Rb_2ZnC\ell_4$ and $(Rb_{.995}K_{.005})_2ZnC\ell_4$. The inclusion of K ions as impurities has two effects at the N-IC transition. Firstly, the neighbouring ions are displaced, producing a random pinning field on the phase mode which might be expected to destroy the long range order (Sham and Patton 1976). Secondly, the impurities produce changes in the force constants which would alter T_C and possibly other properties of the transition if it occurs. Experimentally the transition is observed to be sharp and the width of the Bragg satellites arising from the IC distortion (303.3 K in $Rb_2ZnC\ell_4$, 304.8 K in $(Rb_{.995}K_{.005})_2ZnC\ell_4$) is instrument limited just below T_C in both the pure and impure crystals. This suggests that effectively long range order (>5000 Å) develops in both crystals and that the effect of the random fields is small. In such a case Harris (1974) has shown that systems with random force constants are expected to have the same exponents as those of the analogous pure system provided that the specific heat exponent $\alpha < 0$, as it is for the d = 3 XY model which is expected to describe the N-IC transition. This is in good agreement with our measurements of the static critical exponents β, γ and ν (Table 1) carried out using X-rays (Mashiyama 1981, Mashiyama and Andrews 1983, Andrews and Mashiyama 1983).

Table 1

		β	γ	ν
Experiment	pure	.345(.010)[1]	1.26(.04)	.693(.015)
$RB_2ZnC\ell_4$	mixed	.350(.010)	1.31(.05)	.683(.015)
Theory[2]	d=y XY model	.345(.002)	1.316(.003)	.669(.003)

(1) Mashiyama 1981

(2) Le Guillou and Zinn-Justin 1981

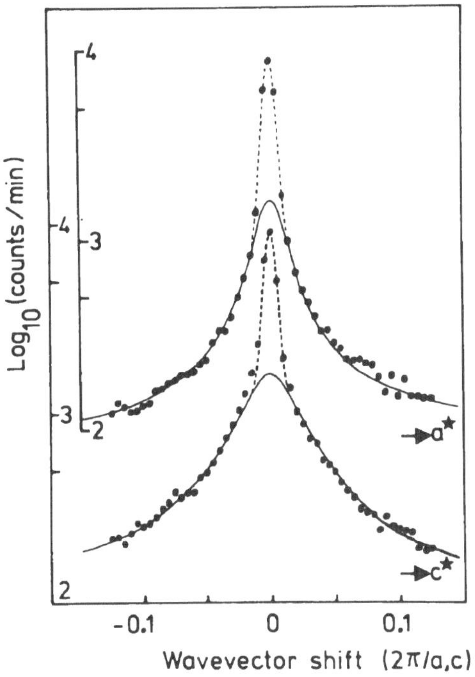

Fig. 1. Scattering around (6, 0, $1-q_s$) showing the diffuse (solid curve) and Bragg (dashed curve) contributions discussed in the text.

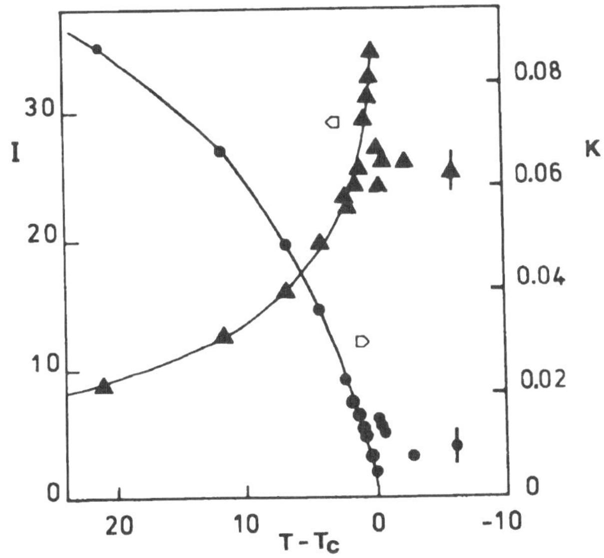

Fig. 2. Measured integrated intensity in arbitrary units (triangles) and deconvolved width (circles) in units of C* of diffuse scattering around (6, 0, $1-q_s$).

Just below T_c the satellite Bragg peaks arising from the IC distortions are superimposed on diffuse scattering (fig. 1) which we attribute to the phason and whose width and integrated intensity assuming a Lorentzian scattering form are shown for $Rb_2ZnC\ell_4$ in fig. 2 as a function of temperature. For temperatures lower than only a few degrees below T_c the diffuse scattering arises largely from the acoustic modes and the intensity and widths of the residual diffuse scattering cannot be determined. Figure 2 shows that the behaviour of the scattering is very different from that expected from systems without a broken continuous symmetry. The intensity does not decrease as rapidly with $|t|$ below T_c as it does for t greater than T_c and the width remains very narrow below T_c. This behaviour is characteristic of systems with a continuous symmetry which is broken by the transition and in this case is direct evidence for the existence of nearly gapless phase modes below T_c. The choice of a Lorentizan scattering function with its implicit small gap ($\lesssim$10 GHz) gave a better fit to the data than a function of the form.

$$I(q) = k_B T(f(q_x^2 + q_y^2) + gq_z^2)^{-1}$$

expected for a gapless phason. This is possibly because the contribution of the amplitude mode was omitted and possibly became the deconvolution of the experimental measurement may not be adequate.

Fig. 3 shows the scattering observed at three different temperatures, one of which is below T_L and the others above T_L. It consists of various orders of satellite reflections superimposed on a large sloping background due to acoustic mode scattering emanating from the primary satellite. The acoustic mode scattering is asymmetric in wavevector shift q from the primary satellite above T_L but not below, the asymmetry being largely independent of temperature between $T_c - 10$ and $T_c - 100$ K. We believe that the origin of the asymmetry is the interference between scattering from the phase mode and the transverse acoustic mode propagating parallel to c and polarised along a, although we have been unable to account in detail for the origin of a coupling constant proportional to $A_0 q^2$, where A_0 is the amplitude of the distortion, a choice which most satisfactorily fits the data.

The temperature dependence of several orders of satellite reflection at wavevector $(6, 0, p + nq_s)$ with $|n| = 1$-5 are shown in figs. 4a,b. As T_L is approached the satellite integrated intensities increase with the exponents given in table 2. Neither the high intensities nor the large values of the exponents can be understood on the basis of the satellites being diffraction harmonics of a simple plane wave distortion. More probably, a description in terms of a broad soliton model is appropriate. We see no indication of a narrow soliton regime observed in NMR experiments within a few

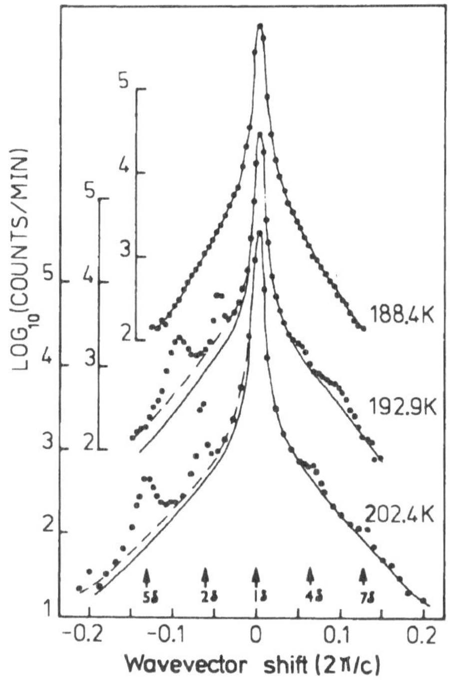

Fig. 3. Scattering along C* about $(6, 0, 1-q_s)$ close to $T_c \sim 190$ K. The solid curves which are symmetrical about zero frequency shift, are guides to the eye. Various orders of satellite reflection are indicated by arrows.

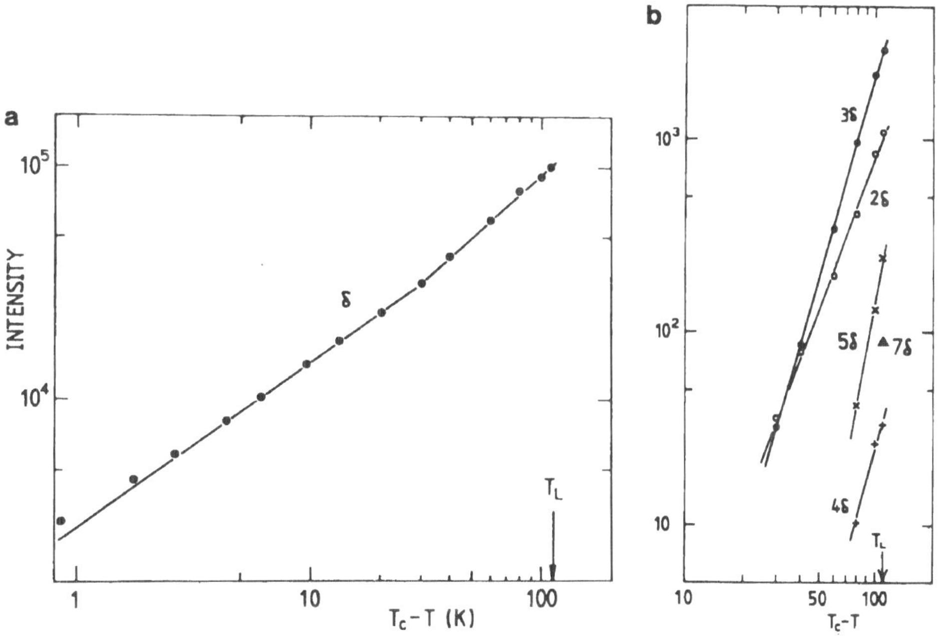

Fig. 4. Integrated intensity of a) the primary and b) several higher order satellites

Table 2

Scattering intensities and effective exponents of satellites at $(6, 0, p + nq_s)$

n	p	I_n/I_1 (193 K)	$d(\log I_n)/d(\log(T_c - T))$ $(30 < T_c - T < 100)$
-1	1	1	0.92 (.05)
2	0	10^{-2}	2.7 (.1)
-3	3	3.10^{-2}	3.6 (.1)
-4	2	5.10^{-5}	3.9 (.5)
5	-1	3.10^{-4}	5.8 (1)
-6	2	$\lesssim 10^{-5}$	-
7	3	6.10^{-5}	-

degrees of T_L (Blinc et al. 1981). In mean field theory and neglecting Debye-Waller factors we would expect the satellite intensities to be largely independent of temperature in such a limit (Cowley 1980).

REFERENCES

Andrews, S.R. and Mashiyama, H., 1983, to be published in J. Phys. C.

Bak, P. and Emery, V.J., 1976, Phys, Rev. Lett. 36, 978.

Blinc, R., Rutar, V., Topic, B., Milia, F., Alexandrova, I.P., Chaves, A.S. and Gazzinelli, R., 1981, Phys. Rev. Lett. 46, 1406.

Cowley, R.A., 1980, Adv. in Physics, 29, 1.

Harris, A.B., 1974, J. Phys. C 7, 1671.

Iizumi, M., Axe, J.D., Shirane, G. and Shimaoka, K., 1977, Phys. Rev. B15, 4382.

Le Guillou, J.C., and Zinn-Justin, J., 1980, Phys. Rev. B21, 3976.

Mashiyama, H., 1981, J. Phys, Soc. Jpn., 50, 2655.

Mashiyama, H. and Andrews, S.R., 1983, J. Phys. C, L247.

McMillan, W.L., 1976, Phys, Rev. B14, 1496.

de Pater, C.J. and van Dijk, C., 1978, Phys. Rev. B19, 4684.

Sham, L.J., and Patton, B.R., 1976, Phys. Rev. B13, 3151.

NEW PHENOMENA IN INCOMMENSURATE BARIUM SODIUM NIOBATE :

MEMORY AND RELAXATION EFFECTS

Gilles Errandonéa, Arlette Litzler, Hervé Savary, Jean-Claude Tolédano, Jacques Schneck, and Jack Aubrée

Centre National d'Etudes des Télécommunications
196 rue de Paris, 92220 Bagneux (France)

We report the observation of two new effects in the incommensurate phase of Barium Sodium Niobate : a relaxation and a memory effect. These very large effects are assumed to be induced by a straong interaction between the incommensurate modulation and slightly mobile defects.

INTRODUCTION

Within the last few years it has become apparent that several effects observed in incommensurate materials cannot be accounted for by phenomenological[1,2] theories of "pure systems" but must be attributed to the presence of defects : thermal hysteresis[3,4,5], suppression of the lock-in transition [6], memory effect[7]...

Barium Sodium Niobate (BSN), $Ba_2NaNb_5O_{15}$, is a very interesting material to study the interaction between defects and modulation since neutrons, X-rays and birefringence measurements[8,9] have shown in it the existence of a large thermal hysteresis ($\approx$ 30°C), one or two orders of magnitude greater than in the most of incommensurate materials. BSN undergoes at $T_I \approx 300°C$ a transition from a tetragonal 4mm crystalline phase towards an incommensurate phase, the modulation wavevector being :

$\vec{k} = (a^*+b^*)(1+\delta)/4 + c^*/2$ with δ decreasing from 0.12 to 0.08, and an incomplete lock-in transition at $T_L \approx 270°C$ towards a nearly commensurate phase ($\delta \approx 0.01$).

The onset of the modulation along a [110] direction below T_I breaks the tetragonal symmetry of BSN into an "average" orthorhombic (mm2) symmetry. Thus certain macroscopic quantities take spontaneous values within the modulated phases and it is possible to study these phases by experimental techniques such as birefringence in the (001) planes, easier to work with than neutrons or X-rays diffraction measurements.

EXPERIMENTAL RESULTS

Four remarkable effects can be observed in different measuring conditions of the birefringence.

1) Limit curves

If we measure the temperature dependence of the birefringence during a thermal cycle between 20 and 350°C (heating run followed by a cooling run) with a fixed heating/cooling rate, we sometimes observe significant differences between the birefringence curves obtained during consecutive cycles. But after several cycles we obtain an accurately reproducible "limit curve". The curve corresponding to a rate of 2.5°C/mn is plotted on Fig.1. It exhibits a thermal hysteresis of $\approx$ 5°C between 20 and 250°C and of $\approx$ 20°C above 250°C. Let us also note that the limit curve is very dependent of the heating/cooling rate. Limit curves related to several rates from 0·2 to 4°C/mn are plotted on Figure 2.

2) Relaxation effect

If we interrupt the thermal cycle on a given point of the heating/cooling run and keep the sample temperature constant, we observe a slow evolution of the birefringence with respect to time (AA' dashed line on figure 1) with the asymptote reaching a constant value.

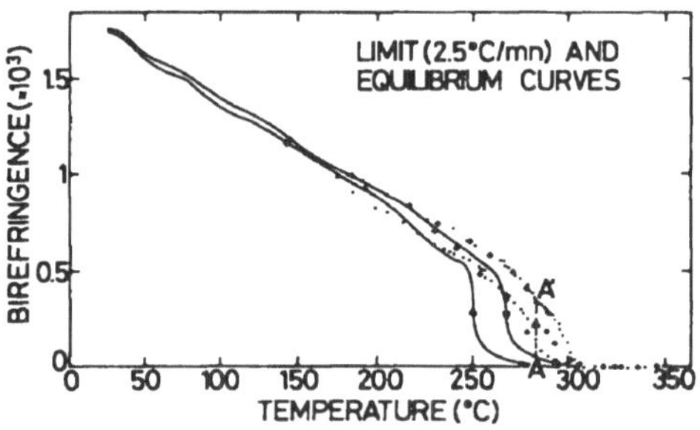

Figure 1 : Continuous line: Birefringence limit curve obtained with an heating/cooling rate of 2.5°C/mn. AA; ; Evolution of the birefringence at fixed temperature. Points and crosses : Equilibrium curve.

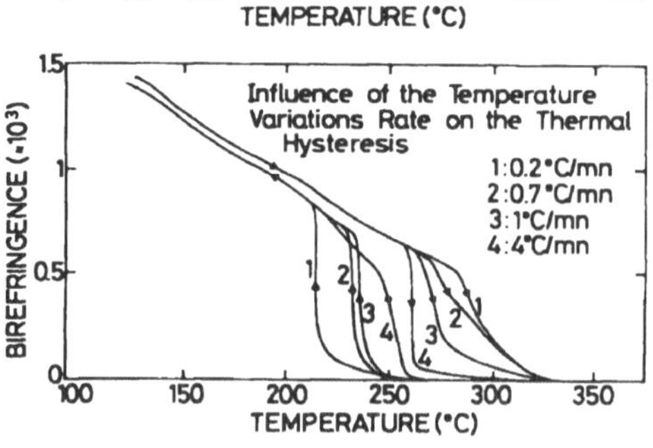

Figure 2 ; Birefringence limit curves related to different heating/cooling rates.

This relaxation effect is negligible below 220°C but becomes very large above 250°C (for example at 280°C the birefringence is multiplied by a factor 8 in 40 hours). An example of the time dependence of the birefringence at 255°C is plotted on Figure 3.

We have tried to fit the experimental data to exponential laws in order to measure the characteristic times of this relaxation process. Fits with only one characteristic time are not good. By contrast, an excellent fit is obtained (see Fig. 3) if we assume the existence of two relaxation-times τ_1 and τ_2 . Results show that the relaxation curves of the birefringence can be characterized by a fast relaxation time τ_1 (of the order of one hour) and a slower one τ_2 (of the order of 10 to 50 hours). To our knowledge this is the first quantitative characterization of a relaxational effect in an incommensurate material. The study of the temperature dependence of τ_1 and τ_2 is in progress.

3) Birefringence equilibrium curve

The time evolution of the birefringence suggests that the limit curves correspond to metastable states, while the final values reached by the birefringence at the end of the relaxation process are related to equilibrium states. The final values are plotted with crosses and points in figure 1. From them we can plot the temperature dependence of the birefringence of BSN at the thermodynamic equilibrium. This "equilibirum curve" coincides with the limit curves of figures 1 and 2 below 250°C, but its variations are smoother and without any inflexion between 250°C and 310°C. In this temperature range, the thermal hysteresis is reduced to about 15°C.

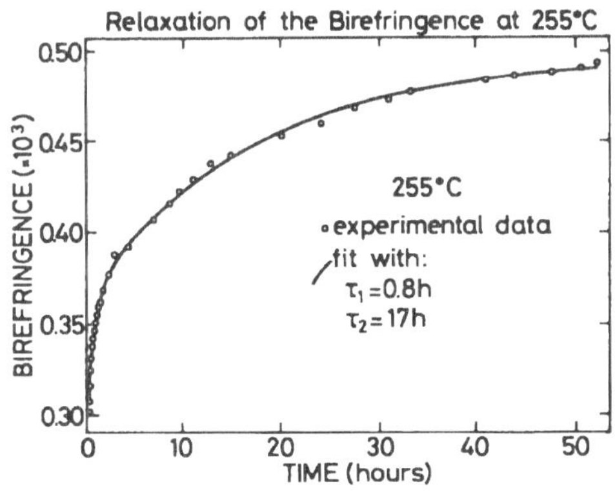

Figure 3 : Relaxation of the birefringence at 255°C. Circles : experimental data. Continuous line : fit to the law: $\Delta n(t)=\Delta n_{\infty}+A_1\exp(-t/\tau_1)+A_2\exp(-t/\tau_2)$.

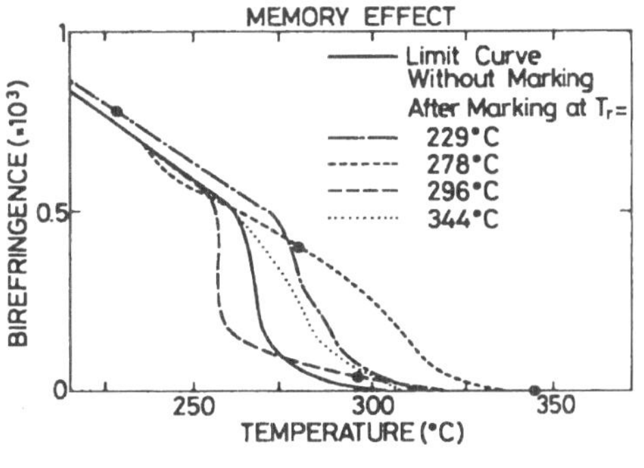

Figure 4 : Influence of annealing on the heating curve of the birefringence

4) Memory effect

We also measured the temperature dependence of the birefringence on a heating run from 20°C to 350°C at 2.5°C/mn, just after a relaxation process at a given temperature T_r followed by a fast cooling run down to 20°C. The resulting birefringence curves are plotted in figure 4. They show a large dependence on the annealing temperature T_r, thus disclosing that the measured sample keeps a memory of its annealing temperature. When $285°C<T_r<310°C$, the birefringence curves exhibit a clear discontinuity near 260°C. When $T_r<285°C$ or $T_r>310°C$ we observe smoother variations than those of the limit curve of figure 1.

This memory effect can also be observed on cooling runs but it is progressively "erased" by successive "reading" thermal cycles (20-350-20°C) which always lead to the "limit curve" described in 1) whatever the annealing temperature. Let us also note that a similar memory effect has already been observed in an other incommensurate material : thiourea[7]. But in the latter material the relative amplitude of the observed effect ($\approx 10^{-3}$) is several orders of magnitude weaker than in BSN.

DISCUSSION

Similar to the interpretation given for thiourea[7], we assume these phenomena to be induced by a strong interaction between the incommensurate modulation and mobile defects. This interaction induces a displacement of the defects, as well as a modification of the amplitude of the modulation in order to reach an equilibrium state minimizing the total energy of the system (modulation + defects).

During the heating or cooling runs, the defects have not enough time to follow the variations of the modulation, therefore BSN samples occupy metastable states. When the temperature is stabilized at T_r, the defects and the modulation relax towards an equilibrium state. If the defects were motionless during the relaxation process, BSN samples should loose the memory of the annealing at T_r on heating above 300°C in the normal high-temperature phase where the incommensurate modulation disappears. This is in contradiction with the experimental results. Thus, the memory effect clearly requires that the defects move during the relaxation process.

At the end of the process, their spatial distribution is in equilibrium with the incommensurate modulation : the sample is "marked". We assume that this marking is not completely erased on cooling the sample to room temperature because the characteristic relaxation times at 20°C are very long (> 72 hours). Therefore, on the next thermal cycle the marked defect distribution favors the stabilization of a modulation compatible with it : the modulation which occured at the relaxation temperature T_r. If $T_r<T_L$

(the incomplete lock-in temperature) the quasi-commensurate phase is stabilized in a wider temperature range than for an unmarked sample and the onset of the incommensurate phase is retarded with the birefringence taking higher values than those of the limit curve. On the contrary, when $T_L < T_r < 310°C$ the birefringence undergoes a discontinuity towards low values near 260°C corresponding to an anticipated onset of the incommensurate phase.

Identification of the nature of the mobile defects

Available samples of BSN always present deviations from the stoichiometric composition mainly consisting in a lack of sodium, leaving sodium vacancies in the structure. Two observations suggest that these vacancies are the main defect involved in the relaxation and memory effects.

First, it has been demonstrated[10] by systematic examination of crystals of different sodium content that these vacancies play a fundamental role in another transition of BSN existing below room temperature and whose characteristic features are close to those of the studied transition. Actually, this other transition is characterized by the onset of a second incommensurate modulation along $[1\bar{1}0]$ while the 300°C transition is related to the $[110]$ direction.

Secondly, structural data[11] have shown that BSN samples are formed by a stack of two kinds of layers. The layers which contain Na^+ ions are filled only to 60 %. Therefore, because of their small size (r_i = 0,95 Å) the Na^+ ions can migrate by hopping from site to site under the influence of the driving modulated potential.

This migration must involve an ionic conductivity of BSN samples in the 250-300°C temperature range whose measurement should confirm the role of the Na^+ vacancies. Preliminary results are in agreement with this statement.

REFERENCES

1. A.P. Levanyuk, D.G. Sannikov, Sov. Phys, Solid State 18, 245 (1976).

2. W.I. Mc Millan, Phys. Rev. B14, 1496 (1976).

3. W.D. Ellenson, S.M. Shapiro, G. Shirane, A.F. Garito, Phys. Rev. B16, 3244 (1977).

4. R.M. Fleming, D.E. Moncton, D.B. McWhan, F.J. DiSalvo, Phys. Rev. Lett. 45, 576 (1980).

5. K. Hamano, Y. Ikeda, T. Fujimoto, K. Ema, S. Hirotsu, J. Phys. Soc. Jpn 49, 2278 (1980).

6. F.J. DiSalvo, J.A. Wilson, Phys. Rve. B12, 2220 (1975).

7. J.P. Jamet, P. Lederer, J. Phys. Lett. 44, 2xx (1983).

8. J. Schneck, J.C. Toledano et al., Phys. Rev. B25, 1776 (1982).

9. J. Schneck, F. Denoyer, Phys. Rev. B23, 383 (1981).

10. J. Schneck, Thesis, Université Paris VI (1982).

11. P.B. Jamieson, S.C. Abrahams et al., J. Chem. Phys. 50, 4352 (1969).

CRITICAL FLUCTUATIONS AND MAGNETIC ORDERING IN PRASEODYMIUM METAL

W.G. Stirling and K.A. McEwen

Institut Laue-Langevin

156X, 38042 Grenoble Cedex, France

The magnetic ordering process in the light rare-earth metal praseodymium displays very unusual features which are not readily explained by currently available theories of magnetic phase transitions.

In the metallic state, Pr^{3+} ions ($4f^2$, $J = 4$) have singlet ground states and the induced moments have an XY-like character as a consequence of the crystalline electric fields acting in the double-hexagonal close-packed crystal structure. Furthermore, analysis of the comprehensive neutron studies of the magnetic excitations in paramagnetic praseodymium[1] demonstrates the long-range oscillatory nature of the RKKY exchange interaction which produces a maximum two-ion coupling at wavevectors around 0.10 to 0.15 τ_{100} (basal plane direction ΓM). Whilst the strength of this bare interaction is just below the critical value for magnetic ordering[2], the polarization, at very low temperatures, of the spins of the Pr^{141} nuclei is apparently sufficient to drive the coupled electron-nuclear system into an ordered state[3]. Alternatively, magnetic ordering can be induced by the application of a suitable uniaxial stress to modify directly the crystal field levels[4].

In this paper we summarize the results of a detailed investigation, using thermal neutron scattering techniques, of the static and dynamic response of praseodymium at low temperatures and under applied uniaxial stress[5]. We emphasize recent experimental results which indicate that the ordering process is more complicated than hitherto suspected. More complete accounts of this work will be published.

Under zero applied stress, a broad magnetic peak[6] at a wavevector of $q_1 = 0.105\ \tau_{100}$ appears in the neutron elastic scattering spectrum, below about 10K. (See, for example, the spectra shown in Fig. 2). This peak has a FWHM of about 0.024 τ_{100}, compared with a typical instrumental width of about 0.007 τ_{100}, and although its intensity grows steadily with decreasing temperature, it does not diverge. Moreover, the width of this peak is essentially independent of temperature. Below approximately 2K, the spectrum contains a second, much sharper, component[7,8] at $q_2 = 0.13\ \tau_{100}$. This "satellite" grows rapidly in intensity below 0.1K, as shown in Figure 1(a). Measurements around a series of reciprocal lattice points[7] have revealed that these satellite reflections may be interpreted as arising from a sinusoidally modulated magnetic structure in which the moments and modulation wavevector are parallel (along the basal plane b-direction). Alternate layers of magnetic moments are coupled antiferromagnetically along the hexagonal axis. Neutron inelastic scattering studies have shown that the incipient soft mode, the longitudinal optic exciton which exhibits a pronounced minimum around the satellite wavevector, does not soften significantly below 7K.

In direct contrast to the results at low temperatures, the application of modest uniaxial stresses (< 2000 bars) produces a marked soft mode behaviour of the LO exciton, accompanied by the growth of <u>both</u> the broad peak at q_1 and the sharper satellite at q_2 [4,9,10]. Even quite small stresses raise the characteristic ordering temperatures to more accessible values. For example, at 150 bars, the transition temperature associated with the long-range ordering (q_2 satellites) is $T_N \cong 6.5$K. An example of the elastic scattering spectrum at 1.43K, under such a stress, is shown in Figure 2.

When higher stresses ($\sim$ 1000 bars) are applied, the LO mode is found to completely condense at (or near) q_2, and the dispersion curves are observed to emanate linearly from this wavevector[9]. These phason-type modes are overdamped for wavevectors closer than $0.04\tau_{100}$ ($0.08\ \text{Å}^{-1}$) to q_2 but are clearly resolved elsewhere. Such behaviour is qualitatively very similar to recent calculations for the dynamics of incommensurable structures[11].

At 980 bars, the transition temperature deduced from the satellite temperature dependence is $T_N \cong 9.5$K. Further, the magnetic satellites are now some 50 times more intense than those measured at very low temperatures with no applied stress, implying a corresponding larger induced magnetic moment. As at low temperatures, the stress-induced structure is sinusoidally modulated, but the modulation-wavevector is stress-dependent. From a value of 0.132 τ_{100} at zero stress, it changes smoothly to 0.122 τ_{100} at 1680 bars; irreversible plastic deformation of the crystal occurs beyond this stress.

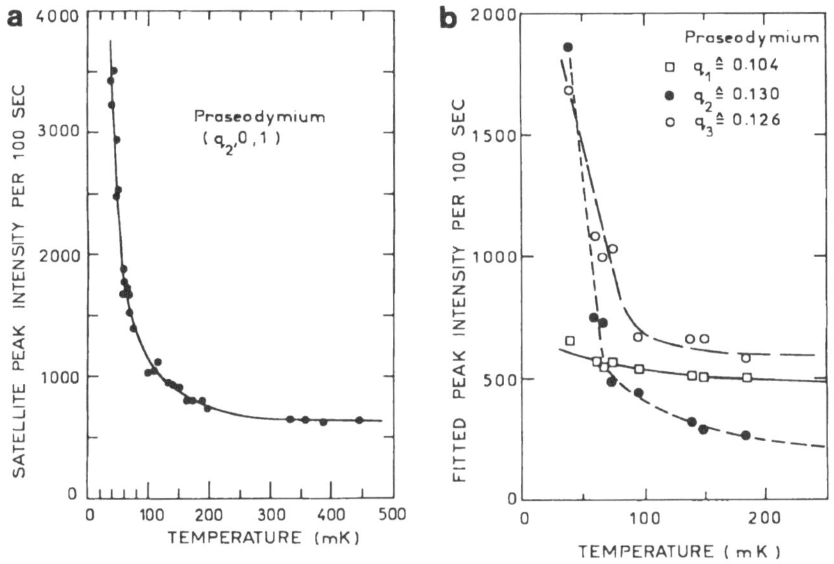

Fig. 1 (a) Satellite peak intensity as a function of temperature.
(b) Fitted peak intensities as a function of temperature for the "3-peak Model" described in the text. The lines are guides for the eye.

Thus, whilst very similar magnetic structures appear at very low temperatures (with zero applied stress) and at significantly higher temperatures under stress, the behaviour of the inelastic response is quite different in the two cases. However, in both cases there are two coexisting magnetic peaks in the elastic scattering spectrum, one of which remains broader than the experimental resolution. At zero stress, the integrated intensity of the broad peak is comparable to that of the satellite peak (see Fig. 2(a), for example).

The temperature dependence of the mean square value of the magnetic moment has also been deduced from recent measurements[12] of the heat capacity of praseodymium below 1K. It is interesting to note that the moment derived in this way is consistent with that calculated from the total integrated intensity of both the q_1 and q_2 peaks, rather than the long-range order peak (q_2) alone.

Indeed, Bjerrum Møller et al.[8] proposed that there is a "great ambiguity in the ordering wavevector" and that true long-range order in unstressed Pr develops only at around 50 mK. Using the somewhat arbitrary criterion of a maximum in $|dC/dT|$, Eriksen et al.[12] determined a transition temperature of 42 mK from their heat capacity measurements. In our most recent experiments, we have therefore reexamined, in more detail, the low temperature elastic scattering from praseodymium. Using a new technique of sample mounting[13], we have largely eliminated earlier problems of thermal contact and accurate thermometry.

As shown in Figure 1(a), there is a particularly large increase in the q_2 peak intensity below 100 mK. Following the procedure described in our earlier paper[7], the spectrum at 40 mK was fitted initially to the sum of two Gaussian functions, with the result shown in the upper left section of Figure 2. Whilst providing a reasonable interpretation of the data, this "2-peak model" has several shortcomings. Firstly, the fitted q_1 peak occurs at a wavevector which is too large, and with a width which is too great, when compared with its high temperature parameters. Secondly, the q_2 peak height is underestimated by the fit, and again the fitted peak is too wide, particularly if it represents long-range order when it should have the instrumental width. Finally the region between the peaks is poorly represented.

The inclusion of a third Gaussian peak in the fitting procedure permits a much improved fit to the data (see Figure 2, lower left). This "3-peak model" has a q_1 peak at the correct wavevector (0.104 τ_{100}), with a reasonable width (FWHM = 0.024 τ_{100}). Now there are two peaks within the "satellite" : one (q_2) with the resolution width (0.007 τ_{100}) and the other between 2 and 3 times wider (0.019 τ_{100}). Although they are close, q_2 (= 0.133 τ_{100}) and q_3 (= 0.129) are apparently not equal as the quality of the fit degrades on constraining these values to be identical.

As a test of this 3-peak model, we have applied it to our data at higher temperatures, using the peak widths obtained at 40 mK. The fitted peak heights are shown, in Figure 1(b), as a function of temperature. It may be noted that the sharp q_2 peak increases particularly rapidly below 100 mK, overtaking the broader q_3 peak. The q_1 peak has a rather slow increase with decreasing temperature.

In a second test of this multiwavevector hypothesis we have applied the 3-peak model to data taken at higher temperatures under uniaxial stress. In the example of Figure 2, upper right, the inadequacy of the 2-peak model is mainly visible in the poor fit to the q_1 peak : as above, it is too wide and at too high a wavevector. The inclusion of the third peak clearly improves the fit (see Figure 2, lower right) with a broad q_3 peak and a resolution limited q_2 peak. It should be stressed that the parameters of all the fits described above were not constrained during the fitting procedure.

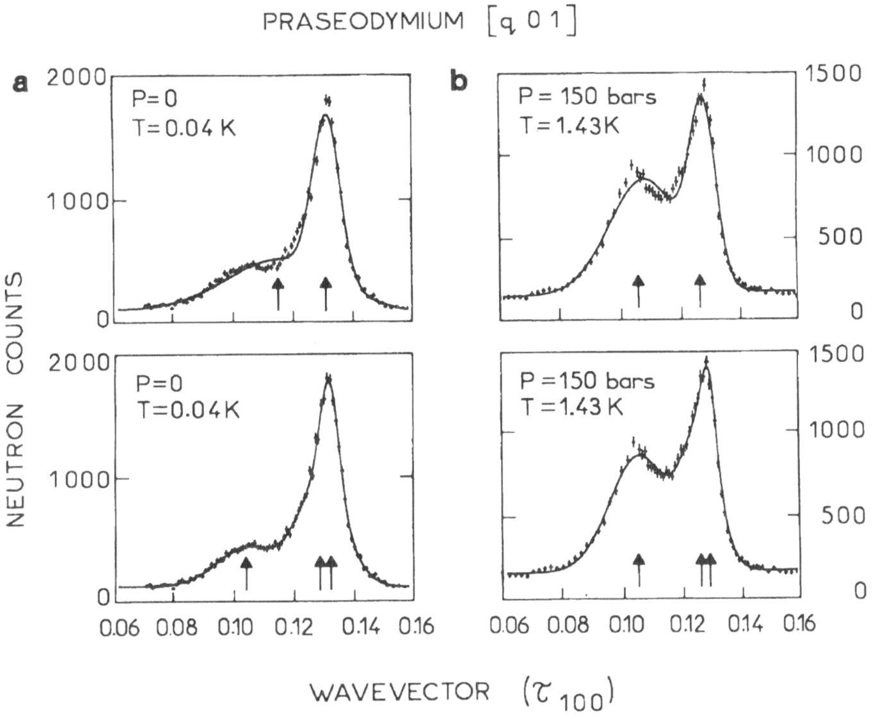

Fig. 2 Examples of elastic scattering spectra at wavevector [q,0,1] ; (a) at zero stress and 40 mK, (b) at 150 bars uniaxial stress and 1.43K. The lines are the result of fitting two Gaussian functions (upper diagrams) or three Gaussian functions (lower diagram), as described in the text.

It seems most appropriate to continue to interpret the q_2 peak as arising from true long-range magnetic order and it is plausible that the broader q_3 peak represents fluctuations of the same order parameter. The rôle of the coexisting q_1 peak remains puzzling. There have been a number of attempts[6] to explain its origin but none deal satisfactorily with the coexistence of this peak and the satellite peaks in the ordered phase of Pr. Recently, Lindgård[14] has developed a correlation theory of XY-like magnets which includes a quadrupolar type coupling between the ground and excited states. This calculation predicts, inter alia, (i) a quasi-elastic central peak in the paramagnetic phase and (ii) increased damping and decreasing intensity of the excitonic mode with a simultaneous increase in the central peak intensity as $T \to T_N$. The experimental situation is, in reality, more complicated, as discussed above. It now seems likely that this mechanism is more appropriate for the q_2/q_3 peaks than for the q_1 peak. Finally, it is interesting to note that the magnitude (in τ_{100} units) of the wavevector q_2 corresponds to the commensurable value 2/15 whereas the q_1 peak is centred at 2/19 and its width encompasses the region from 2/21 to 2/17.

In summary, the general mechanism of the magnetic ordering process in Pr can be understood in terms of (a) the hyperfine-enhanced exchange coupling at very low temperatures and (b) the mixing of crystal field levels by applied stress. However the detailed behaviour contains a number of outstanding puzzles, the most important being the unusual coexistence of broad (short-range order) peaks and long-range order peaks well below the transition temperature in this magnetic system.

ACKNOWLEDGMENTS

Many of our friends have contributed to this work. In particular we thank A. Benoit, S.K. Burke, J. Flouquet, J. Jensen, P.A. Lindgård, A.R. Mackintosh, R. Pynn and C. Vettier for valuable help and advice.

REFERENCES

1. J.G. Houmann, B.D. Rainford, J. Jensen and A.R. Mackintosh, Phys. Rev. B20:1105 (1979).
2. See for example, J. Jensen, J. Physique 40:C5-1 (1979).
3. T. Murao, J. Phys. C16:335 (1983) and references therein.
4. K.A. McEwen, W.G. Stirling and C. Vettier, Phys. Rev. Lett. 41:343 (1978).
5. K.A. McEwen, W.G. Stirling and C. Vettier, in "Crystalline Electric Field Effects in f-electron Magnetism", ed. R.P. Guertin, W. Suski and Z. Zolnierek (Plenum Press, New York), 57 (1982).
6. S.K. Burke, W.G. Stirling, K.A. McEwen, J.Phys. C 14:L967 (1981).
7. K.A. McEwen and W.G. Stirling, J. Phys. C14:157 (1981).
8. H. Bjerrum-Møller, J.Z. Jensen, M. Wulff, A.R. Mackintosh, O.D. McMasters and K.A. Gschneidner, Jr.,Phys. Rev. Lett. 49:482 (1982).
9. K.A. McEwen, W.G. Stirling and C. Vettier, J. Magn. Magn. Mat. 31-34:599 (1983).
10. K.A. McEwen, W.G. Stirling and C. Vettier, Physica B 28 (1983), in press.
11. R. Zeyher and W. Finger, Phys. Rev. Lett. 49:1833 (1982).
12. M. Eriksen, E.M. Forgan, C.M. Muirhead and R.C. Young, J. Phys. F 13:929 (1983).
13. A. Benoit, J. Flouquet, K.A. McEwen and W.G. Stirling, to be published.
14. P.A. Lindgård, Phys. Rev. Lett. 50:690 (1983), and this conference.

SOFT MODE AND CENTRAL PEAK AS A CONSEQUENCE

OF COMPETING INTERACTIONS

Per-Anker Lindgård
Physics Department
Risø National Laboratory
DK-4000 Roskilde Denmark

SUMMARY

Singlet ground state magnets represent an interesting class of competing interaction systems in which a single ion anisotropy DS_z^2 favours for every site a non magnetic ground state $|0\rangle$ while exchange interactions J_0 favour some magnetic structure with a magnetic moment at each site. Approaching the critical temperature T_c from above ordered regions are building up on a background of the singlet ground state matrix. The dynamics of such a system is interesting because excitations characteristic for both regions will be present and change in dominance. The excitation corresponding to the single ground state region is the propagating crystal field excitation at finite frequency. This mode is softening and loosing weight approaching T_c. The fluctuating ordered regions will on the other hand show spin diffusion i.e. a central peak as expected for a Heisenberg magnet. This behaviour is predicted to dominate near T_c and the characteristic critical features are those for a n=2 system. The phase diagram Fig. 1 was calculated for lattice dimensions d=3, 2 and 1. For D equal to zero one has a bicritical point and T_c increases initially for increasing D. The theory presented for the singlet-doublet system, includes correlation effects and calculates the renormalization and damping of the soft mode and the central peak. No details will be given here since a short version has recently been

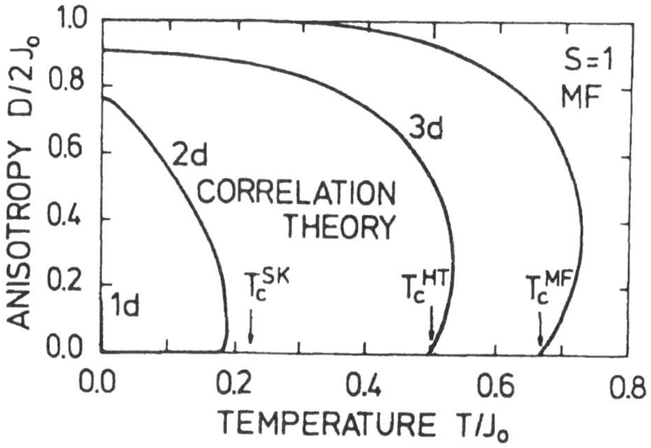

Fig. 1. Phase diagram $T_c(D,J_0)$ for different lattice dimensions d compared with the mean field (MF) result. The theory gives good quantitation agreement with high temperature series expansion results. Notice the bicritical points at D = 0.

published[1] and compared with experiments on Pr[2,3] and $CsFeCl_3$[4,5]. The preliminary results for the effect of applying a magnetic field along the z-direction is calculated. The principal consequence is a relative reduction of the central peak intensity because of the formation of a magnetic component along the field . also the field dependence of the exitonic peak is found to be different from that predicted by the random phase approximation.

REFERENCES

1. P.-A. Lindgård, Phys. Rev. Lett. 50, 690 (1983).
2. P.-A. Lindgård, J. Mag. Mag. Mat. 31-34, 1099 (1983).
3. K.A. McEwen, W.G. Stirling and C. Vettier, Proc. Conf. on Crystal Electric Field and Structural Effects in f-electron Systems, Wroclaw, Poland (1981), eds. R.P. Guertin and W. Suski (Plenum, New York, 1982) and Phys. Rev. Lett. 41, 343 (1978). S.K. Burke, W.G. Stirling and K.A. McEwen, J. Phys. C14, L967 (1981).
4. P.-A. Lindgård, Proc. from VI Yamada conf. (Hakone, Sep. 82), Physica (1983).
5. H. Yoshizawa, W. Kozuke and K. Hirakawa, J. Phys, Soc. Jap. 49. 144 (1980). M. Steiner, K. Kakurai, W. Knop, B. Dorner, R. Pynn, U. Kappek, P. Day and G. Mcleen, Solid State Commun. 38, 1179 (1981).

CRITICALITY AND CROSSOVER IN STUCTURAL PHASE TRANSITIONS

Paul D. Beale

Dept. of Theoretical Physics, Oxford University

1 Keble Road, Oxford, England OX1 3NP

ABSTRACT

Renormalization-group techniques and the exact factorization of the partition function of a particular model in the class of double well Hamiltonians are used to examine the critical behaviour and the high temperature gaussian-Ising crossover behaviour of systems undergoing structural phase transitions. The block probability density function is shown to be consistant with the scaling hypothesis near criticality.

A simple microscopic model which has been used to describe systems undergoing structural phase transitions[1] is described by the classical motion of a set of N particles interacting through the potential energy

$$\beta U_N = \sum_j \beta V(x_j) + \sum_{(ij)} \frac{\beta C}{2} (x_i - x_j)^2. \qquad (1)$$

Here (ij) denotes pairs of nearest neighbour sites on a d-dimensional hypercubic lattice and $\beta = 1/k_BT$. The scalar variables $\{x_i\}$denote the displacement of the particles from the lattice sites along a principle symmetry direction. The site potential energy V(x) has a symmetric double well form with two degenerate minima at $\pm x_0$. The thermodynamics of this system is best described in terms of the dimensionless parameters

$$K^{-1} = T = (\beta C x_0^{\,2})^{-1}, \text{ and } \theta = \frac{4(V(0) - V(x_0))}{C x_0^{\,2}}$$

In the limit $\theta \to \infty$ (the Ising limit) the system is identical to a nearest neighbour Ising model. In the limit $\theta \to 0$ (the displacive limit) the well depth and the critical temperature vanish. In that limit the system is equivalent to a noncritical gaussian or phonon model.

Real-space renormalization-group (RG) results[2] indicate that the critical line $T_c(\theta)$ varies smoothly from zero at the displacive limit to T_c^{Ising} in the Ising limit and that the critical behaviour along the line is Ising-like for all $\theta > 0$. For $T < T_c(\theta)$ the order parameter $\langle x_j \rangle$ takes a nonzero value. For $T \geq T_c(\theta)$ the order parameter vanishes. In addition, there exists a high temperature crossover region where the behaviour of short-range correlations changes qualitatively from being of Ising character to having a distinctly phonon-like behaviour. This crossover temperature $T_x(\theta)$ is not a sharp feature but it can be identified as the temperature at which central peak in the dynamic response function begins to appear above T_c.[2]

The double gaussian[3,4] (DG) model is a particularly useful model in the class of Hamiltonians (1). It is defined by

$$\beta V(x) = \frac{1}{2}\left[\frac{x}{w}\right]^2 - \ln\left[\cosh\left[\frac{xv}{w^2}\right]\right]. \tag{2}$$

One interesting feature of this model is that its partition function Z_N^{DG} can be factorized[3] exactly into a noncritical gaussian model and a long-ranged Ising model, ie.

$$\begin{aligned} Z_N^{DG} &= \int dx_1 \ldots \int dx_N \exp(-\beta U_N) \\ &= \left[\frac{1}{2}\exp\left[-\frac{1}{2}\frac{v^2}{w^2}\right]\right]^N Z_N^G Z_N^I \end{aligned} \tag{3}$$

where

$$Z_N^G = \exp\{-\tfrac{1}{2}\sum_q \ln[1 + 2Kw^2(d - \sum_n \cos(q_n)]\} \tag{4}$$

is the gaussian partition function and

$$Z_N^I = \sum_{\{s_i=\pm 1\}} \exp[\tfrac{1}{2}\sum_i \sum_j K_{ij}\, s_i\, s_j]. \tag{5}$$

is the Ising partition function. The variable q in (4) is summed over the first Brillouin zone of the reciprocal lattice and q_n denotes the nth cartesian coordinate of q. The long-ranged interaction K_{ij} in the Ising partition function is given by

$$K_{ij} = \frac{1}{2}(v/w)^2 N^{-1} \sum_q \frac{\exp(iq\cdot r_{ij})}{1 + 2Kw^2(d - \sum_n \cos q_n)}, \tag{6}$$

where r_{ij} is the displacement vector from site i to site j. For $|r_{ij}| \gg 1$ the interaction has the form

$$K_{ij} \sim |r_{ij}|^{-(d-1)/2} \exp[-|r_{ij}|/R] \tag{7}$$

where R is the range of phonon correlations in the gaussian partition function. In the Ising limit this range of interaction vanishes, ie. the model is equivalent to a nearest neighbour Ising model. In the displacive limit along the critical line R diverges and a mean field theory of the phase transition is accurate except very close to T_c.

From this factorization, rigorous upper and lower bounds for the critical temperature can be derived[5]. The exact asymptotic form of critical temperature near the Ising and displacive limits may also be calculated. The results are

$$T_c(\theta) \approx T_c^{Ising}(1 - \frac{4d+3}{\theta} + \ldots), \text{ for } \theta \gg 1 \tag{8}$$

$$T_c(\theta) \approx A(d)\,\theta + \ldots, \quad \text{for } \theta \ll 1 \text{ and } d > 2; \tag{9}$$

$$T_c(\theta) \ln[T_c(\theta)] \sim \theta + \ldots, \text{ for } \theta \ll 1 \text{ and } d = 2. \tag{10}$$

These results agree with known rigorous results[1,6], RG calculations[2,7] and molecular dynamics simulations[8].

The presence of the high temperature gaussian-Ising crossover line may be investigated by examining the relative contributions of the Ising and gaussian parts of the partition to the free energy. A simple strength criterion argument[5] gives a crossover temperature in good agreement with the RG analysis[2].

The DG model is also useful for calculating the behaviour of the block probability density function $P_L(x)$[9]. This quantity is defined by

$$P_L(x) = \langle [\delta\, x - L^{-d} \sum_{j\epsilon L^d} x_j] \rangle \tag{11}$$

where the sum is only over the lattice sites contained within a block of linear dimension L, and δ is Dirac's delta function. This function gives the probability density that the order parameter averaged over a block of size L will lie between x and x + dx. When the blocksize and the correlation length are sufficiently large, nonuniversal features of the model will be washed out and $P_L(x)$ should take on the universal scaling form[9,10]

$$P_L(x) \approx \frac{L^{\beta/\nu}}{\sigma} \tilde{P}\left(\frac{L^{\beta/\nu}x}{\sigma}, L/\xi\right) \qquad (12)$$

where β and ν are the exponents associated with the order parameter and correlation length, respectively. The behaviour of $\tilde{P}(x,L/\xi)$ has been calculated for $L \ll \xi$ in various dimensions[9,10] and is found to be strongly dependent on the dimensionality.

The block PDF of the DG model can be factorized in much the same way as the partition function[11]. The result of the factorization is that the PDF of the DG model is exactly the convolution of the PDF's of the associated gaussian and Ising models.

$$P_L^{Dg}(x) = \int dx' \, P_{L'}^{G}(x - x') \, P_{L'}^{I}(x'/v) \qquad (13)$$

$P_{L'}^{G}$ is the PDF of the gaussian model and $P_{L'}^{I}$ is the PDF of the long-ranged Ising model. The blocksize L' on the right-hand-side of (13) is given by

$$L' = \begin{cases} R & \text{if } L \ll R \\ L & \text{if } L \gg R \end{cases} \qquad (14)$$

where R is the phonon correlation length. From this factorization one can show that the block PDF of the DG model is consistant with the scaling hypothesis (12) if and only if both $L \gg R$ and $\xi \gg R$. Only when the blocksize and the correlation length are large compared with the phonon correlation range is the scaling hypothesis true. Therefore the phonon correlation length R is the longest non-critical length in double well systems. The crossover from Ising to gaussian behaviour is determined by the relative size of the correlation length and the phonon correlation length.

REFERENCES

1. A.D. Bruce, Adv. in Physics, 29:111 (1980).
2. P.D. Beale, S. Sarker and J.A. Krumhansl, Phys. Rev. B 24:266 (1981). P.D. Beale, Phys.Rev. B 24:6711 (1981).
3. J.H. Chen, M.E. Fisher and B.G. Nickel, Phys.Rev.Lett. 48:630 (1982).
4. G.A. Baker,Jr. and A.R. Bishop, J.Phys. A 15:L201 (1982).
5. G.A. Baker,Jr., A.R. Bishop, K. Fesser, P.D. Beale and J.A. Krumhansl, Phys.Rev. B 26:2596 (1982).
6. J. Bricmont and J.R. Fontaine, preprint.
7. T.W. Burkhardt and W. Kinzel, Phys.Rev. B 20:4730 (1979).

8. T. Schneider and E. Stoll, Phys. Rev. B 13:1216 (1976).
9. A.D. Bruce, T. Schneider and E. Stoll, Phys. Rev. Lett. 43:1284 (1979).
10. K. Binder, Z. Phys. B 43:119 (1981).
11. P.D. Beale, Phys. Rev. B 27:5804 (1983).

ANOMALOUS VIBRATIONAL ABSORPTION NEAR THE ANTIFERROELECTRIC PHASE TRANSITION IN ALKALI CYANIDES

I. Vilfan and R. Pirc

J.Stefan Institute, 61111 Ljubljana, Yugoslavia

F. Lüty

Physics Department, University of Utah
Salt Lake City, Utah 84112, U.S.A.

The critical behaviour of high-frequency anharmonic vibrations coupled to the orientational degrees of freedom in alkali cyanides is discussed.

INTRODUCTION

The purpose of this paper is to discuss the vibrational behaviour of a system of electric dipoles in the critical region of an order-disorder phase transition if the vibrational potential is anharmonic. A typical example of such systems are alkali cyanides like KCN or NaCN.[1,2] The CN^- dipoles in these systems are antiferroelectrically ordered in the low temperature phase. Above the transition temperature T_{c1} the electrically disordered phase, in which the CN^- dipoles are still aligned along the same axis but have head to tail disorder, takes place. Upon further increase in temperature the crystals undergo at T_{c2} another transition to the elastically disordered cubic phase with the CN^- dipoles randomly and dynamically distributed in a multiwell potential with no preferential orientation. The lower, (antiferro)electric transition at T_{c1} is second order, while the second, elastic, transition at T_{c2} is first order.

The vibrational absorption as observed by Raman scattering and infrared absorption experiments can serve as a method for studying the phase transitions in systems of electric dipoles provided the vibrational potential is anharmonic and coupled to the electric fields.[3] The gradual development of the dipolar ordering on cooling is responsible for the formation of a local electric field. While the local field varies randomly from site to site in the electrically disordered phase, it obtains a static component modulated with the wavevector of the dipolar ordering below T_{c1}. Such a local electric field causes the vibrational frequency shift of anharmonic oscillators and fluctuations in vibrational frequency which are observed as broadening of the line.

In this paper we report the results of a calculation of the CN^- stretching line shift and width in alkalí cyanides in the critical region of the electric transition. Since the stretching vibrational frequency is several orders of magnitude higher than the reorientational frequency of the dipoles,[4] the electric field experienced by the dipoles will be considered as static, the dynamics of the dipolar reorientational motion will be ignored.

THE HAMILTONIAN

The model Hamiltonian of the systems of anharmonic oscillators coupled to the dipolar orientational order parameter can be written as

$$H = \sum_i \left[\hbar\omega_0(a_i^+ a_i + \frac{1}{2}) - c(a_i^+ + a_i)^3\right]$$

$$- \frac{1}{2} \sum_{ij} \left[J_{ij}S_iS_j + 2\sqrt{\frac{\hbar}{m\omega_0}}\, J'_{ij}S_iS_j(a_i^+ + a_i)\right] \qquad (1)$$

The first sum represents the Hamiltonian of the localised anharmonic oscillators, a_i^+ and a_i being the creation and annihilation operators, respectively, for the vibrational excitation at the site i, c is the cubic anharmonicity constant. The second sum is the dipolar interaction energy between vibrating dipoles. J_{ij} is the nonperturbed interaction energy between permanent dipoles at the sites i and j, J'_{ij} is the derivative of the dipolar interaction energy with respect to the C-N internuclear separation evaluated at the equilibrium C-N distance, and m is the reduced mass. The pseudospin S_i describes the orientation of the dipole i and is $S_i = \pm 1$ when the electric phase transition is concerned. The magnitude of the dipole moment is to a good approximation linearly related to the C-N separation,[5] J'_{ij} is proportional to J_{ij},

$$J'_{ij} = d\, J_{ij}. \qquad (2)$$

The constant d is equal to the logarithmic derivative of the dipole moment with respect to the C-N separation.

The vibrational frequency of local oscillators in the presence of the electric fields is obtained via the canonical transformation of (1),

$$a_i = a_{io} + \tilde{a}_i \tag{3}$$

with the operator for the equilibrium displacement being equal to

$$a_{io} = \frac{3c}{2\hbar\omega_o} + \frac{d}{\hbar\omega_o}\sqrt{\frac{\hbar}{m\omega_o}} \sum_j J_{ij} S_i S_j . \tag{4}$$

The first term represents the standard anharmonic displacement and the second term displays the effect of the local electric field. The shifted quasiharmonic frequency of the fundamental vibrational transition of the oscillator at the site i is then

$$\omega_i = \omega_o - \frac{15}{4}\frac{c^2}{\hbar^2\omega_o} - \frac{3\,cd}{\hbar^2\omega_o}\sqrt{\frac{\hbar}{m\omega_o}} \sum_j J_{ij} S_i S_j . \tag{5}$$

Here the second term includes the static and dynamic anharmonic frequency shifts. The last term is proportional to the local electric field at the site i and depends thus on all pseudospin orientations.

Individual vibrational frequencies ω_i are undamped in this model, but they vary from site to site due to local fluctuations in the electric field. The essential property of dipolar fields is their long range which makes the spectrum of vibrational frequencies continuously broadened contrary to a discrete spectrum which would appear if J_{ij} were short ranged. If the vibrational lineshape is approximated by a Gaussian distribution, the frequency shift and width are equal to the first and second moment, respectively, of the frequency distribution.

EFFECT OF THE ELECTRIC PHASE TRANSITION ON THE VIBRATIONAL FREQUENCY

In the preceding section we have seen that the local vibrational frequency ω_i is proportional to the electric field at the site i. Below the transition temperature at T_{c1} the vibrational line splits into two lines, the first one being due to the dipoles parallel to the local electric field and the second one to the dipoles antiparallel to the local field. Above T_{c1} the two lines merge into a single line. The field dependent shift Δ of the average frequency is obtained from (5) as

$$\Delta = -\frac{3\,cd}{\hbar^2\omega_o}\sqrt{\frac{\hbar}{m\omega_o}}\sum_j J_{ij}\langle S_i S_j\rangle . \tag{6}$$

Notice that Δ is proportional to the internal energy of the dipolar system per particle, $\langle \mathcal{H}^{DIP}\rangle/N$. On the other hand, the width W turns out to be proportional to the fluctuation of the local electric field. Using (5) with $S_i = 1$ we find

$$W^2 = \langle \omega_i^2\rangle - \langle \omega_i\rangle^2 = \left(\frac{6\,cd}{\hbar^2\omega_o}\right)^2 \frac{\hbar}{m\omega_o}\sum_{jk} J_{ij}J_{ik}\left[\langle S_j S_k\rangle - \langle S_j\rangle\langle S_k\rangle\right] \tag{7}$$

In the shift of the vibrational line thus the complete pseudospin correlation function enters, while in the width of the line the two-point cumulant (connected correlation) of pseudospin operators is present.

In studying the temperature dependence of Δ and W it is important to notice that the dipolar subsystem in the critical region of the electric transition has the behaviour characteristic of Ising systems with dipolar interactions in d = 3 dimensions as the dipoles are antiferroelectrically ordered below T_{c1}. The temperature dependence of the frequency shift Δ is according to (6) given by the behaviour of the internal energy and is

$$\Delta = \Delta(0)\left[1 - A_{\pm}\, t\,|t|^{-\alpha}\right] \qquad t \gtrless 0 \tag{8}$$

where $t = T/T_{c1} - 1$, α is the specific heat exponent being ≈ 0.12, A_+ (for $t > 0$) and A_- (for $t < 0$) are the nonuniversal constants with the universal ratio $A_+/A_- \approx 2^{\alpha-1} \approx 0.54$.[6] The critical behaviour of Δ is governed by the exponent $1 - \alpha$ above and below the transition temperature, as shown in Fig. 1.

The critical behaviour of the vibrational linewidth W follows from (7) taking into account the critical dependence of the pseudospin cumulants:

$$W = W(0)\left[1 - A_+\, t^{\,1-\alpha}\right] \qquad t > 0 \tag{9a}$$

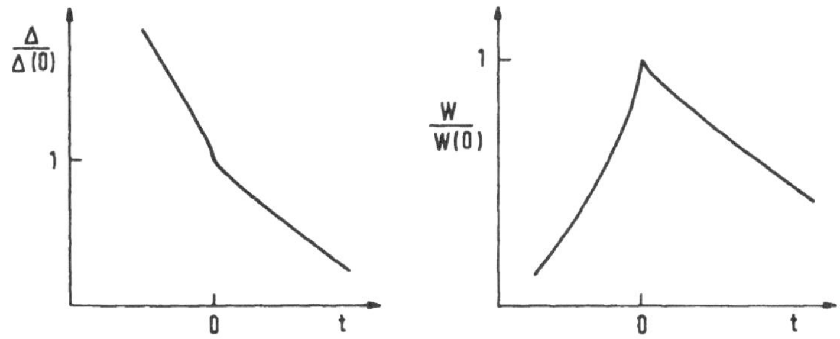

Fig. 1. Schematic temperature dependence of the vibrational frequency shift Δ and width W in the critical temperature interval.

$$W = W(0)\left[1 - B(-t)^{2\beta}\right] \qquad t < 0, \qquad (9b)$$

Here the order parameter exponent is $\beta = 0.32$ and B is a nonuniversal constant. The temperature dependence of W above T_{c1} is governed by the specific heat exponent, while below T_{c1} the order parameter exponent predominates, and has a cusp at T_{c1}, as seen in Fig. 1.

The detailed analysis of critical behaviour of anharmonic vibrations including their dispersion will be published elsewhere.

REFERENCES

1. D.Durand, L.C. Scavarda do Carmo, A.Anderson, and F.Lüty, Phys. Rev. B 22, 4005 (1980).
2. F.Lüty, in International Conference on Defects in Insulating Crystals (Riga, USSR, 1981).
3. L.C.Scavarda do Carmo, F.Lüty, T.Holstein, and R.Orbach, Phys. Rev. B 23, 3186 (1981).
4. H.T. Stokes, T.A. Case, and D.C. Ailion, Phys. Rev. Lett. 47, 268 (1981).
5. J.E. Gready, G.B. Bacskay, and N.S. Hush, Chem. Phys. 24, 333 (1977).
6. E.Brezin, J.C. le Guillou, and J. Zinn-Justin in: "Phase transitions and Critical Phenomena", edited by C.Domb and M.S. Green (Academic Press, New York, 1975) Vol. VI.

A PLENITUDE OF COMMENSURATE PHASES IN SIMPLE MODELS

Michael E. Fisher

Baker Laboratory, Cornell University

Itihaca, New York 14853, U.S.A.

SUMMARY

Various bulk physical systems exhibit a number of distinct complex commensurate modulated phases : particularly notable is the magnetic material CeSb, which exhibits over a dozen such phases[1]. One of the simplest models capable of displaying analogous behavior is the axial next-nearest neighbor Ising model or ANNNI model. In d dimensions the model consists of (d-1)-dimensional layers of Ising spins ($s_i = \pm 1$) with nearest neighbor ferromagnetic couplings, $J_o > 0$, within layers but competing ferromagnetic, J_1, and <u>anti</u>-ferromagnetic, $J_2 = -\kappa J_1 < 0$, axial couplings between spins <u>in</u> first and second-neighboring layers. The ground state is ferromagnetic for $\kappa < 1/2$; of (2,2) or $\langle 2 \rangle$, antiphase character (two layers 'up', two 'down' periodically) for $\kappa > 1/2$, which matches the zero field ground state of CeSb; but infinitely degenerate for $\kappa = 1/2$. The high-temperature paramagnetic phase terminates[2] in a ferromagnetic critical line for $\kappa < \kappa_L \simeq 0.27$ but is bounded by a modulated critical line for $\kappa > \kappa_L$ with a critical wave-vector $q_c(\kappa)$, increasingly monotonically from zero at $\kappa = \kappa_L$. Thus $[\kappa = \kappa_L, T = T_c(\kappa_L)]$ is a Lifshitz point which, in fact, matches that found in MnP surprisingly closely[3]. At intermediate temperatures Monte Carlo work[4] is informative but the technique proves incapable of resolving definitely the fine features of the phase diagram.

At low temperatures it has proved possible[5,6] to construct systematic expansions in powers of $w = \exp(-2J_o/k_BT)$ about <u>all possible</u> ground states and to carry the expansions where necessary <u>to all</u> orders. For d > 2 these expansions show that ($\kappa = 1/2$, $T = 0$) is a <u>multiphase point</u> from which springs an infinite cascade of distinct, spatially modulated phases of layer structure defined by $< 2^{j-1}\, 3 >$ with $j = 1,2,3,\ldots$ and with (mean) wavevectors $\bar{q} = \pi j/(2j+1)a$ (where the spacing between lattice planes is a). The free energy, phase boundaries, and corresponding domain wall energies (or tensions) can be given explicitly at low temperatures. On approaching the commensurate "melting" line, $\kappa = \kappa_\infty(T)$, of the $< 2 >$ phase, the wave vector varies quasicontinuously as $1/|\ln[\kappa_\infty(T) - \kappa]|$ and displays a "devil's top step" just as the $< 2 >$ phase locks in. This variation corresponds to an exponential repulsion between "walls", or "solitons", or "discommensurations", represented by separated bands of three adjacent, coaligned layers. Such a repulsion and reciprocal logarithmic variation is predicted by one-dimensional models at $T = 0$ and by mean-field treatments which neglect statistical fluctuations[7].

A similar low-temperature analysis can be given for the p-state chiral clock models[8] in which the Ising spins are replaced by Potts variables, $n_i = 0,1,2,\ldots(p-1)$, and the axial coupling is taken as $-J_1 \cos[2\pi(n_i - n_{i+1} - \kappa)/p]$, where κ is now the chirality parameter (which vanishes within layers). For $p \geqslant 3$ and $d > 2$ one again finds an infinite sequence of phases[9,10], $< 12^j >$ and $< 32^{j-1} >$ springing from a multiphase point at ($\kappa = 1/2$, $T = 0$) but for $p = 3$ all but five of the phases terminate at low-temperature triple points, $T_{t,j}$ which approach zero as the order, j, of the phases increases to infinity : consequently, at any fixed T, one has only a 'harmless' staircase of wavevectors with no devil's top step[11].

The effect of a magnetic field on the ANNNI model can be studied by mean field theory[12] and by extending the systematic low-temperature expansion technique[13]. At intermediate temperatures, below $T_c(\kappa)$, however, the low temperature expansions are inadequate and it is hard to determine the true phase diagram. For $d \geqslant 3$, nonetheless, it seems likely that the layer-by-layer mean field theory initiated and developed by Bak[14] provides a qualitatively (and even semiquantitatively) correct account of the complex nature of the phase diagram with, probably, first a complete and then an incomplete devil's staircase of wavevectors occurring sufficiently close to the phase boundary[15]; but it remains a difficult challenge to substantiate this conjecture by more rigorous methods.

ACKNOWLEDGEMENTS

In the work summarized above I have benefitted from interactions with Per Bak, David A. Huse, Walter Selke, Julia M. Yeomans and Jacques Villain. Support has been provided by the National Science Foundation, in part through the Materials Science Center at Cornell University, and by the John Simon Guggenheim Memorial Foundation.

REFERENCES

1. For review see B. Lebech, J. Appl. Phys. 52 : 2019 (1981) and the lecture by R. Currat in these Proceedings.
2. S. Redner and H.E. Stanley, Phys. Rev. B 16:4901 (1977) and J. Phys. C10:4765 (1977).
3. See the lectures by Y. Shapira in these Proceedings.
4. W. Selke and M.E. Fisher, Phys. Rev.B20:257 (1979).
5. M.E. Fisher and W. Selke, Phys. Rev. Lett. 44:1502; 45:E148 (1980); Phil. Trans. Roy. Soc. 302:1 (1981).
6. For a short review of the results and the theoretical technique see M.E. Fisher, J. Appl. Phys. 52:2014-18 (1981).
7. See e.g. F.C. Frank and J.H. van der Merwe, Proc. Roy. Soc. 198:205 (1949) and articles in "Solitons in Condensed Matter Physics", Eds. A.R. Bishop and T. Schneider (Springer-Verlag, Berlin, 1978).
8. S. Ostlund, Phys. Rev. B24:398 (1981); D.A. Huse, Phys. Rev. B 24:5180 (1981); see also the overview below by the author in these Proceedings.
9. J.M. Yeomans and M.E. Fisher, J. Phys.C 14 :L835 (1981)
10. J.M. Yeomans, J. Phys. C 15:7305 (1982) and these Proceedings.
11. Such behavior was originally suggested for the ANNNI model by J. Villain and M. Gordon, J. Phys. C 13:3117 (1980).
12. See S.R. Salinas in these Proceedings.
13. V.L. Pokrovsky and G.V. Uimin, J. Physique (1983); J. Smith and J.M. Yeomans in these Proceedings.
14. J. von Boehm and P. Bak, Phys. Rev. Lett. 42:122 (1979); P. Bak and J. von Boehm, Phys. Rev. B21:5297 (1980); M.H. Jensen and P. Bak, Phys. Rev. B27 (May, 1983) and in these Proceedings.
15. See also M.E. Fisher and D.A. Huse in "Melting Localization and Chaos", eds. R. Kalia and P. Vashishta (North-Holland, New York, 1982) pp. 259-293.

COMMENSURABILITY, CHAOS, AND THE DEVIL'S STAIRCASE

Per Bak

Physics Department
Brookhaven National Laboratory
Upton, New York 11973 U.S.A.

M. Hoegh Jensen

H. C. Orsted Institute
Universitetsparken 5
Copenhagen, Denmark

INTRODUCTION

These lectures deal with the effects of the discrete lattice on modulated structures in solid state physics. The modulated structure could be a periodic lattice distortion, a sinusoidal or helical magnetic structure, staging of alkali metals in graphite intercalation compounds, polytypism in crystal formation, or some other periodic arrangement.

Modulated structures are often discussed within a continuum picture, where the lattice constant plays no role in determining the periodicity. The wave vector varies in a smooth way when some parameter is changed, and the period will generally be incommensurable with that of the crystal lattice. The discreteness of the lattice causes some interesting qualitative features not contained in the continuum models to be discussed here:

- Commensurability. The lattice acts as a periodic potential on the modulated structures which favors periodicities which are commensurable with the lattice constant. The wave vector may jump between several, or even an infinity of commensurate values. Incommensurate phases may or may not exist between the commensurate ones[1].

- Pinning. The transition from commensurate (C) to incommensurate (I) structures is brought about by formation of local defects ("solitons") in the commensurate structure. The solitons are subjected to a periodic field of the lattice (the Peierls potential) which tends to "pin" the solitons to the lattice. If the solitons are pinned in a regular way, the resulting modulated phase will be high-order commensurate. If they are pinned in an irregular way (which they might well be in a real system), we will have

- Chaos. It may seem surprising that stable chaotic structures may appear in systems with no intrinsic quenched disorder. A simple example: consider the simple Frenkel-Kontorowa model (Fig. 1) consisting of a periodic array of particles, connected with harmonic springs, in a periodic potential. If the spring constant is weak enough, clearly there exist stable configurations in which the particles are distributed in a more or less random way among the potential minima. The chaotic states that we will be dealing with in the following are the more sophisticated pinned soliton states described above. The situation is quite analogous to the chaos in dynamical systems (with no intrinsic disorder) to be discussed in the lectures by Grossmann: "Chaos in time" may set in as a consequence of the interplay between an intrinsic fundamental periodic motion and (for instance) an external periodic perturbation in a dynamical system[2,3]. In our case "Chaos in space" is caused by the interplay between the intrinsic period of the modulated structure and the fundamental lattice constant. In fact, we shall see that the external configurations of models such as the Frenkel-Kontorowa model exhibit infinite series of bifurcations with universal exponents similar to those found by Feigenbaum for dynamic systems[4].

Rather than discussing these various phenomena in general we shall concentrate our efforts on some specific simple models which I believe are quite representative:

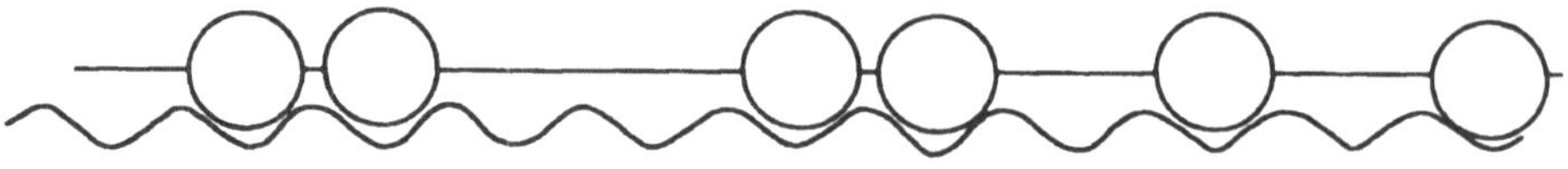

Fig. 1 "Chaotic" structure in the Frenkel-Kontorowa model.

- The 1d Ising model with long range antiferromagnetic interactions in a magnetic field. It has been suggested that the model describes the "staging" phenomenon in intercalation compounds[5] or neutral-ionic transitions in organic charge transfer compounds[6]. As the field is varied the periodicity of the spin arrangement jumps between an infinity of commensurate values, with no incommensurate phases between (Fig. 6). This phenomenon is known as the complete devil's staircase. The width of the commensurate steps obeys a simple scaling relation which is described by the fractal dimension of the staircase. The discussion of the 1d Ising model follows the work of R. Bruinsma and one of the authors[7].

- The 3d Ising model with competing interactions[8] (ANNNI model). This model exhibits a complicated pattern of commensurate and incommensurate phases (Fig. 2). The model has been studied extensively (Ref. 1 and references therein), and here we shall restrict ourself to demonstrate how the mean field theory of the model can be written as a four-dimensional mapping similar to those which have been applied to dynamic systems. Numerical studies of the map reveal pinned, unpinned and irregular chaotic soliton arrangements in addition to simple commensurate and incommensurate phases[9]. Near the C phases, the I phase vanish due to pinning of solitons. The ANNNI model may describe complicated magnetic structures like those found in CeSb[10], which will be discussed in the lectures by Currat.

- The discrete ϕ^4 model. This model describes an array of particles in a double well potential. The model has an infinite series of bifurcations leading to chaos, characterized by the same universal numbers as those of the Hénon model[11]. A slight modification of the potential causes the bifurcations and the chaos to disappear.

THE 1d ISING MODEL WITH LONG RANGE ANTIFERROMAGNETIC INTERACTIONS

Since Onsager solved the 2d Ising model decades ago it may seem surprising that anybody bothers with the ground state of the 1d model today. Consider first a few real physical systems: Fig. 3a illustrates the phenomenon of staging in graphite intercalation compounds. The broken lines represent layers of, for instance, alkali atoms intercalated between graphite layers. The "staging" is the number of graphite layers between successive intercalated layers, and depending upon the conditions under which the compound is formed, stagings between 1 and 11 may occur. Fig. 3a shows a stage 3 compound. Safran[5] has proposed the following phenomenological model. Associate a variable σ_i

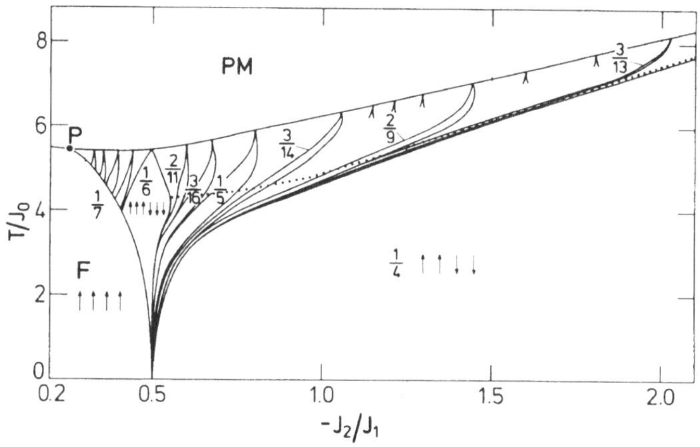

Fig. 2 Phase diagram of the 3d Ising model with competing interactions. (Bak and Von Boehm, Ref. 8).

to each space i between graphite layers, such that $\sigma_i = 1$ if an alkali layer is present and $\sigma_i = 0$ if it is not, and a long range power law repulsive interaction V(i-j) between layers. The energy of a configuration of filled and empty layers is then given by the expression

$$H = \sum_i \mu\sigma_i + \sum_{ij} V(i-j)\sigma_i\sigma_j , \quad (1)$$

where μ is a chemical potential.

Another system to consider is a mixed-stack organic charge transfer compound undergoing a neutral-ionic transition (Ref. 6). Fig. 3b shows schematically the structure of such a system. The +'s indicate donor molecules which have transferrred electrons to occupied acceptor molecules (the -'s). The O's indicate neutral acceptor or donor ions. Let I be the donor ionization energy, A the electron affinity of the acceptor, and

V(i-j) the (exponential) repulsion between ionized layers. Then the energy of the compound may be written

$$H = \sum_i (I-A)p_i + \sum_{ij} V(i-j)p_i p_i \tag{2}$$

Both (1) and (2) can trivially be transformed to the 1d Ising model in a field.

$$H = -\sum_i HS_i + \sum_{ij} J(i-j)(S_i + \tfrac{1}{2})(S_j + \tfrac{1}{2}),\ S_i = \pm\tfrac{1}{2} \tag{3}$$

so a solution to (3) will provide insight into the systems described above. We have written the Hamiltonian in an asymmetric way where only up spins interact.

In the following we shall follow Bak and Bruinsma[7]. The only assumption which will be made is that the interaction J is convex: $J(n+1) + J(n-1) - 2J(n) > 0$ for all n. The interaction could thus be either a power law or an exponential interaction. It is not difficult to see that for $H < 0$ the ground state is the one with all spins down since this would minimize both terms in (3). Also, it is easy to see that for $H > 2\sum_i J(i)$ the state with all spins up is stable since it would then cost energy to flip a single spin from + 1/2 to − 1/2. The problem then is to determine the stable structure for

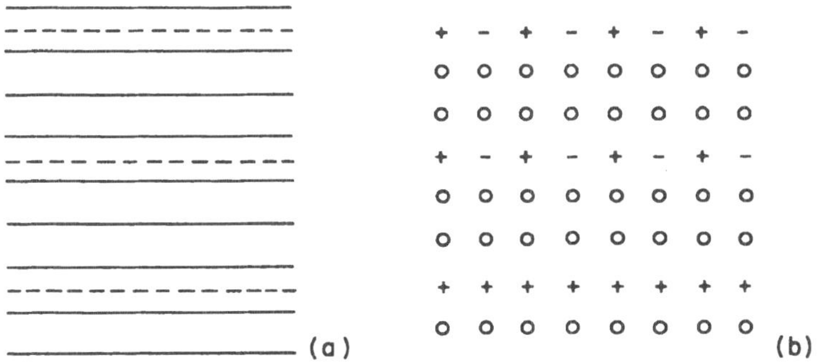

Fig. 3(a) Graphite intercalation compound (schematic). The full lines represent graphite layers, the broken lines the intercalated atoms. b) Charge-transfer compound (schematic).

$$0 < H < 2\sum_i J(i) \tag{4}$$

The analysis of the ground state of the model has two steps. First, for a given ratio of up spins to down spins, q, we find the configuration which minimizes the energy (3). Then we determine the range of fields for which the structure with a given q is stable.

Let there be N^+ up spins and $N^- = N-N^+$ down spins, and let X_i^p denote the distance between the i'th up spin and the p'th nearest up spin in the positive direction. For the configuration in Fig. 4, $X_1^1 = 3$, $X_1^2 = 4$, $X_1^2 = X_1^1 + X_1^2 = 5$, etc. A little thought will convince the reader that

$$\sum_{i=1}^{N^+} X_i^p = pN \tag{5}$$

The average value of X_i^p is $pN/N^+ = p/q$. This quantity is not an integer in general. Assume that $\langle X_i^p\rangle$ is between the integers r_p and r_p+1

$$r_p \leqslant \langle X_i^p\rangle < r_p+1 \;, \tag{6}$$

and that we have generated a configuration for which

$$X_i^p = r_p \text{ or } X_i^p = r_p+1 \tag{7}$$

for all i, and all p. Suppose we want to decrease one of the distances $X_{i'}^p$, from r_p to r_p-1. In order to fulfil (5) we must compensate by increasing another p'th nearest up-spin distance $X_{i''}^p$ from r_p to r_p+1. This will cause the total p'th nearest neighbor energy to go up by an amount

$$\Delta E_p = J(r_p-1) + J(r_p+1) - 2J(r_p) > 0. \tag{8}$$

Similarly, if we increase a distance X_i^p from r_p+1 to r_p+2 we must decrease another distance from r_p+1 to r_p, causing an increase of the energy by an amount

$$\Delta E_p = J(r_p+2) + J(r_p) - 2J(r_p+1) > 0 \tag{9}$$

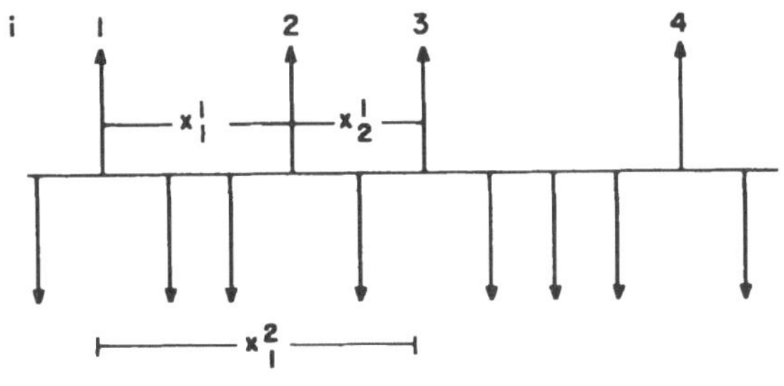

Fig. 4 Spin configuration of the Ising chain. Definitions of X_i^p.

By continuing this line of thinking one realizes that it is possible to reach any configuration fulfilling (5) through configurations with increasing energies. Therefore, our starting configuration minimizes the energy. But is it possible to find a configuration for which (7) holds? Indeed, if we choose the position X_i of the i'th up spin as

$$X = \mathrm{Int}(\frac{i}{q}) \tag{10}$$

then

$$\begin{aligned} X_i^p &= \mathrm{Int}(\frac{i+p}{q}) - \mathrm{Int}(\frac{i}{q}) \\ &= \mathrm{Int}(\frac{p}{q}) \text{ or } \mathrm{Int}(\frac{p}{q}) + 1. \end{aligned} \tag{11}$$

Fig. 5 shows the stable configurations for q = 1/3 and q = 10/23. Note that in general (10) will give an incommensurate ordered structure for fixed number of up-spins in the limit $N \to \infty$.In a scattering experiment, one would observe Bragg peaks at incommensurate (I) positions despite the fact that we have discrete spins on a discrete lattice. Hence, one can certainly not rule out ab initio the possibility of having I phases stable as the field in varying.

Next, let us consider the stability of the commensurate phase with q = m/n. This phase is stable as long as no energy can be gained by flipping one up-spin down, or one down-spin up, and rearrange the structure to minimize the energy. When one down-spin is flipped, the number of p'th nearest neighbor interactions which are r_p will increase by r_p+1, and the number of p'th nearest neighbor interactions which are r_p+1 will decrease by r_p for the equations (5) and (7) to hold with one up spin added. Summing over all p'th nearest neighbor energies, we find the total gain in energy:

$$\begin{aligned} U(\downarrow\uparrow)\binom{m}{n} = & \ H \\ & + (r_1+1)J(r_1) - r_1 J(r_1+1) \\ & + (r_2+1)J(r_2) - r_2 J(r_2+1) \\ & \ \vdots \\ & + nJ(n-1) - (n-1)J(n) \\ & \ \vdots \\ & + 2nJ(2n-1) - (2n-1)J(2n) \end{aligned} \tag{12a}$$

where $r_m = n$ has been inserted. Similarly, the energy gained by flipping one up-spin down is

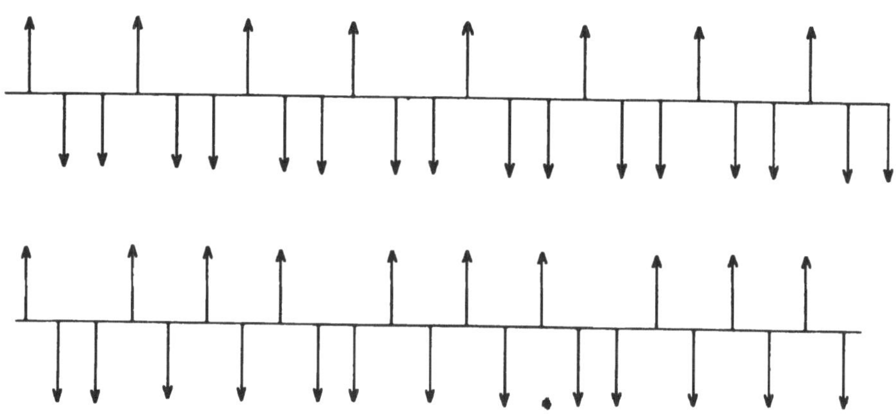

Fig. 5 Stable spin configurations for q = 1/3 and q = 10/23.

$$U(\uparrow\downarrow)(\tfrac{m}{n}) = - H - (r_1+1)J(r_1) + r_1J(r_1+1) - (r_2+1)J(r_2) + r_2J(r_2+1) - (n+1)J(n) + nJ(n+1) - (2n+)J(2n) + 2nJ(2n+1) \tag{12b}$$

The stability interval is the difference between the value of H for which $U(\downarrow\uparrow) = 0$ and the value of H for which $U(\uparrow\downarrow) = 0$:

$$\Delta H(\tfrac{m}{n}) = \sum_k kn(J(kn+1) + J(kn-1) - 2J(kn)). \tag{13}$$

This expression is positive for all m and n because of the convexity condition. Note also that the stability does not depend on the numerator m.

So, every single commensurate phase, $q = m/n$, is stable for a finite range of values of the magnetic field. Hence, $q(H)$ is a devil's staircase. The total width of all the commensurate phases is

$$\begin{aligned} \Delta H &= \sum_{m,n} (m,n \text{ irreducible}) \sum_k kn(J(kn+1) + J(kn-1) -2J(kn)) \\ &= \sum_{m,n} (\text{all } m,n)\ [n(J(n+1) + J(n-1) - 2J(n))] \\ &= \sum_n n(n-1)[J(n+1) + J(n-1) - 2J(n)]. \end{aligned} \tag{14}$$

Let us count the number of times the interaction $J(p)$ enters this expression. It appears for $n = p+1$, $n = p-1$ and $n = p$, so the number becomes $-2p(p-1) + p(p+1) + (p-1)(p-2) = 2$, and

$$\Delta H = \sum_p 2J(p) ,$$

i.e. the C phases fill up the whole interval (4), and the Devil's staircase is complete. The I phases can thus only be stable for a set of H values of zero measure, and I phases will not be observed in an experiment. Figure 6 shows the devil's staircase for $J(n) = 1/n^2$. Note the self-similarity ("scaling") of the staircase under magnification.

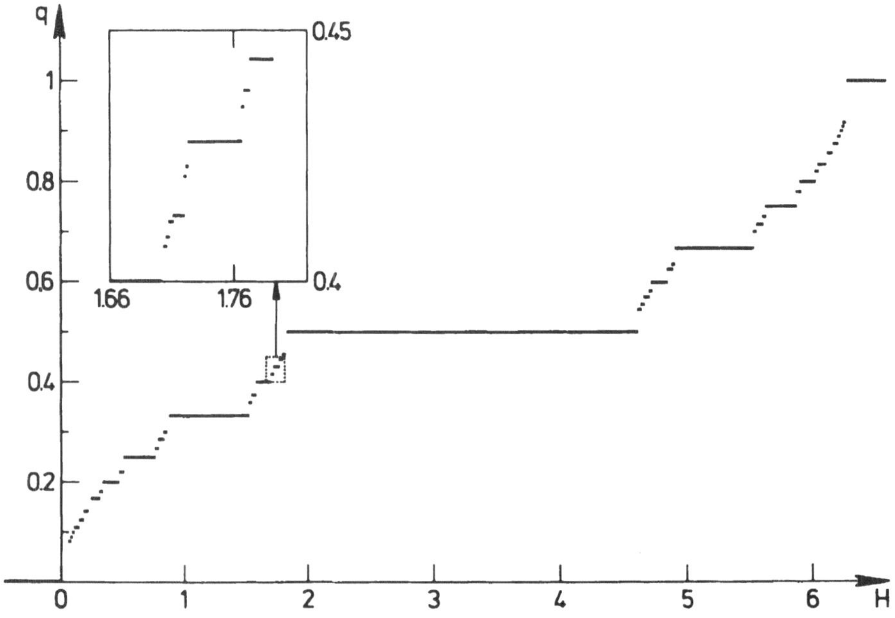

Fig. 6 The complete Devil's staircase for $J(n) = 1/n^2$.

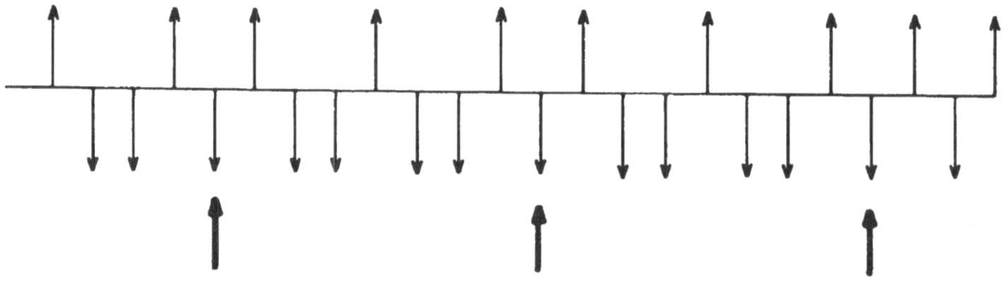

Fig. 7 Lowest excited state for $q = 1/3$ with three solitons.

Let us briefly consider the lowest lying excitations whose energy we have just calculated. Fig. 7 shows the lowest excited state for $q = 1/3$ which has one more up spin. Note the formation of 3 localized defects. For an infinite system these defects (or solitons) would be infinitely far apart. It is not difficult to see that the excess energy of this configuration is in fact given by (12a) with $n = 3$. Note that one spin has been flipped, and three solitons are formed. Hence the solitons have fractional spin $S = 1/3$. Solitons with any rational spin can be formed in this way. There is a very close connection between the fractional spins found here (which are defects or discommensurations in a spin system) and the fractionally charged objects found by Su, Schrieffer and Heeger[12] in a model of polyacetylene, which are discommensurations in a charge-density-wave system. Clearly, one can not isolate fractional spins by breaking the chain between the solitons. We shall see later in connection with the ANNNI model that such defects generally play an important role in determining the stability of C phases with respect to I phases.

The soliton states, and all incommensurate states belong to the Cantor set of measure zero which is complementary to the stability intervals for the commensurate phases. The scaling properties of the staircase are described by the fractal dimension[13] or Hausdorf dimension, D, of the staircase or, more precisely, the fractal dimension of the complementary Cantor set. Let N_s be the number of steps which are wider than a given scale s. Then

$$D = \frac{\ln N_s}{\ln 1/s} \qquad (15)$$

If we choose s equal to the width of a C phase with denominator n, N_s is the number of rational numbers with denominator < n, i.e. $N_s \sim n^2$. For a power law interaction, $J(i) \sim i^{-\alpha}$, Eq. (14) yields $\Delta H(\frac{m}{n}) \sim s \sim n^{-\alpha-1}$, hence

$$D = \frac{\ln n^2}{\ln n^{\alpha+1}} = \frac{2}{\alpha+1} \qquad (16)$$

The dimension is thus between 0 and 1 for $\alpha > 1$, as it should be for a set of zero measure. The fractal dimension gives information about the amount of additional phases which will be observable for a given increase of the resolution. Of course, one can not "prove" in an experiment that a staircase is complete since this would require infinite resolution. What one can do is to investigate the scaling properties and see if this is consistent with a fractal dimension between 0 and 1. For exponential interactions, D = 0. In fact there exist systems (such as CeSb[10]) which exhibit many commensurate phases. However, the number (∿8) is too small for a meaningful scaling analysis to be performed.

What are the implications for the intercalation compounds of the results found here? If the staging is indeed described by the Safran model (1), there should exist fractional staging in addition to the integer stages discussed by Safran. Fig. 8 shows an example of a fractional staging with q = 2/5 together with the corresponding spin structure. The stability interval of the 2/5 phase should be the same as for the integer 1/5 phase, but this would probably be taking the model too seriously. The application of the model to neutral-ionic transitions is not well-established, so it seems speculative to expect any significance of our results in this case. The exponential interaction would apply that only very few phases be observable.

THE ANISOTROPIC ISING MODEL WITH COMPETING INTERACTIONS

The 1d Ising model which was solved in the preceding section has commensurate phases only. The 3d ANNNI model is much richer

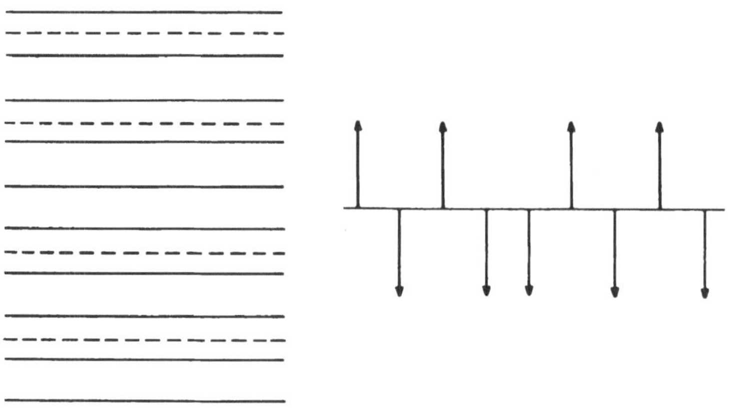

Fig. 8 Fractional staging (q = 2/5) in intercalation compound, and the corresponding spin configuration.

and shows essentially all the various features that we are interested in: Incommensurate phases, commensurate phases, pinned and unpinned solitons, and chaotic metastable states. Unfortunately, it can not be solved rigorously.

In the ANNNI model, Ising spins are situated on a cubic lattice and interacting with ferromagnetic nearest neighbor interactions J_1 and next nearest neighbor interactions J_2 in the uniaxial Z-direction, and a ferromagnetic interaction J_0 in the perpendicular direction.

Figure 2 shows the phase diagram for the model, derived by a combination of analytical methods and brute force numerical work. The phase diagram consists of a complicated network of commensurate and incommensurate phases[8]. An infinity of phases come together at $T = 0$, $J_2/J_1 = -0.5$[8]. As the temperature is increased, more commensurate phases appear. In general not even the mean field theory can be solved (which should be perfectly clear at the end of this section!) but near T_c it can. Near the critical line the widths of the commensurate phases $q = m/n$ go to zero as $(T_c-T)^{(n-2)/4}$ for n even, $(T_c-T)^{(2n-2)/4}$ for n odd. This result was derived by calculating the stability of the C phases with respect to local defects like the ones found for the 1d-long range Ising model. In the limit $T \to T_c$ all possible C phases will eventually become stable; however their total width goes to zero as $T \to T_c$, so, in contrast to the situation for the 1d model the devil's staircase will not be complete near T_c. There is room for I phases between the infinity of C phases. Critical fluctuations will change the exponents describing the vanishing of the C phases near T_c[14], but they are not expected to change the topology of the phase diagram. Near $T = 0$ only commensurate phases exist as discussed in the lecture by Fisher. The mean field theory can be solved in this limit and gives a result which is in agreement with the low temperature expansion of Fisher and Selke[15]. Since I phases exist near T_c, but not near $T = 0$ they must disappear in between, but how and where does this happen?

Mean field theory as a 4d mapping

The mean field free energy for the Ising model is

$$F = - \sum_{\langle ij \rangle} J_{ij} M_i M_j + T \sum_i \int_0^{M_i} \tanh^{-1} \sigma d\sigma \tag{17}$$

$$= - \sum_{\langle ij \rangle} J_{ij} M_i M_j + \sum_i \frac{T}{2} \left[(1+M_i) \ln(1+M_i) + (1-M_i) \ln(1-M_i) \right]$$

where J_{ij} is the appropriate interaction between spins on sites i and j, and M_i is the magnetization $\langle S_i \rangle$ of the spin at site i. Our problem is "simply" to minimize this expression with respect to the M_i's. Since J_0 couples spins within planes ferromagnetically, we shall assume that M_i depends only on the uniaxial coordinate. The condition for extremum, $\partial F/\partial M_i = 0$, leads to the MF equations

$$\begin{aligned} T \tanh^{-1} M_i &= 4J_0 M_i \\ &+ J_1(M_{i+1} + M_{i-1}) \\ &+ J_2(M_{i+2} + M_{i-2}) \end{aligned} \tag{18}$$

where M_i is now the magnetization of a spin in the i'th plane. Eq. (18) is a relation between the magnetization in five consecutive layers. If we know the magnetization in the four layers i-2, i-1, i, i+1 we can thus find the magnetization in the layer i+2 by the relation

$$\begin{aligned} M_{i+2} &= f(M_{i-2}, M_{i-1}, M_i, M_{i+1}) \\ &= - M_{i-2} - \frac{J_1}{J_2}(M_{i-1} + M_{i+1}) - \frac{4J_0 M_i}{J_2} + \frac{T}{2J_2} \ln \frac{1+M_i}{1-M_i} \end{aligned} \tag{19}$$

This defines a mapping

$$T : (M_{i-2}, M_{i-1}, M_i, M_{i+1}) \to (M_{i-1}, M_i, M_{i+1}, f(M_{i-2} \ldots M_{i+1})) \tag{20}$$

By iterating (19) or, equivalently, iterating the mapping (20) one may in principle generate all possible solutions to the MF theory. Usually one needs the orbit which minimizes the free energy, and if a starting point is chosen at random it is extremely unlikely that it belongs to the desired trajectory.

Among the orbits generated by successive iterations are limit cycles with $T^N V = V$, describing commensurate phases of order N. A limit cycle is mathematically stable (to be distinguished from physically stable!) if a neighborhood around V is transformed to the same neighborhood around V. If the trajectory diverges under repeated iteration, the limit cycle fixed point is said to be mathematically unstable. The stability can be determined by studying the eigenvalues of the linearized mapping DT^N at a limit cycle point. If all eigenvalues are on the complex unit circle, the limit cycle is mathematically stable, otherwise not. The problem is that the physically stable configurations which correspond to (local) minima of (17) are all mathematically unstable[16]. This hampers a numerical search for stable orbits tremendously, since the orbits can only be followed during a finite number of iterations.

Incommensurate phases

The most general incommensurate structure can be written

$$M_n = g(qn+\alpha)$$

where

$$g(x) = g(x+2\pi)$$
$$g(x+\pi) = -g(x) \qquad (21)$$
$$g(x) = g(-x) \; .$$

Here q is the irrational wave vector (or winding number) of the I phase. The energy is independent of the phase α which can be chosen at will. The soft mode associated with this continuous symmetry is usually called the phason. The orbit corresponding to the structure (21) consists of the points

$$V_n = (g(qn+\alpha),\ g(q(n+1)+\alpha),\ g(q(n+2)+\alpha),\ g(q(n+3)+\alpha)). \qquad (22)$$

Since q is irrational this defines a one-dimensional orbit which is filled up ergodically. The orbits consist of all points of the form $(g(x), g(x+q), g(x+2q), g(x+3q))$ with $0 < x < 2\pi$. Incommensurate phases are thus described by one-dimensional orbits. At first sight, one has to search a space of dimension four to find the 1d orbit. However, the orbit (22) passes through points of specially high symmetry. For instance, choosing $\alpha = -5q/2$ the initial point can be written

$$V_1 = (g(\tfrac{-3q}{2}),\ g(\tfrac{-q}{2}),\ g(\tfrac{q}{2}),\ g(\tfrac{3q}{2}))$$
$$= (b,a,a,b) \; , \qquad (23)$$

and we have reduced the search to a 2 parameter problem. Fig. 9 shows an I orbit slighty below T_c, and the corresponding spin configuration, which is almost sinusoidal. The stable wave vector is simply the winding number of the orbit which has the smallest free energy.

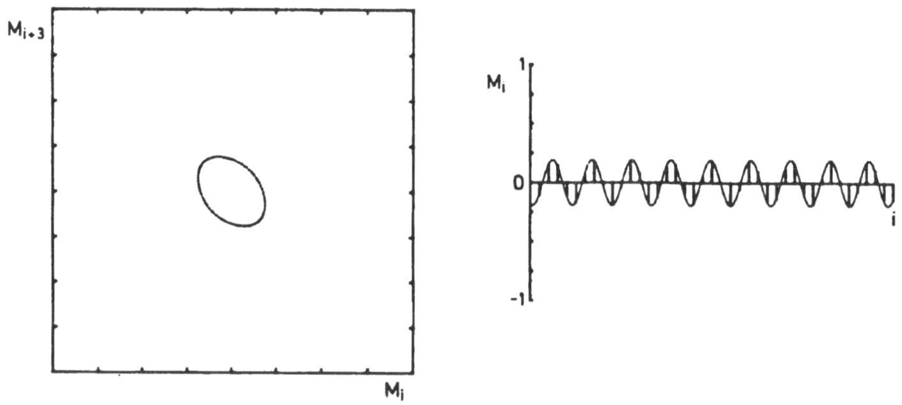

Fig. 9 One-dimensional incommensurate trajectory found numerically for $J_2/J_1 = -2$, $T = 8.05\ J_0$.

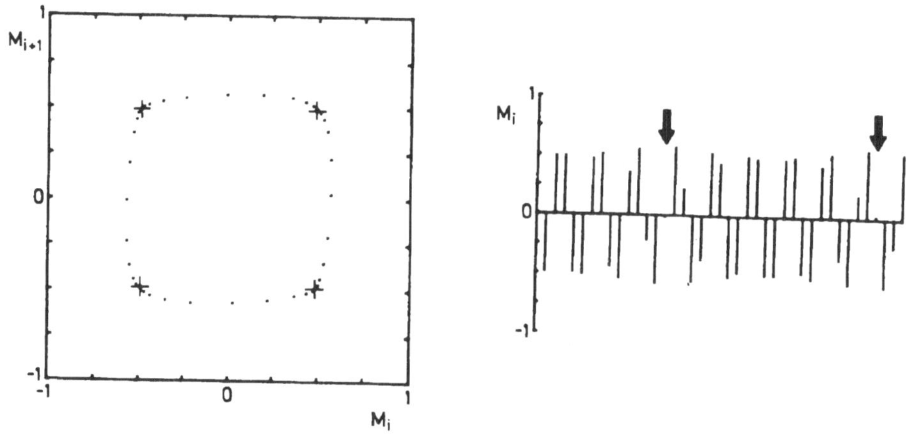

Fig. 10 Unpinned soliton lattice near the CI transition.

Pinning of solitons near the CI transition

The low order C phase with q = 1/4 fills up a substantial fraction of the phase diagram. Fig. 10 shows the trajectory and the corresponding spin structure for an incommensurate phase near the lock-in transition. It is an unpinned regular lattice of defects which separate almost commensurate regions. The solitons have different shapes since there is still a continuous phase associated with a sliding of the spin structure (or the soliton lattice). Compare with the defect structure Fig. 7.

Fig. 11 shows the free energy as a function of the wave vector, F(q), for a temperature which is very close to the lock-in transition. The most surprising feature is that when the wave vector becomes too close to the commensurate one, (1/4-q)<0.015, the I phases cease to exist. We shall argue later that this is due to pinning of solitons. The energies of the C phases q = 7/30, q = 4/17 and q = 1/4 fall approximately on a straight

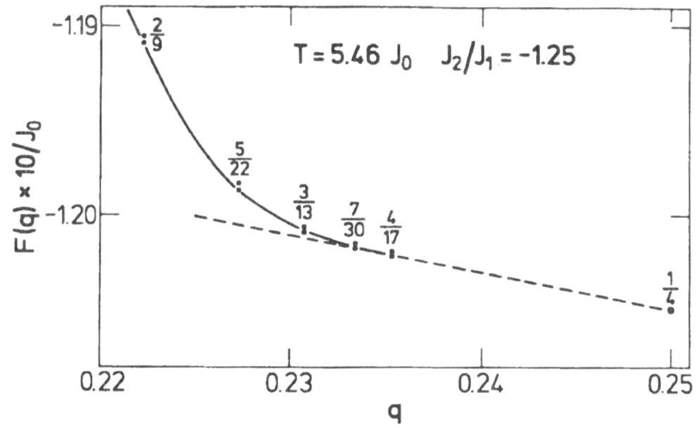

Fig. 11 Free energy vs. wave vector near the CI transition. The two dots at each C value indicate the energies of the pinned and the "depinned" soliton configurations.

line. This means that the excess free energy in the "I" phase is proportional to the number of solitons, so the slope can be considered the soliton energy. In Fig. 11 the slope (starting from the $q = 1/4$ phase) is positive and the C phase is stable. At a slightly higher temperature the soliton energy becomes negative and the I phase is stable.

To understand the origin of the disappearance of the I phases we must consider the effects of pinning of the solitons to the lattice. An I phase always has a Goldstone mode associated with the phase α in Eq. (21), and a shift of α will not change the energy. A C structure is generally locked to the lattice, and a shift of α will cost energy: this is in fact what is meant by pinning. When the high-order C phase takes the form of a soliton lattice it is favorable for the solitons to be pinned such that the spin at the center of the soliton assumes the maximum value.

Figure 12 shows commensurate phases for $\alpha = -5q/2$ and for $\alpha = -5q/2 + \pi/2$. In both cases the configuration is a soliton lattice. In the former case the spin assumes its largest values at the center, and in the latter unfavorable case the magnetization is zero at the centers indicated by arrows. The width of the solitons is 3 lattice units. Compare again with the soliton lattice in Fig. 7, which could also be classified as a pinned soliton structure. The energies of the "pinned" and "depinned" configurations are indicated by two dots at each commensurate phase in Fig. 11. The pinning energy is the difference between these two energies. The pinning energy is proportional to the soliton density: Fig. 13 shows the pinning energy per soliton which indeed is independent of the number of solitons,

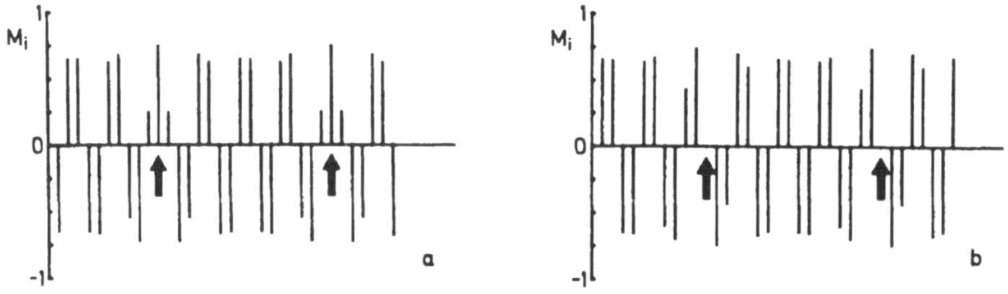

Fig. 12. Pinned and "depinned" soliton structures for $q = 4/17$.

justifying the interpretation of the pinning energy as a soliton pinning energy.

Bak and Pokrovsky[17] have argued that the I phase should disappear, and the soliton lattice became pinned in a regular or (most probably) irregular way when the soliton distance ℓ becomes so large that the interaction between solitons can not overcome the pinning energy. The soliton interaction energy, E_{int}, is the deviation from the straight line, per soliton, of the free energy, Fig. 11, which is also shown in the log plot in Fig. 13. The straight line indicates an exponential interaction as could be expected from analytical considerations[8]. The pinning takes place above the critical value ℓ_c at which $E_{int} \approx E_{pin}$. From the figure we find $\ell_c \sim 14$ lattice constants corresponding to a critical wave vector $q_c \sim 0.232$. As can be seen on Fig. 11 this is in good agreement with the point beyond which incommensurate phases could not be located numerically.

In Fig. 2 we have indicated very schematically the line beyond which there are no I phases. Probably an I phase will cease to exist every time a C phase is approached, when the distance between the appropriate solitons becomes too large. The

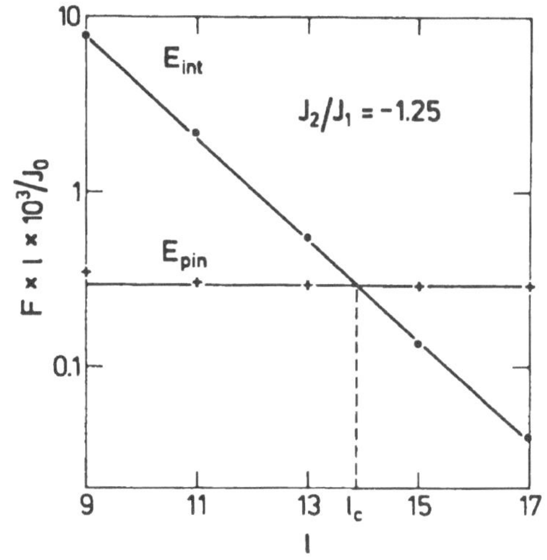

Fig. 13 Pinning energy E_{pin} and interaction energy E_{int} per soliton vs the distance between solitons.

critical line (shown by dots) must in principle be very irregular or "fractal"[13] since it must approach T_c at every rational point. Above the line, the I phases have finite measure so they are indeed important in the upper part of the phase diagram. At low temperatures there are only commensurate phases separated by first order transitions. In our terminology, the whole low temperature region belongs to the pinned regime. We cannot rule out small first-order transitions between commensurate and incommensurate phases.

Let us return to the regime where the incommensurate phases do not exist. The considerations presented here establish the states in this regimes to be pinned soliton arrangements. Assuming the picture of exponentially interacting solitons to be valid, the theory for the 1d long range Ising model can be applied diretly, and the true ground state is always a high-order C phase. As the parameters are varied, the system will lock into the infinity of phases forming the devil's staircase.

In the pinned regime we were able to locate not only higher-order C phases, but also irregular curves which seem to have 1d character. The spin configuration (Fig. 14) is an array of randomly pinned solitons. It is very difficult to locate the various metastable and stable configurations numerically in the pinned regime, and to compare their energies in order to detemine the ground state. This is because of the extremely weak interactions between solitons, which rapidly becomes lower than the numerical accuracy. From a physical point of view we are dealing with pinned, noninteracting solitons. In a diffraction experiment one would not see Bragg peaks at the average q-vector but

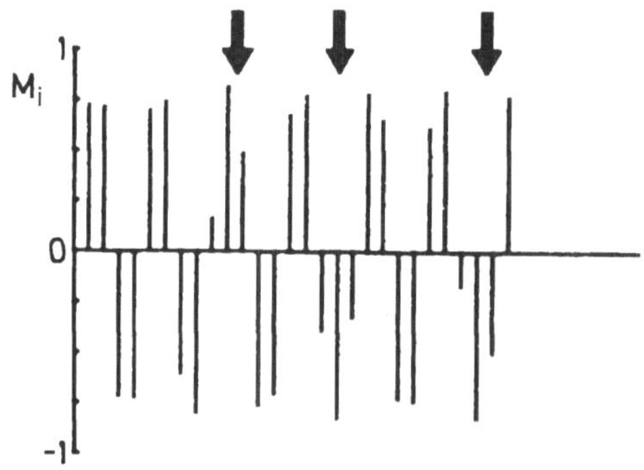

Fig. 14 Chaotic pinning of solitons.

smeared peaks corresponding to a random "chaotic" configuration of solitons, since the system can not diffuse to the real ground state through the high Peierls pinning barriers in a finite amount of time.

We therefore suggest to call this regime the chaotic phase. The situation corresponds in many ways to the one for spin-glasses: there are several almost degenerate metastable states separated by high energy barriers, and the system will not generally be in the ground state. Here, we can identify the metastable states as pinned soliton states; generally one can not identify these states in the various spin-glass models.

THE DISCRETE ϕ^4 MODEL

As a final example showing the connection between chaos in discrete lattice models and in dynamical systems, let us consider the discrete ϕ^4 model which has an infinite series of geometrically converging bifurcations leading to chaos[11]. The discrete ϕ^4 model is an array of particles, connected with springs, in a double-well potential:

$$E = \sum_n \frac{1}{2}(\phi_n - \phi_{n-1})^2 + \frac{1}{4} a(\phi_n^2 - 1)^2 \tag{24}$$

The stable configurations are found by differentiating (24) with respect to ϕ_n:

$$(\phi_{n+1} - \phi_n) - (\phi_n - \phi_{n-1}) = a\phi_n(\phi_n^2 - 1) , \tag{25}$$

which can be written as a 2d area-preserving mapping

$$T : \begin{pmatrix} \phi_n \\ \phi_{n-1} \end{pmatrix} \rightarrow \begin{pmatrix} \phi_{n+1} = a\phi_n(\phi_n^2-1) + 2\phi_n - \phi_{n-1} \\ \phi_n \end{pmatrix} \tag{26}$$

In addition to regular C and I solutions the equations have chaotic solutions[17,11]. Fig. 15 shows trajectories generated by numerical iterations for a = 8/9. The narrow closed islands represent unstable modulated C states, and the scattered points belong to a chaotic trajectory.

Now, focus the attention on the elliptic fixed point (0,0) in the middle. Figs. 16a and 16b show plots for a = 3.98 and a = 4.02. A bifurcation of the elliptic fixed point into two elliptic fixed points takes place between these two values, and the fixed point itself becomes hyperbolic. Note that the fixed point which bifurcates is elliptic, i.e. mathematically stable, and hence physically unstable. No direct application of the phenomenon to real physical systems is obvious.

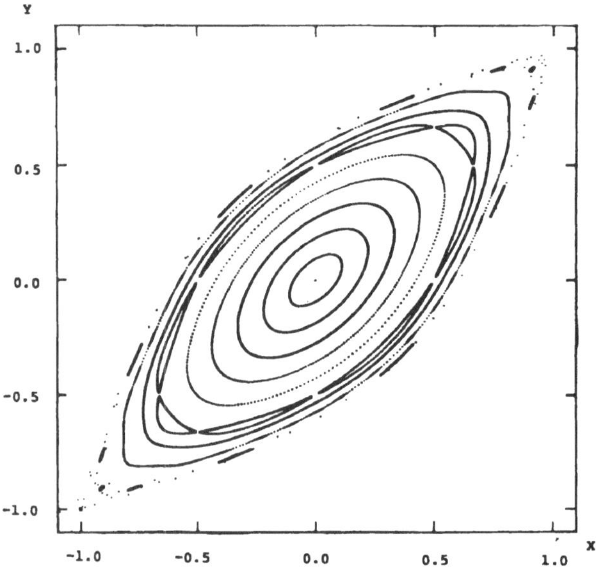

Fig. 15 Orbits of the discrete ϕ^4 mapping for a = 8/9.

Figs. 16c and 16d show trajectories in the neighborhood of one of the symmetric bifurcated two-cycle points for a = 4.995 and a = 5.005. At a = 5 the fixed point becomes marginally stable. The "starfish" is shown to illustrate the richness of the mapping.

Figs. 16e and 16f show plots for a = 5.98 and a = 6.02. At a = 6 the two-cycle splits into two two-cycles and not into a four-cycle as in a regular bifurcation. We call this phenomenon the "banana split." Figs. 16g and 16h (a = 6.24 and a = 6.245) show a bifurcation from a two-cycle to a four cycle. With increasing values of a, an infinite series of bifurcations, converging at a = 6.274515•••, seems to take place.

On the basis of the numerical calculations we can determine the Feigenbaum convergence number

$$\delta = \lim_{k\to\infty} \frac{a_{k-2}-a_{k-1}}{a_{k-1}-a_k}\ , \tag{27}$$

where a_k is the value of a at the k'th bifurcation. We find $\delta = 8.721096\cdots$. We also calculate the number

$$\alpha = \lim_{k\to\infty} \frac{\delta_{k-1}}{\delta_k} = 4.0180\cdots \qquad (28)$$

where δ_k is the distance between the elliptic fixed points which separated at the k-1'th bifurcation. These values are, up to the numerical accuracy, the same as for other two-dimensional area preserving mappings, as for instance the Hénon mapping[18], but they are different from the numbers found by Feigenbaum[4] for the "dissipative" case. This supports the hypothesis that α and δ are the same for 2d conservative ("area preserving") maps.

Chaotic solutions, bifurcations, etc. seem to be almost universal features of mappings such as (20) and (26). However, if the ϕ^4 potential is replaced by the slightly different potential[19,20]

$$V(\phi) = (1 - \frac{2}{a})\ln(1 - \frac{a}{2}\phi^2) - \phi^2 , \qquad (29)$$

the discrete equations corresponding to (25) can be solved analytically. Hence, there is no chaos. In fact, the solutions to (25) with the potential (29) are identical to the solutions for the ϕ^4 potential in the continuum approximation. They are expressed in terms of elliptic functions as for instance $\phi_n = \phi_0 \mathrm{sn}(nq)$.

Fig. 17 shows the ϕ^4 potential and the potential (29) for a = 1.6, and Fig. 18 shows the results of numerical iterations in the two cases. Note that the irregular chaotic solutions have been replaced by smooth 1d trajectories, representing incommensurate states. These curves are filled up ergodically as the iteration proceeds. No pinning of solitons takes place as can be seen by inserting the single soliton solution

$$\phi_n = \tanh(\sqrt{\frac{a}{2}}\, n - \alpha). \qquad (30)$$

The energy is independent of the position α of the soliton, so the potential (29) is pinning free.

It should be borne in mind, however, that the potential (29) is quite pathological and was selected in a careful way. In general, one would expect pinned and exponentially interacting solitons in commensurate or almost commensurate structures.

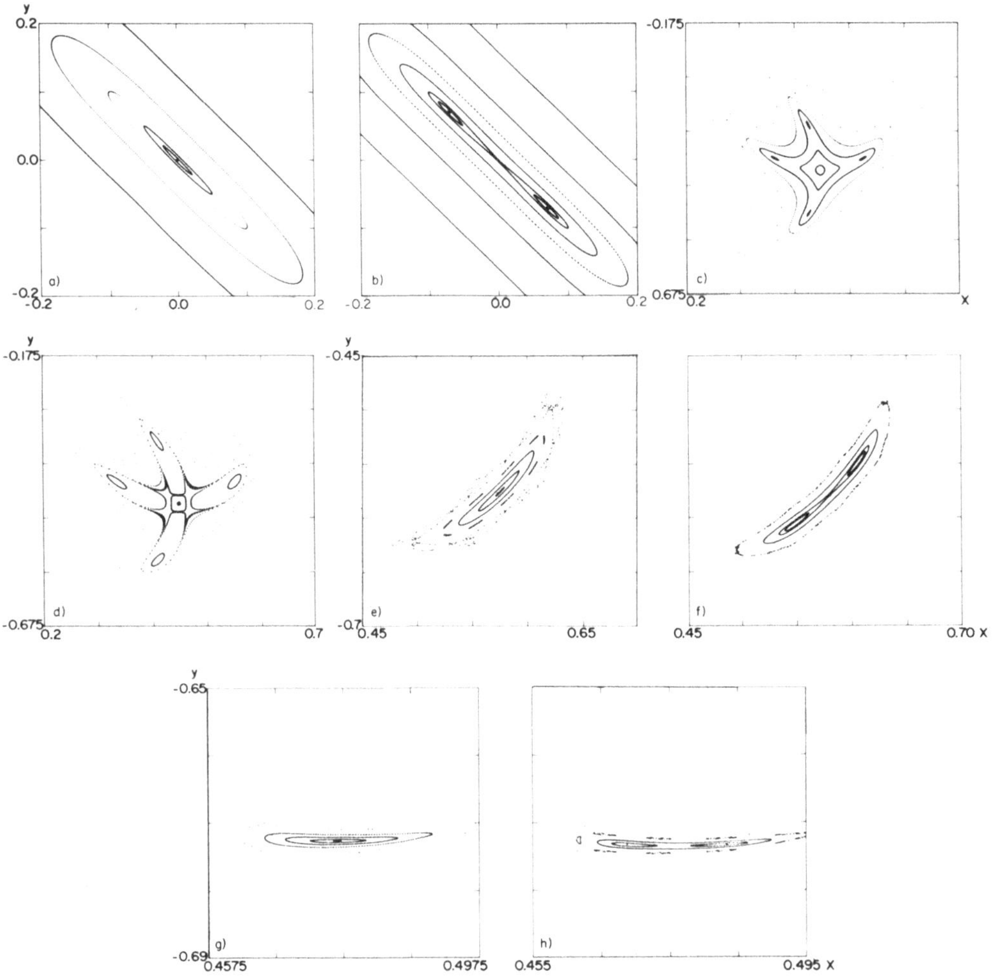

Fig. 16 Trajectories determined by numerical iterations for increasing values of a. The scale of the plots is varying.

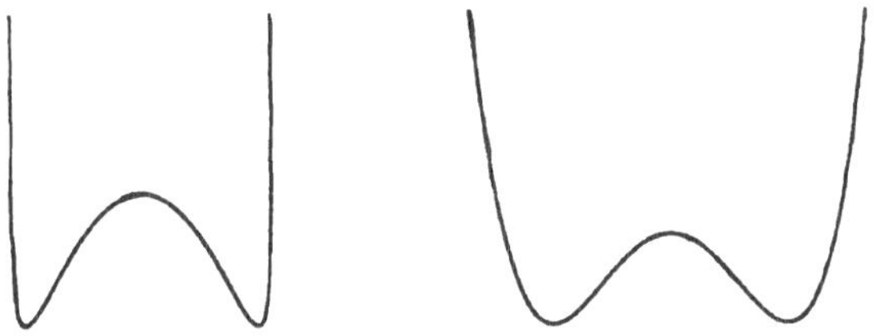

Fig. 17 The ϕ^4 potential and the potential (29) for a = 1.6.

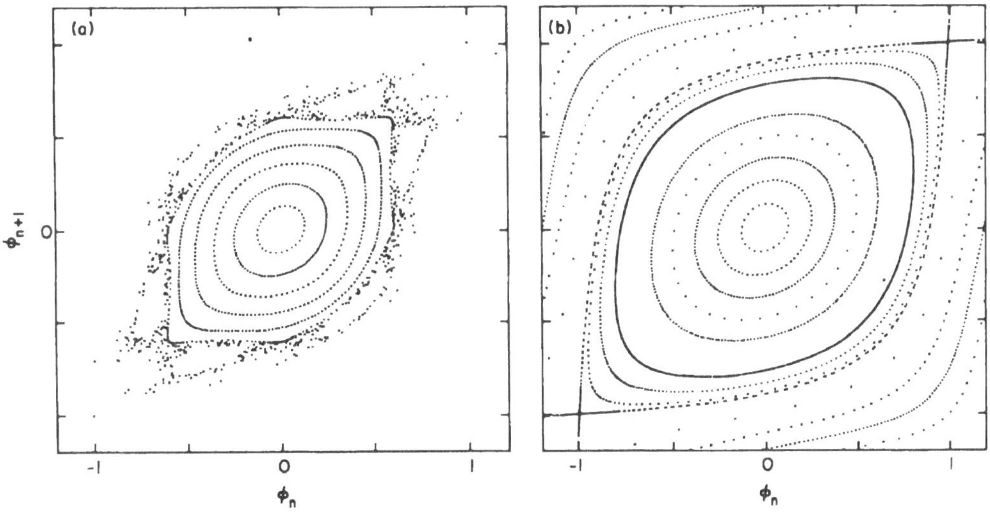

Fig. 18 Trajectories for the ϕ^4 potential and the potential (29) for a = 1.6.

REFERENCES

1. For a review, see for instance P. Bak, Rep. Prog. Phys. 45, 587 (1982).

2. M. J. Feigenbaum, L. P. Kadanoff, and S. J. Shenker, Physica 5D, 370 (1982); D. Rand, S. Ostlund, J. Sethna and E. Siggia, Phys. Rev. Lett. 49, 132 (1982).

3. M. Hoegh Jensen, P. Bak, and Tomas Bohr, preprint.

4. M. J. Feigenbaum, J. Stat. Phys. 19, 25 (1978).

5. S. Safran, Phys. Rev. Lett. 44, 937 (1980).

6. J. Hubbard and J. B. Torrance, Phys. Rev. Lett. 47, 1750 (1981).

7. P. Bak and R. Bruinsma, Phys. Rev. Lett. 49, 249 (1982); Phys. Rev. B (to be published).

8. P. Bak and J. von Boehm, Phys. Rev. B 21, 5297 (1980).

9. M. Hoegh Jensen and P. Bak, Phys. Rev. B (to be published).

10. P. Fischer, B. Lebech, G. Meier, B. D. Rainford, and O. Vogt, J. Phys. C 11, 345 (1978). J. Rossat-Mignod, P. Burlet, J. Villain, H. Bartholin, T. S. Wang, D. Florence, and O. Vogt, Phys. Rev. B 16, 440 (1977).

11. P. Bak and M. Hoegh Jensen, J. Phys. A 15, 1893 (1982).

12. W. P. Su, J. R. Schrieffer, and A. J. Heeger, Phys. Rev. Lett. 42, 1698 (1978).

13. B. Mandelbrot, "Fractals: Form, chance, and dimension" (Freeman, San Francisco, 1977).

14. A. Aharony and P. Bak, Phys. Rev. B 23, 4770 (1981).

15. M. E. Fisher and W. Selke, Phys. Rev. Lett. 44, 1502 (1980).

16. T. Janssen and J. A. Tjon, J. Phys. A.

17. P. Bak and V. L. Pokrovsky, Phys. Rev. Lett. 47, 958 (1981).

18. T. C. Bountis, Physica D 3, 577 (1981).

19. M. Hoegh Jensen, P. Bak and A. Popielewicz, preprint.

20. V. H. Schmidt, Phys. Rev. B 20, 4397 (1979).

The work at Brookhaven National Laboratory was supported by the Division of Materials Sciences U.S. Department of Energy under contract DE-AC02-76CH00016.

THE COMPLETE DEVIL'S STAIRCASE AND UNIVERSALITY OF MODE LOCKING STRUCTURES IN DISCRETE MAPPINGS

M. Høgh Jensen, Per Bak* and Tomas Bohr

H.C. Ørsted Institute
Universitetsparken 5
DK-2100 Copenhagen Ø, Denmark

ABSTRACT We study mode locking in certain discrete mappings. The mode locking traces up a complete devil's staircase of fractal dimension $D \sim 0.87$. This exponent is universal for a large class of functions and we expect to find similar behaviour in systems with two competing frequencies.

The transition to chaos has been intensively studied by means of discrete maps. Feigenbaum[1] found universal properties in the bifurcation route and recently a transition through quasi-periodic behaviour has attracted much attention.[2,3,4] This transition is realized by varying two frequencies and may be studied by the "circle map"

$$f(\theta) = \theta+\Omega- \frac{K}{2\pi} \sin(2\pi\theta) \quad . \tag{1}$$

The ratio between the two frequencies is determined by the winding number

$$W(K,\Omega) = \lim_{n\to\infty} \frac{f^n(\theta)-\theta}{n} \quad . \tag{2}$$

When the winding number approaches the Golden mean Shenker[2] found that the mapping exhibits unusual scaling behaviour at the point, $K = 1$, where chaos sets in. This scaling has also been treated within a renormalization group technique.[3,4]

---

*Now at: Brookhaven National Laboratory, Upton, N.Y. 11973.

A competition between two frequencies can be realized when a system with a characteristic frequency is perturbed by a varying external frequency. When the ratio of the frequencies approaches simple rational values one observes resonances which eventually become weaker as the ratio approaches irrational values. A resonance is stable in a finite range of the external frequency so sudden jumps between different resonances will occur as this frequency is varied. Such behaviour has been observed in great detail in the I-V characteristic of a shunted Josephson junction.[5]

By studying the mapping (1) we encounter a simple mechanism for this type of mode locking phenomena. The function (1) must be considered as a Poincaré-mapping of the phase-space trajectory of a particular dynamical system. If $\Omega = 0$, a progressive iteration of (1) will converge to a one-cycle, i.e. a fixed point; when $\Omega \neq 0$ the behaviour in general becomes more complicated. In fact it has been shown for $0<K<1$ that the winding number (2) takes all rational values each locked in a finite interval of Ω.[6] However, these intervals do not fill up the whole Ω-axis so there is room of finite measure for solutions with irrational winding numbers.

We have studied the critical point, $K = 1$, at which the inverse of the mapping (1) still exist but has a cubic singularity at $\theta = 0$.[7] The stability interval (in Ω) of a particular winding number $W = P/Q$ is denoted $\Delta\Omega(P/Q)$. Fig.1 shows intervals $\Delta\Omega$ larger than 0.0015; the inset shows intervals larger than 0.00015.

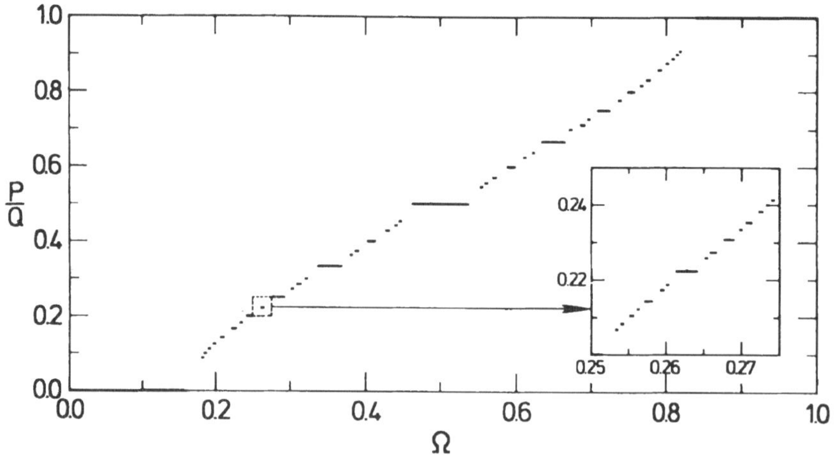

Fig. 1. The complete devil's staircase of the critical circle map (1). Stability intervals of periodic solutions with $\Delta\Omega>0.0015$ are shown. The inset is magnified 10 times and shows intervals with $\Delta\Omega>0.00015$.

We have numerically determined all intervals $\Delta\Omega(P/Q)$ with $Q \leq 50$ to a precision of 10^{-6} and found evidence that all rational winding numbers are present in finite stability intervals. The plateaux thus trace up a devil's staircase. The staircase is said to be complete if the intervals fill up the whole Ω-axis. To investigate the completeness we consider the fraction of $0 \leq \Omega \leq 1$ which is not occupied by plateaux larger than a particular scale. As decreasing length scales we arbitrarily choose $r_Q = \Delta\Omega(1/Q)$. First, the total length $S(r_Q)$, of plateaux larger than r_Q is measured. The holes between the plateaux now have the total length $1-S(r_Q)$ and we define the number $N(r_Q)$ as this length measured on the scale r_Q, $N(r_Q) = \dfrac{(1-S(r_Q))}{r_Q}$

Fig.2a shows a plot of $\log N(r_Q)$ versus $\log \dfrac{1}{r_Q}$ and the straight line indicate a power law

$$N(r) \sim \left(\frac{1}{r}\right)^D \tag{3}$$

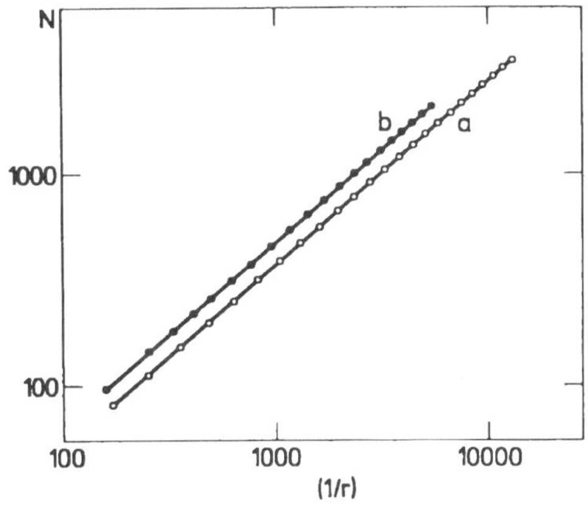

Fig. 2. Plot of $\log N(r)$ versus $\log (\frac{1}{r})$ for a): the critical map (1); b): the map (4) with $a = -0.8$.

From the figure one finds $D = 0.8669$, so the length of the holes, $1-S(r) \sim r^{1-D}$, will vanish when the length scale, r, goes to zero showing that the staircase is complete; i.e. the space between the plateaux is of zero measure. This space is therefore a Cantor set and D is the fractal dimension[8] of the Cantor set.

When K exceeds 1 the stability intervals will overlap so the winding number depends on the starting point and this might cause bysteresis effects. Also, for K>1, some periodic solutions will bifurcate and eventually turn into chaos. Thus the fractal dimension of the staircase at the critical point is a characteristic index for the mode locking at the transition to chaos. To test the universality of this result we have studied a class of mappings with a cubic singularity (K=1)

$$f(\theta) = \theta+\Omega- \frac{K}{2\pi}(\sin 2\pi\theta + a \sin^3 2\pi\theta) \ , \qquad (4)$$

which are monotonic for $-\frac{4}{3} \leq a \leq \frac{1}{6}$. Indeed, for these mappings the staircase also seems to be complete with fractal dimension $D \sim 0.87$. On this basis we have conjectured that $D = 0.87$ is a universal index for cubic critical mappings. Fig.2b is a plot of $\log N(r)$ versus $\log \frac{1}{r}$ for (4) with $a = -0.8$; the slope is by linear regression determined to 0.8652.

We believe that this observation is of experimental importance to a large class of systems with a characteristic frequency, which is perturbed by an oscillation. The functions (4) are one-dimensional Poincaré mappings of the phase-space trajectories. The class we consider consists of monotonic increasing functions with the lowest (a cubic) singularity. As soon as the function exhibits a local extremum bistability and bifurcations will occur.

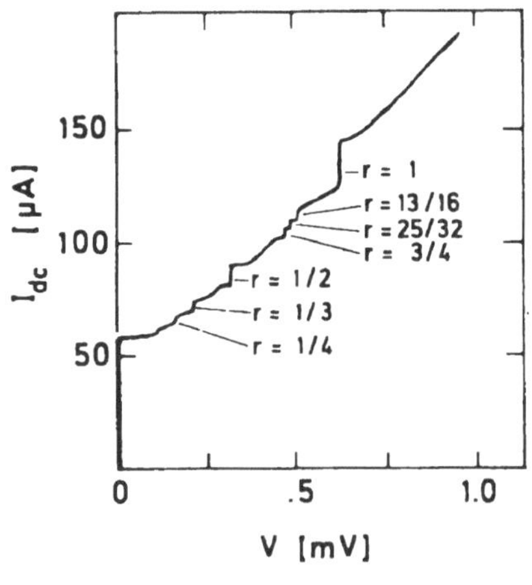

Fig. 3. Experimental I-V curve for a Josephson junction irradiated by 295 GHz microwaves (Ref.5).

In a Josephson junction a staircase with steps to Q = 32 has been found[5] (Fig. 3) and it should indeed be possible to check if the scaling behaviour is described by the universal index D = 0.87 at the critical point where hysteresis sets in. Also a Rayleigh Bénard experiment perturbed by up-down oscillations in the gravitational field[9] might show similar behaviour.

This work has been supported in part by the Danish Natural Science Research Council.

REFERENCES

1. M.J. Feigenbaum, J.Stat.Phys. 19:25 (1978); 21:669 (1979).
2. S.J. Shenker, Physica 5D:405 (1982).
3. M.J. Feigenbaum, L.P. Kadanoff and S.J. Shenker, Physica 5D: 370 (1982).
4. D. Rand, S. Ostlund, J. Sethna and E. Siggia, Phys.Rev.Lett. 49:132 (1982); S. Ostlund, D. Rand, J. Sethna and E. Siggia, Santa Barbara Preprint (1982).
5. V.N. Belykh, N.F. Pedersen and O.H. Soerensen, Phys.Rev.B 16: 4860 (1977).
6. M.R. Herman, Lecture Notes in Mathematics 597:271 (1977).
7. M. Høgh Jensen, Per Bak and Tomas Bohr, H.C. Ørsted Institute Preprint (1983).
8. B.B. Mandelbrot, "Fractals: Form, Chance, and Dimension", Freeman, San Francisco (1977).
9. A. Libchaber, private communication.

LOW TEMPERATURE ANALYSIS OF THE ANNNI MODEL IN A FIELD

Jonathan Smith and Julia Yeomans

Department of Physics
The University
Southampton S09 5NH, U.K.

Recently Fisher and Selke[1] have shown that, at low temperatures, the axial next nearest neighbour Ising (ANNNI) model exhibits an infinite sequence of long wavelength commensurate phases. In this paper we discuss the changes in the phase diagram under the influence of a uniform external field.

The ANNNI model in a field is defined by the Hamiltonian

$$\mathcal{H} = - J_0 \sum_{nn}^{\perp} S_i S_j - J_1 \sum_{nn}^{//} S_i S_j - J_2 \sum_{nnn}^{//} S_i S_j - H \sum_i S_i \qquad (1)$$

where the spins, $S_i = \pm 1$, occupy the sites i of a d-dimensional lattice. Along a singled-out, axial direction, //, the nearest neighbour (nn) coupling, J_1, competes with an antiferromagnetic next nearest neighbour (nnn) coupling, $J_2 < 0$. In the (d-1) perpendicular directions, $\perp$, the spins interact through ferromagnetic nearest neighbour bonds, $J_0 > 0$. Hence, at low temperatures, for $d > 2$, the model is made up of ferromagnetically aligned layers coupled by competing interactions: here we shall be mainly concerned with the effect of these interactions on the relative spin orientations along the axial direction.

The low temperature configurations of the model comprise a sequence of bands, groups of consecutive layers with the same spin value terminated by layers of opposite spin. For example, the configuration

$$\ldots \downarrow\downarrow\uparrow\uparrow\downarrow\downarrow\uparrow\uparrow\uparrow\downarrow\downarrow\uparrow\uparrow\downarrow\downarrow\uparrow\uparrow\uparrow \ldots \qquad (2)$$

is constructed from a basic unit of three two-bands followed by one three-band. We denote this <2$\bar{2}$2$\bar{3}$> where, if necessary, bars are used to indicate that the spins lie antiparallel to the magnetic field ($S_i = -1$). Useful parameters which will be used later are

$$\kappa \equiv \tfrac{1}{2} + \delta = -J_2/|J_1| \text{ and } \eta = H/|J_1|. \tag{3}$$

In zero field, H = 0, a reversal in the sign of J_1 and alternate spins along the competion direction, //, leaves the Hamiltonian unchanged. Hence we may take $J_1 > 0$ without loss of generality. At zero temperature, for $\kappa < \frac{1}{2}$, the ordering in the axial direction is ferromagnetic, < ∞ >; however, for $\kappa > \frac{1}{2}$, the second neighbour antiferromagnetic interaction dominates and the ground state is the antiphase structure, <2>. At the multiphase point, ($\kappa=\frac{1}{2}$, $\eta=0$), however, all structures with bands of width $\kappa \geq 2$ are degenerate at zero temperature.

Fisher and Selke[1] have analysed the phase diagram of the ANNNI model in the vicinity of the zero-field multiphase point using a combination of low temperature series and linear programming techniques. They show that, as the temperature is increased, an infinite sequence of phases <2^k3>, k = 0,1,2 ..., spring from the multiphase point interpolating between the ferromagnetic, <∞>, and antiphase, <2>, states. The extent of successive phases decreases geometrically with increasing k. The corresponding sequence of phases for $J_1 < 0$ is easily shown to be <1^k2>, k = 1,2, 3 ... .

We now present the results of a similar low temperature analysis of the ANNNI model in a field. The first step is to determine the ground-state. To do this we follow Pokrovsky and Uimin[2] in introducing configuration variables

$$p = (\uparrow\uparrow), \qquad q = (\downarrow\downarrow), \qquad r = (\uparrow\uparrow\uparrow), \qquad t = (\downarrow\downarrow\downarrow) \tag{4}$$

where, for example, (↑↑) denotes the number of pairs of consecutive layers with spin, $S_i = +1$, per layer. The energy of a general layered configuration is then

$$E = -\tfrac{1}{2}q_\perp J_0 + J_1 - J_2 - H(p-q) - 2(J_1 - 2J_2)(p+q) - 4J_2(r+t) \tag{5}$$

where $q_\perp$ is the number of nearest neighbours within a layer. The ground-state is determined by minimising (6) subject to the constraints

$$p \geq r \geq 0, \quad q \geq t \geq 0, \quad 3p + q - 2r \leq 1, \; 3q + p - 2t \leq 1 \tag{6}$$

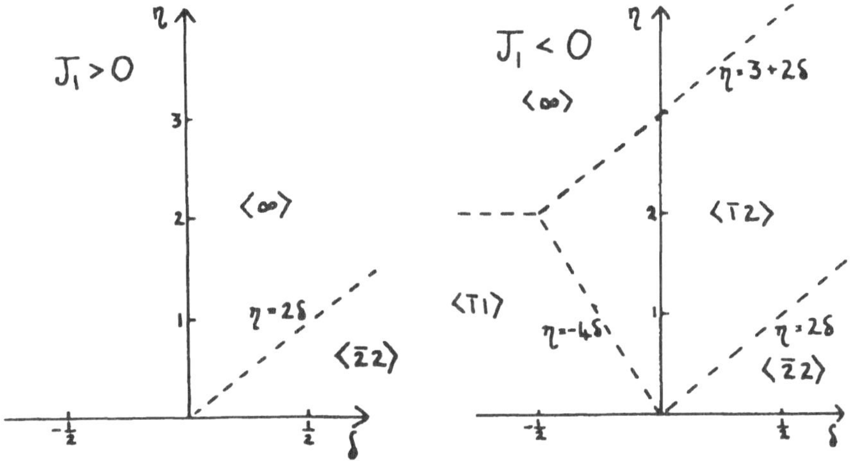

Figure 1: Ground state of the ANNNI model in a field.

This is a simple linear programming problem in four dimensions. The results, for $J_1 > 0$ and $J_1 < 0$, are shown in Figure 1. The stable zero-temperature phases are separated by multiphase lines on which the ground-state is highly degenerate. The allowed zero-temperature states for each of the multiphase lines are listed in Table 1.

Table 1. Degeneracy of the multiphase lines at zero temperature.

Degenerate line		Neighbouring phases		Allowed bands on degenerate line
$J_1>0$	$\eta=2\delta$	$<\infty>$	$<\bar{2}2>$	$2,3,4\ldots\ ;\bar{2}$
$J_1<0$	$\eta=2\delta$	$<\bar{1}2>$	$<\bar{2}2>$	$2\ ;\ \bar{1}\ ,\ \bar{2}$
	$\eta=-4\delta<2$	$<\bar{1}2>$	$<\bar{1}1>$	$1\ ,\ 2\ ;\ \bar{1}$
	$\eta=3+2\delta>2$	$<\bar{1}2>$	$<\infty>$	$2,3,4\ldots\ ;\bar{1}$
	$\eta=2,\delta<-\frac{1}{2}$	$<\bar{1}1>$	$<\infty>$	$1,\infty;\ \bar{1}$

To examine which of these persist to finite temperature it is necessary to construct a low temperature expansion about all possible ground states. The expansion must be taken to all orders where necessary to confirm the existence of increasingly long-wavelength phases. Fisher and Selke[4] have shown that this is feasible as it is possible to extract the important terms at each order of the series expansion. The method is described in detail in their paper and the extension to $H \neq 0$ is given in Smith and Yeomans [3]. Therefore here we restrict ourselves to an outline of the results obtained.

For ferromagnetic nearest neighbour interactions in the axial direction there is a single multiphase line $\eta=2\delta$. At finite temperatures in the vicinity of this line an infinite sequence of commensurate phases $\langle\bar{2}(2\bar{2})^k 3\rangle$, k=0,1,2..., interpolates between the $\langle\infty\rangle$ and $\langle 22\rangle$ phases. The phases are separated by first order boundaries and their widths decrease geometrically with increasing k. The sequence of phases does not persist to arbitrarily high fields: near $\delta=\frac{1}{2}$, $\eta=1$, each phase becomes unstable to the ferromagnetic phase through a first order transition. Thus the model exhibits a geometrically converging sequence of triple points where the phases $\langle\infty\rangle$, $\langle\bar{2}(2\bar{2})^{k-1}3\rangle$ and $\langle 2(2\bar{2})^k 3\rangle$, k = 1,2,3... , coexist.

For antiferromagnetic nearest neighbour interactions we must consider three multiphase lines. For $\eta=2\delta$ a sequence of phases $\langle\bar{1}2(\bar{2}2)^k\rangle$, k=0,1,2,..., interpolates between the phases $\langle 12\rangle$ and $\langle\bar{2}2\rangle$. In this case, however, there is no sequence of triple points and all the phases persist to arbitrarily high fields.

A first-order low temperature series calculation performed in the vicinity of the lines $\eta=4\delta$ and $\eta= 3+2\delta > 2$ indicates that the behaviour here is quite different. In each case a single, discontinuous transition is seen between the neighbouring phases with no interpolating sequence.

The ANNNI model in a field has also been studied by Yokoi et al.[4] for $J_1 > 0$ using a numerical mean-field technique. Their results are in agreement with ours to within the accuracy of their numerical techniques.

Pokrovsky and Uimin[2] have studied the ANNNI model in a field using a perturbative high temperature expansion for both $J_1 > 0$ and $J_1 < 0$. Their results agree with ours except near the line $\eta=3 + 2\delta > 2$ where they find a sequence of long-wavelength commensurate phases to be stable.

REFERENCES

1. Fisher M.E. and Selke W. 1980 Phys.Rev.Lett. 44 1502
1981 Phil.Trans.R.Soc. 302 1

2. Pokrovsky V.L. and Uimin G.V. 1982 J.Phys.C. Solid St.Phys. 42 L11
1982 J.E.T.P. to be published

3. Smith J. and Yeomans J.M. 1982 J.Phys.C. Solid St.Phys. 15 L1053
1983 submitted to J.Phys.C.

4. Yokoi C.S.O., Coutinho Filho M.D. and Salinas S.R. 1981 Phys. Rev. B24 4047.

FIELD BEHAVIOR OF SPIN MODELS WITH COMPETING INTERACTIONS

S.R. Salinas

Instituto de Física
Universidade de S. Paulo - C.P. 20516
01000 São Paulo, SP, Brazil

Our interest in the study of spin hamiltonians with competing interactions has been stimulated by recent measurements for the field-temperature phase diagram of compounds which exhibit modulated phases[1,2]. In the first part of this talk we report mean-field calculations[3] for the global T-p-H phase diagram of the ANNNI model (where $p=-J_2/J_1$ gives the strength of the axially competing interactions). At high temperatures we obtained several analytic results, but at low temperatures the modulated phases in a field were studied numerically. The analysis in the neighborhood of the critical surface was also supplemented[4] by an ε-expansion renormalization-group treatment which will be reported in part II. We finally talk about an XY spin hamiltonian with competing interactions[5] which is suitable for explaining the Lifshitz point in MnP.

I. MEAN FIELD TREATMENT OF THE ANNNI MODEL

Consider the ANNNI model on the sc lattice with ferromagnetic nearest-neighbor couplings ($J_0>0$ in the xy plane, and $J_1>0$ in the z direction) and competing antiferromagnetic second-neighbor couplings ($J_2<0$) in the z direction. The mean-field free energy may be written in terms of the set of values $\{m_z\}$, where m_z is the average magnetization per spin in layer z, given by the system of N coupled equations

$$H + 4J_0 m_z + J_1(m_{z-1}+m_{z+1}) + J_2(m_{z-2}+m_{z+2}) = kT \tanh^{-1} m_z . \quad (1)$$

In Fig. 1 we depict the global phase diagram which is obtained from Eq.(1). If we write $m_z=m_0+\delta m_z$, where m_0 is the paramagnetic solution, $H + \hat{J}(0)\, m_0 = kT \tanh^{-1} m_0$, with $\hat{J}(q) = 4J_0 + 2J_1 \cos q +$

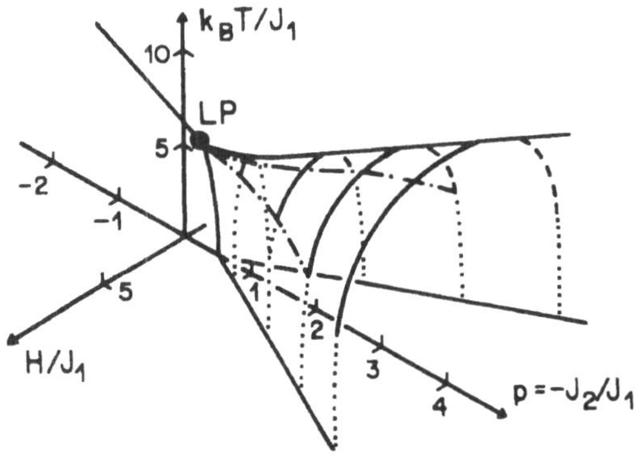

Fig. 1. Global phase diagram of the ANNNI model (for $J_0=J_1$).

$2J_2\cos 2q$, it is easy to Fourier-analyse Eq.(1) and obtain an expression for the critical surface, $kT/(1-m_0^2) = \max_q \hat{J}(q) = \hat{J}(q_c)$, which is bounded by two lines of tricritical points. It should be noticed that these lines, as well as the ferro-modulated first order line, meet tangentially at the H=0, p=1/4 Lifshitz point (LP). At a certain higher value of p the tricritical lines end at multicritical points of higher order, beyond which lines of critical and double critical end-points are expected to occur. If we write the expansion in harmonic components $\delta m_z = M_0+M_1\cos(q_cz+\phi_1) + M_2\cos(2q_cz+2\phi_2)+\ldots$,

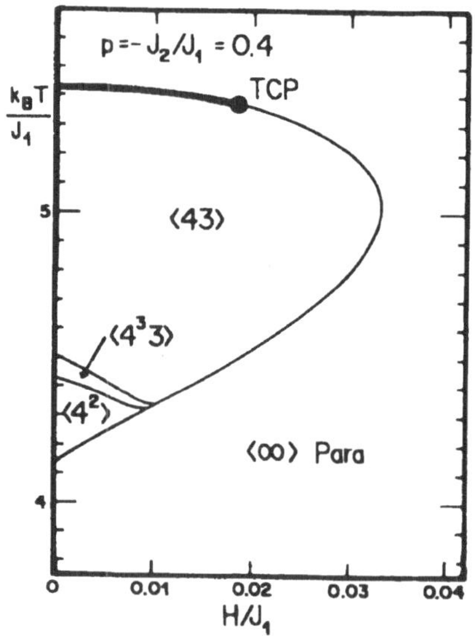

Fig. 2. Phase diagram in the H-T plane (for p=0.4).

it is possible to obtain asymptotic expressions for the amplitudes M_n. In zero field $M_n=0$ for n even. For $H\neq 0$, however, $M_0 \sim M_1^2$, and $M_n \sim M_1^n$ for $n \geq 2$, where M_1^2 vanishes with (T_c-T). For truly incommensurate values of q_c, the phase ϕ_1 is arbitrary (with $\phi_n=\phi_1$), but for commensurate critical wave vectors the value of ϕ_1 is determined by the minimization of the free energy. It is necessary to take into account higher harmonic components of δm_z to obtain correct expressions for the tricritical lines and the asymptotic dependence of the first order ferro-modulated line.

At lower temperatures, inside the modulated region, we were forced to treat Eq.(1) numerically. We considered solutions with periodicities L up to 20 lattice spacings, and examined wave vectors of the form $q=2\pi K/L$, where $0 \leq K < L \leq 20$. Among the various solutions found for a given point in the T-p-H space, only that one which minimizes the free energy is physically relevant. Of course, our results are limited to the main commensurate phases, and it is to be understood that in between them there may exist other commensurate or incommensurate phases. As an example, we show in Fig. 2 a typical phase diagram in the T-H plane, for $1/4<p<1/2$. The notation <43> means that 4 planes with spins predominantly parallel are followed by 3 planes with spins predominantly antiparallel to the field. The ground state is ferromagnetic, and the modulated phases are limited to a hump-shaped region which is stable for lower fields. The second order transition line (heavy tracing) ends at the tricritical point TCP. This figure bears a striking similarity to the electric field-temperature phase diagram of the ferroelectric compound Thiourea which exhibits several modulated phases. A quantitative comparison with the data for Thiourea, however, demands either a more complex spin hamiltonian or some improvements in the simple meanfield approach (which are presently being carried out by C. Yokoi).

II. RENORMALIZATION-GROUP TREATMENT OF THE ANNNI MODEL

Previous RG studies of the ANNNI model in zero field have shown that the para-modulated transition is XY-like due to the spin fluctuations about the wave vectors $+\vec{q}_c=(0,0,\ldots,+q_c)$ and $-\vec{q}_c=(0,0,\ldots,-q_c)$, which correspond to the maxima of $\mathcal{J}(\vec{q})$ on a d-dimensional sc lattice. The presence of H introduces many couplings between these critical fields, associated with the main harmonic components of the magnetization, and the non-critical fields, corresponding to the higher harmonic components. Since these couplings are ultimately responsible for driving the system to the tricritical point, we paid special attention to their behavior under the RG iterations. Our calculation is closely related to Nelson and Fisher's analysis of metamagnetism. However, in the present case, besides the non-critical uniform spin field, the non-critical second harmonic spin field is also relevant to the analysis.

According to the usual procedures of the RG technique, we use the weighting factor $\exp(-aS^2/2-bS^4/4)$ to write the reduced hamiltonian

$$\bar{H}[S_q] = \frac{1}{2}\int u_2(q)\, S_q S_{-q} + w\int S_q S_{q'} S_{-q-q'} + u\int S_q S_{q'} S_{q''} S_{-q-q'-q''} , \qquad (2)$$

where $u_2(q) = a+3bM^2-\beta\hat{J}(q)$, $w=bM$, $u=b/4$, and M is given by $-\beta H + [a-\beta\hat{J}(0)]M + bM^3 = 0$. In Eq.(2) the spin variables have already been shifted to eliminate the field dependent linear terms. Now we consider the spin variables S_q with wave vectors in the neighborhood of 0, $\pm 2q_c$, $\pm 3q_c$,.... It is then convenient to introduce a new set of variables, given by $S_q=\sigma_{0q}$ and $S_{\pm nq_c+q} = \sigma^1_{nq} \pm i\sigma^2_{nq}$, in terms of which we have

$$\bar{H} = \frac{1}{2}\int r_0\sigma_0\sigma_0 + \frac{1}{2}\int (r_1+q^2)\sum_\alpha \sigma_1^\alpha\sigma_1^\alpha + \frac{1}{2}\int r_2 \sum_\alpha \sigma_2^\alpha\sigma_2^\alpha + w_0\iint[\sigma_0\sigma_1^1\sigma_1^1 + \sigma_0\sigma_1^2\sigma_1^2] + w_2\iint[\sigma_2^1\sigma_1^1\sigma_1^1 - \sigma_2^1\sigma_1^2\sigma_1^2 + 2\sigma_2^2\sigma_1^2\sigma_1^1] + u\iiint \sum_{\alpha,\beta} \sigma_1^\alpha\sigma_1^\alpha\sigma_1^\beta\sigma_1^\beta , \qquad (3)$$

where the momentum conserving subscripts have been omitted, and we have not included higher order contributions and higher harmonic components which are found to be irrelevant to the analysis of the critical behavior. In Eq.(3) w_0, $w_2 \alpha w$, and $r_j \alpha\ a+3bM^2-\beta\hat{J}(jq_c)$, for $j=0,1,2$. Now it is straightforward to carry out a perturbative expansion about $\varepsilon=4-d$, with the ansatz r_1, $u=O(\varepsilon)$, and $w_0,w_2=O(\sqrt{\varepsilon})$. The recursion relations for r_1 and $x_j=w_j^2/r_j$ $(j=0,2)$ determine eight fixed points, four of which are not accessible in the case of the ANNNI model. There is a trivial crossover for $H\neq 0$, as in the case of the Ising metamagnet, but now both fixed points are XY-like. The tricritical behavior is governed by a Gaussian-like fixed point, given by $r_1^*=0$, $x_0^*=x_2^*=u^*$, which can be reached when $u=(x_0+x_2)/2$, with $x_0,x_2\neq 0$. For $u<(x_0+x_2)/2$, with $x_0,x_2\neq 0$, there is a critical run away leading to a first order transition.

III. ANISOTROPIC XY MODEL FOR MnP

Detailed measurements for the field-temperature phase diagram of MnP in the vicinity of the LP have been reported by Dr. Shapira. With H applied along the intermediate y direction, we proposed a qualitative interpretation of these measurements on the basis of the model hamiltonian

$$H = -\frac{1}{2}\sum_{\ell,\ell'} J(\ell-\ell')\vec{S}_\ell\cdot\vec{S}_{\ell'} - D\sum_\ell [(S_\ell^x)^2-(S_\ell^y)^2] - H\sum_\ell S_\ell^y , \qquad (4)$$

where $D>0$ and $\vec{S}=(S^x,S^y)$. The exchange parameters are ferromagnetic in the xy planes and include the effects of competing interactions along the hard z direction. A mean-field treatment of this model system can be performed along the lines described for the ANNNI model. However, in order to produce a LP in the H-T plane, it is necessary to assume that $\hat{J}(q)=\hat{J}(0)-\alpha_2 q^2-\alpha_4 q^4+\ldots$, with $\alpha_4>0$, and that $\hat{J}(0)$ and α_2 are smooth functions of T and H, with $\alpha_2=0$ at the LP. With these assumptions it is possible to obtain expressions for the transition lines, which meet tangentially at the LP, and for several thermodynamic quantities of experimental interest. The uniform transverse susceptibility χ^x, for example, obeys the usual Curie -Weiss law across the para-ferro transition line; however, it is continuous and shows a finite cusp across the para-fan line. As one moves along this line, χ^x diverges as $|\Delta T|^{-2}$ for $T\to T_L$. The longitudinal susceptibility is given by the constant 1/4D throughout the ferromagnetic phase, and shows a discontinuous behavior across the transition lines. All these features seem to be in qualitative agreement with the experiments reported by Dr. Shapira.

A renormalization-group analysis of hamiltonian (4) has been carried out along the lines described in the last section. It should be noticed that, since D>0, the spin field S_q^y is non-critical and may be integrated to produce an effective hamiltonian where only the S_q^x components are present. After a few iterations of the RG, we obtain a one-component (n=1), uniaxial (m=1), hamiltonian of the same form which is produced by the simple ANNNI model. Therefore, the LP in MnP should exhibit the characteristic multicritical behavior of the uniaxial one-component case, as it is indeed suggested by the experimental results.

REFERENCES

1. Y. Shapira, C. C. Becerra, N. F. Oliveira Jr., and T. S. Chang, Phys. Rev. B24,2780 (1981).
2. J.P. Jamet, J. Phys. (Paris) 42,L123 (1981).
3. C. S. O. Yokoi, M. D. Coutinho-Filho, and S. R. Salinas, Phys. Rev. B24,4047 (1981).
4. M. D. Coutinho-Filho, C. S. O. Yokoi, and S. R. Salinas, J. Magn. Magn. Mater. 31-34,1265 (1983).
5. C. S. O. Yokoi, M. D. Coutinho-Filho, and S. R. Salinas, Phys. Rev. B24,5430 (1981).

PHASE TRANSITIONS IN CRYSTALS WITH COMPETING INTERACTIONS

T. Janssen and J.A. Tjon

Institute for Theoretical Physics, Nijmegen, Holland
Institute for Theoretical Physics, Utrecht, Holland

INTRODUCTION

Incommensurate phase transitions have mainly been studied by means of phenomenological Landau theory. Recently one has considered also microscopic models (refs.1-3) and from these it may be understood that these crystal phases in insulators are a consequence of the occurrence of competitive interactions, each favouring a different periodicity. The resulting frustration shows itself by the formation of a phase without space group symmetry. Its structure may be described as a periodic distortion of an ordinary crystal. The main features of the phase diagram are explained in such a model. Application of similar models to specific materials [4)] has given a fairly nice agreement with experiment. We shall briefly discuss here the results and will consider in particular the changes caused by an external (electric) field.

MODELS

As a model a one-dimensional chain of classical particles is considered. Each particle has one degree of freedom denoted by x_n. There is a harmonic coupling between first and second neighbours. The potential energy in the model is

$$V = \sum_n \{\frac{\alpha+2\beta+3\delta}{2} x_n^2 + \frac{1}{4}x_n^4 + (\beta+2\delta)x_n x_{n-1} + \delta x_n x_{n-2}\}. \qquad (1)$$

The anharmonic term is necessary to keep the potential energy bounded. The somewhat unusual choice of parameters stems from an interpretation of x_n as the difference $x_n=u_n-u_{n-1}$ of displacements of adjacent particles in an equidistant array. Then V is the potential energy of a linear chain with harmonic and anharmonic springs:

$$V = \sum_n \{\frac{\alpha}{2}(u_n-u_{n-1})^2+\frac{1}{4}(u_n-u_{n-1})^4+\frac{\beta}{2}(u_n-u_{n-2})^2+\delta(u_n-u_{n-3})^2\}. \quad (2)$$

However, x_n may also be interpreted as, for example, the torsion angle of a molecule at site n. For having a phase transition at least one of the parameters should be negative, a value $\neq 0$ of δ is required for an incommensurate phase transition.

The equilibrium configuration of the system is found from the requirement $\partial V/\partial_{x_n} = 0$. This gives

$$(\alpha+2\beta+3\delta)x_n+x_n^3 + (\beta+2\delta)(x_{n-1}+x_{n+1})+\delta(x_{n-2}+x_{n+2}) = 0. \quad (3)$$

Analytical solutions of these coupled nonlinear equations can be found under the restriction to periodic solutions of low period (N=1,2,3,4). Numerically one can determine periodic solutions with a period that is limited by the computer capacity. A third way to generate (periodic and nonperiodic) solutions is to use the fact that eq.(3) is a recurrence relation: for arbitrary values of x_1,x_2,x_3,x_4 one may calculate a configuration (for δ=0 already from x_1,x_2).

In refs. 1-3 the different solutions are discussed. Only stable solutions are retained. Stability here means linear stability with respect to small displacements. The stable solutions with the lowest energy are indicated in Fig.3a as a function of the parameters α and δ. Incommensurate phases correspond to the hatched regions. For $\delta>1/6$ the ground state may be incommensurate.

A simple argument shows that one can consider the parameter α as effectively dependent on temperature: an increase in temperature corresponds to an increase in α. However, one can also use a mean-field approximation to study the thermodynamic properties in a model.

PHASE DIAGRAMS

In mean field theory the local potential energy is

$$V_n = \frac{\alpha+2\beta+3\delta}{2}x^2+\frac{1}{4}x^4+ (\beta+2\delta)x(\bar{x}_{n-1}+\bar{x}_{n+1})+\delta x(\bar{x}_{n-2}+\bar{x}_{n+2}), \quad (4)$$

where $\bar{x}$ is the thermal average of x. This quantity can be calculated in a self-consistent way from

$$\bar{x}_n = \int x \exp[-\beta V_n(x)]dx \,/\, \int \exp[-\beta V_n(x)]dx. \quad (5)$$

The solutions of these self-consistent equations correspond to stationary points of the free energy:

$$F = -\Sigma_n\{(\beta+2\delta)\bar{x}_n\bar{x}_{n-1}+\delta\bar{x}_n\bar{x}_{n-2}+kT\ \ln Z_n\}. \quad (6)$$

One solution is the trivial one: $\bar{x}_n=0$. Other solutions may be found numerically. Just as in the T=0 case there are many possible solutions for values of T and the parameters α,β,δ for which the trivial solution is unstable. By comparing the free energies of (stable) solutions, one determines the ground state and constructs the phase diagram. The result for fixed value α and for β positive or negative is given in Fig.1. Comparison with Fig.3a shows that indeed α may be considered as a T dependent quantity.

In the phase diagrams one may distinguish several regions. For high T the thermodynamic ground state has the structure of an undistorted equidistant array. For not too large values of δ the ground state is again commensurate at T=0. For $\delta>1/6$ the high temperature structure becomes unstable with respect to a deformation with wave vector determined by δ. The instability is preceded (for T decreasing towards T_i) by the formation of a soft mode. Because for $\delta<0.9$ the T=0 state is commensurate there is a second phase transition, the lock-in transition, at T_c for $1/6<\delta<0.9$ ($\beta=-1$) or for $1/2<\delta<0.9$ ($\beta=+1$).

In the incommensurate phase there are two different regions. For temperatures close to T_i the displacement wave is more or less sinusoidal. Its wave vector changes little and is that of the soft mode. A dynamical analysis shows that in this region there is a vibration mode which can be described as a shift of the modulation wave. The frequency of this "phason" is zero.

Near the lock-in transition the structure consists of commensurate domains separated by narrow regions, the discommensurations, such that the over all wave vector is incommensurate. In this region the wave vector of the modulation changes rapidly. Moreover, the phason frequency rises steeply. There are new excitations which correspond to localized vibrations.

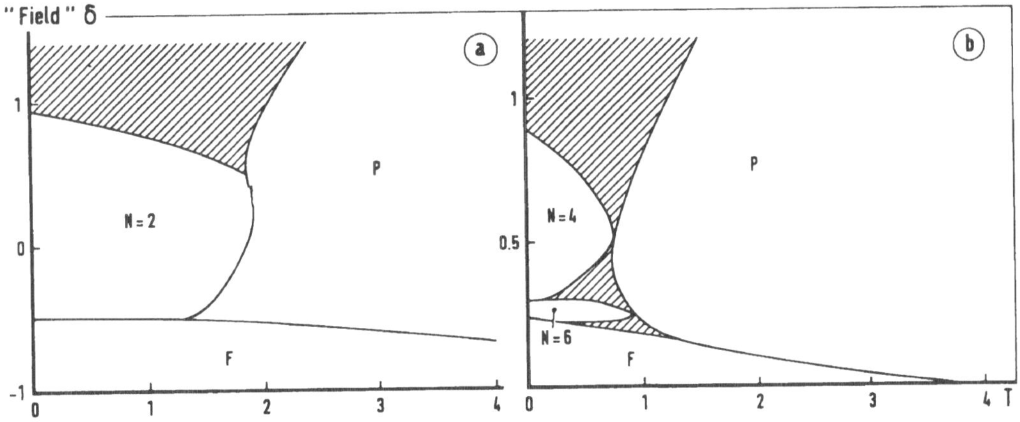

Fig.1 Phase diagrams in mean field approximation for the chain of eq.(1). a. Antiferroic coupling ($\beta=+1$), b. Ferroic first neighbour coupling ($\beta=-1$). Fixed values for α(0 in case b, -2 in case a).

Close to the lock-in tran[illegible]n a new phenomenon occurs. Because there the free energies of several solutions of the self-consistent equations are nearly degenerate the real structure is a random distribution of discommensuration distances. In fact one has entered the chaotic state.

The phase diagram shows several multicritical points. A conspicuous one lies at the point where the basic P-phase, the ferroic F-phase and the incommensurate I-phase coincide. Another one occurs for $\delta=0.5$, $\beta=-1$, where the P-phase, the I-phase and a commensurate period four phase meet. The former one disappears if one introduces an external field, the latter persists as discussed in the next section.

THE PHASE DIAGRAM IN AN EXTERNAL FIELD

When the variable x_n is coupled to an external field E, the phase diagram should be studied in the E-parameter too. An example is the coupling of the (spontaneous or induced) dipole moment to an electric field. Such phase diagrams as function of an external electric field have been measured, for instance, for thiourea $SC(NH_2)_2$[5], $NaNO_2$[5] and $RbLiSO_4$[7]. If the coupling between x_n and E is linear the potential energy (1) gets an additional term $\Sigma_n Ex_n$. In a similar way as discussed above one can determine the stationary points and their stability. Because one has seen that the parameter α may be interpreted as temperature, for convenience the structures with the lowest potential energy have been calculated (instead of the thermodynamic ground state in mean field approximation). The results are shown in Fig.2 as a function of α and E for fixed values $\beta=-1$, $\delta=0.2298$.

The transition from the P- and F-phases to the I-phase may be determined from a stability analysis. For constant solutions $x_n=a$ the equilibrium condition is

$$(\alpha+4\beta+9\delta)a + a^3 + E = 0. \tag{7}$$

The stability follows from the dispersion of the vibrations:

$$m\omega_k^2 = \alpha+2\beta+3\delta+3a^2+2(\beta+2\delta)\cos(k)+2\delta\cos(2k). \tag{8}$$

As a function of α (which corresponds to T) there is a soft mode for $k=k_o=\cos^{-1}(-(\beta+2\delta)/4\delta)$. This soft mode, however, does not go to zero when E exceeds a critical value. One sees from (8) that both P- and F-solutions develop a soft mode at the same wave vector, which is independent of the field E. In order to find an E dependent wave vector of the soft mode one has to take into account anharmonic terms in the neighbour interaction. If one adds to V a term with $(\gamma/2)\Sigma(x_n x_{n-1})^2$ the equilibrium condition becomes

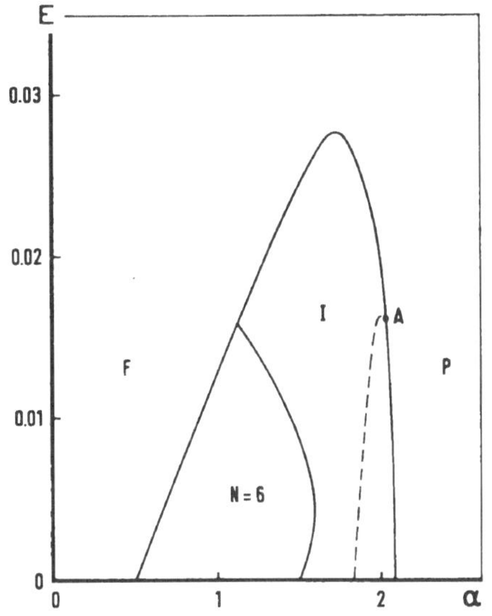

Fig.2 Phase diagram in the E-α(T) plane for β=-1, δ=0.2298. A denotes the end point of a line of continuous transitions. Dashed line: instability of F-phase.

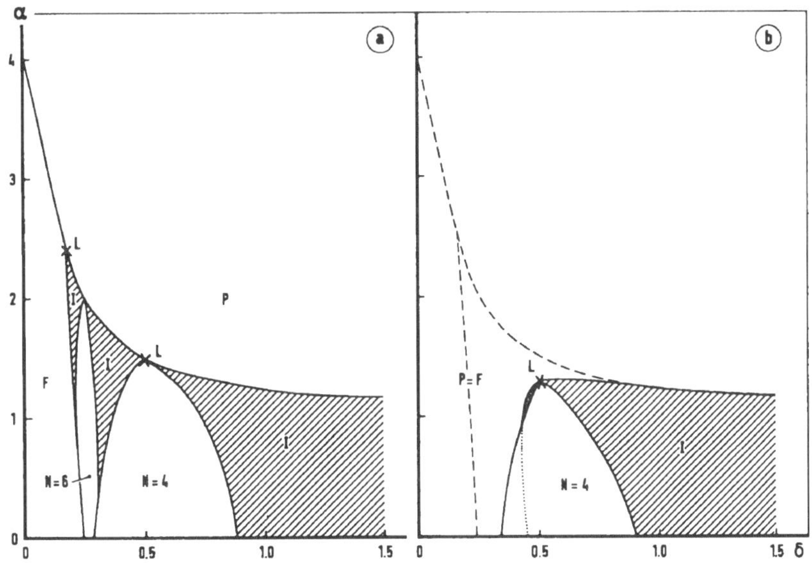

Fig.3 Phase diagram in the α-δ-plane. Indicated are the para-phase (P), the ferroic phase (F), commensurate phases with period N = 4 or 6. The hatched regions indicate the incommensurate I-phase. (a. E=0, b. E=0.75). Dashed curve: E=0 lines, dotted line: instability of the F-phase.

$$(\alpha+4\beta+9\delta)a + (1+2\gamma)a^3 + E = 0 \tag{9}$$

and there is a soft mode at wave vector $k=k_o=\cos^{-1}[-(\beta+2\delta+2\gamma a^2)/4\delta]$. Because a^2 depends on E, the wave vector does so too, as has been observed, e.g. in $NaNO_2$. In first non-vanishing order the deviation of k_o from its value at E=0 is quadratic in the field.

The E-α-diagram of Fig.2 shows an incommensurate phase between the ferroelectric and the paraelectric phases. These become identical for $E>E_c=0.028$. In the modulated phase there is a lock-in transition to a period six solution. For E>0.016 this lock-in transition is suppressed. The transition from the P-phase to the I-phase is continuous for E<0.016. Therefore, the point A is the end of a line of second order transitions.

In Fig.3 the α-δ phase diagrams are compared for E=0 and E≠0. For E≠0 the F-I-P point is eliminated and the I-phase is pushed towards higher values of δ. The Lifshitz point between the P- , I- and N=4 phases, however, moves over a line and still exists for E>0.

REFERENCES

1. T. Janssen and J.A. Tjon, Phys. Rev. B25, 3767 (1982)
2. T. Janssen and J.A. Tjon, J. Phys. A16, 673 (1983)
3. T. Janssen and J.A. Tjon, Incommensurate crystal phases in mean field approximation, preprint (1983)
4. K. Parlinski and K.H. Michel, Crystalline phases of thiourea, (preprint 1982)
5. J.P. Jamet, J. Physique-Lettres 42, L-123 (1981)
6. D. Durand et al., Neutron diffraction study of $NaNO_2$ in an applied electric field, (preprint 1983)
7. H. Mashiyama et al., J. Phys. Soc. Jap. 46, 1959 (1979)

COMMENSURATE MELTING AND DOMAIN WALLS IN SURFACE PHASES

Michael E. Fisher

Baker Laboratory
Cornell University
Ithaca, New York 14853, U.S.A.

SUMMARY

Certain adsorbed surface systems display commensurate ordered phases which, as temperature T, or chemical potential, ζ, (controlled by the vapor pressure) vary, may melt via a continuous transition. Examples of particular interest are the $\sqrt{3} \times \sqrt{3}$ hexagonal phases of helium on graphite[1] and krypton on graphite[2] and the 3×1 rectangular phase of atomic hydrogen on the (110) surface of iron[3,4]. Both these types of ordered phase exhibit $p = 3$ distinct but fully equivalent types of domain, say, A, B and C. This fact can be represented by an internal, global symmetry, Y_p, which, for $p = 3$, is the same as that of the Potts model. On this basis it has been suggested[5,6] that the melting of such phases should be in the universality class of the 3-state Potts model with, in particular, a specific heat exponent $\alpha = \frac{1}{3}$ as, indeed, is confirmed for one coverage of helium on graphite[1]. However, the real physical symmetry is lower than the product, $Y_p \times L$, of the internal symmetry and the lattice symmetry, L, which describes the simple Potts model[7,8]. This reduced symmetry can be seen in a physically instructive way[8,9] by examining the structure of the walls separating domains in various relative arrangements : thus in the rectangular 3×1 phases one finds that the [+] walls, A|B, B|C and C|A (oriented normal to the direction in which the adsorbate lattice constant is 3 times the substrate lattice constant) are all equivalent but differ in structure from the complementary [-] walls A‖C, C‖B and B‖A; likewise in the hexagonal phases one finds two physically distinct types of wall[8]. The [+] walls and [-] walls will have free energies or 'tensions' $\Sigma_+(T,\zeta)$ and $\Sigma_-(T,\zeta)$ which, in general, are unequal : by contrast, in the Potts model the corresponding walls are all equivalent and have the same surface tension.

The identification of the lowered symmetry suggests the existence of melting transitions of a new, chiral character : however, one must ask if the corresponding chiral perturbation is relevant, in which case the Potts transition point will be a multicritical point, accessible only for some special chemical potential, say $\zeta = \zeta_0$, or if the perturbation is irrelevant, in which case a continuous section of the transition line should remain of Potts character[8]. Even in the latter case the transition line might acquire a chiral nature at a sufficiently large value of the symmetry breaking perturbation. In fact, by considering the scaling properties of the experimentally observable structure factor, $S(q;T,\zeta)$ of the fluid near a chiral melting transition, one can identify a special signature[8,9], namely the equality of the exponents $\bar{\beta}$ and ν, which appears to have been seen in the fluid-to-commensurate transition around 97°K in krypton on graphite[2].

The relevance of the chiral perturbation for rectangular $p \times 1$ phases has also been studied by generalizing the Potts model to a chiral clock model[7,10] in which the Potts coupling between neighbouring 'spins' n_i (= 0, 1, 2) along one lattice axis is replaced by the term

$$- J \cos \left[2\pi(n_i - n_j + \kappa)/p\right] \qquad \text{(with } p = 3)$$

in which κ represents the chirality : thus for $0 < \kappa < \frac{1}{2}$ the sequence of states 012012... has lower energy than the mirror sequence 210210... but the ferromagnetic pairs 00, 11, and 22 still have lowest energy. For $p = 4$ the crossover exponent for κ can be determined exactly as $\phi = \gamma - \nu$ for all dimensionality[11]; this is positive so that κ is relevant. For $p = 3$ heuristic arguments[8,9] suggest that chirality is again relevant. This conclusion has been confirmed for $d = 2$ dimensions by extrapolation of high-temperature series expansions for $(\partial^2\chi/\partial q \partial\kappa)_{q=\kappa=0}$ and of low-temperature expansions for $(\partial\Sigma_+/\partial\kappa)_{\kappa=0}$, which yield $\lambda = 0.19 \pm 0.06$.[12] Thus a chiral commensurate melting transition is anticipated for adsorbed 3×1 phases. A preliminary investigation suggests that chirality will not be relevant at the Potts point for hexagonal $\sqrt{3} \times \sqrt{3}$ phases but, as explained, this does not preclude a chiral transition above some finite value of the symmetry breaking[8,9].

Even granting chiral crossover it is not certain that the new exponents will actually differ from the pure Potts exponents, $\alpha = \frac{1}{3}$, $\nu = \frac{5}{6}$, etc. Nevertheless the chiral critical fluctuations should differ significantly from the ordinary $p = 3$ Potts critical fluctuations and hence yield new exponents. The difference arises from a domain wall wetting transition which must take place as the melting line is approached : the tension of the favoured type of wall, say [+], must approach zero at melting; before it vanishes it will fail to satisfy the stability condition $2\Sigma_+ > \Sigma_-$. When this happens an unfavoured [-] wall, say $A \| C$, undergoes wetting by interpolation of a B domain and decomposes into the form A|B|C with two [+] walls. As a result of this wetting transition (which can be found explicitly in the chiral clock model[12]) only one type of wall or microdomain boundary survives near the transition so modifying the fluctuations. In a low-temperature 'solid-on-solid' limit the wetting transition is continuous[8,12] with

$$(2\Sigma_+ - \Sigma_-) \sim (T_W - T)^2 \qquad \text{for } T \to T_W-$$

but this behaviour is expected to be universal since it arises directly from the properties of random walks which represent the domain wall wandering[9,13].

ACKNOWLEDGEMENTS

Active collaboration with David Huse and Anthony Szpilka played a vital role in the researches reported here. The support of the National Science Foundation, in part through the Materials Science Center at Cornell University, is gratefully acknowledged.

REFERENCES

1. M. Bretz, Phys. Rev. Lett. 38:501 (1977).
2. D.E. Moncton, P.W. Stephens, R.J. Birgeneau, P.M. Horn G.S. Brown,Phys. Rev. Lett. 46:1533 (1981).
3. F. Boszo, G. Ertl, M. Grunze and M. Weiss, Appl. Surf. Sci. 1:103 (1977).
4. R. Imbihl, R.J. Behm, K. Christmann, G. Ertl and T. Matsushima, Surface Sci. 117:257 (1982).
5. S. Alexander, Phys. Lett. 54A:353 (1975).
6. E. Domany, E.K. Riedel and M. Schick, Phys. Rev. B18:2209 (1978).
7. S. Ostlund, Phys. Rev. B 24:398 (1981) and 23:2235 (1981).
8. D.A. Huse and M.E. Fisher, Phys. Rev. Lett. 49:793 (1983).
9. D.A. Huse, Phys. Rev. [submitted].
10. D.A. Huse, Phys. Rev. B 24:5180 (1981).
11. D. Huse and M.E. Fisher, J. Phys. C 15:L585 (1982).
12. D.A. Huse, A.M. Szpilka and M.E. Fisher, Physica [submitted].
13. M.E. Fisher and D.S. Fisher, Phys. Rev. B 25:3192 (1982).

THE ROLE OF DOMAIN WALLS IN THE CRITICAL BEHAVIOUR OF THE THREE-DIMENSIONAL P-STATE ASYMMETRIC CLOCK MODEL

Julia Yeomans

Department of Physics
The University
Southampton SO9 5NH, U.K.

INTRODUCTION

The behaviour of systems with competing interactions has recently provided considerable interest[1]. Experiments on, for example, cerium antimonide [2], silicon carbide [3] and graphite intercalation compounds have shown that a fine balance between different interactions can lead to a compound exhibiting many different stable thermodynamic phases as the external parameters are varied[5,6].

A simple spin model which mirrors this behaviour is the p-state asymmetric clock model [7,8]. As the parameter controlling the competition between chiral and ferromagnetic ordering is varied this model displays, for dimensions $d > 2$, infinite sequences of long wavelength phases, separated by first order phase boundaries, springing from a unique multiphase point at zero temperature [9,10].

Our aim in this paper is to discuss two different approaches to the determination of the low temperature phase diagram of the p-state asymmetric clock model for all values of p. We first present results obtained using a low temperature series technique [6,10]. The advantage of this treatment is that the results are exact at low temperatures. However, low temperature series calculations give little physical insight into the reasons for the occurrence of the long wavelength phases.

Therefore we look again at the phase diagram of the model using a low temperature mean field theory developed by Villain and Gordon [11]. This allows us to map the system onto a set of interacting domain walls. The mean field phase diagram is in excellent agreement with that resulting from low temperature series and considering a "wall" picture of the model allows us to gain insight into the reasons for the appearance of the phase sequences.

The p-state asymmetric clock model is described by the Hamiltonian [7,8]

$$\mathcal{H} = -J_0 \sum_{\langle ij \rangle}^{\perp} \cos \frac{2\pi}{p}(n_i - n_j) - J \sum_{\langle ij \rangle}^{//} \cos \frac{2\pi}{p}(n_i - n_j + \Delta) \qquad (1)$$

where the n_i are integer variables taking values from 0 to p-1 and the summations are over nearest neighbours on a d-dimensional hypercubic lattice in a singled out axial (//) or in the (d-1) perpendicular ($\perp$) directions respectively. For $0 \leq \Delta < \frac{1}{2}$ the ground state is ferromagnetic. For $\frac{1}{2} < \Delta \leq 1$, however, axially connected spins order chirally. $\Delta = \frac{1}{2}$ is a multiphase point at which the ground state is infinitely degenerate. At this point each ground state comprises a sequence of bands, consecutive layers of spin value n_i preceded by a layer of spin $(n_i - 1)$ and followed by a layer of spin $(n_i + 1)$. For example, a possible ground state for p = 4 is

$$\ldots 0011122233000111223333000 11222\ldots \qquad (2)$$

Here the repeating sequence consists of a two-band followed by two three-bands.

Our aim is to discover which of the ground state phases persist to finite temperature. To describe the states we follow Fisher and Selke [6] in introducing the notation $\langle n_1 n_2 \ldots \; n_m \rangle$ to represent a basic repeating sequence of m bands of length n_1, n_2 $\ldots n_m$. (2) can then be written as $\langle 233 \rangle$ or $\langle 23^2 \rangle$.

LOW TEMPERATURE SERIES

We first describe the phase diagrams which result from studying the p-state asymmetric clock model using low temperature series. Results for p = 3 are exact to all orders [9]; results for p > 3 follow from a numerical calculation of the important terms in the series to at least tenth order [10].

Consider first p = 3 [9]. For $\Delta = \frac{1}{2}$ symmetry demands that the $\langle 2 \rangle$ phase is stable [8]. Between the ferromagnetic phase and $\langle 2 \rangle$ the model locks into an infinite sequence of commensurate phases $\langle 2^n 3 \rangle$, n = 0,1,2 ... The phase boundaries are first order and the width of successive phases decreases exponentially with increasing n. For $\Delta > \frac{1}{2}$, a similar sequence $\langle 12^n \rangle$, n = 1,2,3..., interpolates between $\langle 1 \rangle$ and $\langle 2 \rangle$. As the temperature is raised the long wavelength phases become stable.

Qualitatively similar phase diagrams are seen for higher

values of p [10]. However, with increasing p more phase sequences become stable. The phase diagram for p = 4 is shown in Figure 1. New phases $<12^{n-1}\ 12^n>$ intercalate between $<12^{n-1}>$ and $<12^n>$, n=1, 2,3 ..., and < 4> and $<2^{n-1}32^n3>$, n = 1,2,3..., appear for $\Delta < \frac{1}{2}$. All phases now remain stable throughout the range of validity of the low temperature series. For p = 5, for $\Delta > \frac{1}{2}$, new sequences $<12^{n-1}\ 12^{n-1}\ 12^n>$ and $<12^{n-1}\ 12^n\ 12^n>$, n = 1,2,3 ..., become stable. In each case the intercalating phase is closely related to its neighbours.

Note that, for $\Delta > \frac{1}{2}$, only one- and two-bands are allowed. A surprising feature of the results is that the value of p for which a given phase $<\nu>$ first appears depends only on the number of one bands, n_1, in the sequence ν through the expression

$$p = n_1 + 2 \ . \tag{3}$$

The low temperature series calculations give no insight into the physics which leads to this relation and hence we have studied the p-state asymmetric clock model using a mean-field theory [11].

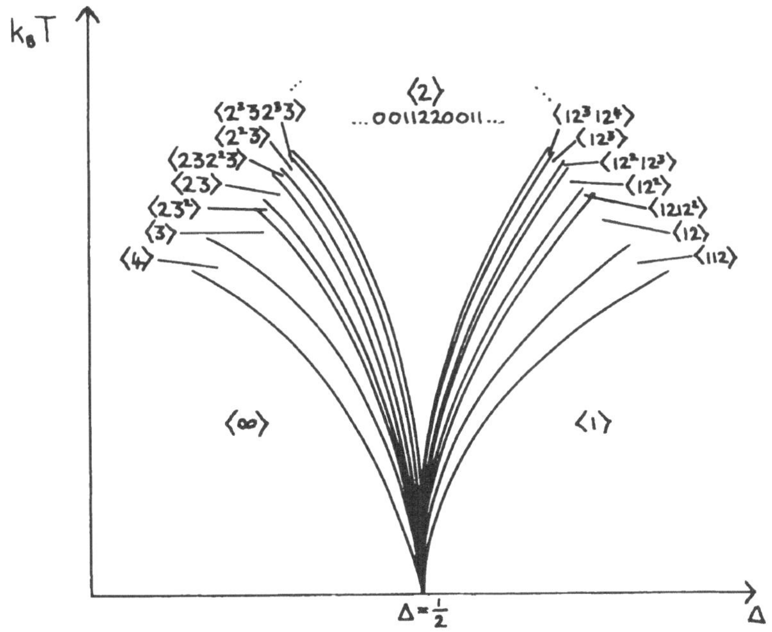

Fig.1. Schematic phase diagram of the asymmetric clock model for p = 4.

MEAN FIELD THEORY

The mean-field theory [11] reduces the d-dimensional Hamiltonian of the asymmetric clock model to the one-dimensional problem of a series of interacting domain walls. This is done by treating the in-layer interactions using mean-field theory and considering the chiral interaction to act between the mean spin in neighbouring layers. The Hamiltonian (1) then becomes

$$\mathcal{H} = -\frac{J}{2}\sum_i \{ e^{\frac{2\pi i\Delta}{p}} \langle e^{\frac{2\pi i n_i}{p}} \rangle \langle e^{\frac{-2\pi i n_{i+1}}{p}} \rangle + e^{\frac{-2\pi i\Delta}{p}} \langle e^{\frac{-2\pi i n_i}{p}} \rangle \langle e^{\frac{2\pi i n_{i+1}}{p}} \rangle \} \qquad (4)$$

Writing

$$\langle e^{\frac{2\pi i n_i}{p}} \rangle = (1-\delta_i)\, e^{\frac{2\pi i n_i}{p}} \equiv \phi_i - u_i \qquad (5)$$

and adding a quadratic potential term to force the deviations, $\delta_i = |u_i|$, of the layer spin from its value at zero temperature, ϕ_i, to remain small gives a free energy

$$\mathcal{F} = A\sum_i |u_i|^2 - \frac{J}{2}\{e^{\frac{2\pi i\Delta}{p}} \sum_i (\phi_i - u_i)(\phi^*_{i+1} - u^*_{i+1}) + e^{\frac{-2\pi i\Delta}{p}} \sum_i (\phi^*_i - u^*_i)(\phi_{i+1} - u_{i+1}) \} \qquad (6)$$

We stress that the expression (6) has been derived using the following approximations:
(i) a mean-field approximation within the ferromagnetically coupled layers
(ii) a low temperature approximation - $|u_i|$ must remain small
(iii) fluctuations in the phase of the layer spin have been ignored.

Standard techniques [11] permit (6) to be written in terms of an effective long-range interaction between the layer variables ϕ_i.

$$\mathcal{F} = -\sum_{i\neq j} \gamma(i-j)\phi_i\phi_j^* \qquad (7)$$

where $\gamma(j) = \lambda\, e^{-\kappa|j|}\, e^{\frac{-2\pi i j\Delta}{p}}$, (8)

$$\lambda = \frac{1}{J\sinh\kappa}\ , \qquad \cosh\kappa = \frac{A}{J}\ . \qquad (9)$$

By considering the relative values of ϕ_i and ϕ_j in successive spin-bands the free energy (7) can be rewritten in terms of a series of ν interacting walls at positions x_q [11]

$$\mathcal{F} = \nu \left\{ \left(1 - e^{\frac{-2\pi i}{p}}\right) w_1 + \left(1 - e^{\frac{2\pi i}{p}}\right) w_1^* \right\}$$

$$+ \sum_{r=1}^{\infty} \sum_{q} 4\sin^2\frac{\pi}{p} \left\{ e^{\frac{-2\pi i r}{p}} U(x_{q+r} - x_q) + e^{\frac{2\pi i r}{p}} U^*(x_{q+r} - x_q) \right\} \qquad (10)$$

where w_1 and $U(j)$ are related to $\gamma(j)$ by

$$w_1 = \sum_{k=1}^{\infty} k\gamma(-k) \quad , \quad U(j) = \sum_{k=1}^{\infty} kU(-j-k) \ . \qquad (11)$$

It is apparent from (10) that two terms contribute to the free energy. The first term represents the free energy of ν non-interacting walls. The second term is the contribution to the free energy which results from interaction between walls: the summations are over all walls q and over rth neighbour interactions.

Note that the free energy is an oscillatory function of Δ. It is this which allows the incomplete devil's staircase, an infinite sequence of phases separated by first order phase transitions, to appear in the phase diagram of the p-state asymmetric clock model. The oscillatory behaviour is a direct consequence of the chiral nature of the Hamiltonian.

ANALYSIS OF THE MEAN FIELD FREE ENERGY

To analyse the free energy (10) we expand it in terms of the variable $e^{-\kappa}$ which is small for low temperatures. The phase diagram can then be determined inductively from a consideration of successively higher order terms in the series expansion [6]. It is helpful to follow Fisher and Selke [6] in defining the structural variables, ℓ_ν, as the number of spin sequences ν per spin which appear in a given structure. For example, for the sequence (2)

$$\ell_1 = 0,\ \ell_2 = \frac{1}{8},\ \ell_3 = \frac{1}{4},\ \ell_{23} = \frac{1}{8},\ \ell_{33} = \frac{1}{8},\ \ell_{32} = \frac{1}{8}\ . \qquad (12)$$

Defining α_i by

$$\Delta = \tfrac{1}{2} - (\alpha_1 e^{-\kappa} + \alpha_2 e^{-2\kappa} + \alpha_3 e^{-3\kappa} + \dots) \qquad (13)$$

we obtain to lowest order

$$\frac{\mathcal{F}}{N} = \left(\sum_{k=1}^{\infty} \ell_k \, (4s\alpha_1 - 8s^2) + 8s^2 \ell_1 \right) e^{-2\kappa} + o(e^{-3\kappa}) \qquad (14)$$

where $s = \sin \frac{\pi}{p}$ and $c = \cos \frac{\pi}{p}$. This expression is minimised by the ferromagnetic phase, $<\infty>$, for $\alpha_1 > 2s$, by $<2>$ for $-2s < \alpha_1 < 2s$ and by the chiral phase, $<1>$, for $\alpha < -2s$. On the phase boundary between $<1>$ and $<2>$, $\alpha = -2s$, all phases which contain only one- and two-bands have the same free energy to order $e^{-2\kappa}$. Therefore, to lift the degeneracy we must consider higher order terms in the expansion [6].

In the vicinity of this boundary the free energy can be written as a linear function of the structural variables

$$\frac{\mathcal{F}}{N} = (\ell_1 + \ell_2) \left\{ \left(1 - e^{\frac{-2\pi i}{p}}\right) w_i + \left(1 - e^{\frac{-2\pi i}{p}}\right) w_1^{*} \right\} + \sum_{\mu} c_\mu \ell_\mu \qquad (15)$$

where the ℓ_μ are a subset of the ℓ_ν chosen so that they are linearly independent. Hence linear programming theory can be used and the phase diagram can be built up, order by order, through the following inductive steps [6]:

(i) the lowest order phase which can intercalate in the vicinity of a given phase boundary is determined. The new phase is always closely related to its neighbours, for example, on the boundary between $<12^k>$ and $<12^{k+1}>$ the candidate for stability is $<12^k12^{k+1}>$.

(ii) the stability of the new phase, μ, is investigated. This depends on the sign of the coefficient c_μ of ℓ_μ in the expression (15) for the free energy. If c_μ is positive the appearance of the new phase ($\ell_\mu \neq 0$) can only increase the free energy and the existing boundary will remain stable to all orders of the expansion. If c_μ is negative, however, the new phase will become stable in the vicinity of the boundary and the analysis must recommence about the two new boundaries generated.

A careful comparison of (10) and (15) shows that for all phases which can appear in the phase diagram of the p-state asymmetric clock model for $\Delta > \frac{1}{2}$ the coefficient of ℓ_μ is given to leading order by

$$c_\mu = - 8s^2 \cos \frac{\pi n_1}{p} e^{-\kappa n_\mu} \qquad (16)$$

where n_μ is the number of spins in the sequence μ. Hence we see that the stability of a given phase $<\mu>$ depends only on the number of one-bands, n_1, in μ. We shall return to consider the expression (16) in more detail but first we use it to predict the phase sequences appearing in the mean-field treatment of p-state asymmetric clock model and compare the results to those obtained using low temperature series [10].

RESULTS OF THE MEAN-FIELD APPROXIMATION

Despite the approximations inherent in the mean-field theory the resultant phase diagrams are in excellent agreement with those obtained using low temperature series. For p = 3 the stable phase sequences are identical. Successive phases become exponentially narrower (by a factor $e^{-\kappa}$) and are separated by first order phase boundaries. No cut-off of the longer wavelength phases is seen.

As the value of p is increased new phase sequences appear in the same order as predicted by the low temperature series expansion [10]. However, consideration of (16) shows that the value of p at which a given phase first becomes stable is

$$p = 2n_1 + 1 \ .$$

For $p = 2n_1$, $c_\mu = 0$, and it is necessary to look at the next order of the expansion to check the stability of the new phase. This has not been done generally but checks of the lower order phases (for example whether <112> appears for p = 4) suggest that the new phase does become stable to this order. Its width will, however, be reduced by a factor $e^{-\kappa}$.

Hence the mean-field phase diagram for $p = p_0$ contains the same phase sequences as the low temperature phase diagram, which we assume to be exact, for $p = \lfloor p_0/2 \rfloor + 2$ where $\lfloor \ \rfloor$ denotes the integer part. The close similarities gives us confidence that the underlying physics is treated correctly by the mean-field theory.

We return to a consideration of the expression (16). For the sequence $\nu = \langle 1\ \tilde{\nu}\ 2 \rangle$ the dominant term arises from the free energy due to interaction between the walls which bound the phase sequence $1\ \tilde{\nu}\ 1$. If this interaction is attractive the phase $\langle 1\ \tilde{\nu}\ 2 \rangle$ is not stable and vice versa. The interaction depends only on the number of one-bands, n_1, in $\tilde{\nu}$. Hence the two bands seem to provide an unimportant "background" term in the determination of the stability of the phases. The interaction between the one-bands is more likely to be repulsive for larger values of p and for smaller values of n_1.

CONCLUSION

In summary we have shown that the p-state asymmetric clock model exhibits infinite sequences of long wavelength commensurate phases as the parameter, Δ, governing the competition is varied. Results from low temperature series [10] and a mean-field theory [11] have been compared and shown to be qualitatively very similar. The series method has the advantage of being exact at low temperatures whereas the mean-field theory allows us to gain a greater insight into the role of domain walls in the stability of the phase diagram. In particular the interaction between one-bands is important in determining which phases become stable for a given value of p.

I should like to thank J.Villain for suggesting the application of the mean-field theory to the asymmetric clock model.

REFERENCES

1.Bak P. 1982 Rep.Prog.Phys. 45 587
2.Rossat-Mignod J., Burlet P., Batholin H., Vogt O and Lagnier R. 1980 J.Phys.C. Solid St.Phys. 13 6381
3.Jepps N.W. and Page T.F. 1983 to be published in J.Cryst.Growth
4.Clarke R. 1980 in "Ordering in two-dimensions", ed. S.Sinha, North Holland, New York p.53
5.Bak P. and von Boehm J. 1980 Phys.Rev. B21 5297
6.Fisher M.E. and Selke W. 1981 Phil.Trans.Roy.Soc. 302 1
7.Huse D.A. 1981 Phys.Rev. B24 5180
8.Ostlund S. 1981 Phys.Rev. B24 398
9.Yeomans J.M. and Fisher M.E. 1981 J.Phys.C. Solid St.Phys. 14 L835
10.Yeomans J.M. 1982 J.Phys.C. Solid St.Phys. 15 7305
11.Villain J. and Gordon M.B. 1980 J.Phys.C. Solid St.Phys. 13 3117.

INTERFACIAL ADSORPTION IN MULTI-STATE MODELS

Walter Selke

Institut für Festkörperforschung der
KFA Jülich, 5170 Jülich 1,
Federal Republic of Germany

Interfaces in multi-state models, such as Potts models and higher spin Ising models, may be introduced by fixing the variables at opposite boundaries in two different states. Using, in particular, Monte Carlo techniques it is seen that at the interface more of the non-boundary states are generated than is the case in the absence of interfaces[1,2,3], see Fig.1. This phenomenon, the interfacial adsorption of non-boundary states, resembles the appearance of a wetting layer in a two-fluid system in equilibrium with its vapour phase[4]. Other realizations might be expected in adsorbate systems described by Potts models.

The critical behaviour of the interfacial adsorption depends crucially on the class of the bulk transition (first-order, second-order, tricritical, Potts-like), as is found from the analysis, using Monte Carlo methods, of two-dimensional Potts models [1,2] and the two-dimensional Blume-Capel (BC) model[3], a special case of the Blume-Emery-Griffiths (BEG) model.

The Blume-Capel model is described by the Hamiltonian

$$H = -J \sum_{<ij>} S_i S_j + D \sum_i S_i^2 \qquad S_i = 1, 0, -1$$

where $<ij>$ indicates summation over nearest neighbor spins on a square lattice of size NxN; J and D are taken to be positive. The model is known to display a second-order transition for $0 \leq D/J < (D/J)_t$, a tricritical point at $(D/J)_t \approx 1.931$[5] and a first-order transition in the range $(D/J)_t < D < 2$.

The Hamiltonian of the q-state Potts model is

$$H = -J \sum_{<ij>} \delta_{S_i, S_j} \qquad S_i = 1, 2, \ldots q$$

The summation extends over all nearest neighbour pairs of sites i,j; J is positive. In two dimensions, the phase transition is known[6] to be second order for $q \leq 4$, and of first order for higher values of q. It is interesting to note that the three-state Potts transition point is a special tricritical point in the BEG model[3].

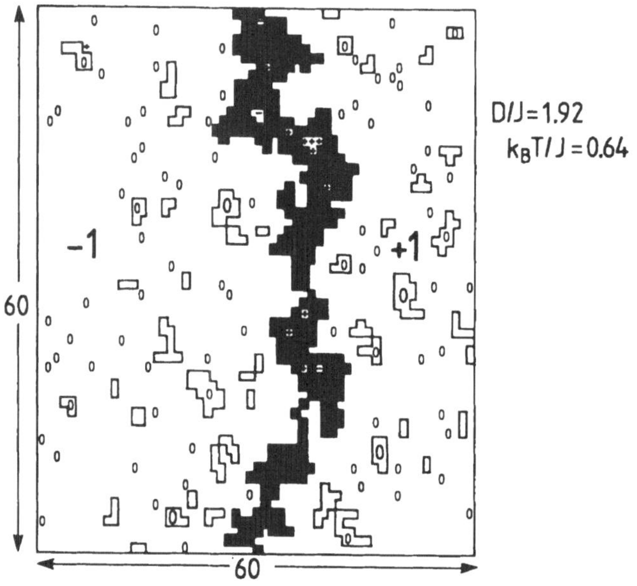

Fig.1 Typical Monte Carlo equilibrium configuration close to the tricritical point of the Blume-Capel model. The non-boundary states (0) adsorbed on the interface are shown blackened.

To introduce an interface into the BC model the spins on opposite boundaries are fixed in the two different non-zero states, say, -1 on the left side boundary (S_L=-1) and +1 on the right hand boundary (S_R=+1). For the Potts models we set S_L=1 and S_R=2. At zero temperature the interface is a straight line for the Potts, models (and the BC model with D/J<1) separating "1" and "2" ("1" and "-1") domains. For D/J>1, however, the ground state energy is lowered by the insertion of a S_i=0 monolayer at the interface. By examining typical equilibrium Monte Carlo configurations[1,2,3] it is seen that an excess of the non-boundary states is generated at the interface (this observation provides also the clue for the modification of the interface approximation of Müller-Hartmann and Zittartz to treat three-state models[1,3,7]). The interfacial adsorption occurs in a layer-like fashion in the BC model in contrast to the more droplet-like adsorption in Potts models as expected on the basis of single spin-flip considerations. In general, the boundary states form small islands in the adsorbed layers or droplets. In the case of $q \geq 4$, the large droplets are usually composed of clusters of different non-boundary states.

To describe this phenomenon quantitatively we define the net adsorption per unit length of the interface as

$$W_n = \frac{1}{N} \sum_{n,i} (<\delta_{n,S_i}>_I - <\delta_{n,S_i}>_{1:1})$$

The sum is taken over all lattice sites, i, and non-boundary states n (0 for the BC model; 3,...q for the Potts models); the < >-brackets denote thermal averages and the subscripts I (1:-1 in the BC case; 1:2 in the Potts case) and 1:1 refer to systems with and without ($S_L = S_R = 1$) interfaces, respectively. W_n may be also considered as a measure of the thickness of an idealized adsorption layer at the interface, which contains only non-boundary states.

The non-boundary states generated at the interface involve an energy cost, which is compensated by a gain in entropy. This compensation is clearly seen in the temperature dependence of W_n:W_n increases up to its maximum value close to the bulk transition temperature, T_c (approaching T_c as N goes to infinity), and then decreases sharply. The latter sharp decrease is consistent with the vanishing of the interface above T_c, which we know must occur in the thermodynamic limit.

To study the critical properties of the net adsorption W_n we recorded its critical temperature dependence, $W_n(t=(T_c-T)/T_c)$ for fixed size N(=40,60) as well as the size dependence of the maximal adsorption, $W_n^{max}(N)$, and of the adsorption at T_c, $W_n(T_c,N)$; $6 \leq N \leq 100$. The results for the BC model can be summarized in the following form:

(i) at the continuous transition:

$$W_n^{max}(N) = W_n^{max}(\infty) - c(D/J)/N$$

where $W_n^{max}(\infty)$ is finite. Accordingly, the temperature dependence is non-critical.

(ii) at the tricritical point ($D/J \approx 1.931$):

$$W_n^{max} \sim \ln N \text{ or } N^a \text{ with } a<0.2$$

Note that the slope. $d(\ln W_n)/d(\ln N)$ decreases appreciably with increasing N giving an upper bound for a: similary for ω (see below). If there is a power-law behavior, then the exponent is probably very small. The temperature dependence is also quite likely to be of logarithmic type

$$W_n(t) \sim \ln t \text{ or } t^{-\omega} \text{with } \omega<0.15$$

(iii) at the first-order transition:
A power law divergence is found

$$W_n^{max} \sim N^a \quad \text{with} \quad a=0.7\pm0.05$$

$$W_n(t) \sim t^{-\omega} \quad \text{with} \quad \omega=0.33\pm0.03$$

The power-law divergence at the first-order transition is of particular interest. It contradicts the common belief that no critical exponents may characterise such transitions, because of the finite correlation length. It should be noted that logarithmic divergencies at first-order transitions have been found in different connections[8,9] using mean-field or Landau theory. - In our case, a qualitative argument can be given[10], which yields $\omega=1/3$ for a model similar to a three-state model of SOS-type[11].

In the three- and four-state Potts model W_n is found to diverge when T_c is approached as

$$W_n \sim t^{-\omega_q} \quad q = 3,4$$

At the bulk critical temperature one obtains

$$W_n(T_c,N) \sim N^{a_q} \quad q = 3,4$$

The exponents, as determined from Monte Carlo data, are consistent with the ones following from a scaling argument[2] (based on the observation that the non-boundary states observed at the interface are components of the order parameter)

$$\omega_q = \nu_q - \beta_q \quad \text{and} \quad a_q = 1-(\beta_q/\nu_q)$$

where ν_q and β_q are the known[6] bulk exponents. The same scaling relations have been pointed out in the context of adsorption on a wall of a two-phase system[12].

It is very interesting to study higher state (q>4) Potts models to check, whether the exponents a and ω found at the first order transition of the BC model are "universal" in two dimensions. However, preliminary Monte Carlo data for $q=6$, 10 and 20 indicate that it is difficult to establish the critical behaviour convincingly. In particular, we studied $W_n(T_c,N)$ for $6\leq N\leq 60$. In all cases, $d(\ln W_n)/d(\ln N)$ showed noticable curvature (the slope decreases with increasing N), presumably because the asymptotic behaviour has not been reached for these sizes[13] due to the large, albeit finite correlation length at T_c, which seems to decrease rather slowly with larger values of q. Note that the effective exponent a^q_{eff}, determined in the range $20 \leq N \leq 60$, tends towards the BC value ($a=0.7\pm0.05$): We obtain $a^q_{eff} \approx 0.95$ for q=6, 0.85 for q=10 and 0.65 for q=20. (Note also that on the basis of this limited set of data we cannot even rule out a finite maximal adsorption for $N \to \infty$ governed possibly by the correlation length). $d(\ell n\ W_n)/$ $/d(\ell n\ t)$ shows an even more pronounced curvature (e.g. for N=40, q=6) and we did not try to estimate ω. At any rate, the true critical behaviour should be clarified in a later, careful study using large lattice sizes and/or values of q.

I wish to thank Julia Yeomans, Werner Pesch and David Huse for productive collaborations on many of the results reported in this contribution.

References

1. W. Selke and W. Pesch, Interface properties of the three-state Potts model in two dimensions, Z. Physik B47: 335 (1982)
2. W. Selke and D.A. Huse, Interfacial adsorption in planar Potts models, Z. Phys. B50: 113 (1983)
3. W. Selke and J. Yeomans, Interface properties of the two-dimensional Blume-Emery-Griffiths model, J. Phys. A: in press (1983)
4. M. Moldover and J. Cahn, An interface phase transition: complete to partial wetting, Science 207: 1073 (1980); B. Widom, Structure of the $\alpha\gamma$ interface, J. Chem. Phys. 68: 3878 (1978)
5. D.P. Landau and R.H. Swendsen, private communication
6. F.Y. Wu, The Potts model, Rev. Mod. Phys. 54: 235 (1982)
7. T. Temesvari, Interface method for the antiferromagnetic three-state Potts model on a square lattice, J. Phys. A15: L625 (1982)
8. J. Lajzerowicz, Domain wall near a first order transition: role of elastic forces, Preprint (1982)
9. R. Lipowsky, Critical phenomena at first-order bulk transitions, Phys. Rev. Lett. 49: 1575 (1982)
10. D.A. Huse, private communication
11. D.B. Abraham and E.R. Smith, Surface-film thickening: an exactly soluble model, Phys. Rev. B26: 1480 (1982)
12. M.E. Fisher and P.G. de Gennes, Phénomènes aux parois dans un mélange binaire critique, C.R. Acad. Sci. Paris B287: 207 (1978)
13. K. Binder, Static and dynamic critical phenomena of the two-dimensional q-state Potts model, J. Stat. Phys. 24: 69 (1981)

VORTEX PINNING IN THIN SUPERFLUID FILMS

J.L. McCauley, Jr.

Department of Physics
University of Houston
Houston, Texas 77004, USA

In a thin superfluid film, a region of reduced superfluid density will attract a vortex. Consider a closed region of size $\sim b$ over which the superfluid density vanishes locally and consider a random distribution of such sites with density $N_p \ll b^{-2}$. When the surrounding superfluid is set into motion, such sites will create and annihilate free vortices[1]. Because circulation is conserved, the annihilation of a vortex by a neutral site will cause that site to become "charged" i.e., to take on the circulation of the vortex and similarly for vortex creation. Hence, both charged and neutral sites must be treated simultaneously in the dynamics. This is the phenomenon of 'vortex pinning' in the two dimensional limit.

Neutral sites exert a dipolar attraction upon a free vortex while charged sites are similar to vortices that are fixed in space. We have calculated the rates of vortex creation and annihilation due to the combined effects of pinning sites and bulk processes. The results are in good agreement with the observed persistent current decays for ^{4}He films where measurements have been reported for eight different film coverages, with the average film thickness d ranging from 6 to 10 monolayers[2]. If we denote as y_0 the saddle-point separation for a vortex pair in the bulk, then when $b/y_0 \ll 1$ we obtain our previous limit where bulk processes dominate and which is correct for $d \sim 6$ [3]. When $b/y_0 \gg 1$, we retrieve the result predicted by Ambegaokar et al.[4] due to an edge effect ($b = \infty$) which is correct if $d \sim 10$ [1]. Previously, no first principles theory could describe all the decays in the region $6 < d < 10$ without serious violations of fundamental physics [3]. The present theory agrees with experiment if we assume that (i) the size b increase rapidly with film coverage, and (ii) the site density is constant but the density of bound vortex pairs in the bulk (roughly, the fugacity of ref. 4) increases rapidly as ^4He coverage is reduced below d ~ 8. Physically, we regard the parameter b as an effective site size whose scale is determined by the average film thickness d.

Our results can be extrapolated by a renormalization group method to the Kosterlitz-Thouless critical point where the resulting predictions may help to understand differences between thermal conductivities obtained from ^4He on different substrates[5,6].

REFERENCES

1. J.L. McCauley, Jr. and C.W. Allen, Phys. Rev. B25 (1982) 5680
2. D.T. Ekholm and R.B. Hallock, Phys. Rev. B21 (1980) 3902
3. J.L. McCauley, Jr. Phys. Rev. Lett. 45 (1980) 467
4. V. Ambegaokar, B.I. Halperin, D.R. Nelson and E.D. Siggia, Phys. Rev. B21 (1980) 1806
5. J. Maps and R.B. Hallock, Phys. Rev. Lett. 47 (1981) 1533
6. R.A. Joseph and F.M Gasparini, Phys. Rev. B26 (1982 3982

THEORY OF CRITICAL PHENOMENA IN RANDOM SYSTEMS

Amnon Aharony

Department of Physics and Astronomy
Tel Aviv University
69978 Tel Aviv, Israel

I. INTRODUCTION

The aim of this series of lectures is to review some of the theoretical ideas which are used in the study of critical phenomena in random systems. Emphasis is put on phase diagrams which involve *multicritical points*, and on crossover behavior in their vicinity.

II. DILUTE FERROMAGNETS

Consider first the simple dilute ferromagnet, with the Hamiltonian

$$\mathcal{H} = - \sum_{<ij>} J_{ij} \vec{S}_i \cdot \vec{S}_j \quad , \tag{1}$$

where $\vec{S}_i$ is an m-component classical spin vector and the nearest neighbor exchange coefficient J_{ij} is either zero (with probability 1-p) or J>0 (with probability p). We expect the transition temperature T_c to decrease with decreasing p, and to vanish at and below the *percolation threshold* p_c (below which there exist only finite clusters of sites connected by a non-zero exchange). The phase diagram is thus expected to have the qualitative shape shown in Fig.1.

The diagram in Fig.1 contains two special *multicritical points*. One of these represents the "pure" critical behavior, at p=1. The other represents the *percolation threshold* p_c, at T=0. We start with a discussion of the crossover near the former.

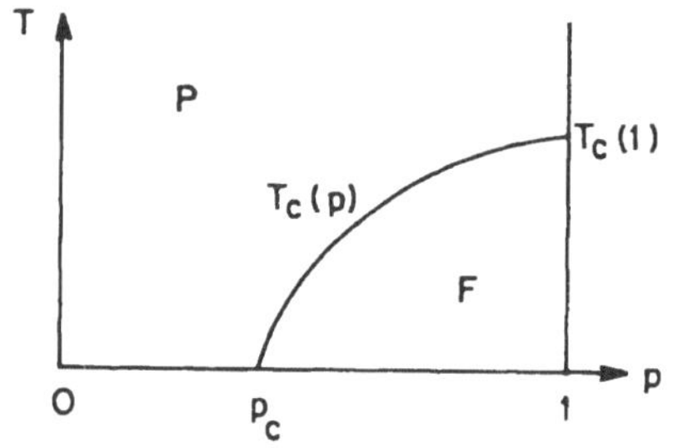

Fig. 1. Phase diagram of dilute Ferromagnet. P = paramagnet; F = ferromagnet.

The critical properties as T approaches $T_c(1)$ on the line p=1 are those of the non-random m-component spin model. As soon as p<1, the transition at $T_c(p)$ may have modified critical properties. One can write a *crossover scaling form* for the singular free energy density near p=1, $T=T_c(1)$,

$$F_s(T, p) = |t|^{2-\alpha} f(q/|t|^{\phi_J}) , \quad (2)$$

where q=1-p, $t=[T-T_c(1)]/T_c(1)$ and ϕ_J is the crossover exponent. If $\phi_J>0$ then the randomness is "relevant", and one must search for new critical properties.

There are three simple ways to identify ϕ_J, all giving $\phi_J=\alpha$. The heuristic Harris criterion[1] compares the fluctuations of the local transition temperature, ΔT_c, within a volume ξ^d (where ξ is the magnetic correlation length and d is the dimensionality),

$$\Delta T_c/T_c \propto \xi^{-d/2} , \text{ with } (T-T_c)/T_c \propto \xi^{-1/\nu} .$$

The transition will remain sharp and ξ will behave as $(T-T_c)^{-\nu}$ only if $\Delta T_c \ll (T-T_c)$, i.e. $(d/2) > (1/\nu)$. Using the hyperscaling relation $d\nu=2-\alpha$, this happens if $\alpha<0$. Fluctuations in T_c are thus irrelevant if $\alpha<0$, and a modified behavior is expected if $\alpha>0$.

A more quantitative way is based on expanding F_s in powers of $\Delta_J = [(\delta J_{ij})^2]_{av} = [(J_{ij} - [J_{ij}]_{av})^2]_{av}$, where $[J_{ij}]_{av} = pJ$ is the configurational average of J_{ij}.[2] Since $\Delta_J=p(1-p)J^2=pq\,J^2$, the singular term in $(\partial F_s/\partial\Delta_J)$ is proportional to that in $(\partial F_s/\partial q)$. At q=0, this derivative should have a term which scales as $|t|^{2-\alpha-\phi_J}$, from which ϕ_J may be identified. A straightforward calculation yields

$$\left.\frac{\partial(-\beta F_s)}{\partial(\beta^2\Delta_J)}\right|_{\Delta_J=0} = \frac{1}{2N}\sum_{<ij>}(<(\vec{S}_i\cdot\vec{S}_j)^2>_o - <\vec{S}_i\cdot\vec{S}_j>_o^{\,2}) \quad , \qquad (3)$$

where $\beta=1/k\,T$ and $<\ >_o$ denotes a thermal average with respect to the non-random (p=1) Hamiltonian. Since $<\vec{S}_i\cdot\vec{S}_j>$ behaves (for nearest neighbors) as the energy density, $|t|^{1-\alpha}$, one readily identifies $\phi_J=\alpha$.

Eq.(3) yields not only the crossover exponent, but also the *magnitude* of the term linear in Δ_J in the expansion of F_s. This term should become comparable to the non-random free energy density before a crossover to a modified behavior can be felt (when $\alpha>0$).

A third method is based on the *"replica trick"*.[3] The average free energy density is written as

$$-\ \beta F = \frac{1}{N}[\ln Z\{J_{ij}\}]_{av} \quad , \qquad (4)$$

where

$$Z\{J_{ij}\} = \mathrm{Tr}\exp(\beta\sum_{<ij>} J_{ij}\,\vec{S}_i\cdot\vec{S}_j) \ . \qquad (5)$$

One now replaces $\ln Z$ by $\lim_{n\to 0}(Z^n-1)/n$, and averages over the distribution of J_{ij}'s. The result is

$$[Z^n]_{av} = \mathrm{Tr}\,\pi_{<ij>}\,[p\exp(\beta J\sum_\alpha \vec{S}_i^{\,\alpha}\cdot\vec{S}_j^{\,\alpha}) + q] \quad , \qquad (6)$$

where the trace is now over the n (replicated) m-component spins $(\vec{S}_i^1, \ldots \vec{S}_i^n)$. Eq.(6) is then written as

$$[Z^n]_{av} = \mathrm{Tr}\,\exp\,[-\beta\, H_{eff}\,\{\vec{S}_i^{\,\alpha}\}] \quad , \tag{7}$$

and the effective Hamiltonian $H_{eff}\,\{S_i^{\alpha}\}$ is analyzed using the usual techniques. The limit $n\to 0$ is taken only at the end.

Near the "pure" point, βH_{eff} may be expanded in powers of Δ_J. To leading order,

$$-\beta H_{eff} = \beta \sum_{\alpha=1}^{n} \sum_{<ij>} [J_{ij}]_{av} \; (\vec{S}_i^{\,\alpha}\cdot\vec{S}_j^{\,\alpha}) +$$

$$+ \frac{1}{2}\beta^2\Delta_J \sum_{\alpha,\beta} \sum_{<ij>} (\vec{S}_i^{\,\alpha}\cdot\vec{S}_j^{\,\alpha})\;(\vec{S}_i^{\,\beta}\cdot\vec{S}_j^{\,\beta}) \; . \tag{8}$$

In the limit $\Delta_J\to 0$, βH_{eff} separates into n decoupled m-component terms (each equivalent to the original "pure" system). Thus, $(\vec{S}_i^{\,\alpha}\cdot\vec{S}_j^{\,\alpha})$ scales (e.g. under the renormalization group) as the energy density of the pure system. Rescaling lengths by b, the energy density scales as $b^{-(1-\alpha)/\nu}$, and thus Δ_J scales as $b^{d-2(1-\alpha)/\nu}=b^{\alpha/\nu}$, implying that $\phi_J=\alpha$.[3]

In three dimensions, the Heisenberg (m=3) and XY (m=2) models have $\alpha<0$. We therefore expect the randomness to be *"irrelevant"*, (the argument in f, Eq.(2), becomes small when $t\to 0$), and the critical behavior at $p<1$ is the same as that of the "pure" system. In the Ising case (m=1), $\alpha\simeq 0.12$ and we expect crossover to a modified behavior. In the language of the renormalization group, the Hamiltonian with $p<1$ (or $\Delta_J>0$) "flows" to a new, "random" fixed point. This "random" fixed point has been identified in $d=4-\varepsilon$ dimensions,[4-6] where one finds a fixed point at which $\Delta_J=O(\varepsilon^{\frac{1}{2}})$, and

$$\alpha = -\,(6\varepsilon/53)^{\frac{1}{2}} \;+\; O(\varepsilon) \quad . \tag{9}$$

Since $|\alpha|$ is very small, the crossover from the "pure" to the "random" behavior is expected to be rather slow: the argument $(q/|t|^{\phi_J})$ in Eq.(2) varies by a factor 10 only if $|t|$ varies by a factor $\sim 10^8$! One will therefore always observe only "effective" critical exponents, which will vary with $|t|$.

In contrast to the weak quantitative changes in critical exponents, which will not be easy to observe, we now wish to draw attention to a strong *qualitative* change in the behavior of the *structure factor*, observable via neutron scattering. The structure factor, $S(\vec{q})$, is defined as the Fourier transform of the spin correlation function $[\langle S_i S_j \rangle]_{av}$, or as $[\langle \hat{S}(\vec{q})\ \hat{S}(-\vec{q}) \rangle]_{av}$, where $\hat{S}(\vec{q})$ is the Fourier transform of S_i. This may be written as[7]

$$S(\vec{q}) = \chi(\vec{q}) + M^2\ \delta(\vec{q}) + C^{(s)}(\vec{q}) \quad , \tag{10}$$

where $M^2\ \delta(\vec{q})$ is the usual Bragg peak,

$$\chi(\vec{q}) = [\langle (\hat{S}(\vec{q}) - \langle \hat{S}(\vec{q}) \rangle)\ (\hat{S}(-\vec{q}) - \langle \hat{S}(-\vec{q}) \rangle)]_{av} \tag{11}$$

is the response function (or the susceptibility) to a $\vec{q}$-dependent field $h(\vec{q})$, and $C^{(s)}(\vec{q})$ is the Fourier transform of $[(\langle S_i \rangle - M)\ (\langle S_j \rangle - M)]_{av}$, i.e.

$$C^{(s)}(\vec{q}) = [\langle \hat{S}(\vec{q}) \rangle \langle \hat{S}(-\vec{q}) \rangle]_{av} - M^2\ \delta(\vec{q}) \quad . \tag{12}$$

Clearly, $C^{(s)}(\vec{q}) \equiv 0$ for the non-random system, when $\langle \hat{S}(\vec{q}) \rangle \equiv M\ \delta(\vec{q})$. Expanding $\langle \hat{S}(\vec{q}) \rangle$ in powers of δJ_{ij}, and averaging, it is easy to check that for $\vec{q} \neq 0$, to leading order in Δ_J, one has[7,8]

$$C^{(s)}(\vec{q}) = 4\beta^2\ \Delta_J M^2 \langle \hat{S}(\vec{q})\ \hat{S}(-\vec{q}) \rangle_0^2 \quad . \tag{13}$$

In the Ornstein-Zernike approximation, $\chi(\vec{q})$ and $\langle \hat{S}(\vec{q})\ \hat{S}(-\vec{q}) \rangle_0$ may be written ($q \neq 0$) as Lorentzians, of the form $(|t|^\gamma + q^2)^{-1}$. Thus, the Lorentzian is supplemented by a squared Lorentzian whenever $M \neq 0$! This qualitatively changes the q-dependence of $S(q)$ in the ordered phase of any dilute ferromagnet! This modified behavior should be searched for in neutron scattering experiments.

As we shall see below, a squared Lorentzian is expected whenever the system is subject to random ordering fields, of the form $\sum_i h_i S_i$. If one replaces S_i in Eq.(1) by $(\langle S_i \rangle + \delta S_i)$, then one indeed finds a term linear in the fluctuation δS_i, whose coefficient is proportional to $\sum_j \langle S_j \rangle \delta J_{ij}$. This may be considered as an effective random field, h_i, with $[h_i^2]_{av}$ proportional to $M^2 \Delta_J$.

III. PERCOLATION

We now turn to the second multicritical point in Fig.1, i.e. the *percolation threshold* $p = p_c$, $T=0$. At $T=0$, all the spins which belong to a cluster (connected to each other by at least one non-zero exchange J_{ij}) point in the same direction. The spontaneous magnetization per spin is therefore equal to the probability of a spin to belong to the infinite cluster, $P_\infty(p)$, and the susceptibility is proportional to the cluster average of s^2, where s is the number of sites in the cluster.

There are several ways to show that the critical exponents associated with the transition at p_c are the same as those of the q-state *Potts model*, in the limit $q \to 1$.[9] In the Potts model, each site i can be in one of q states, $\alpha_i = 1,2,\ldots,q$, and the interaction between sites has the form $-J\,\delta_{\alpha_i \alpha_j}$.

Beginning with Eq.(1) for the Ising case (m=1), ($S_i = \pm 1$), and replacing $S_i S_j$ by $(S_i S_j - 1)$, the replicated Hamiltonian of Eq.(6) may be written as

$$[Z^n]_{av} = \mathrm{Tr} \prod_{<ij>} e^{h_{ij}} , \tag{14}$$

where[10]

$$e^{h_{ij}} = p\, e^{K \sum_\alpha (S_i^\alpha S_j^\alpha - 1)} + q . \tag{15}$$

and $K = \beta J$. The sum $\sum_\alpha (S_i^\alpha S_j^\alpha - 1)$ is equal to 0 if $S_i^\alpha \equiv S_j^\alpha$ for all α's. Otherwise, it will be negative. In the limit $T \to 0$, or $K \to \infty$, a negative power of e^K vanishes. Thus,

$$e^{h_{ij}} \xrightarrow[K \to \infty]{} p \prod_\alpha \delta_{S_i^\alpha S_j^\alpha} + q ,$$

or

$$h_{ij} \longrightarrow - \ln q \left(\prod_\alpha \delta_{S_i^\alpha S_j^\alpha} - 1\right) . \tag{16}$$

Since each S_i^α has two states, the set $\{S_i^\alpha\}$ has 2^n states and h_{ij} represents the 2^n-state Potts model. The limit $n \to 0$ yields the 1-state Potts model. It should be noted that the same result is obtained for any dilute magnetic system (and not just for the Ising example shown here).

At low temperatures, when K is finite (but large), the high symmetry of the Potts model is broken. The magnitude of the leading symmetry breaking term (i.e. the first correction in Eq.(15)) is of order e^{-2K}. We thus expect the crossover to T>0 to be of the scaling form

$$F_s(T, p) = |p - p_c|^{2-\alpha_p} \tilde{f}\left(e^{-2K} / |p-p_c|^{\phi_p}\right) . \tag{17}$$

Indeed, such a form results from an analysis of the anisotropic 2^n-state Potts model in mean field theory and in $d=6-\varepsilon$ dimensions.[10] That calculation also yields $\phi_p=1$, to all orders in ε.[11] The exponent ϕ_p determines the shape of the critical line $T_c(p)$ near p_c, $\exp(-2K_c) \sim (p-p_c)^{\phi_p}$.

A geometrical heuristic way to understand Eq.(17) and to estimate ϕ_p follows from the *"links and nodes"* model.[12,13,14] In this model, the backbone of the infinite cluster is a superlattice made of nodes, separated by a distance of the order of the connectedness length $\xi_p \propto (p-p_c)^{\nu_p}$. The "one-dimensional" links between these nodes are made of L nearest neighbor bonds, with $L \propto (p-p_c)^{-\zeta}$ (See Fig.2a). Along each link, the spin correlations propagate in a one-dimensional way. Decimation of the L spins within a link creates a renormalized coupling constant K_L, with

$$\tanh K_L = (\tanh K)^L \quad . \tag{18}$$

For very large K, we may approximate tanh K by $(1-2e^{-2K})$, and find that $e^{2K_L} \simeq L\, e^{2K}$. The free energy of the superlattice is a function of K_L, and thus of $L\, e^{-2K} \propto e^{-2K}(p-p_c)^{-\zeta}$. We therefore identify $\phi_p=\zeta$.

In the simple model of Fig.1a, one must have $L>\xi_p$, and there-

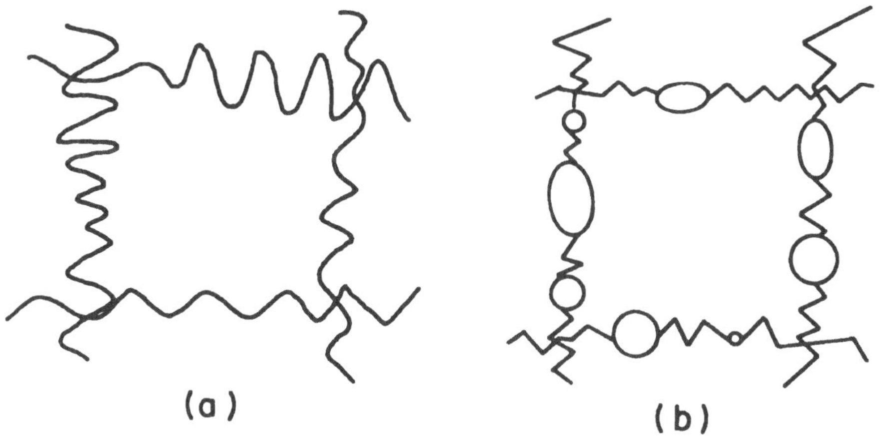

Fig. 2. "Links and nodes" model. (a) The original Skal-Shklovskii-de Gennes model. (b) The modified Stanley-Coniglio "link-node-blob" model.

fore $\zeta > \nu_p$. If one accepts that $\phi_p=1$ in all dimensions,[11] ϕ_p cannot be equal to ζ in two dimensions, when $\nu_p=4/3$.[15] One therefore needs a modified model. A possible modification, suggested by Stanley and discussed in detail by Coniglio,[16,17] adds within each link "blobs" of spins which are multiply-connected (Fig.2b). If two spins are connected by two bonds in parallel, then the effective coupling constant between them is 2K. Since $\tanh 2K \simeq 1-2e^{-4K}$, such a factor will not contribute to the renormalized coupling constant K at the leading order e^{-2K}. To this order, the power L on the RHS of Eq.(18) must be replaced by L_1, i.e. the number of singly connected bonds along the link (cutting of such a bond breaks the connection between the ends of the link). The number of these "cutting bonds", L_1, may be much smaller than L. Writing $L_1 \propto (p-p_c)^{-\zeta_1}$, we thus identify $\phi_p=\zeta_1$. Coniglio[18] recently presented a general proof of the result $\zeta_1=1$, in any dimension, confirming that $\phi_p=1$.

The above argument is valid only if $e^{-2KL} \simeq L_1 e^{-2K} << 1$. As we approach p_c at a fixed temperature, L_1 increases, $L_1 e^{-2K}$ is no longer small and the expansion of tanh K is no longer justified. Moreover, the number of bonds within the "blobs" is of order $L >> L_1$. One can show that $L \propto \xi^{D_B} \propto (p-p_c)^{-D_B\nu_p}$, where D_B is fractal dimensionality of the backbone,[19,20] $D_B=d-\beta_B/\nu_p$ (d is the Euclidean dimensionality, and the probability of a bond to belong to the backbone behaves as $(p-p_c)^{\beta_B}$). In addition to $L_1 e^{-2K}$, we expect "higher order" terms, like $L_2 e^{-4K}$. L_2 is the number of pairs of bonds such that the cutting of both breaks the connection between the ends of the link. Writing $L_2 \propto (p-p_c)^{-\zeta_2}$, this term behaves as $e^{-4K}(p-p_c)^{-\zeta_2}$. We expect that $\zeta_2 > \zeta_1$. As soon as $L_2 e^{-2K} > L_1$, i.e.

$$(p - p_c)^{\zeta_2-\zeta_1} < e^{-2K} , \tag{19}$$

these terms will dominate and the scaling with a single variable, Eq.(17), will have to be modified.[21] Eq.(19) represents an additional line in Fig.1, above which these modifications become important.

Consider now the spin-spin correlation function $\langle S_0 S_r \rangle$. For length scales $r>\xi_p$, this function is described by the random Ising model, as discussed in the previous section. (The effective coupling in this Ising model should be K_ξ). On length scales $r<\xi_p$, the two spins are within a single "link". The correlation function should therefore not depend on ξ_p. In the spirit of the Coniglio model, the function $\langle S_0 S_r \rangle$ is now given by

$$\langle S_0 S_r \rangle = (\tanh K)^{L_1(r)} \simeq e^{-L_1(r)/\xi_1} ,$$

where $\xi_1 = \frac{1}{2}e^{2K}$ is the one-dimensional Ising correlation length, while $L_1(r)$ is the number of singly connected bonds within a distance r. We now generalize the Stanley-Coniglio picture, by adding the assumption that the geometry is *self-similar* on length scales $r<\xi_p$. Noting that $L_1(\xi_p) \propto \xi_p^{\zeta_1/\nu_p}$, we thus write $L_1(r) = A_1 r^{\zeta_1/\nu_p}$ and obtain

$$<S_o S_r> \simeq \exp(-A_1 r^{\zeta_1/\nu_p}/\xi_1) \quad . \tag{20}$$

Clearly, the structure factor $S(q)$ obtained from Eq.(20) will be quite different from the usual Lorentzian one; at large q we expect that $S(q) \propto \xi_p^{-\zeta_1/\nu_p} / q^{2+\zeta_1/\nu_p}$.[21]

Once the "blobs" are taken into account, in a self-similar way, the exponent in Eq.(20) may have the additional term $A r^{\zeta_2/\nu_p}/\xi_1^2$, which dominates at the intermediate length scales $\xi_1^{\nu_p/(\zeta_2-\zeta_1)} < r < \xi_p$.[21] At $p=p_c$, $\xi_p=\infty$ and this new term dominates for large r, thus affecting the small-q dependence of $S(q)$. The detailed q-dependence of $S(q)$ at p_c is therefore much more complex than hitherto expected.

Thus far we discussed only the Ising case. At very low temperatures, decimation of the one-dimensional m-component spin model yields

$$y_m(K_L) = [y_m(K)]^L \quad , \tag{21}$$

where $y_m(K)$ is the nearest neighbor one-dimensional spin correlation function, instead of Eq.(18).[22,23] Using $y_m(x) \simeq 1-a/x$ one finds that the free energy depends on the scaled variable $K_L^{-1} \simeq L/K = LT/J$. Unlike the Ising case, adding two bonds in parallel yields a factor $y_m(2K) \simeq 1-a/2K$, and thus contributes to *leading* order in $T \propto 1/K$. One easily checks that to this order, $K_L \simeq K/L_R$, where L_R is the total resistance of the link (assuming each bond has a unit resistance).[23,17] Writing $L_R \propto (p-p_c)^{-\zeta_R}$, Eq.(17) is replaced by

$$F_s(T,p) = |p-p_c|^{2-\alpha_p} \tilde{f}(T / |p-p_c|^{\zeta_R}) \quad . \tag{22}$$

Again, as p_c is approached this result breaks down and one must consider terms of order $L_2 T^2$, with $L_2 \propto \xi^{\tilde{\zeta}_2}$.[21]

Up to this point we emphasized the "link-node-blob" model. An opposite extreme models the backbone by a *self-similar fractal structure*, such as the Sierpinski gasket shown in Fig.3.[20] In this model there exist no quasi-one-dimensional links, and loops appear at all length scales. One can solve spin models exactly on such models, and use the results to learn on their behavior on the real backbone. For the Ising model, the resulting correlation length behaves (for d=2) as[24] $\xi \propto \exp(4\ e^{4K})$. The difference between this result and the one following from Eq.(20), i.e. $\xi \propto \xi_1{}^{\nu_p/\zeta_1}$, or from the modified $\xi \propto \xi_1{}^{2\nu_p/\zeta_2}$, comes from the different geometrical assumptions. A combined model remains to be constructed.

Before concluding this section, we mention briefly several extensions of the phase diagram of Fig.1 and of the percolation multicritical point. First, Fig.1 represents the phase diagram only for *short range interactions*. As the range of the interactions, R, increases, p_c decreases. For infinite R, p_c decreases to zero and the point $T=p=1/R=0$ is a new multicritical point.[25-27] The critical region of the random Ising model transition, described in the previous section, shrinks to zero as p and $T_c(p)$ approach zero. For dipolar interactions, spin glass ordering may occur at sufficiently small p.[26] At higher p, random dipolar Ising systems exhibit unusual singularities, e.g. $\chi \propto t^{-1} \exp(D|\ell n t|)^{\frac{1}{2}}$.[6]

Second, we should note that near p_c, $T_c(p)$ is very low. One therefore might expect *quantum effects* to become important. It is well known that a *transverse magnetic field* Γ lowers the transition temperature of the quantum Ising model. At sufficiently large Γ, $T_c(\Gamma)$ is lowered down to zero. The corresponding multicritical point has in d dimensions the critical properties (as function of Γ) of a spatially anisotropic (d+1)-dimensional classical Ising model (as function of T).[28] Dilution of the quantum model is equivalent to the introduction of randomness in d out of these (d+1) dimensions. Such a randomness, with infinite correlations along one axis, is highly relevant. The upper critical dimension is shifted up to five, and new critical properties have been predicted.[29] Varying p and Γ we may then reach a new zero-temperature multicritical point, at which both the percolation and the quantum phenomena are important.

Third, some of the impurities may be *annealed*, i.e. in thermal equilibrium with the spins. A simple model for such a situation is the Blume-Emergy-Griffiths spin-1 model.[30] A large chemical potential of the annealed impurities, Δ, may turn the Ising transition first order, via a *tricritical point*. Since the specific heat diverges strongly at the "pure" tricritical point ($\alpha_p=\frac{1}{2}$ at d=3), one expects quenched dilution to cause a fast crossover to a "random" tricritical behavior. This random behavior has not yet been identified (theoretically or experimentally). Again, the T-p-Δ phase diagram may exhibit novel interesting phenomena. In particular, one expects a sequence of many phase transitions as Δ is increased for fixed $p>p_c$ and T=0.[31]

Fig. 3. Three stages of the self similar Sierpinski gasket.

IV. DILUTE ANTIFERROMAGNETS

If the exchange coefficient J is negative, and if the lattice has two interpenetrating sublattices, then the ground state is *antiferromagnetic* (the spins on the two sublattices have opposite directions). In the absence of a magnetic field, the problem can be transformed into the one discussed above by a simple gauge transformation which "flops" the spins on one sublattice. We thus recover the phase diagram of Fig. 1. However, it turns out that these results are drastically changed in the presence of an external field.[32]

The unit cell of the simple antiferromagnet contains two sites (one from each sublattice). Denoting the corresponding spins by $\vec{S}_{i_1}$ and $\vec{S}_{i_2}$, one may define the local uniform magnetization $\vec{M}_i=\frac{1}{2}\langle\vec{S}_{i_1}+\vec{S}_{i_2}\rangle$ and the local staggered magnetization $\vec{M}_i^+=\frac{1}{2}\langle\vec{S}_{i_1}-\vec{S}_{i_2}\rangle$. The exchange coupling [Eq.(1)] now generates three types of terms, $\mathcal{J}_{ij}^+\,\vec{M}_i\cdot\vec{M}_j$, $\mathcal{J}_{ij}^-\vec{M}_i^+\cdot\vec{M}_j^+$ and $\mathcal{J}_{ij}^*\,\vec{M}_i\cdot\vec{M}_j^+$. The last term is absent in pure (translationally invariant) antiferromagnets, because of the local symmetry between the two sublattices (i.e. between $\vec{M}_i^+$ and $-\vec{M}_i^+$). However, this symmetry is broken locally when the J_{ij}'s are random. Thus, $\tilde{J}_{ij}^*$ may be non-zero although its configurational average $[\tilde{J}_{ij}^*]_{av}$ vanishes.

A uniform external magnetic field $\vec{H}$ induces a non-zero magnetization M_i. The random coupling to $\vec{M}_i^\dagger$ now generates a local random staggered field, of order $\vec{h}_i = \sum_i \tilde{J}_{ij}^* \vec{M}_i$. At low fields, $\vec{M}_i \propto \vec{H}$ and thus the strength of the local random field, $\vec{h}_i$, is proportional to δJ_{ij} and to $\vec{H}$, and one has $[\vec{h}_i^{\,2}]_{av} \propto H^2 \Delta_J$. As mentioned above, such random fields generate new (Lorentzian squared) terms in the structure factor. As we shall see below, such terms have crucial effects on the existence of long range order at low dimensionalities.

Much attention has been devoted in recent years to multicritical phenomena in *anisotropic antiferromagnets.*[33] For strong anisotropies, a strong field turns the Ising transition first order via a *tricritical point.* For low anisotropies, the field causes a spin flop transition and a *bicritical* phase diagram. At high dimensionalities, such phase diagrams are also reproduced in the presence of random fields (such as those generated by the uniform field, as explained above).[34,35] Moreover, for intermediate anisotropies one can expect a splitting of the bicritical point into the tricritical point and a critical end point at a new multicritical point.[36] However, many of the transition lines in these phase diagrams are expected to disappear at low dimensionalities, due to the absence of long range order in the presence of random fields.[37]

V. RANDOM FIELDS

Consider now the Hamiltonian

$$\mathcal{H} = - J \sum_{\langle ij \rangle} \vec{S}_i \cdot \vec{S}_j - \sum_i \vec{h}_i \cdot \vec{S}_i \; , \qquad (23)$$

with $[\vec{h}_i]_{av}=0$, $[\vec{h}_i^{\,2}]_{av}=\Delta_h$. As we saw above, such random fields may be generated in many realistic situations (more will be mentioned below). The first question one should ask concerns the relevance of such terms. One can follow each of the methods described in Sec.II, i.e. expand in Δ_h or use replicas, to show that [2,34] $(\partial\chi/\partial\Delta_h)_{\Delta_h=0}$ contains the term $\chi_o^{\,2}$, and therefore that the crossover exponent ϕ_h near the "pure" behavior is equal to the susceptibility exponent, γ. Since γ is usually larger than unity, a fast crossover to a new behavior is expected.

At high dimensionalities, the nature of this new behavior has been analyzed using renormalization group and diagrammatic methods. In the simple Gaussian model, one can show (e.g. by the methods mentioned above) that the Lorentzian propagator $1/(r+q^2)$ is replaced, when $\Delta_h \neq 0$, by $[1/(r+q^2) + A\Delta_h/(r+q^2)^2]$ (r is the inverse susceptibility).[38] Expanding about this Gaussian model, integrals which involved $1/(r+q^2)$ will now involve $1/(r+q^2)^2$, and thus diverge at dimensionalities higher by 2. This led to the general proofs[38,39] that the critical behavior in the presence of random fields in d dimensions is the same as that of the corresponding "pure" system in (d-2) dimensions. These proofs have recently been rephrased in terms of supersymmetry.[40]

These diagrammatic expansions work only near the *upper critical dimensionality*, which is now shifted to six. They are therefore useless at dimensionalities $d<4$. The study at low dimensionalities has been concentrated on the identification of the *lower critical dimensionality*, d_ℓ, below which there exists no long range order. A simple heuristic argument, due to Imry and Ma,[41] gave $d_\ell=2$ for the Ising case (m=1) and $d_\ell=4$ for $m\geq 2$. The argument checks the stability of the ferromagnetic ground state against the formation of a domain, of linear size L, of opposite spins. Since the domain contains L^d spins, the bulk gain due to the random fields will be of order $(\Delta_h L^d)^{\frac{1}{2}}$. For m=1, the loss due to the new boundary is of order JL^{d-1}. Clearly, the domains are favored and long range order is destroyed for $d<2$. For $m\geq 2$, the domain wall energy is of order JL^{d-2} and one finds $d_\ell=4$.

For $m\geq 2$, the result $d_\ell=4$ has been supported by a spin-wave low-temperature renormalization group calculation in $d=4+\varepsilon$ dimensions.[42,39] It is also confirmed by exact results in the limit $m\to\infty$.[43,44] A simple way to derive this result, and to obtain information on the behavior for $d<d_\ell$, is based on low temperature scaling arguments.[45] For $d<d_\ell$, the spontaneous magnetization M is expected to be zero for all $\Delta_h>0$, while $M\neq 0$ at $\Delta_h=0$ (provided d is above the lower critical dimensionality of the "pure" system, $d_\ell{}^0$, and $T<T_c{}^0$). Therefore, M has a discontinuity as function of Δ_h when $\Delta_h\to 0$. This is similar to the temperature dependence of M in the one-dimensional Ising model, when M=0 for T>0 and M=1 at T=0. If we write $M \propto \Delta^{\beta_\Delta}$, then a discontinuity implies that $\beta_\Delta=0$.

Consider now the spin-spin correlation function $\langle\vec{S}_i\cdot\vec{S}_j\rangle$ (without subtracting $\langle\vec{S}_i\rangle\cdot\langle\vec{S}_j\rangle$). This function has an infinite range at $\Delta_h=0$, and we expect its range (related to the flopped domain sizes) to diverge as $\Delta_h\to 0$, $\xi\propto\Delta_h^{-\nu_\Delta}$. Writing the usual scaling

$$[\langle\vec{S}_i\cdot\vec{S}_j\rangle]_{av} = r_{ij}^{-(d-2+\eta_{ij})} f(r_{ij}/\xi) \quad , \tag{24}$$

the scaling relation $\beta_\Lambda= \nu_\Lambda(d-2+\eta_\Lambda)$ implies that $\eta_\Delta=2-d$. This immediately yields the scaled *structure factor*,

$$S(q,\Delta) = \xi^d \tilde{f}(q\xi) \quad . \tag{25}$$

This scaling form has yet to be checked experimentally.

The discontinuity in M at $\Delta=0$ implies that a uniform field h should scale, under a renormalization group length rescaling by b, via $(h/T)'=b^d(h/T)$.[46] Defining the Edwards-Anderson spin glass order parameter by $Q=[\langle\vec{S}_i\rangle^2]_{av}$, we also expect a discontinuity in Q as $\Delta_h\to0$. This implies that $(\Delta_h/T^2)'=b^d(\Delta_h/T^2)$. For the case $m\geq2$, spin wave arguments yield $T'=b^{2-d}\,T$. It is now easy to combine these and show that $\Delta_h'=b^{4-d}\Delta_h$, i.e. $d_\ell=4$ (Δ_h is irrelevant at low temperature and $d>4$).

In the Ising case, the low temperature spin wave argument is replaced by the capillary wave interface argument.[47] This yields $T'=b^{1-d}\,T$. Accepting the assumption that Q is discontinuous as $\Delta_h\to0$, this yields $\Delta_h'=b^{2-d}\Delta_h$, i.e. $d_\ell=2$,[45] in agreement with the Imry-Ma argument.

The capillary wave interface argument considers the energy of a (d-1)-dimensional interface between Ising up- and down-spins (without "bubbles" and "overhangs"). Denoting the height of this interface by $f(\vec{x})$, the exchange energy is written as[47]

$$\mathcal{H} = \beta J \int d^{d-1}x \sqrt{1+(\nabla f)^2} \quad . \tag{26}$$

Adding a random field, the interface energy must be modified by

$$\mathcal{H}_h = \beta \int d^{d-1}x \int_0^{f(\vec{x})} h(\vec{x},z)\,dz \quad . \tag{27}$$

If the variation of $h(\vec{x},z)$ with z is slow, one may replace the integral over z by $\tilde{h}(\vec{x})\,f(x)$, where $\tilde{h}(\vec{x})$ is some average of $h(\vec{x},z)$. This is linear in f, with a random coefficient. As such, it must shift the dimensionality in diagrammatic expansions by two. The usual $(1+\varepsilon)$ expansion[47] is thus mapped on an expansion near three dimensions, and one identifies $d_\ell=3$.[48,49]

On the other hand, a fast variation of the sign of $h(\vec{x},z)$ implies that the integral over z is of order $\tilde{h}_1(\vec{x})\,|f(\vec{x})|^{\frac{1}{2}}$. One can then count powers and obtain $d_\ell=2$.[50,51,52]

The value of d_ℓ for m=1 remains unclear at present, and more studies are necessary to identify it unambiguously. In any case it is clear that random field systems at d=2 and d=3 have very special new properties.

VI. RANDOM ANISOTROPIES

Consider now the Hamiltonian[53]

$$H = -J \sum_{<ij>} \vec{S}_i \cdot \vec{S}_j - D \sum (\hat{n}_i \cdot \vec{S}_i)^2 , \tag{28}$$

where $\hat{n}_i$ is a unit vector of random direction. Similar effects are expected whenever there arise random off-diagonal exchange interactions, e.g. of the form $J_{ij}^{xy} S_i^x S_j^y$.[54]

The relevance of D may again be checked by expansion or by replicas. One finds that $\phi_{D^2}=2\phi_a-2+\alpha$, where ϕ_a is the spin anisotropy crossover exponent.[33] For d=n=3, $\phi_{D^2}\simeq 0.37$ and one expects new behavior.

Assuming that there exists long range order in the x-direction, $\langle S_i^x \rangle = M$, the random anisotropy generates a random transverse field $h_i^y = DM\, n_i^x n_i^y$ (or $M\sum_j J_{ji}^{xy}$). This, in turn, destroys long range order for d<4.[55] The behavior of random anisotropy systems at d<4 remains unclear. To lowest order in D^2, the equation of state has been found to have the form[56]

$$h/M = t + M^2 + A\, D^2\, (h/M)^{(d-4)/2} , \tag{29}$$

implying that the zero field susceptibility $\chi=(M/h)_{h\to 0}$ remains infinite for all t<0. Higher order terms (yet to be calculated) may change this result and yield a finite susceptibility. If χ is finite at $D\neq 0$ then low temperature scaling arguments[45] imply that $\chi \propto D^{-4/(4-d)}$. Cubic anisotropy may turn the transition to first order.[54]

VII. COMPETING ANISOTROPIES

Consider now an alloy of two magnets, which tend to order along the z-axis or in the XY plane. Under the simplest conditions, this alloy will be described by the additional term[57]

$$H_A = -\sum_{<ij>} D_{ij} (2\, S_i^z S_j^z - S_i^x S_j^x - S_i^y S_j^y) , \tag{30}$$

where D_{ij} assumes a positive or negative value at random. As a function of the relative concentration (of $D_{ij}>0$ versus $D_{ij}<0$), one expects ordering of S^z or of S^x and S^y (see Fig. 4a). The transition into the former phase will be that of a random Ising model, and that into the latter will be the same as for the pure XY model (since $\alpha_{xy} < 0$).

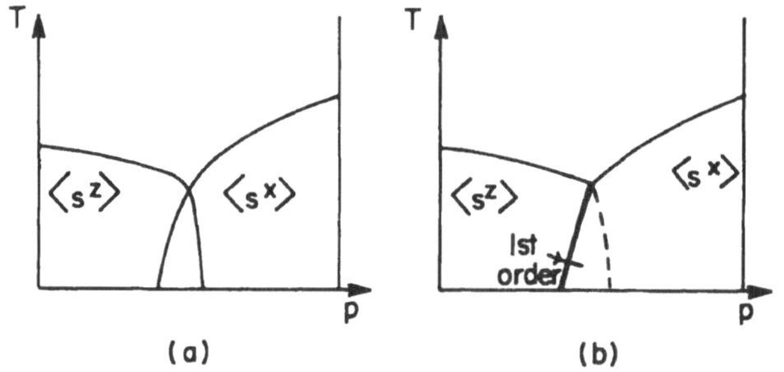

Fig. 4. Phase diagram of an alloy with competing anisotropies. (a) The simple case of Eq.(30). (b) With the additional interaction $S^zS^x[(S^x)^2 - 3(S^y)^2]$.

The stability of the *tetracritical point* (at which all four critical lines meet) can again be checked using the methods described above. In particular, the replica method yields a term $[D_{ij}^2]_{av}$ $(S_i^{z\alpha}S_j^{z\alpha})$ $(S_i^{x\beta}S_j^{x\beta})$. Since the two operators here scale as the energies of the Ising and XY models, we find that $[D_{ij}^2]_{av}$ scales as b^λ, with[3] $\lambda=d-(1-\alpha_1)/\nu_1-(1-\alpha_2)/\nu_2 = \frac{1}{2}(\alpha_1/\nu_1+\alpha_2/\nu_2)$. Since α_1 and α_2 are negative at the random fixed points, we find that the terms which couple the z component to the x and y components are irrelevant. One thus expects a *decoupled tetracritical point.*[57] As shown in Fig.4a, this implies that the four lines meet at the tetracritical point with finite angles between them. Since $|\alpha_1|$ and $|\alpha_2|$ are small, the crossover to this asymptotic decoupled behavior may be rather slow.

It has recently been noted,[59] that many of the mixed antiferromagnets in which such a phase diagram is expected have additional terms, e.g. $w\ S_i^z\ S_i^x\ [(S_i^x)^2 - 3(S_i^y)^2]$. Such terms are relevant near the (m=3) decoupled point: S_i^z scales as an Ising magnetization, with $b^{-\beta_1/\nu_1}$, while $[(S_i^x)^3 - 3\ S_i^x(S_i^y)^2]$ scales with $b^{\lambda_{Potts}}$, where λ_{Potts} reflects the relevance of the Potts symmetry breaking term.[60] Thus, w scales with $\lambda_w = \lambda_{Potts} - \beta_1/\nu_1 \simeq 0.33$. Since λ_w is not very large, the crossover to new behavior in the disordered phase is probably slow. However, the situation in the *ordered* phase is very different: ordering of S^x (or S^y) generates an effective uniform field on S^z, "smearing" the transition into the mixed phase (Fig.4b). Alternatively, ordering of S^z generates a Potts term in S^x and S^y, turning the transition into the mixed phase first order (Fig.4b).[60]

Many of these systems also have *random anisotropy* like terms, e.g. $A_{ij}\, S_i^z\, S_j^x$, with $[A_{ij}]_{av} = 0$.[61] Now, ordering of S^x will generate a *random field* on S^z, destroying (for $d \leq d_c$) the further transition into the phase in which S^z would order. Similarly, ordering of S^z will generate a Potts model with a random field, in which long range order is also destroyed at sufficiently low d.[62] The detailed behavior of the alloy at low temperatures is thus strongly tied together with the properties of systems with random fields.

It is worth emphasizing that an alloy with competing anisotropies allows one to study anisotropic antiferromagnets with various values of the anisotropy. Adding a uniform magnetic field one may thus explore the bicritical-tricritical properties mentioned above, in Sec.III.

VIII. CONCLUSION

The detailed critical properties of random systems involve various novel considerations: geometrical structures, exact expressions for crossover exponents, absence of long range order, etc. I hope that this review gave the flavour of some of the relevant considerations, and will stimulate further thought into the many open problems.

The work summarized here benefitted greatly from discussions and collaborations with S. Fishman, Y. Gefen, Y. Imry, S. Ma, E. Pytte and Y. Shapir. It was supported in part by grants from the U.S.-Israel Binational Science Foundation and from the Israel Academy of Sciences and Humanities.

REFERENCES

1. A. B. Harris, J. Phys. C7, 1671 (1974).
2. Y. Shapir and A. Aharony, J. Phys. C14, L905 (1981).
3. See e.g. A. Aharony, in "Phase Transitions and Critical Phenomena," ed. C. Domb and M. S. Green, Academic, New York (1976), Vol.6, p.357.
4. D. E. Khmelnitzkii, Sov. Phys. JETP 41, 981 (1976).

5. C. Jayaprakash and H. J. Katz, Phys. Rev. B16, 3987 (1977).
6. A. Aharony, Phys. Rev. B13, 2092 (1976).
7. G. Grinstein, S. Ma and G. F. Mazenko, Phys. Rev. B15, 258 (1977).
8. A. Aharony, to be published.
9. P. W. Kasteleyn and C. M. Fortuin, J. Phys. Soc. Japan Suppl. 26, 11 (1969).
10. M. J. Stephen and G. S. Grest, Phys. Rev. Lett. 38, 567 (1977).
11. D. J. Wallace and A. P. Young, Phys. Rev. B17, 2384 (1978).
12. A. Skal and B. I. Shklovskii, Sov. Phys. Semicond. 8, 1029 (1975).
13. P. G. de Gennes, J. Phys. Lett. (Paris) 37, L1 (1976).
14. T. C. Lubensky, Phys. Rev. B15, 311 (1977).
15. e.g. D. Stauffer, Phys. Reports 51, 1 (1979).
16. H. E. Stanley, J. Phys. A10, L211 (1977).
17. A. Coniglio, Phys. Rev. Lett. 46, 250 (1981).
18. A. Coniglio, J. Phys. A15, 3829 (1982).
19. S. Kirkpatrick, in "Les Houches Summer School on Ill Condensed Matter," ed. R. Balian, R. Maynard and G. Toulouse, North Holland, Amsterdam (1979).
20. Y. Gefen, A. Aharony, B. B. Mandelbrot and S. Kirkpatrick, Phys. Rev. Lett. 47, 1771 (1981).
21. Y. Gefen, Y. Kantor and A. Aharony, to be published.
22. e.g. H. E. Stanley, Phys. Rev. 179, 570 (1969).
23. R. B. Stinchcombe, J. Phys. C12, 2625 (1979).
24. Y. Gefen, B. B. Mandelbrot and A. Aharony, Phys. Rev. Lett. 45, 855 (1980).
25. D. Stauffer and A. Coniglio, Z. Phys. 38, 267 (1980).
26. M. J. Stephen and A. Aharony, J. Phys. C14, 1665 (1981).
27. A. Aharony and D. Stauffer, Z. Phys. 47, 175 (1982).
28. M. Suzuki, Phys. Lett. 34A, 94 (1974).
29. S. N. Dorogovtsev, Phys. Lett. 76A, 169 (1980); D. Boyanovsky and J. L. Cardy, Phys. Rev. B26, 154 (1982).
30. M. Blume, V. J. Emery and R. J. Griffiths, Phys. Rev. A4, 1071 (1971).
31. U. Stein and A. Aharony, to be published.
32. S. Fishman and A. Aharony, J. Phys. C12, L729 (1976).
33. e.g. M. E. Fisher, A.I.P. Conf. Proc. 24, 273 (1974).
34. A. Aharony, Phys. Rev. B18, 3318 (1978).
35. A. Aharony, Phys. Rev. B18, 3328 (1978).
36. S. Galam and A. Aharony, J. Phys. C13, 1065 (1980); 14, 3603 (1981).
37. R. J. Birgeneau and A. N. Berker, Phys. Rev. B26, 3751 (1982).
38. A. Aharony, Y. Imry and S. Ma, Phys. Rev. Lett. 37, 1364 (1976).
39. A. P. Young, J. Phys. C10, L257 (1977).
40. G. Parisi and N. Sourlas, Phys. Rev. Lett. 43, 744 (1979).
41. Y. Imry and S. Ma, Phys. Rev. Lett. 35, 1399 (1975).
42. R. A. Pelcovits, Phys. Rev. B19, 465 (1979).

43. P. Lacour-Gayet and G. Toulouse, J. Phys. (Paris) 35, 425 (1974).
44. H. M. Hornreich and H. G. Schuster, Phys. Rev. B26, 3929 (1982).
45. A. Aharony and E. Pytte, Phys. Rev. B (in press).
46. B. Nienhuis and M. Nauenberg, Phys. Rev. Lett. 35, 477 (1975).
47. D. J. Wallace and R. K. P. Zia, Phys. Rev. Lett. 43, 803 (1979).
48. E. Pytte, Y. Imry and D. Mukamel, Phys. Rev. Lett. 46, 1173 (1981).
49. H. Kogon and D. J. Wallace, J. Phys. A14, L527 (1981).
50. I first learnt this from discussions with Y. Shapir in 1979.
51. G. Grinstein and S. Ma, Phys. Rev. Lett. 49, 685 (1982).
52. J. Villain, J. Phys. Lett. (Paris) 43, L551 (1982).
53. R. Harris, M. Plischke and M. J. Zuckerman, Phys. Rev. Lett. 31, 160 (1973).
54. A. Aharony, Solid State Commun. 28, 667 (1978).
55. R. A. Pelcovits, E. Pytte and J. Rudnick, Phys. Rev. Lett. 40, 476 (1978).
56. A. Aharony and E. Pytte, Phys. Rev. Lett. 45, 1583 (1980).
57. A. Aharony and S. Fishman, Phys. Rev. Lett. 37, 1587 (1976).
58. S. Fishman and A. Aharony, Phys. Rev. B18, 3507 (1978).
59. D. Mukamel, Phys. Rev. Lett. 46, 845 (1981).
60. A. Aharony, K. A. Müller and W. Berlinger, Phys. Rev. Lett. 38, 33 (1977).
61. P. Wong, P. M. Horn, R. J. Birgeneau, C. R. Safinya and G. Shirane, Phys. Rev. Lett. 45, 974 (1980).
62. D. Blankschtein, Y. Shapir and A. Aharony, to be published.

CRITICAL BEHAVIOUR OF DISORDERED ISING SYSTEMS:

SUPPRESSION OF THE SPECIFIC HEAT DIVERGENCE AT T_c

Giancarlo Jug

Department of Theoretical Physics

1 Keble Road, Oxford OX1 3NP, England

Recently, it has become possible to use field theoretic renormalization group (RG) techniques for studying critical phenomena directly in two and three dimensions (2, 3D). High-order perturbative calculations for the 3D M-vector model have yielded very accurate predictions for the critical exponents.[1]

Random-exchange magnetic models lend themselves to this kind of RG studies since no reliable evaluation of the critical exponents is available from conventional (high-T series, MonteCarlo,...) and ϵ-expansion methods.

I have applied[2,3] fixed-dimension RG techniques to study the critical properties of the spin system with Hamiltonian $H=-\Sigma_{ij}J_{ij}p_ip_j\underline{S}_i\underline{S}_j$ ($\underline{S}_i$: M-component classical spin variable; J_{ij}: short-ranged ferromagnetic interaction; p_i: uncorrelated random site variables with distribution $P(p_i)=p\delta(p_i-1)+(1-p)\delta(p_i)$; $c=1-p$: concentration of non-magnetic impurities). The model's Landau-Ginzburg-Wilson Hamiltonian has the form:

$$H[\underline{\phi},\sigma]=\int dx\left[\frac{1}{2}(m_0{}^2-\sigma)\underline{\phi}^2+\frac{1}{2}(\nabla\underline{\phi})^2+\frac{\lambda}{4!}(\underline{\phi}^2)^2\right], \qquad (1)$$

where $\underline{\phi}(x)$ is the magnetization density field, $m_0{}^2\propto T-T_c{}^0(p)$ and $\sigma(x)$ is a quenched random field which simulates the effect of disorder and for which $[\sigma(x)\sigma(x')]_{av}=\lambda_2\delta(x-x')/3$. $T_c{}^0(p)=pT_c{}^0(p=1)$ is the mean field critical temperature, $p\lambda_1$ is the pure ϕ^4 exchange coupling, whereas $\lambda_2=(1-p)\lambda_1$ is the impurity coupling. Near T_c, the model's free energy is:

$$F=\int D\sigma \exp[-\tfrac{3}{2}\lambda_2^{-1}\int dx\ \sigma^2(x)]\ \ln\int D\underline{\phi}\ \exp-H[\underline{\phi},\sigma].$$

Evaluating the quenched average by means of the n=0 replica trick, an effective Hamiltonian is introduced, with density:[4]

$$L[\underline{\phi}^{\alpha}]=\Sigma_{\alpha=1}^{n}[\tfrac{1}{2}m_0{}^2\underline{\phi}^{\alpha 2}+\tfrac{1}{2}(\nabla\underline{\phi}^{\alpha})^2+\tfrac{\lambda}{4!}{}_1(\underline{\phi}^{\alpha 2})^2]-\tfrac{\lambda}{4!}{}_2[\Sigma_{\alpha}\underline{\phi}^{\alpha 2}]^2.$$

The above expression is the Lagrangian of a field theory, the asymptotic behaviour of which gives, for n=0, the critical properties of the original spin model. This behaviour is extracted from the coefficients $\beta_1(u_1,u_2)$, $\beta_2(u_1,u_2)$, $\eta(u_1,u_2)$ and $\theta(u_1,u_2)$ of the Callan-Symanzik equations appropriate for the above Lagrangian. I have calculated these functions directly in 2 and 3D to second order in renormalised perturbation theory.[2] Due to the asymptotic nature of these series expansions, a simple two-variable resummation technique has had to be used to make the calculation meaningful. The zeros (u_1^*,u_2^*) of the resummed β functions correspond to the RG fixed points and the eigenvalues b_1, b_2 of the matrix $\partial\beta_i(u_1^*,u_2^*)/\partial u_j$ account for their stability. Random-exchange critical behaviour corresponds to a fixed point with $u_1^*\geq 0$, $u_2^*\geq 0$ and $b_1\geq 0$, $b_2\geq 0$.
I have observed random critical behaviour only for the random Ising model (M=1). This is in agreement with the (generalised[5]) Harris criterion:[6] random (as opposite to pure) critical behaviour characterises a disordered spin system only if the pure system's specific heat $C(T,p=1)$ diverges as $T\to T_C$.

Accordingly, the random 3D Ising model presents new critical exponents. The exponents η and ν are obtained from the resummed η and θ functions through $\eta=\eta(u_1^*,u_2^*)$, $\nu^{-1}=2+\theta(u_1^*,u_2^*)$, the other exponents following from the ordinary scaling laws. My estimate[2] for the various critical exponents and for the correction to scaling exponent Δ_1 is reported in Table 1, first row. In the second row of this table, an independent RG evaluation of the same exponents by Newman and Riedel[7] is also reported. The two sets of results compare well, except for the magnetic exponent η.[8] In the third row, very recent experimental results[9] for the random critical exponents are reported. Neutron scattering and linear birefringence measurements have been performed on the randomly diluted 3D Ising antiferromagnet $Fe_{0.5}Zn_{0.5}F_2$ in zero field and for $10^{-3}<\tau<10^{-1}$ $(\tau=|T-T_C|/T_C)$. Although these measurements do not show the expected pure to random crossover behaviour and do not probe the random critical region ($\tau<2\times10^{-3}$ for p=0.5) deeply enough,

TABLE 1: Critical Exponents of the Random 3D Ising Model

	η	ν	α	γ	Δ_1
1)	0.031	0.68	-0.04	1.34	0.31
2)	0.015	0.70	-0.09	1.39	0.29
3)	-	0.73±3	-0.09±3	1.44±6	0.50
4)	0.031±4	0.6300±15	0.1100±45	1.241±2	0.498±20
5)	0.031±4	0.708±31	-0.124±10	1.394±59	0.559±45

1)Present Author[2]; 2)Newman & Riedel[7]; 3)Experiment[9]; 4)Pure Model[1]; 5)Annealed Model[1+10]

the semi-quantitative agreement with the theoretical predictions represents a success for the RG theory of critical phenomena.
In particular, both theory and experiment indicate clearly that the main effect of random exchange is the suppression of the pure system's specific heat divergence at T_c. This can be understood in terms of the effective random field $\sigma(x)$ in Eq.(1). This field couples to the spin energy density $\phi^2(x)$; hence, one expects $\sigma(x)$ to destroy the fluctuations of $\phi^2(x)$ at criticality and to give rise to a suppression of the divergence of $C(T,p)$ at T_c.
In the last two rows of Table 1 I report, for comparison, the pure Ising exponents[1] and the annealed random Ising exponents as deduced from the pure ones using Fisher's renormalization.[10] Qualitatively, annealing and quenching change the pure exponents in the same way, though in the former case the effect on the specific heat is more dramatic.

In the case of the random 2D Ising model disorder is a marginal perturbation, as follows from the Harris criterion[6] or from the equivalence[11] $\phi=\alpha_P$ between the pure to random crossover exponent ϕ and the pure system's specific heat exponent α_P (=0 for this model). Thus, the disordered model has the critical exponents of the pure one and impurity-induced logarithmic corrections appear. Since only the specific heat exponent is mean field in value, $C(T,p)$ is the one thermodynamic function to develop a logarithmic correction. By solving exactly the corresponding RG equation, I have found[3] a new specific heat singularity of the form:

$$C(T,p) \sim |\ln|T-T_c(p)||^{1-\mu} \qquad \text{if } \mu\neq 1,$$
$$C(T,p) \sim \ln|\ln|T-T_c(p)|| \qquad \text{if } \mu=1. \qquad (2)$$

μ is a new critical exponent, rather difficult to calculate perturbatively expanding from d=4.

However, μ is defined all along the disorder marginality line $\alpha_P(M,d)=0$, where Eq.(2) holds good, and in $d=4-\epsilon$ dimensions I have found:

$$\mu(4-\epsilon)=2-\epsilon+O(\epsilon^2)\gtrsim 1.$$

Also, using resummed perturbation theory (badly convergent in 2D) to second order, I have found: $\mu(3D)\simeq 1.2$ and $\mu(2D)\sim O(1)$. Clearly, non-perturbative techniques should be used to calculate $\mu(2D)$ and verify the expected result: $\mu(2D)\gtrsim 1$. This would agree with the observed behaviour for $d\geq 3$ and with the generally expected cusp-like behaviour of the specific heat at T_C resulting from the introduction of exchange randomness into the system. Such behaviour is also suggested by the known[10] specific heat $|\ln|T-T_C(p)||^{-1}$ singularity of the analogous annealed random 2D Ising model.

References

1. G.A.Baker, B.G.Nickel, M.I.Meiron, Phys.Rev. B17, 1365 (1978); J.C.LeGuillou, J.Zinn-Justin, ibid.21, 3976 (1980)
2. G.Jug, Phys.Rev. B27, 609 (1983) (Rap.Comm.)
3. G.Jug, Phys.Rev. B27, April 1 (1983) (Rap.Comm.)
4. G.Grinstein, A.Luther, Phys.Rev. B13, 1329 (1976)
5. A.Aharony, Phys.Rev. B13, 2092 (1976)
6. A.B.Harris, J.Phys. C7, 1671 (1974)
7. K.E.Newman, E.K.Riedel, Phys.Rev. B25, 264 (1982)
8. However, observe that Riedel's RG technique is unreliable for the evaluation of η
9. R.J.Birgeneau, R.A.Cowley, G.Shirane, H.Yoshizawa, D.P.Belanger, A.R.King, V.Jaccarino, BNL preprint 32489 (1983) (unpublished)
10. M.E.Fisher, Phys.Rev. 176, 257 (1968)
11. A.Aharony, in "Phase Transitions and Critical Phenomena", edited by C.Domb and M.S.Green (Academic, New York 1976), Vol.6. See also G.Jug, C.E.I.Carneiro, Oxford preprint TP64/82 (unpublished)

DISORDER, RANDOM FIELDS AND COMPETING INTERACTIONS IN ANTIFERROMAGNETS

R.A. Cowley

Department of Physics
University of Edinburgh
Edinburgh, EH9 3JZ
Scotland

R.J. Birgeneau

Department of Physics
Massachusetts Institute of Technology
Cambridge
Mass. 02129, USA

G. Shirane and H. Yoshizawa

Brookhaven National Laboratory
Upton
New York 11973, USA

ABSTRACT

Neutron scattering studies of disordered antiferromagnets have proved to be a very profitable way of studying random systems. Several recent examples are selected and include a detailed study of the phase transition of a $d = 3$ Ising system showing a well defined transition with properties different from these of a pure $d = 3$ Ising system. Much of the article is then concerned with the effect of a random field on the ordering and phase transitions. It is shown that random fields do have a large effect on the critical properties and in practice destroy the long range order in the good Ising systems with $d = 2$ and $d = 3$, although not in a nearly Heisenberg-like system. These results are compared with current theories and the discrepancies discussed. Finally measurements on a system with competing interactions are discussed and shown to be strongly influenced by the random fields produced in that system.

INTRODUCTION

The fluorides and chlorides of the 3d transition metals are ideal materials on which to study the properties of disordered systems. Transition metal ions such as Mn^{++}, Co^{++}, Fe^{++} and Mg^{++} are chemically very similar but magnetically very different. Consequently it is possible to grow single crystals in which the magnetically very different ions are distributed at random over the transition metal sites, so that the systems are chemically quite uniform but magnetically very disordered. The magnetic properties can then be studied in detail by various experimental techniques, but particularly detailed information is provided by neutron scattering techniques. These results often provide very stringent tests of the theories of disordered systems because the magnetic interactions in these systems are well known, and often of short range, and so are very similar to the theoretical models for which calculations are most often performed. Secondly by choosing different magnetic ions and different crystal structures a variety of different models can be studied — Ising or Heisenberg-like magnetic interactions — magnetic systems in which the interactions are largely confined to, $d = 1$, lines of ions, or $d = 2$, planes of ions, or throughout the whole crystal when the dimensionality $d = 3$.

At an earlier Geilo school in 1979, we[1] reviewed some of the measurements which had been made on these systems to study the properties of the excitations and phase transitions of disordered systems. In this article, we shall not repeat that review but concentrate on the progress which has been made since 1979. All of the work reviewed earlier was on antiferromagnetic materials containing either two different magnetic ions, or a magnetic ion and a non-magnetic ion. There is then no doubt about the nature of the long range order, if any, because there is no competition between the different interactions. Since 1979 a large part of the work has been the extension of these studies into systems, where there are competing interactions and the nature of the ground state is by no means so clear or even understood. A large part of this article describes the experiments which have been performed on these more complicated systems. Since the nature of the ground state in these systems is still not yet understood, there has been comparatively little progress made on the study of the excitations; consequently we concentrate on the properties of the phase diagram and the nature of the phase transitions.

Before describing the work performed on systems with competing interactions, we describe one experiment on the phase transition in a relatively simple disordered antiferromagnet. The reason for describing this is that it sheds light on one of the questions raised in our earlier review — namely whether the phase transition in a disordered system is sharp or rounded. This experiment is described in the next section.

The remainder of this article is then a discussion of the results found so far for systems with competing interactions. The third section describes the effect of applying a random field to the systems so that there is a competition between the random field and the antiferromagnetic exchange interactions. Because the strength of the random field can be altered continuously, this problem is in many ways the most amenable of the systems with competing interactions for detailed study and has attracted the most experimental effort recently. Another type of system with competing interactions are those with random anisotropies so that the anisotropy favours one spin direction on some of the ions and a different direction on the other ions. These systems are described in the fourth section. Finally we should mention measurements made on systems with mixtures of ferromagnetic and antiferromagnetic exchange constants which, in principle at least, should give insight into the behaviour of the metallic spin glasses. One system which has been studied in considerable detail is that of $Eu_xSr_{1-x}S$, and the results for this system will be described by Coles in his lectures. Another very interesting system is $Rb_2Mn_xCr_{1-x}C\ell_4$ where the Mn ions are coupled antiferromagnetically and the Cr ions ferromagnetically, and the magnetic ions lie in sheets. Although there is evidence for a $d = 2$ spin glass phase in this system, detailed studies have not yet been performed and so we do not discuss it further here.

Most of the measurements which will be described, have been performed with neutron scattering techniques. Except when explicitly stated these were performed at the High Flux Beam Reactor at the Brookhaven National Laboratory. They were mostly performed with a two-axis spectrometer and an incident neutron energy of 14 meV and with pyrolytic graphite used as both monochromator and as a filter to suppress the higher order contaminent neutrons in the incident beam. The critical scattering observed close to the phase transition was analysed by convoluting the appropriate functional form with the experimental resolution and fitting the parameters of that assumed form to give a good fit to the experimental results.

PHASE TRANSITIONS WITHOUT COMPETING INTERACTIONS

The nature of the phase transitions in disordered systems with exchange constants of the same sign has been the subject of a considerable amount of work. Harris[2] showed that if the specific heat exponent, α, of the pure system was negative then the critical phenomena would be very similar (have the same universal behaviour) as that of the corresponding pure systems. In contrast if α was positive the disorder would have a larger effect which might lead either to a first order transition or to a continuous transition but with different exponents from those of the analogous pure systems. These results have also been obtained using renormalisation group theory[3] and explicit predictions made for some of the exponents[4] which differ from those of the pure system.

Experimental work on these systems has not until recently been able to confirm these results. Although these predictions agreed with the measurements[5] on the mixed d = 2 Ising system $Rb_2Ni_{0.5}Mn_{0.5}F_4$ for which $\alpha = 0$, measurements on three dimensional systems[6] gave a smeared transition which made the determination of the exponents very difficult if not illegitimate. It was impossible to ascertain whether the smearing was due to macroscopic concentration fluctuations or would arise in a homogeneous random crystal.

Recently measurements[7] of the critical behaviour have been performed on a sample of $Fe_{0.5}Zn_{0.5}F_2$. FeF_2 is an antiferromagnet in which the S = 2 magnetic ions interact with Heisenberg-like exchange interactions, but the proximity of nearby crystal field levels causes a large single ion anisotropy. This anisotropy produces a large gap (6.5 meV) in the spin wave spectrum[8] so that the critical behaviour of FeF_2 should be that of the d = 3 Ising model and is found to be that experimentally[9].

The sample of $Fe_{0.5}Zn_{0.5}F_2$ was found to be of excellent quality with a mosaic spread of less than 0.02 degrees and a spread in T_N of less than 2×10^{-3} T_N. These properties enabled a very detailed study of the critical properties of the disordered d = 3 Ising model to be made. This is a particularly interesting system because α is positive for the pure d = 3 Ising model and so the critical properties of the disordered crystal are expected to be different from those of pure FeF_2.

The critical scattering was measured using neutron scattering techniques and was analysed assuming that the scattered intensity was given by

$$I(\vec{Q}) = \frac{AT}{\kappa^2 + q^2} + C\delta(\vec{q}) \qquad (1)$$

where $\vec{Q}$ is the wavevector transfer and $\vec{q}$ is the difference between $\vec{Q}$ and the nearest reciprocal lattice vector as measured in reciprocal lattice units. The parameters A and κ were fitted to the observed critical scattering and the results for κ, and for κ^2/A are shown in fig. 1.

The results for the inverse correlation length and amplitude were fitted to power law forms;

$$\kappa(\tau) = \kappa_0^+ \tau^{\nu_+} \qquad T > T_N$$

$$\kappa(\tau) = \kappa_0^- \tau^{\nu_-} \qquad T < T_N$$

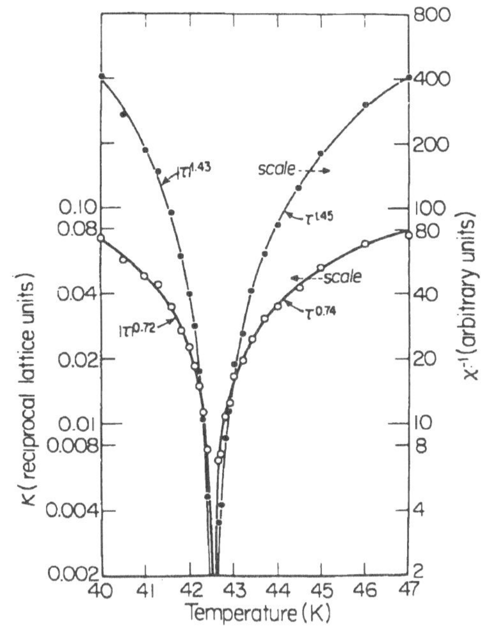

Fig. 1. Inverse correlation length κ and inverse staggered susceptibility κ^2/A for $Fe_{0.5}Zn_{0.5}F_2$ above and below $T_N = 42.50$ K. The solid lines are fits to single power laws and $\tau = (T - T_N)/T_N$.

with $\tau = |T - T_N|/T_N$, and it was found that within error $\nu_+ = \nu_-$. In table 1, we collect the best estimates of the exponents for the correlation length, ν, the susceptibility, γ, from these experiments and of the specific heat, α, from measurements of the birefringence. Also collected in table 1 are the approximate amplitude ratios; we note that the exponents for the disordered and pure systems differ only slightly while the amplitude ratios are very significantly different. Table 1 also shows current theoretical estimates of these quantities.

This experiment very strongly suggests that the theory of the phase transitions in disordered systems is correct in predicting that even when the exponent α is negative for the pure system, the

Table 1. Exponents and Amplitude Ratios for Pure and Random d = 3 Ising Models

	Pure Ising		Random Ising	
	Theory[10]	FeF_2 [9]	Theory[11]	$Fe_{0.5}Zn_{0.5}F_2$ [7]
Susceptibility				
γ	1.24	1.38±0.08	1.39	1.44±0.06
Amplitude Ratio	5.1	6.1 ±1.0	1.7	2.2 ±0.1
Correlation Length				
ν	0.63	0.67±0.04	0.70	0.73±0.03
Amplitude Ratio	0.53	0.43±0.07	0.83	0.73±0.02
Specific Heat				
α	0.11	0.11±0.005	-0.09	-0.09±0.03
Amplitude Ratio	0.51	0.54±0.02	-0.5	1.6 ±0.3

transition is sharp but that the critical properties are altered. The smearing observed in earlier experiments[6] is most probably due to macroscopic fluctuations in the concentrations and hence to a varying T_N throughout the crystal.

ORDERING AND RANDOM FIELDS

Theoretical Introduction

The effect of random fields on the ordering of magnetic systems was first discussed by Imry and Ma[12]. They considered the Hamiltonian for a system of n-component spins as

$$H = - J \sum_{ij} \vec{S}_i . \vec{S}_j - g\mu_B \sum_i \vec{H}_i . \vec{S}_i \qquad (2)$$

where the sum over i and j is restricted to nearest neighbours and the fields $\vec{H}_i$ are randomly directed. If it is to be energetically favourable for the system with long range order to break up into randomly oriented domains, the gain in the random field energy must outweigh the energy needed to create the domain walls. In a system of dimensionality d, the gain of random field energy in creating a domain of size L is proportional to the net magnetic field $L^{d/2}$.

In a system for which $n > 1$, the energy in creating a smooth domain wall is proportional to L^{d-2} and consequently the gain in magnetic field energy destroys the long range order for $d < d_c = 4$. In Ising systems for which $n = 1$, the domain wall energy is larger and proportional to L^{d-1}, and so Imry and Ma predicted that $d_c = 2$ for Ising systems.

Since this pioneering work of Imry and Ma there have been many theoretical treatments of this problem. Briefly renormalisation group theory and diagrammatic expansions[13] can be performed about the upper critical dimensionality, $d = 6$. A particularly appealing approach was that of Parisi and Sourlas[14] who showed by using supersymmetry that the universal properties of a system in a random field were the same as those of the corresponding pure system but in two less dimensions. This result is believed to be an exact result of the symmetry of the problem for $d > 4$, but its status for $d < 4$ is uncertain. This dimensionality shift is consistent with the arguments of Imry and Ma for the lower critical dimension of systems with more than one ($n > 1$) spin component, but is inconsistent with their predictions for $n = 1$ systems because the lower critical dimension of Ising systems in the absence of a random field is one and the dimensionality shift would then predict that in a random field, $d_c = 3$.

Since this work there have been several attempts to calculate d_c and to understand the origin of this discrepancy between the results of supersymmetry and simple domain arguments. Briefly it[15] was suggested that the calculation of the domain wall energy by Imry and Ma was incorrect because it assumed there were smooth domain walls, whereas if the domain walls were rough they might better be able to take advantage of the random field energy. Calculations[15] of this effect of domain wall roughening suggested that the effect could increase d_c from 2 to 3. This approach has, however, been criticised because it assumed that the domain wall energy was an analytic function of its position and could be expanded in a Taylor series. Arguments[16] based on a different model of the domain wall energy gave a singular dependence of the energy on the location of the domain wall and again suggested $d_c = 2$. Consequently, at present the lower critical dimension of the Ising model in a random field is an unsolved problem.

The dimensionality shift from supersymmetry also gives[15] a direct prediction for the form of the scattered intensity. The correlation function of a d dimensional Ising model is given for $T > T_c$ by;

$$\langle S_i S_j \rangle \sim R^{(1-d)/2} \exp(-\kappa R),$$

where R is the distance between the spins S_i and S_j. In the presence of a random field the dimensionality shift immediately predicts that

$$<S_i \, S_j> \sim R^{(3-d)/2} \exp(-\kappa R),$$

when this form is Fourier transformed to give the scattering cross-section we obtain

$$I(\vec{Q})/|f(\vec{Q})|^2 = \frac{A}{\kappa^2 + q^2} + \frac{B}{(\kappa^2 + q^2)^2}, \qquad (3)$$

where $f(\vec{Q})$ is the magnetic form factor of the ions, A and B are constants and κ is related to the size of the domains. Based on the known behaviour of κ for $d \leqslant d_c$ in systems without a random field it is then predicted[15] that

$$\kappa \sim <H_i^2>^x$$

where $x = 1/(d_c - d)$ for $d < d_c$

$$\kappa \sim \exp(-a/<H_i^2>) \quad \text{for } d = d_c, \qquad (4)$$

where a is a constant. If for small κ and $T < T_c$, the intensity (3) is to go over smoothly into the Bragg peak which arises from true long range order, then

$$B = <M>^2 \, \kappa^{4-d}, \qquad (5)$$

where M is the magnetisation of the system, while the constant A must tend to zero as $\kappa \to 0$.

Equation (3) together with equation (5) are consistent with the scaling relations obtained by Aharony and Pytte[17] who also suggest that at the T_c of the system in the absence of the random field, then

$$\kappa \sim <H_i^2>^{\nu/\gamma},$$

and

$$I(\vec{Q}) \sim <H_i^2>^{-1}, \quad \text{when } \vec{q} = 0. \qquad (6)$$

For several years these results[12] did not attract the attention of experimentalists because of the difficulty of producing controlled random fields. It was then pointed out by Fishman and Aharony[18] that the application of a uniform field to a dilute antiferromagnet produces a random staggered field, which is then a random field conjugate to the antiferromagnetic order parameter. Fishman and Aharony initially considered an antiferromagnet with only bond disorder, whereas the experimentally realisable systems have site disorder. In the case of site disorder there are two contributions[19] to the random field. The first arises from the randomness in the magnetic moments from one cell to the next. In a site diluted antiferromagnet only one sub-lattice may be occupied in one cell whereas in another cell the other sub-lattice may be occupied, leading to a random field proportional to the applied field. In addition to the randomness in the moments the field will also produce a nearly uniform magnetisation. Due to the randomness in the exchange interactions this magnetisation will produce a random fluctuation in the local staggered mangetic field as originally described by Fishman and Aharony. This contribution is proportional to the magnetisation and hence to the product of the applied uniform field and the uniform susceptibility. Since the susceptibility is temperature dependent this can lead to a temperature dependence in the connection between the strength of the uniform field and the strength of the random field.

The above description has recently been criticised by Cardy[20] who argues that the strength of the random field is always proportional to the magnetisation. We believe his arguments to be in error, but further work is needed to establish the exact connection between the random field model and the realisation by a site disordered antiferromagnet in a uniform field. For example, the theory assumes that the random fields are uncorrelated, whereas there are certainly short range correlations between the random fields produced in the site diluted systems although the correlation range is much less than $1/\kappa$ in most of the experiments.

Finally it is unfortunate that the Fishman and Aharony suggestion for a realisation of random fields is applicable only to Ising systems. The application of a magnetic field to an ordered antiferromagnet with a continuous symmetry immediately causes a domain realignment in favour of the domains in which the spin directions are perpendicular to the applied field (the spin-flop phase)[21]. In most systems, there are then no components of the random fields parallel to the spin directions and the behaviour is very similar to that of a spin-flop phase in a pure antiferromagnet. In systems of sufficiently low symmetry there may be random interactions of the form $S_i^{\parallel} S_j^{\perp}$; such systems may allow study of the XY model in a random field.

Neutron Scattering Results

We shall describe the results which have been obtained by applying a magnetic field to four different antiferromagnetic systems. The $Rb_2Co_xMg_{1-x}F_4$ system is an example of a two-dimensional antiferromagnet with essentially Ising interactions[22]. The Co^{2+} ions have a Kramer's doublet as their ground state and the interactions between the effective spins $S = \frac{1}{2}$ are anisotropic due to the anisotropic effects of the crystal field on the Co ions. The ions are arranged on the sites of a two dimensional square lattice and one plane of magnetic ions is sufficiently well separated from the neighbouring planes that the inter-plane magnetic interactions are negligible compared with the intra-plane interactions.

The other materials are all of the rutile structure in which the magnetic ions are arranged on a three dimensional lattice of the body centred tetragonal type. We have studied the crystals $Co_xZn_{1-x}F_2$, $Fe_xZn_{1-x}F_2$ and $Mn_xZn_{1-x}F_2$. All of these materials are antiferromagnets with predominantly exchange interactions between only the nearest neighbour antiferromagnetic ions, but the interactions between the different magnetic ions are quite different. In CoF_2 the interactions are anisotropic because of the crystal field effects discussed above and result in the material having a large spin wave gap and exhibiting Ising critical behaviour[23,6]. The interactions in FeF_2 were discussed in the preceding section, and this again is a good example of an Ising system[8] but the anisotropy results from single ion terms in the Hamiltonian instead of anisotropic exchange. Finally in MnF_2 the exchange interactions are predominantly of Heisenberg character, but the weaker magnetic dipolar forces in the tetragonal rutile structure give rise to a small anisotropy[24]. Consequently the MnF_2 system does belong to the Ising universality class like FeF_2 and CoF_2, but the anisotropy is very much smaller in the MnF_2 system.

A preliminary account[25] of the measurements on the Co salts and on the Mn salt[26] have already been published and full papers on the $Rb_2Co_xMg_{1-x}F_4$ system[27] and $Co_xZn_{1-x}F_2$ system[28] have been submitted for publication. Consequently we shall here review only the essential features of these measurements.

One of the difficulties in working with Ising systems at low temperatures is the problem of knowing whether the system is in thermodynamic equilibrium. In the experiments on the Co and Fe salts, it was found that the results at low temperatures were dependent upon the way in which the state was prepared. For example, if the sample was cooled to low temperatures and the field then applied, the scattering was different from that obtained when the sample was cooled from above T_N in a constant field. This showed that the spins were indeed frozen at low temperatures and could not then respond to changes in the field. Reproducible results were obtained if the field was always changed at temperatures above T_N and the sample then cooled while keeping the field constant. Most of the results were therefore taken using this procedure. Although there is no guarantee that this does result in the achievement of thermodynamic equilibrium, it is the accepted way of studying spin glasses and has the merit of being a well defined procedure. We shall return to the problem of the freezing again later.

When the Co and Fe systems are cooled in the presence of a magnetic field, the scattering in the neighbourhood of the magnetic Bragg reflection is dependent upon the field. In particular it becomes progressively broader with increasing field, as shown in figs. 2 and 3. This increase in the width shows directly that the long range magnetic order is being destroyed by the application of a magnetic field. The nuclear Bragg peaks were unchanged so that the effect is not due to a change in the crystal structure. These results show directly that a random field does appear to destroy the long range order in both d = 2 and d = 3 Ising models and so the lower critical dimension d_c is apparently larger than 3.

The temperature dependence of the scattering is illustrated in fig. 4, where it is seen that the width of the scattering increases steadily with increasing temperature and that the peak intensity steadily decreases.

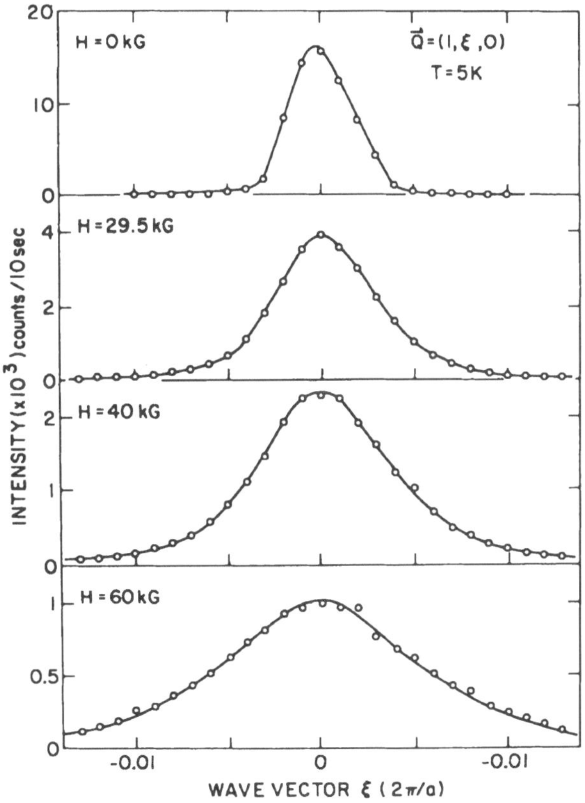

Fig. 2. The neutron scattering observed in $Rb_2Co_{0.7}Mg_{0.3}F_4$, a d = 2 Ising system, as a function of applied field.

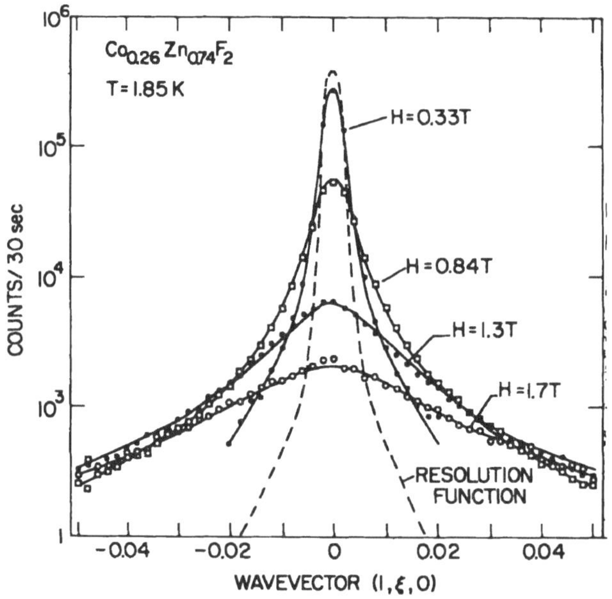

Fig. 3. The neutron scattering observed in $Co_{0.26}Zn_{0.74}F_2$, a d = 3 Ising system, as a function of applied field.

Initially the experimental results were fitted to a variety of different functional forms. The scattering profiles shown in figs. 2-4 are unusual in that the wings are more intense than would be expected from a Gaussian spread of domains, but can also not be described by the usual Lorentzian form which has proved so successful in describing critical phenomena, As shown by the agreement between the solid lines and the experimental results in figs. 3 and 4, a good description of the results was obtained using eqn. (3), and by fitting the parameters A, B and κ to the experimental data. Although we cannot rule out the possibility that other functional forms would fit as well,[27,28] we do consider that this agreement provides support for the dimensionality shift analysis described above.

The results obtained for the inverse correlation length, κ, as a function of temperature for various different fields are shown in figs. 5-7. The results show that there is a monotonic increase in κ with increasing field, but that the effect of the field is at least relatively larger at low temperatures. On cooling in an applied field, κ decreases in the neighbourhood of T_N but becomes almost constant at a temperature somewhat below T_N. This temperature

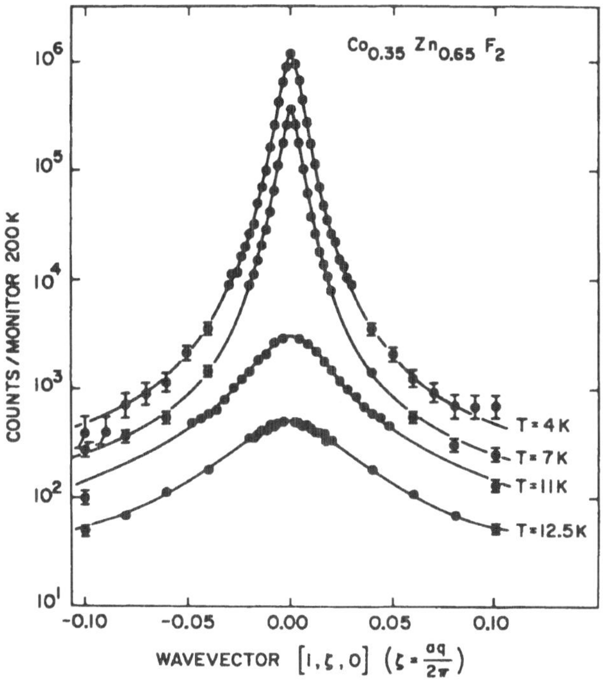

Fig. 4. The neutron scattering observed from $Co_{0.35}Zn_{0.65}F_2$ as a function of temperature in a field of 3.5 T.

decreases with increasing field. Of particular interest is the behaviour of the inverse correlation length as a function of field and at low temperatures. This is shown in figs. 8 and 9. In the case of $Rb_2Co_{0.7}Mg_{0.3}F_4$, the d = 2 system, the inverse correlation length varies as the applied field to the power $2x = 1.6\pm0.2$. For $Fe_{0.35}Zn_{0.65}F_2$, the data at small fields do not give a reliable estimate of κ because of experimental problems due to extinction. Fitting the data at larger fields H > 1.5 T we find that the inverse correlation length varies as the applied field to the power $2x = 2.1\pm0.2$. In contrast similar experiments on $Co_{0.35}Zn_{0.65}F_2$ gave the power $2x = 3.6\pm0.2$. It is difficult to reconcile these results with the predictions given in eqn. (4). The results for $Rb_2Co_{0.7}Mg_{0.3}F_4$ and $Co_{0.35}Zn_{0.65}F_2$ are more or less consistent with eqn. (4) if $d_c \approx 3.4$, which is higher than all the theoretical estimates, but since the $Fe_{0.35}Zn_{0.65}F_2$ system yields a different exponent and a value of $d_c \approx 3.9$, any conclusion must be treated with caution.

The behaviour of the amplitudes A and B is very much that expected from eqns. (3) and (5). At low temperatures it is found in $Rb_2Co_{0.7}Mg_{0.3}F_4$ that the amplitude of the Lorentzian squared, B,

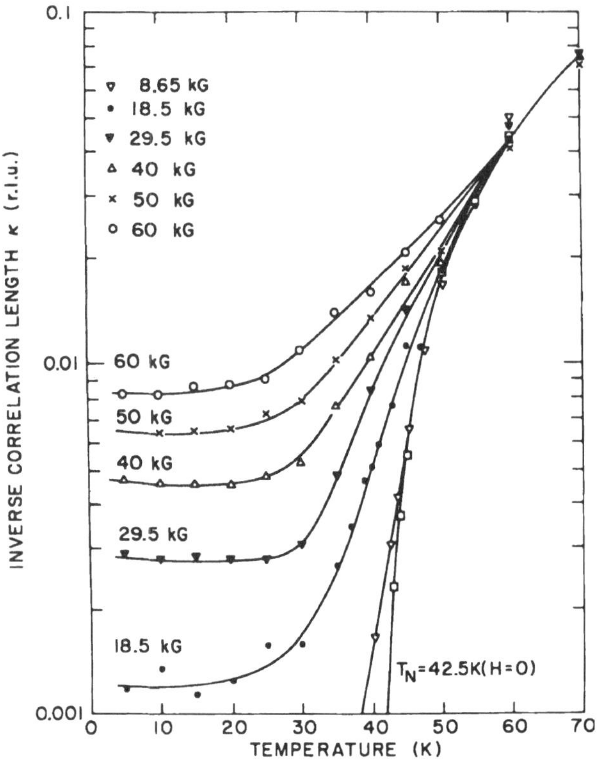

Fig. 5. The inverse correlation length κ as a function of temperature for various different applied fields in $Rb_2Co_{0.7}Mg_{0.3}F_4$.

is proportional to κ^2, and in $Co_{0.35}Zn_{0.65}F_2$ and $Fe_{0.35}Zn_{0.65}F_2$ that B is proportional to κ, in agreement with eqn. (5). In the latter two systems the amplitude of the Lorentzian A is negligible at small fields. The behaviour of the amplitudes in the sample $Co_{0.26}Zn_{0.74}F_2$ is somewhat different. This sample has a concentration of Co ions which is very close to the percolation concentration of 0.24. On applying fields above 1.2 T, it was found that the scattering could be described by a Lorentzian profile without including the Lorentzian squared term. Also for these fields the results were independent of the prior history of the sample. The system was then a normal paramagnet, and the relatively large field had suppressed the characteristic features of the random field behaviour. In this case as the field was increased, B/κ decreased to zero at 1.2 T, and the amplitude A increased with increasing field to 1.2 T and at higher fields remained almost constant.

At temperatures well above T_N, the scattering was well described

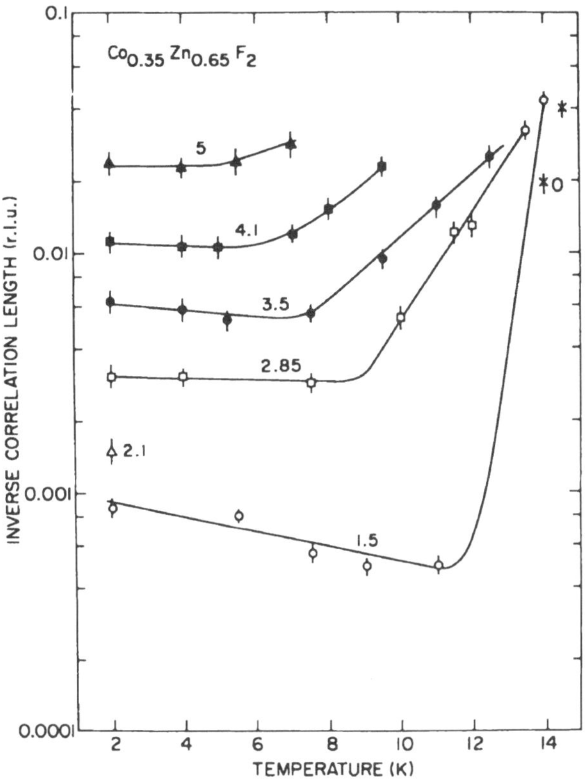

Fig. 6. The inverse correlation length κ as a function of temperature for various applied fields in $Co_{0.35}Zn_{0.65}F_2$.

by the normal Lorentzian profile for every system as shown at 12.5 K in fig. 4 for $Co_{0.35}Zn_{0.65}F_2$. On cooling in a field, the profile changes continuously from a Lorentzian to a largely Lorentzian squared profile. Unfortunately, when eqn. 3 is then fitted to the results, the values of the parameters, A, B and κ, are highly correlated, and it is difficult to obtain reliable estimates. Nevertheless we show in figs. 10 and 11 the behaviour of B/κ^2 for $Rb_2Co_{0.7}Mg_{0.3}F_4$ and B/κ for $Co_{0.35}Zn_{0.65}F_2$. They both behave very much like the square of the order parameter as predicted by eqn. 5, and a similar behaviour is found for $Fe_{0.35}Zn_{0.65}F_2$. The amplitude of the Lorentzian steadily decreases with decreasing temperature although its behaviour close to T_N is difficult to determine. It is worthy of note that the amplitude of the Lorentzian squared is not zero at all temperatures above the break in the κ/T curves shown in figs. 2-4, but is significant for all temperatures below T_N.

The behaviour of the scattering at T_N was studied in

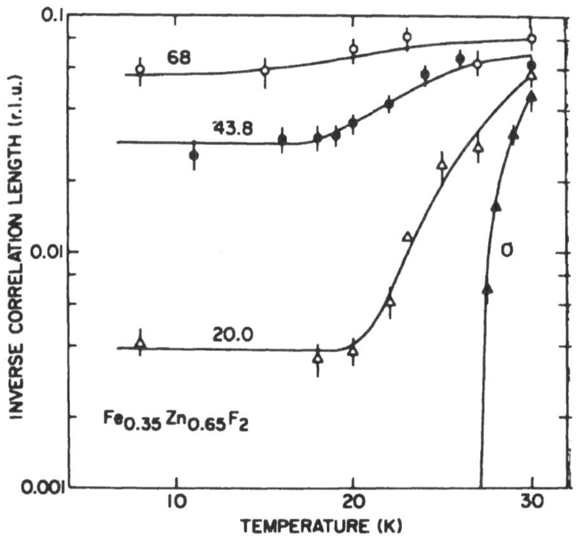

Fig. 7. The inverse correlation length κ as a function of temperature for various applied fields (kG) in $Fe_{0.35}Zn_{0.65}F_2$.

$Fe_{0.35}Zn_{0.65}F_2$, and the results for κ are illustrated in fig. 12. The inverse correlation length κ varies as the applied field to the power 0.8±0.1, and a similar analysis for the peak intensity yields it varying as the applied field to the power -1.6±0.2. As the critical exponent, η, is small for the d = 3 Ising model, these two results are consistent with one another but are not in good agreement with the predictions of scaling theory, eqn. (6), unless the random field is proportional to the applied field to the power 0.8.

Although many of the results described above are more or less consistent with the behaviour expected for an Ising system in a random field, there is one glaring discrepancy. The lower critical dimension is predicted to be variously 2 or 3, whereas a direct interpretation of our results suggests d_c is larger than 3. Because of this discrepancy and the implications it has for the theory as well as for the effect of random fields on the results of other experiments on three dimensional systems, we must consider the problem of the freezing at low temperatures in more detail. In particular it is possible that our results are not characteristic of the ground state of the system, but that the freezing prevents us from determining the properties of the ground state. We have performed two sets of experiments to investigate this possibility. Firstly in $Fe_{0.35}Zn_{0.65}F_2$ we performed several experiments in which the

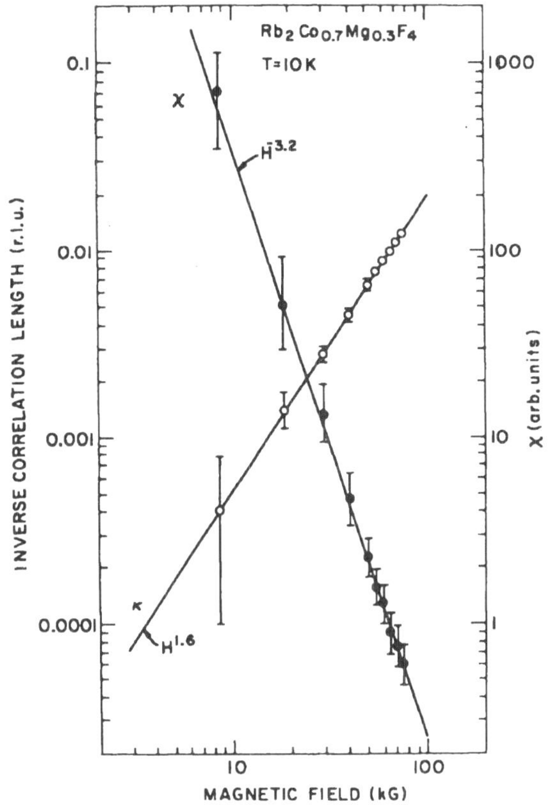

Fig. 8. The magnetic field dependence of the inverse correlation length and susceptibility in $Rb_2Co_{0.7}Mg_{0.3}F_4$.

field was varied while the temperature was held constant. These showed that at a temperature of 8.0 K, the spins were frozen for fields, $0 < H < 7.0$ T. At 16.0 K, however, the spins were able to follow the varying field, on a time constant of several minutes, for magnetic fields as low as 2 T. Since the scattering for fields of 2 T is almost independent of temperature below 20 K, fig. 7, this experiment strongly suggests that the spins are in equilibrium for at least part of the temperature range over which κ is almost constant in fig. 7. The break in behaviour of the κ versus temperature curve is not due to freezing at about 20 K. If however the equilibrium behaviour of the system is such that in a field of 2 T, κ is independent of temperature between 20 K and 15 K and then decreases, freezing may well prevent us from observing this decrease. This behaviour although not impossible seems unlikely to us. It was shown that $Rb_2Co_{1-x}Mg_xF_4$ was in equilibrium at all fields for $T \gtrsim 30$ K and at least for fields greater than 4 T down to 20 K.

The second set of experiments were measurements on the $Mn_xZn_{1-x}F_2$ system. In this system the spin wave gap is much smaller

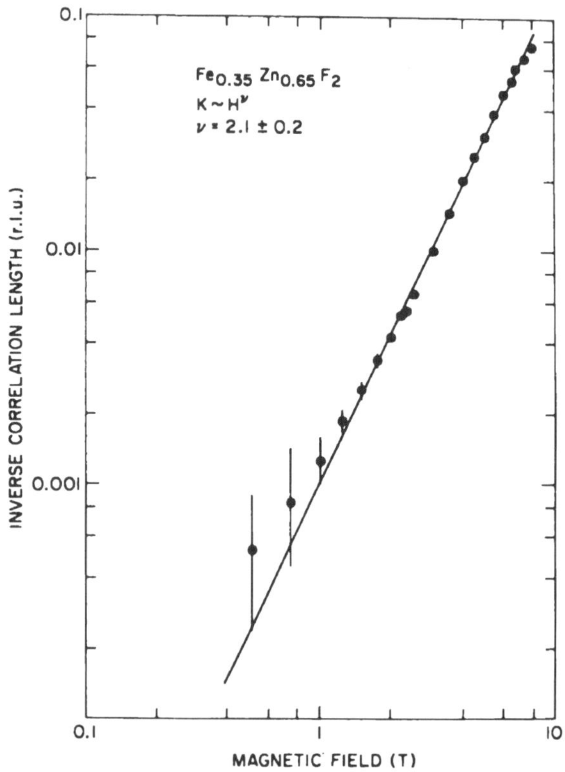

Fig. 9. The magnetic field dependence of the inverse correlation length in $Fe_{0.35}Zn_{0.65}F_2$ at 8.0 K.

and so it is to be expected that the system will remain in a thermodynamic equilibrium state down to lower temperatures. The system is, however, not a good Ising system and in particular the system has spin $S = \frac{5}{2}$ rather than $\frac{1}{2}$, and furthermore the anisotropy arises from the rather complex dipolar interactions; typically for the range of fields studied the random field energy is larger than the anisotropy.

The initial measurements were performed at Chalk River Nuclear Laboratories on a sample of $Mn_{0.78}Zn_{0.22}F_2$. The results[26] were surprising in that at low temperatures the magnetic Bragg reflection was unchanged in width on cooling in a magnetic field as shown in fig. 13. This shows that in $Mn_{0.78}Zn_{0.22}F_2$, long range order is not destroyed by a field unlike the cases discussed above. There was however evidence of a random field state described by a Lorentzian squared cross-section, eqn. (3), but this occurred only within about 2 K of T_N which in this case was 48 K.

More detailed measurements have now been made on a more dilute sample of $Mn_{0.65}Zn_{0.35}F_2$. For a field of about 5 T, the spin wave

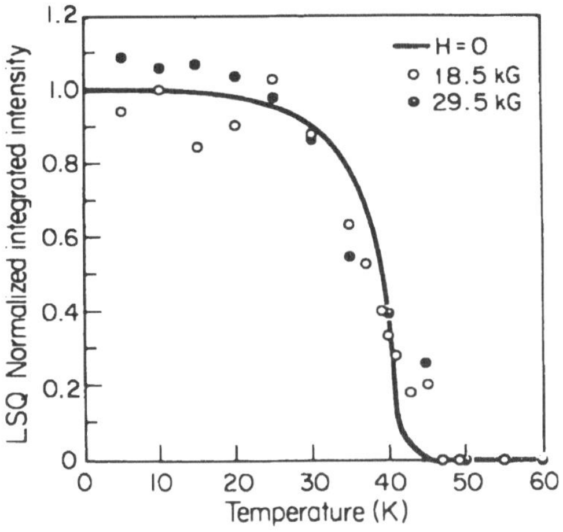

Fig. 10. The integrated intensity of the Lorentzian squared in $Rb_2Co_{0.7}Mg_{0.3}F_4$, B/κ^2, as a function of temperature for three different applied fields.

frequency becomes zero and the sample undergoes a spin-flop transition, as shown in fig. 14, to a phase in which the spins are aligned perpendicular to the c-axis. The paramagnetic phase is characterised by a Lorentzian scattering profile, and the random field phase by a Lorentzian squared profile. In the long range ordered phase the inverse correlation length, κ, is smaller than our resolution which means that κ is less than 0.0005 r.l.u. or that the domain size is in excess of 2000 lattice spacings. Finally in fig. 15 we show a plot of κ against temperature for various applied fields. The behaviour is clearly very different from that occurring in the other samples, figs. 5-7. The inverse correlation length continuously decreases with decreasing temperature and does not show the temperature independent behaviour of the other systems. We do not understand the origin of this difference in the behaviour — whether it results from the freezing although our experiments on $Rb_2Co_{1-x}Mg_xF_4$ and $Fe_{0.65}Zn_{0.35}F_2$ make this appear to be unlikely, or whether it is because $Mn_{0.35}Zn_{0.65}F_2$ is anomalous because it is not a good Ising model, has a large spin $\frac{5}{2}$, and is close to a spin-flop instability. Only further work can elucidate these questions as a function of temperature for various different applied fields.

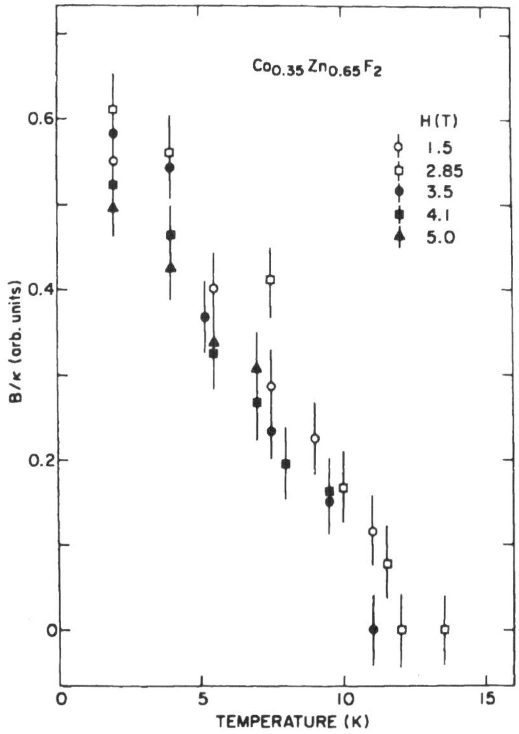

Fig. 11. The integrated intensity of the Lorentzian squared in $Co_{0.35}Zn_{0.65}F_2$, B/κ, as a function of temperature for different applied fields.

Other Measurements and Conclusions

The application of a magnetic field to a dilute antiferromagnet has been used to study the effect of random fields using several techniques other than neutron scattering. The first measurements[29] were susceptibility measurements on $Gd_{0.97}La_{0.03}AlO_3$ which showed that the application of a random field produced a rounding of the transition and were interpreted as showing a destruction of a long range order. Similar susceptibility measurements have been made on the $Rb_2Co_xMg_{1-x}F_4$ systems[30] with x = 0.8, 0.7 and 0.6 and on the $Mn_xZn_{1-x}F_2$ system[31] with x = 0.45 and 0.6. The results on both of these systems also show a rounding of the transition when the field is applied. The $Mn_xZn_{1-x}F_2$ system[32] was also studied for x = 0.94 by thermal expansion measurements and they too found a rounding of the transition was produced by the applied field.

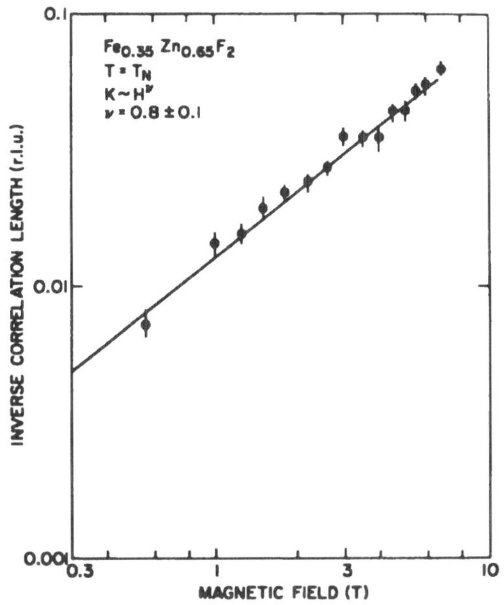

Fig. 12. The inverse correlation length κ for $Fe_{0.65}Zn_{0.65}F_2$ as a function of field for $T = T_N$.

In contrast to these measurements specific heat and linear birefringence measurements[33] which also yield the specific heat exponent show firstly that the critical fluctuations are different in the presence of a field and secondly that the transition does remain sharp. Recent measurements on a sample of $Fe_{0.6}Zn_{0.4}F_2$ gave the results shown in fig. 16 and the transition is clearly extremely sharp. Analysis shows that the temperature derivative of the birefringence is a logarithmic function of $(T - T_N)$, which is the behaviour expected for a $d = 2$ Ising and non-random system. Similar results although of not such high precision were found[33] for $Mn_xZn_{1-x}F_2$.

The results of the susceptibility measurements are consistent with the neutron scattering measurements. There is no doubt that random fields do drastically modify the behaviour of the critical fluctuations close to T_N. There is no doubt that care must be taken in studies of critical phenomena to avoid the presence of random fields because they are relevant and do produce very large effects. In nearly all of the measurements the random field energy/site has been much less than kT_N, or the exchange interactions but there is no doubt that they have caused large effects.

The results of the specific heat measurements are more surprising because they do yield a surprisingly sharp peak as shown in

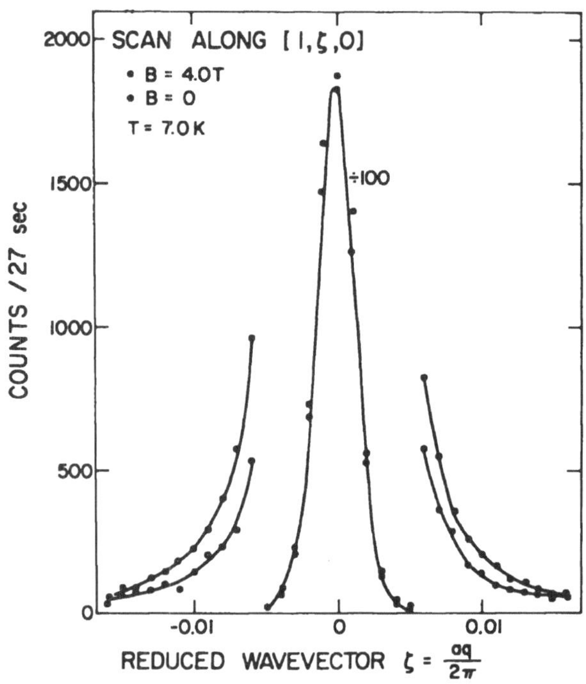

Fig. 13. The neutron scattering intensity observed along the line (1, ζ, 0) at 7.0 K in $Mn_{0.78}Zn_{0.22}F_2$ for a field of 4 T (open circles) and without a field (closed circles).

fig. 16. Not surprisingly the sharpness of this peak has been interpreted as evidence that the phase transition in the d = 3 Ising model is not destroyed by the presence of a random field. The neutron scattering measurements have been performed on the same system, $Fe_xZn_{1-x}F_2$, however, and they show unambiguously that the systems do not have long range antiferromagnetic order. Consequently if there is a sharp transition in the presence of a field it is not a transition between a paramagnetic phase and an antiferromagnetic phase, it must be a transition to a disordered state with a presently unknown order parameter.

The measurements which are sensitive to a particular type of order, susceptibility and neutron scattering both show a smearing of the properties close to T_N, whereas the specific heat measurements are especially sensitive to the local properties. Whether the domains caused by the random field are sufficiently large that the ordering of the spins within one domain can account for the specific

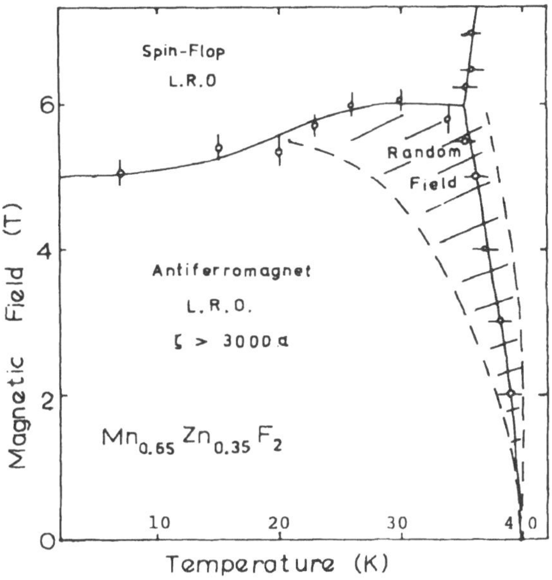

Fig. 14. The phase diagram of $Mn_{0.65}Zn_{0.35}F_2$ in a field.

heat results, or whether the specific heat results do signify a new type of ordered phase only more experiments and theory will tell.

In conclusion there is no doubt that the presence of random fields does have a large effect on the critical fluctuations close to T_N. They appear to destroy the long range order in $d = 3$ Ising systems although there is the possibility that the low temperature freezing of the spins may be giving problems in the interpretation of the measurements. The behaviour of the $Mn_xZn_{1-x}F_2$ system with a small gap arising from the dipolar anisotropy is different from that of the good Ising models ; $Co_xZn_{1-x}F_2$ and $Fe_xZn_{1-x}F_2$.

Further theoretical work is needed to understand the changes in the critical fluctuations close to T_N, and to understand the freezing of the spins at lower temperatures in addition to further work on the problem of the lower critical dimension for the Ising model in a random field.

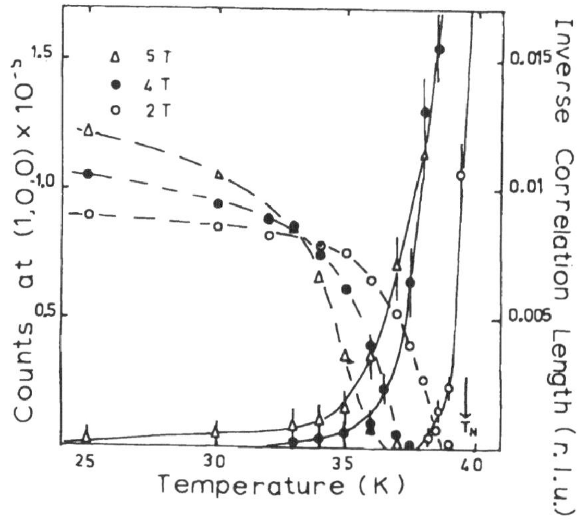

Fig. 15. Inverse correlation length in $Mn_{0.65}Zn_{0.35}F_2$ as a function of temperature for various different applied fields.

COMPETING ANISOTROPY

Systems with competing anisotropies are ones in which the spin direction of one component wishes to align along one crystallographic direction whereas for the other component it is energetically favourable for the spin to align along a different direction. There is then a competition between the anisotropy energy which favours different directions for the spins and the exchange energy which favours parallel or anti-parallel spin directions. There are several examples of systems in which this competition can be studied. The most carefully studied[35,36] is the $Fe_xCo_{1-x}C\ell_2$ system in which the single ion anisotropy energy of the Fe ion favours a spin alignment along the c axis, whereas the anisotropic Co exchange favours a spin direction perpendicular to the c-axis. Other systems are $K_2Co_xFe_{1-x}F_4$ [37], where in contrast, the anisotropic exchange between the Co ions favours alignment along the c-axis whereas the single ion Fe anisotropy favours alignment perpendicular to the c-axis and $K_2Mn_xFe_{1-x}F_4$ where the dipolar forces favour the alignment along the c-axis[38].

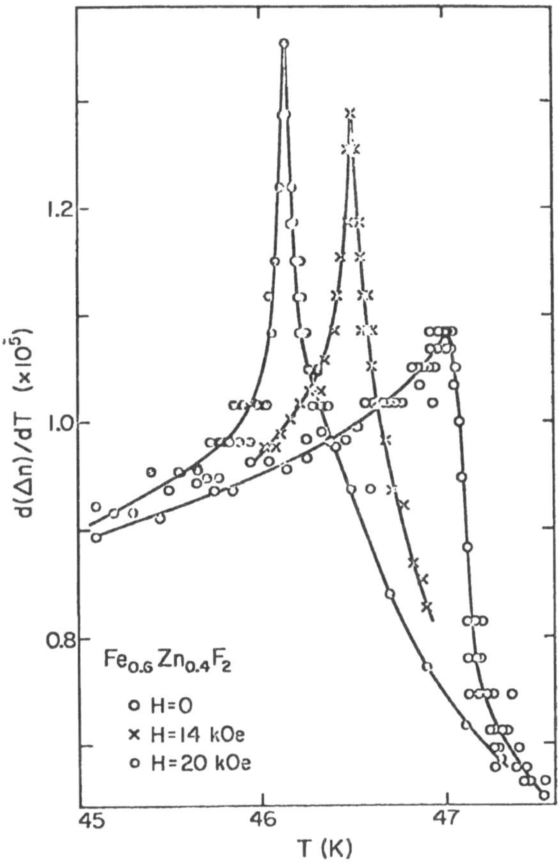

Fig. 16. Measurements[34] on the temperature derivative of the linear birefringence in $Fe_{0.46}Zn_{0.4}F_2$.

The theory of these systems within mean field[39] theory suggests that between the phases in which the spins are aligned parallel or perpendicular to the c-axis, there is a phase in which both the parallel and perpendicular components are ordered. The critical phenomena was then discussed by Fishman and Aharony[40] who suggested that the phase boundaries associated with the ordering of these two components of the spin were decoupled and intersected at a decoupled tetracritical point, and furthermore that the critical behaviour associated with each of the phase lines was that appropriate for the ordering of either the parallel or perpendicular components.

Although the presence of an intermediate phase has been confirmed in several different systems, detailed information on the critical phenomena is available[35,36] on only one system, $Fe_xCo_{1-x}C\ell_2$. Largely, this is because the experiments require very good single crystals with a very homogeneous concentration. In fig. 17. we show measurements of the phase diagram in $Fe_xCo_{1-x}C\ell_2$, showing the phases with spins ordered parallel to the c-axis, perpendicular to the c-axis and a mixed phase. The qualitative agreement with mean field theory is satisfactory. The predictions of Fishman and Aharony suggest that the phase boundary lines are decoupled and are continuous in slope at the tetracritical point. The experimental results clearly suggest that there is a discontinuity in shape, particularly in the line describing the perpendicular ordering.

The behaviour of the phase transitions was also different from that predicted as shown in fig. 18. The scattering for $\vec{Q} = (0,0,8.8)$ is the critical scattering associated with the ordering of the perpendicular component, whereas that for $\vec{Q} = (0.985,0,\bar{1})$ is largely associated with that of the parallel component. Clearly fig. 18 shows that whereas the transition at higher temperature is sharp, that at the lower temperature is rounded. Similar evidence[36] for this was found in studies of the temperature dependence of the ordering. There is no doubt that the ordering of $S_{\parallel}$ and $S_{\perp}$ is coupled. There have been several suggestions put forward[41,36] to explain this discrepancy between experiment and theory. The most appealing[36] is that the smearing is a result of the effect of the random fields produced as a result of the ordering at the first transition.

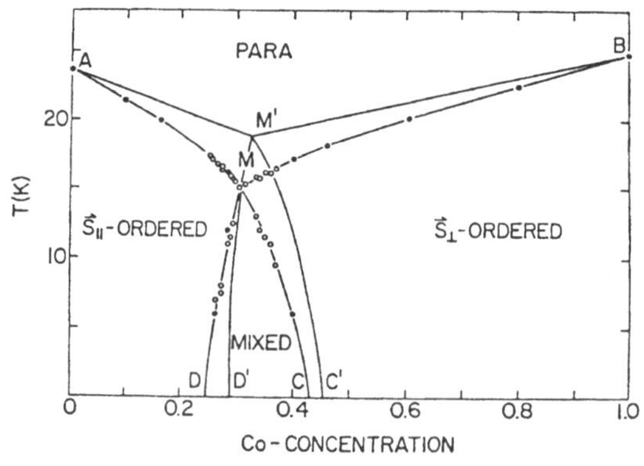

Fig. 17. The phase diagram of $Fe_xCo_{1-x}C\ell_2$ from ref. 36. The lines connecting the dashed points are the results of mean field theory and the experimental points are from susceptibility (solid) or neutron scattering (open).

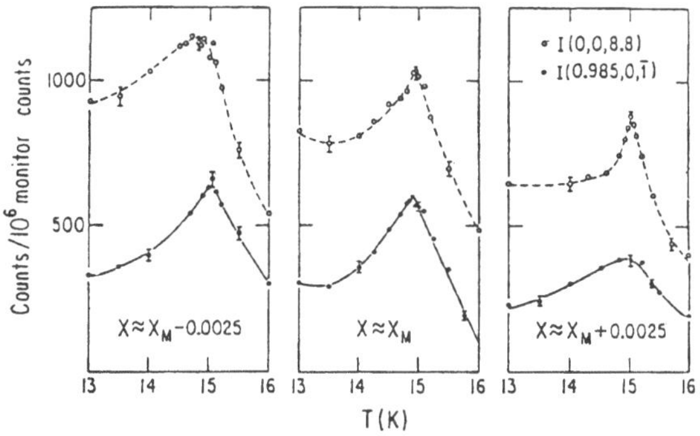

Fig. 18. The neutron scattering intensity for two different wave-vector transfers in $Fe_xCo_{1-x}C\ell_2$ for different concentrations near the tetracritical point with $x = x_M$.

The symmetry of $Fe_xCo_{1-x}C\ell_2$ allows a general exchange interaction between two ions separated by $\vec{R}_{ij}$ of the form

$$\mathcal{H} = - J\, \vec{S}_i . \vec{S}_j + \frac{K}{R^2} (\vec{R}_{ij} . \vec{S}_i)(\vec{R}_{ij} . \vec{S}_j)$$

$$+ \frac{G}{R} \vec{S}_i . (\vec{S}_j \times \vec{R}_{ij}) \ .$$

The last term in this expression gives rise to a coupling between $S_\parallel$ and $S_\perp$ even for interactions between ions in the magnetic sheets. If either $S_\parallel$ or $S_\perp$ orders, this term generates a field on the other spin component, and due to the mixed nature of the sample the net direction of this field varies from site to site. This type of interaction was not considered by Fishman and Aharony. As we have seen in the preceding section random fields do drastically alter the critical fluctuations and in neutron scattering measurements appear to smear out the transition. Consequently this provides a ready explanation of the measurements.

If this is indeed the origin of the effect, it should be absent in, for example, the $K_2Co_xFe_{1-x}F_4$ system because the third term in the exchange interaction is forbidden by symmetry. We must await detailed measurements on this type of system to test both the theory and the explanation of the discrepancies in $Fe_xCo_{1-x}C\ell_2$.

CONCLUSIONS

The first conclusion is that the study of disordered insulating magnets is a very rich field for the testing of the theories of the phase transitions in disordered systems. There is a wide variety of systems available, and experiments continue to provide a crucial test of the theories and new challenges for theoretical work.

The second main conclusion concerns the importance of random fields. Measurements have shown that relatively small random fields do produce large and dramatic effects on the ordering and critical phenomena of systems. Although, as described above, a detailed understanding of these effects is not yet available, there is no doubt that they are large and significant. It is therefore of crucial importance in studying phase transitions to ensure that impurities do not introduce random fields. We have seen above that the first detailed study of the competing anisotropy problem gave results which were not in agreement with the theory and the discrepancies probably resulted from random field effects. Similarly[42] discrepancies between theory and experiment on tricritical points have been explained as arising from these effects. There is no doubt that they are also of importance in the study of competing interactions in metallic spin glasses and they probably also account for many of the puzzling results obtained on structural phase transitions. It is clearly of considerable importance to understand the effect of random fields in more detail, and studies of insulating disordered magnets will probably be the most fruitful line of experimentation.

ACKNOWLEDGMENT

We are grateful for collaboration and discussions of the random field problem with A. Aharony, D.P. Belanger, W.J.L. Buyers, H.J. Guggenheim, M. Hagen, V. Jaccarino, H. Ikeda, A.R. King, S. Satija and D. Wallace.

The work was supported by the National Science Foundation under contract DMR 79-23203, the Division of Basic Energy Sciences under contract DE-AC02-76CH0016 and by the Science and Engineering Research Council.

REFERENCES

1. R. A. Cowley, R.J. Birgeneau and G. Shirane, Excitations and Phase Transitions in Random Anti-Ferromagnets in 'Ordering in Strongly Fluctuating Condensed Matter Systems', T. Riste ed., Plenum, New York (1980).

2. A. B. Harris, J. Phys. C 7, 1671 (1974).
3. G. Grinstein and A. Luther, Phys. Rev. B13, 1329 (1976).
T. C. Lubensky, Phys. Rev. B11, 3580 (1975).
4. D. E. Khmelnitskii, Soviet Phys. JETP, 41, 981 (1976).
5. R. J. Birgeneau, J. Als-Nielsen and G. Shirane, Phys. Rev. B16, 280 (1977).
6. G. M. Meyer and O. W. Dietrich, J. Phys. C 11, 1451 (1978).
R. A. Cowley and K. Carneiro, J. Phys. C 13, 3281 (1980).
7. R. J. Birgeneau, R. A. Cowley, G. Shirane, H. Yoshizawa, D. P. Belanger, A. R. King and V. Jaccarino (to be published).
8. M. T. Hutchings, B. D. Rainford and H. J. Guggenheim, J. Phys. C3, 307 (1970).
9. M. T. Hutchings, M. P. Schulhof and H. J. Guggenheim, Phys. Rev. B5, 154 (1972), G. Ahlers, A. Kornblit and M. B. Salamon, Phys. Rev. B9, 3932 (1974).
10. L. C. Le Guillou and J. Zinn-Justin, Phys. Rev. B21, 3976 (1980).
11. S.A. Newlove (to be published).
K.E. Newman and E.K. Riedel, Phys. Rev. B25, 264 (1982).
G. Jug, Phys. Rev. B27, 609 (1983).
12. Y. Imry and S. Ma, Phys. Rev. Lett. 35, 1399 (1975).
13. G. Grinstein, Phys. Rev. Lett., 37, 944 (1976).
14. G. Parisi and N. Sourlas, Phys. Rev. Lett. 43, 744 (1979).
15. E. Pytte, Y. Imry and D. Mukamel, Phys. Rev. Lett. 46, 1173 (1981), D. Mukamel and E. Pytte, Phys. Rev. B25, 4779 (1982) H. S. Kogon and D. Wallace, J. Phys. A 14, L527 (1981).
16. G. Grinstein and S. Ma, Phys. Rev. Lett. 49, 685 (1982).
17. A. Aharony and E. Pytte (to be published).
18. S. Fishman and A. Aharony, J. Phys. C 12, L729 (1979).
19. P. Z. Wong, S. Von Molnar and P. Dimon, J. Appl. Phys. 53, 7954 (1982).
20. J. L. Cardy (unpublished).
21. A. Aharony, Phys. Rev. B18, 3328 (1978), Experiments on $KMn_xNi_{1-x}F_3$ by M. Hagen have confirmed this result experimentally.
22. H. Ikeda and M. T. Hutchings, J. Phys. C 11, L529 (1978).
23. P. Martel, R. A. Cowley and R. W. H. Stevenson, Can. J. Phys. 46, 1355 (1968).
24. O. Nikotin, P. A. Lindgard and O. W. Dietrich, J. Phys. C 2, 1168 (1969).
25. H. Yoshizawa, R. A. Cowley, G. Shirane, H. J. Guggenheim and H. Ikeda, Phys. Rev. Lett. 48, 438 (1982).
26. R. A. Cowley and W. J. L. Buyers, J. Phys. C 15, L1209 (1982).
27. R. J. Birgeneau, H. Yoshizawa, R. A. Cowley, G. Shirane and H. Ikeda (to be published).
28. M. Hagen, R. A. Cowley, S. Satiya, G. Shirane, H. Yoshizawa, R. J. Birgeneau and H. J. Guggenheim (to be published).
29. H. Rohrer, J. Appl. Phys. 52, 1708 (1981).
30. H. Ikeda, J. Phys. C 16, L21 (1983).
31. H. Ikeda and K. Kibuta (to be published).
32. Y. Shapiro, J. Appl. Phys., 53, 1931 (1982).

33. D. P. Belanger, A. R. King and V. Jaccarino, Phys. Rev. Lett. 48, 1050 (1982).
34. D. P. Belanger, A. R. King, V. Jaccarino and J. Cardy, (to be published).
35. P. Wong, P. M. Horn, R. J. Birgeneau, C. R. Safinga and G. Shirane, Phys. Rev. Lett. 45, 1974 (1980).
36. P. Wong, P. M. Horn, R. J. Birgeneau, and G. Shirane, Phys. Rev. (in press).
37. K. Fendler, W. Lehmann, R. Weber and U. Dürr, J. Phys. C 30, L533 (1982).
38. L. Bevaart, E. Frikee and L. J. de Jongh, Phys. Rev. B18, 3376 (1978), ibid B19, 4741 (1979).
39. P. A. Lindgard, Phys. Rev. B14, 4074 (1976), ibid B16, 2168, (1978).
40. S. Fishman and A. Aharony, Phys. Rev. B18, 3507 (1978).
41. D. Mukamel, Phys. Rev. Lett. 46, 845 (1981).
42. R. J. Birgeneau and A. N. Berker, Phys. Rev. B26, 3751 (1982).

THE ORIGINS AND INFLUENCES OF THE SPIN-GLASS PROBLEM

B.R. Coles

Blackett Laboratory
Imperial College
London SW7

In recent years there has been a burst of interest in what is called the spin-glass problem. This concerns the nature of the magnetic state of systems containing a distribution of magnetic-moment-carrying atoms distributed with some degree of randomness (generally on a small fraction of the atomic sites of a metal crystal, the other site being occupied by non-magnetic atoms), such that there exist both parallel and antiparallel couplings between moments. The ground state clearly lacks long-range magnetic order, but the nature of that state and of the transition to it from the high temperature paramagnetic state is still not clear. This topic is of interest not only as a scientific problem, but also as a good example of the way in which new sub-fields of study in solid state physics emerge and it is as such an example that it is considered here. A rather more complex story could be constructed for the dilute alloy (Kondo) problem, where the interactions between moments can be ignored.

In these notes I have not tried to review the whole history of the spin-glass problem, leading up to the present state of our understanding of random magnetic alloys without long-range order. I have merely tried to show the components of experimental work that provided different types of stimulus at different times to the attempts that were made to construct satisfactory theoretical models. I have also tried to show the part played by interactions between people of different original interests and the influence of local or even national interests. Since the "spin-glass problem" is by no means solved, I have not attempted to bring the story up to date.

First, it should be made clear that the term "spin-glass" is used here as something to describe a type of experimental behaviour, a syndrome of which the various symptons may not all be "present" (as the medical men would put it) in a particular system or experiment. The most significant of these symptoms are a sharp cusp in the susceptibility in very small A.C. or D.C. magnetic fields, time dependent and remanent magnetizations below the temperature of the cusp, displaced hysteresis curves in more concentrated field-cooled alloys, extra contributions to the specific heat which are approximately linear in temperature and very weakly concentration dependent below the susceptibility cusp temperature and falling off slowly (with only a very rounded maximum) at higher temperatures, a magnetic scattering term in the electrical resistivity which increases more slowly, or falls, above the temperature of the susceptibility cusp and, where the Mossbauer effect can be observed, hyperfine fields which appear rather rapidly near the cusp temperature as the specimen is cooled in zero field. In current, as opposed to historical, discussions the term is used sometimes to describe a theoretical construct - a ground state in which magnetic moments are frozen without long range magnetic order that is separated from a paramagnetic state by some novel type of phase transition. I do not aim here to trace the emergence of such concepts, partly because the papers of Adkins and Rivier (1974), Edwards and Anderson (1975), which introduced order parameters, mark a fairly clear recognition of the phase transition aspects of the problem. About an earlier theoretical construct - a distribution of internal fields, usually called P(H) - I shall make some comments, because this concept played an important part in early interactions between experiment and theory.

The first symptom of the spin-glass syndrome was a striking resistivity effect [a maximum at low temperatures in $\rho(T)$] in dilute random alloys of transition metals in simple metals (Gerritsen and Linde, 1951); but from the outset the study of rather dilute alloys meant that the spin-glass effects - those due to the spin freezing when kT became less than the coupling energy of the spins to one another - were almost inextricably entangled with effects due to the scattering of conduction electrons by individual solute spins, effects that later were referred to as the Kondo effect. It was not at first clear that the resistance maximum, the temperature of which was strongly composition dependent, was the result of a competition between a negative $d\rho/dT$ from single impurity (Kondo effect) scattering and a positive $d\rho/dT$ from the freezing-out of spin-disorder scattering, rather like that below the Curie point of long-range magnetically ordered gadolinium. [It should be remembered that not only would it be another 13 years before the Kondo effect was explained, but also that it would not be until the work of Kasuya (1956) and de Gennes and Friedel (1957) that a spin-disorder scattering explanation of the resistivity of local moment materials like Gd or Au_3Mn with long range order would replace the, for them, inappropriate $s \rightarrow d$ scattering theory of Mott (1936). For a review see Coles (1958)].

EARLY WORK ON MAGNETIC PROPERTIES

What is surprising in retrospect is that, in spite of efforts of Korringa and Gerritsen (1953) to use two-level concepts associated with the moment on the impurity, no measurements of the low temperature magnetic properties of such alloys were carried out before the work on Cu-Mn of Owen, Browne, Arp and Kip (1957) and that of Schmitt and Jacobs (1957) which followed Schmitt's effort (1956) to use the onset of ferromagnetic interactions to give Korringa and Gerritsen's two levels. The work of Owen et al, which included ESR, susceptibility and Cu^{60}NMR measurements on alloys of 0.03-11% Mn in Cu as well as some work on other alloys, produced a striking impact on both experimental and theoretical efforts, and it is ironic that it seems in no way to have been triggered by the transport property anomalies. John Owen tells me that the susceptibility measurements which yielded the first observation of a maximum in $\chi(T)$ were made to help in the interpretation of the ESR data he had obtained, and that those measurements were made when he (accustomed to study in Oxford 3d ions in insulating crystals) suggested to Kittel and the Berkeley group where he was a visitor that Cu Mn alloys were more likely to be productive of ESR effects than the Li Mg alloys that had been suggested as a subject of study by Kittel. [It is not clear to me whether the Berkeley interest in Li Mg was in fact in an impurity resonance possibility or in the effect of Mg on the alkali-metal conduction electron resonance which had been a subject of intensive study in Kittel's group. It should be remembered that the period of such studies (1952-1954) predated not only Friedel's pioneering work (1956) on 3d virtual bound states in dilute metallic alloys, but also Friedel's 1954 attack on the problem of the screening of simple heterovalent impurities in monovalent metals (e.g. Cu-Zn or Li-Mg)]. At GE also*, where Roland Schmitt was working, the Cu-Mn work lay at the intersection of two areas of interest, his own efforts to understand the Dutch work on transport property anomalies in dilute Mn alloys and the involvement of Bean and Jacobs in the effects of competing magnetic interactions in producing anomalous magnetic remanence effects which were discussed in terms of exchange anisotropy. This latter aspect was one pursued later in more concentrated

* An important coupling between Berkeley and GE was provided by J.C. Fisher, on leave at the former from the latter, who was acknowledged for helpful discussions in the Owen et al papers, and by R.W. Schmitt for suggesting that ferromagnetic couplings could provide the split levels required by the Gerritsen & Korringa approach. In a footnote to this paper, Schmitt acknowledges a private communication from Owen indicating that their recent experiments at Berkeley suggested the dominance of antiferromagnetism.

Cu-Mn alloys by Kouvel (1961) at GE, yielding features such as the shifted hysteresis loops given by field cooling. Kouvel (1959) had first studied atomically disordered alloys near the Ni_3Mn composition, and since the ordered alloys were known to be ferromagnetic, it was natural to seek an explanation of their behaviour in terms of a competition between antiferromagnetic interactions of the sort to be expected for nearest neighbour Mn atoms and the ferromagnetic interactions, which dominated in the ordered alloys, between Ni neighbours or Ni-Mn neighbours. Although the Cu-Mn alloys were recognized by Kouvel (1960) as lacking the very strong bias towards ferromagnetism of Ni-Mn, it was clear from the time-dependent effects in the magnetization found in Sheffield by Street (1960) and from early neutron-scattering studies that they were not straightforward antiferromagnets. Work on Cu-Mn alloys had begun at Sheffield (Street and Smith, 1959) because of an interest in possible anomalous behaviour of elastic properties at antiferromagnetic phase transitions to match that seen in ferromagnets. Measurements were made on 40 and 50% Mn alloys after such effects had been found in both antiferromagnetic (>70% Mn) γ alloys and in two phase ($\alpha + \gamma$) alloys in which the α -manganese was the antiferromagnetic component and the γ phase contained $\sim$ 40% Mn, and the relationship between the peculiar magnetization behaviour of the 40% alloy and that found by Owen et al in the more dilute alloys was recognized. The ideas of fluctuations in composition or degree of atomic order that produced regions in which different interactions dominated fed back into the spin-glass story later when arguments revolved around the existence of comparatively concentrated transition-metal clusters coupling to one another through a more dilute matrix. The postulate of interactions of different signs within and between clusters was encapsulated by Beck (1971) in the term mictomagnetism. (The concept of clusters was also very influential in the programme of experimental studies at Grenoble in the 1960s (see Tournier and Weil, 1962); since Tournier, who had been greatly stimulated by the Friedel and Blandin (Paris) theories of magnetic (single-impurity) virtual bound states coupling by RKKY interactions (see below), was also inevitably swayed by the authority of Néel, in whose work on a range of materials the idea of fine particles had played an important role, cf. Néel (1962). Discussions of fairly dilute random alloys in terms of clusters were complicated by the fact (not fully recognized in the early 1960s) that there are two distinct ways in which the magnetic behaviour of fairly well separated solute transition metal atoms can be modified when they form clusters of near neighbours. In the first effectively non-magnetic individual solute atoms can thereby gain moments (an idea applied later by many workers to the onset of ferromagnetism in Cu-Ni); in the second well established moments on individual solute atoms can couple together in clusters to act cooperatively - at least over some range of temperatures - as superparamagnetic entities.

INTERACTIONS VIA CONDUCTION ELECTRONS

Following the exciting susceptibility results at Berkeley in 1956, it was recognized that this work was closely related both to the anomalous transport properties (by then being studied at GE also) and to the role of conduction electrons in providing an oscillatory local spin - local spin coupling. This concept had been, for nuclear spins, a Berkeley product (Rudermann and Kittel, 1954), but strangely enough the magnetic evidence for "some sort of antiferromagnetism" in Cu-Mn alloys was first ascribed to some type of superexchange through Cu $3d^{10}$ cores dominating a ferromagnetic local spin-conduction electron interaction, the belief that the latter should be large and show up in g-shifts and Knight shifts being a sort of reversion to the first order Zener (1951), Fröhlich-Nabarro (1940) interaction. This was soon corrected, however, by Hart (1957) and by Yosida (1957a), the latter having joined Kittel from the country in which Kasuya had correctly recognized the role of the local spin-conduction electron interaction in Gd and in which Kondo was later to use that interaction in its full implications for the explanation of the resistivity minimum*.

Thus, after the detailed theoretical work of Yosida, it was clear that the long-range oscillatory (Rudermann-Kittel-Kasuya-Yosida) interaction could explain the fall at low temperatures in sufficiently concentrated Cu-Mn alloys)more than ~ 0.05% Mn) in both the susceptibility and the electrical resistivity. It was emphasized by both Yosida (1957b) and the present writer (Coles, 1958) that the resistivity minimum seemed to be a separate effect not due to the magnetic "ordering". I do not have space here to discuss the effect of the worldwide interest in the Kondo problem and the need to purify it of interaction effects on the build-up of information relevant to the spin-glass problem.

THE SPECIFIC HEAT PROBLEM

The next wave of theoretical interest was stimulated by specific heat measurements on Cu-Mn and related alloys. The first of these were carried out by G.L. Booth and F.E. Hoare (unpublished) in Leeds before 1957, by de Nobel and du Chatenier (1958 and 1959) in Leiden (where the original resistance anomalies had been seen) and by P.L. Smith in Oxford (unpublished) on a specimen John Owen took back from

---

* It is rather remarkable that little notice had been taken of earlier work in th USSR on the s-d or s-f interaction and no significant use was made of the Hamiltonian of Zener, Kasuya and Yosida between 1956/7 and Kondo's paper of 1964, except for the important paper of Abrikosov & Gorkov (1960) on the effect of magnetic impurities on superconductors. (The suggestion that it was at the source of this effect was first made by Herring (1958)).

Berkeley in 1956. The impact on theory, however, seemed to come mainly from the paper of Zimmerman and Hoare (1960). This described work carried out at the Ford Motor Co. Laboratories in Dearborn, where Hoare had been a visitor, coupling his previous interest in the specific heat of alloys with that of the group of people (Zimmerman, Arrott and Goldman), who had moved there from the Carnegie Institute of Technology in Pittsburgh, where the alloy specific heat interest had been mainly in Cu-Ni, which seemed to possess no local moments. (Both Hoare and Booth and the Carnegie Tech. group had made specific heat measurements on manganese itself, and it is interesting to note that Kouvel had carried out his postgraduate work with Hoare at Leeds on the susceptibilities of paramagnetic transition metals, a programme stimulated by Stoner's interest in band magnetism).

The striking feature of the Cu-Mn and Ag-Mn alloys with more than about 0.5% Mn was the presence of an anomalous term in the specific heat (over and above the lattice and electronic terms of pure copper) which was, in the liquid He temperature range, approximately linear in T and roughly independent of Mn concentration. This was clearly associated with the magnetic indications of some sort of "ordering", but had characteristics totally different from those to be expected of long-range antiferromagnetic (or ferromagnetic) ordering. Furthermore, Arrott had been a co-author with Sato and Kikuchi of a paper (Sato, Arrott and Kikuchi 1959), which suggested that long-range order in dilute alloys required a concentration of at least 10-15% of solute. (This percolation concentration has become a separate field of active research).

In a number of laboratories (including that of Friedel in Paris, where Blandin extended his theoretical investigations from the single impurity with a magnetic virtual bound state to interacting impurities) it was recognized that the general features of the specific heat would follow from a broad distribution of molecular fields such that its breadth, and hence the temperature corresponding to some broad maximum in the excess specific heat, would increase with concentration. The total number (as distinct from the fraction) of moments lying in very small molecular fields, and thus able to be excited by a small rise in temperature, would then be roughly independent of concentration. The basic theoretical problem was the construction of this field distribution, and the first and most imaginative suggestion was that of Overhauser (1960) at the Ford laboratories, who suggested that a spin-density wave of the sort he had contemplated as a ground state for alkali metals could be stabilized in the conduction band of copper by its interactions with moments on solute Mn atoms. This was in a sense the first suggestion of a special type of phase transition* in such materials. (It is

* It is ironic that Overhauser (private communication) was first led to the idea of spin density waves in simple metals in a search for a mechanism for a superconducting ground state.

interesting that just this type of "purchase" of a spin density wave by solute moments was later shown to take place in Y-Gd alloys, although preceded at concentrations below 2% by a spin-glass of the conventional Cu-Mn type). Marshall (1960), then at Berkeley, however, claimed that the necessary distribution of effective fields could be generated in Cu-Mn by an RKKY interaction between the solute moments without any special modification of the conduction band states and without any particular ordering of the moments : Blandin (1961) independently reached similar conclusions, while in addition recognizing that the "exchange" operating between local moment and conduction electrons was an effective one, arising from the mixing of local 3d states and conduction band states that created the virtual bound state. The most detailed construction of the effective field distribution on a statistical basis was that of Klein and Brout (1963) using an Ising model. (More recently it has been recognized that the excitation of individual spins in local fields is as incomplete an account of the excitations for the spin glass as for the pure ferromagnet, but there is not yet agreement on the character of the low-lying excitations in the former).

During this period (1956-1965) the ground state of the not-too-dilute magnetic alloy was regarded as in some sense antiferromagnetic, in that a specimen cooled in zero field possessed both parallel and antiparallel interactions but no imbalance between "up" and "down" spins. At the time of Blandin's first discussion of the Cu-Mn alloys (Blandin and Friedel, 1959) the specific heat data had not been published; but that paper spoke of spin freezing, and introduced percolation ideas by recognizing that with strong parallel nearest neighbour interactions a random f.c.c. solid solution would become ferromagnetic (possess an infinite cluster) above about 16% solute. The randomness recognized for the dilute alloys was primarily that of the distribution of the spins over the points of the host lattice. "...we can imagine a sort of antiferromagnetism ... there will be atoms with spin parallel or antiparallel. One can thus think of antiferromagnetism where each spin is fixed but at positions which are random in space", (Blandin, thesis 1961). "[at low temperatures] the atomic disorder and hence the spatial disorder of the moments is still present", Coles (1958). There later developed, however, an awareness that the ground state would also be characterized by spin orientations that were more generally random, while being frozen and time-independent. Thus Tournier (1965) wrote of "a distribution of spin directions [that] is such as to minimize the energy. Each spin is certainly blocked, but there does not have to be a privileged direction of antiferromagnetism". Such an approach was particularly helpful in enabling one to picture a biassing of spin orientations (with consequent remanence) on cooling a specimen in a field, together with the slow decay of such remanence.

SUSCEPTIBILITY CUSPS

During the 1960s a considerable expansion took place in the number of alloy systems in which spin-glass character was found. (Some of the work on more concentrated alloys has been mentioned above). There were now dilute rare-earth systems* as well as dilute 3d systems (although some of those studying the displacement of superconductivity by magnetism failed to recognize the spin-glass character of their magnetic "ordering"), systems in which a strong Kondo demagnetisation of the solute in the dilute limit delayed the onset of the magnetic freezing and systems in which the host was itself a transition metal (the similarity of Mo-Fe to Au-Fe as far as resistivity behaviour had been noted in 1962). There were, in fact, signs of this whole area of study being in danger of becoming a sort of alloy taxonomy, when the situation was transformed and the interest of the theoreticians reawakened by the striking behaviour demonstrated by Cannella and Mydosh (1972) in the very low field and low frequency A.C. susceptibiliites of apparently well trodden systems like Au-Fe and Cu-Mn. They found very sharp cusps (rapidly rounded by superposed D.C. fields) at temperatures which in the former system agreed closely with "ordering" temperatures indicated by the hyperfine fields of Mossbauer effect measurements. (It had long been recognized that the latter (Violet and Borg, 1966) yielded much sharper indications of ordering than the specific heat and little indication of the unsplit component expected from the more naive distribution of effective field models for the specific heat. They had also, however, misled some workers into postulating ferromagnetism at concentrations where it was later shown that no spontaneous magnetization existed). Just as the observations of the first rare earth spin-glass susceptibility anomalies were triggered by a different interest (impurities in superconductors), so the first transition-metal spin glass susceptibility cusps were discovered in alloys examined because of their relevance to an interest in critical phenomena in transport properties at magnetic phase transitions.

Interest in critical phenomena had been growing rapidly in the sixties and the manifestation of such effects in the resistivity of pure nickel was first studied in 1967. It seems likely that the choice of Au-Fe as an alloy system for such studies (Mydosh, Kawatra, Budnick, Kitchens and Borg, 1968) was influenced by Borg's Mössbauer studies of this system. The variations in the character of $\rho(T)$ with composition led Cannella and Mydosh (1972) to make A.C. susceptibility measurements. These revealed the sharp cusps which directly stimulated the theorists (Adkins and Rivier, 1974; Edwards and Anderson,

---

* The first striking low field susceptibility maxima were in fact seen in La-Gd (Finnemore et al 1965, 1968) and related alloys (Guertin Crow and Parks 1966) to which attention had been drawn by the then rapidly expanding interest in magnetic impurities in superconductors.

1975) to formulate order parameters. Some of the early theoretical work was dominated by the need to construct mathematically tractable models and methods, and influenced (see Anderson, 1973) by the ideas of localization induced by randomness, but an important input to the theory was also the recognition (Sherrington and Southern 1975) that many of the alloy systems showing spin glass character could be made ferromagnetic by changes of composition or degree of atomic order. There has been, in return, an active period in which experimentalists have sharpened the tools they use to distinguish between short-range and long-range magnetic order (Murani, 1974) and to examine critical character near percolation limits (Sarkissian, 1981).

A glance at the proceedings of recent conferences on magnetism will show the enormous stimulus of the first fully fledged spin glass theories (especially those of Edwards and Anderson (loc. cit.) and of Sherrington and Kirkpatrick, 1975), but in contrast to the now rather settled look of the Kondo problem, the spin glass field is both active and unsettled. Since my main theme has been the impact of experiment in stimulating theory it should be confessed in closing that one of the few major effects of recent experimental work (especially neutron scattering, nuclear magnetic resonance, and other measurements) has been to make unlikely for dilute alloys those theories that invoke cluster moments freezing by dipole interactions or in private anisotropy fields. Strong influences from theory have been encouraging searches for new types of boundaries in J,H,T phase diagrams, but the wealth of data accumulating on spin glass dynamics from various measurements have yet to fertilize a satisfactory theoretical account.

REFERENCES

Abrikosov, A.A. and Gor'kov, L.P., 1960, Zh. Eksp. Teor. Fiz., 39:1781
Adkins, K. and Rivier, N., 1974, J. de Phys. Coll. C4, 35: C4-237
Anderson, P.W., 1973, in "Amorphous Magnetism", H.O. Hooper and A.M. de Graaf, eds. Plenum, New York
Beck, P.A., 1971, Met. Trans. A.I.M.E., 2 : 2015
Blandin, A., 1961, Thesis, University of Paris
Blandin, A. and Friedel, J., 1959, J. Phys. Radium, 20 : 160
Cannella, V. and Mydosh, J.A., 1972, Phys. Rev. B, 6 : 4220
Coles, B.R., 1958, Adv. in Physics, 7 : 40
Edwards, S.F. and Anderson, P.W., 1975, J. Phys. F., 5 : 965
Finnemore, D.K., Johnson, D.L., Ostenson, J.E., Spedding, F.H. and Beaudry, B.J., 1965, Phys. Rev. A., 137 : 550
Finnemore, D.K., Williams, L.J., Spedding, F.H. and Hopkins, D.C., 1968, Phys. Rev. 176 : 712
Friedel, J., 1956, Canad. J. Phys. 34 : 1190
Fröhlich, H. and Nabarro, F.R.N., 1940, Proc. Roy. Soc. A.175 : 382
de Gennes, P.G. and Friedel, J., 1958, J. Phys. Chem. Solids,4 : 71
Gerritsen, A.N. and Linde, J.O., 1951, Physica 17 : 573

Guertin, R.P., Crow, J.E. and Parks, R.D., 1966, Phys. Rev. Letts. 16: 1095
Hart, E.W., 1957, Phys. Rev. 106 : 467
Herring, C., 1958, Physica, 24 : S184
Kasuya, T., 1956, Prog. Theor. Phys. 16:45, 58
Klein, M.W. and Brout, R., 1963, Phys. Rev. 132 : 2412
Korringa, J. and Gerritsen, A.N., 1953, Physica 19 : 457
Kouvel, J.S., 1959, J. appl. Phys., 30 : 313S
Kouvel, J.S., 1960, J. appl. Phys., 31 : 142S
Kouvel, J.S., 1961, J. Phys. Chem. Solids, 21 : 57
Marshall, W., 1960, Phys. Rev., 118 : 1519
Mott, N.F., 1936, Proc. Roy. Soc. A,, 153 : 699 and 156 : 368
Murani, A.P., 1974, J. Phys. F, 4 : 757
Mydosh, J.A., Kawatra, M.P., Budnick, J.I., Kitchens, T.A. and Borg, R.J., 1969, Proc. L.T. XI (St. Andrews, Scotland) 2:1324
Néel, L., 1962, J. Phys. Soc. Japan, Suppl. B1, 676
de Nobel, J. and du Chatenier, F.J., 1958, Physica, 24 : S175
1959, Physica, 25 : 969
Overhauser, A.W., 1960, J. Phys. Chem. Solids, 13 : 71
Owen, J., Browne, M.E., Arp, V. and Kip. A.F., 1957, 2 : 85
Ruderman, M.A. and Kittel, C., 1954, Phys. Rev. 96 : 99
Sarkissian, B.V.B., 1981, J. Phys. F., 11 : 2191
Sato, H., Arrott, A. and Kikuchi, R., 1959, J. Phys. Chem. Solids, 10 : 19
Schmitt, R.W., 1956, Phys. Rev., 103 : 83
Schmitt, R.W. and Jacobs, I.S., 1957, J. Phys. Chem. Solids, 3 : 329
Sherrington, D. and Kirkpatrick, S., 1975, Phys. Rev. Letts. 35 : 1792
Sherrington, D. and Southern, B.W., 1975, J. Phys. F., 5 : L49
Street, R., 1960, J. appl. Phys., 31 : 310S
Street, R. and Smith, J.H., 1959, J. Phys. Radium, 20 : 82
Tournier, R., 1965, Thesis, University of Grenoble
Tournier, R. and Weil, L., 1962, J. Phys. Soc. Japan, 17 : Suppl. B1, 118
Violet, C.E. and Borg, R.J., 1966, Phys. Rev., 149 : 540
Yosida, K., 1957a, Phys. Rev., 106 : 893
Yosida, K., 1957b, Phys. Rev., 107 : 396
Zener, C., 1951, Phys. Rev., 81 : 440
Zimmerman, J.E. and Hoare, F.E., 1960, J. Phys. Chem. Solids, 17 : 52.

PHASE(?) TRANSITIONS IN SYSTEMS WITH SPIN GLASS AND LONG-RANGE ORDER REGIMES

B.R. Coles

Blackett Laboratory
Imperial College
London SW7

Some of the first spin glasses studied were created by slight modifications of the structural parameters of systems possessing long-range order, and it is of interest to look for general features of systems in which a spin-glass transition intersects a line of Curie or Néel temperatures as the concentration of magnetic sites increases in a random A-B solid solution where A is non-magnetic and B magnetic. Metallic systems are of particular interest since fairly high temperature spin glass transitions can be expected to follow from the RKKY coupling of solute B atoms, but straightforward tests of percolation ideas can be complicated by the presence of two families of interactions, nearest neighbour and RKKY. An added problem in metallic systems is that when strong mixing exists between the conduction electrons of metal A and the part-filled electron shell providing the B moments the B atom may not carry a moment when in dilute solution. (We shall not here discuss the question of whether this situation is better described as a non-magnetic ground state of the Friedel-Anderson approach or of the Kondo problem, but it should be borne in mind that in principle a singlet of the crystal field Hamiltonian, the spin-orbit interaction or even the hyperfine interaction can have an effect which is formally very similar.)

Thus there are in general two types of phase diagram, which are shown in Figure 1. In 1(a) a "good" moment exists on B atoms down to quite low concentrations and the spin-glass line g corresponding to the freezing of these moments is easily located. Archetypal examples of this situation are of course Au-Fe and Cu-Mn with l corresponding to ferromagnetism and antiferromagnetism respectively, but the latter is complicated by the fact that by the time the large percolation concentration of $\sim 65\%$ Mn is reached it is almost certainly incorrect to ignore the itinerant character of the Mn 3d

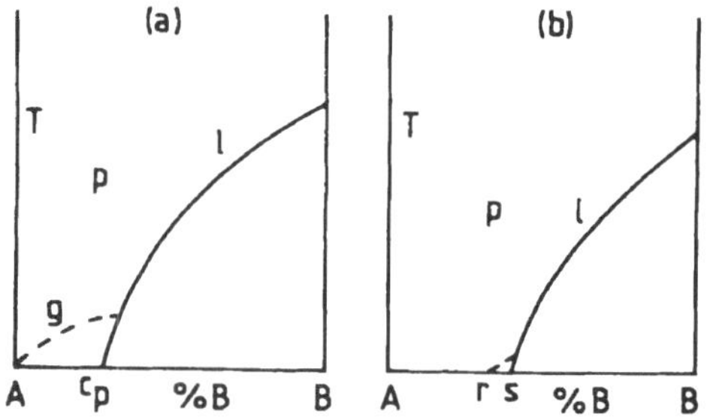

Fig. 1. Types of magnetic phase diagrams for random solid solutions of non-magnetic A with magnetic B.

electrons. There are now examples of this type of diagram for many types of material including $LaAl_2$-$GdAl_2$ (g → ferro Laves phase)[1], Mo-Fe (g → ferro b.c.c.)[2] Y-Gd (g → helix h.c.p.)[3] Pd-Fe (g → ferro in exchange-enhanced host)[4]. In the rare-earth systems both the spin glass and long-range order are sustained by RKKY interactions, but in the 3d systems the nearest neighbour interactions that yield long-range order certainly involve overlap and a fairly clear distinction of these from the RKKY couplings is suggested by the approximation of the critical concentrations in Au-Fe[5] and Cr-Fe[6] to those suggested on percolation arguments. Non-metallic systems with only short-range couplings might be expected to be more straightforward, but, except in special circumstances (e.g. SrS-EuS[7] and CdTe-MnTe[8]), the couplings between finite clusters in the prepercolation range yield spin glass temperatures too low to be conveniently accessible.

In figure 1(b) we show schematically the situation that can arise when a finite concentration, r, of B is required before good moments and spin glass freezing are found. The clearest example of this type is probably Rh-Co[9], but $ZrCo_2$-$ZrFe_2$[10] is also of this type with complications due to non-stoichiometry, and Y-Tb[3] is an example where the dilute non-magnetism is due to a crystal field singlet. The value of r can be quite close to that, s, for the onset of long range order, and in Cu-Ni[11] the behaviour in this composition range is acutely sensitive to chemical clustering. In some such systems based on 4d metals (Rh-Ni and Pd-Ni for example) no spin glass regime has been reported, although they differ very greatly in the critical concentration for ferromagnetism.

There are in principle two distinct ways of studying the character of the order in A-B systems, either by measuring some property as a function of composition (ie lines like 3-4 in Fig. 2(a)) or by examining the temperature-dependence along lines like 1-2 through the transition lines at various compositions. There is a surprising shortage of measurements of the latter type for the specific heat, although, as a zero-field measurement it might be expected to be particularly informative. There are data[12] for certain Co, Zn complex chlorates which show a very rapid loss of sharp critical character along the line 5-6-7, the specific heat being dominated by the short range order long before the critical composition c_p is reached. One should therefore not be surprised that no sharp peak is found in specific heat-temperature curves of spin glasses, and it is not obvious that this is to be used as an argument against phase transitions*. For metallic systems (some semiconducting systems will be discussed later) the only systematic data on specific heats of the 1-2 type are those for Au-Fe[14], and, while there is a very slight rounded maximum at T_c for the 18% Fe alloy, the general behaviour of this alloy is very similar to the cluster glass alloy on the other side of c_p at 15% Fe. (Specific heat results at low temperatures of the 3 - 4 character for this system[15] also show very little change of character with only the weakest suggestion of a maximum at c_p.)

For Au-Fe Sarkissian[16] has made very careful studies of the magnetic response along 1-2 type lines and has shown the vital importance of avoiding demagnetizing field complications. He found

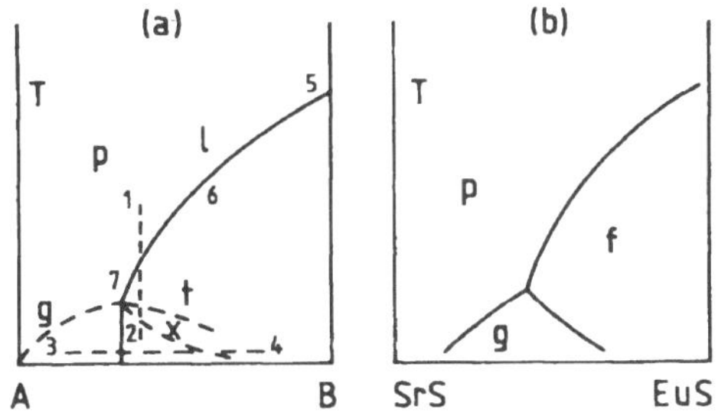

Fig. 2. Types of phase diagram suggested (a) by some theories; (b) for SrS-EuS.

* For a recent discussion of a careful specific heat measurement on a Cu-Mn spin glass see Fogle et al[13].

a very sharp onset of critical character in the A.C. susceptibility, and it would be very interesting to have data of comparable character for an antiferromagnetic system. The results of less detailed small angle neutron scattering studies are in good agreement with magnetic studies both for Au-Fe[17] and Cr-Fe[18]. Since conduction electrons are also a useful probe of magnetic order it would be interesting to have data for these systems like those for Cu-Ni[19], since susceptibility[20] and specific heat[21] studies have also been carried out for this system.

A recent focus of interest in the random magnetic alloy problem has been the possible existence of other phase transitions, either below the spin glass temperature in finite fields or below the Curie temperature in zero applied field. Transition lines of the latter type are shown in Figure 2a and labelled t and x. Attention was first drawn to some line of the X type for AuFe[5] magnetization and A.C. susceptibility results, but there were indications of some such transition in early measurements[22] on the Fe-Al system. Since then a large number of reports has appeared of anomalous behaviour below the Curie temperatures of just ferromagnetic alloys; and, in spite of some attempts to dismiss them as either the trivial onset of normal types of anisotropy or due to the limited dimensionality of the ramified infinite cluster, they have been taken seriously by some workers because of theoretical predictions within Heisenberg mean field theories[23,24,25] of two distinct types of transition, one (t) corresponding to the onset (after longitudinal ferromagnetic ordering at T_c) of a freezing in spin-glass manner of transverse magnetic moment components, and the other to an instability in the molecular field of the spontaneous magnetization of the coupled longitudinal-transverse magnetizations. An effect of the latter sort had first been found theoretically[26] in a classical Ising spin model[27] for spin-glasses with no infinite ferromagnetic cluster but in an applied field, a transition with some formal relationship to the spin-flop transition of a normal antiferromagnet. Very recently Campbell et al[28] have produced fairly convincing evidence of both t and X transitions in a Au 19% Fe alloy, the latter being marked by a sudden onset of strong irreversibility in magnetization curves and the former by the appearance in Mössbauer measurements of an extra frozen component of magnetization rather than just a canting of the static longitudinal component. This question of the character of the low temperature change in just ferromagnetic alloys is by no means settled, however, since other interpretations have included the freezing of co-existent finite clusters[5], the softening of ferromagnetic spin waves[29] and the transition back out of a ferromagnetic state into the phase of spin glass character existing at lower compositions[30], as shown for SrS-EuS in Figure 2(b).

This last interpretation was put forward by Maletta and Convert[30] after observing a decrease in height and increase in width of neutron Bragg peaks in just ferromagnetic SrS-EuS at low temperatures, and was probably influenced by the possibility of such a transition in the Sherrington-Kirkpatrick model[27] of the spin glass for certain ratios of the mean exchange to the spread of exchange values.

This system belongs to a family of semiconducting systems in which a face-centred cubic lattice of magnetic ions is a rich source of frustration. Other systems of this sort are CdTe-MnTe[31] and HgTe-MnTe[32] and it does not yet seem clear whether one can discuss them simply in terms of superexchange via the other ion (as in MnO) or whether a virtual RKKY interaction plays a role. (It seems impossible to ignore the latter in a related system, GeTe-MnTe, which was studied some years earlier[33] and seems to show ferromagnetic character at very low Mn concentrations.) It is clearly desirable to have more experiments on both metallic and semiconducting systems to settle the question of whether a line like 3-4 in Figure 2(a) is in fact crossing a vertical phase boundary. If lines g,t and X are not true phase boundaries, the vertical boundary must exist; if they are the possibility of something like Figure 2(b) is still open.

REFERENCES

1. J.L. Tholence and R. Tournier, J. de Phys. 39:C6-928 (1978)
2. A. Amamou, R. Caudron, P. Costa, J.M. Friedt, F. Gautier and B. Loegel, J. Phys. F,6:2371 (1976)
3. B.V.B. Sarkissian and B.R. Coles, Comm. on Phys. 1:17 (1976)
4. G. Chouteau and R. Tournier, J. de Phys. 32:C1-1002 (1971)
5. B.R. Coles, B.V.B. Sarkissian and R.H. Taylor, Phil. Mag.B 37:489 (1978)
6. S.K. Burke, R. Cywinski, J.R. Davis and B.D. Rainford, J. Phys.F 13:451 (1983)
7. H. Maletta and W. Felsch, Z. Phys. B37:55 (1980)
8. M. Escorne, A. Mauger, R. Triboulet and J.L. Tholence, Physica 107B:309 (1981)
9. B.R. Coles, A. Tari and H.C. Jamieson, in "Low Temperature Physics L.T.13"
10. G. Wiesinger and G. Hilscher, J. Phys. F 12:497 (1982)
11. C.J. Tranchita and H. Claus, Solid State Comms. 27:583 (1978)
12. H.A. Algra, L.J. de Jongh and K.J. Reedijk, Phys. Rev. Letts. 42: 606 (1979)
13. W.E. Fogle, J.D. Boyer, N.E. Phillips and J. Van Curen, Physica 107B:633 (1981)
14. J.W. Loram and K. Mirza, to be published
15. D.G. Dawes and B.R. Coles, J. Phys. F9 : L215 (1979)
16. B.V.B. Sarkissian, J. Phys. F11:2191 (1981)
17. A.P. Murani, S. Roth, P. Radhakrishna, B.D. Rainford, B.R. Coles, K. Ibel, G. Goeltz and F. Mezei, J. Phys. F6:425 (1976)
18. S.K. Burke and B.D. Rainford, J. Phys. F13:471 (1983)

19. J.B. Sousa, M.R. Chaves, M.F. Pinheiro and R.S. Pinto, J. Low Temp. Phys. 18:125 (1975)
20. H. Boghossian, B.V.B. Sarkissian and B.R. Coles, in "Physics of Transition Metals 190", P. Rhodes, ed. Inst. of Physics, Bristol (1981)
21. Z. Chen and J.W. Loram, Physica 107B:101 (1981)
22. R.D. Shull, H. Okamoto and P.A. Beck, Solid State Comms. 20:863 (1976)
23. G. Toulouse, J. de Phys. Lettres, 41:L447 (1980)
24. M. Gabay and G. Toulouse, Phys. Rev. Letts. 47:201 (1981)
25. D.M. Cragg, D. Sherrington and M. Gabay, Phys. Rev. Letts. 49: 158 (1982)
26. J.R.L. de Almeida and D.J. Thouless, J. Phys. A 11:983 (1978)
27. D. Sherrington and S. Kirkpatrick, Phys. Rev. Letts. 35:1792 (1975)
28. I.A. Campbell, S. Senoussi, F. Varret, J. Teillet and A. Hamzic, to be published
29. C.R. Fincher, S.M. Shapiro, A.C. Palumbo and R.D. Parks, Phys. Rev. Letts. 45:474 (1980)
30. H. Maletta and P. Convert, Phys. Rev. Letts. 42:108 (1979)
31. R.R. Galazka, S. Nagata and P.H. Keesom, Phys. Rev. B22:3344 (1980)
32. S. Nagata, R.R. Galazka, G.D. Khattak, C.D. Amarasekara, J.K. Furdyna and P.H. Keesom, Physica 107B:311 (1981)
33. R.W. Cochrane, M. Plischke and J.O. Ström-Olsen, Phys. Rev. B9: 3013 (1974)

TIME EFFECTS OF THE FIELD COOLED MAGNETIZATION OF A Au(Fe) SPIN GLASS

P. Nordblad, P. Svedlindh, L. Lundgren and O. Beckman

Division of Solid State Physics, Institute of Technology
University of Uppsala, Box 534, S-751 21 Uppsala, Sweden

We have studied the field cooled (FC) as well as the zero field cooled (ZFC) magnetization of a Au - 8 at% Fe spin glass using a SQUID magnetometer.

The FC measurements were performed by cooling the sample in a static field (0.1 - 1300 Gauss) in temperature steps of appr. 0.1 K and the change of the magnetization with time (up to 1000 secs) was recorded at constant temperature. The relative change of the magnetization with time at 100 sec is shown in Fig. 1 at various applied fields. (Fig. 2 shows the field dependence of the spin glass freezing

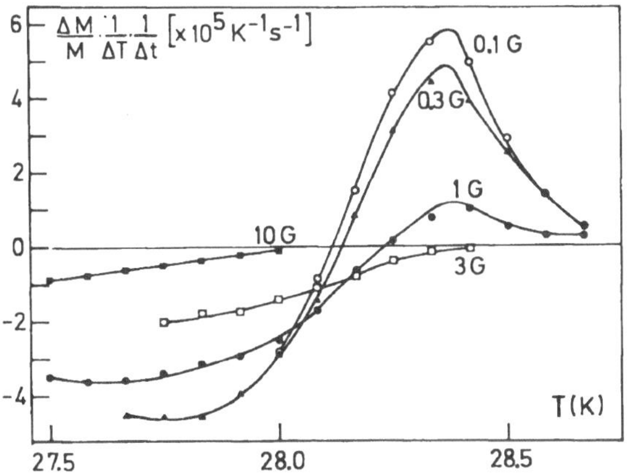

Fig. 1. Time dependence of the FC magnetization at 100 sec.

temperature, T_g, defined as the 'cusp' of the FC magnetization). In the low field regime (< 3 Gauss) there is an increase of the magnetization with time at temperatures above T_g. For temperatures below T_g the magnetization decreases with time. At applied fields above 3 Gauss there is no detectable change of the magnetization with time above T_g. However, at some temperature below T_g the magnetization will decrease with time. In Fig. 2 we have plotted the 100 sec FC line in an H-T diagram representing the onset of a decrease of the magnetization with time at an observation time of 100 sec as found from the graphs in Fig. 1.

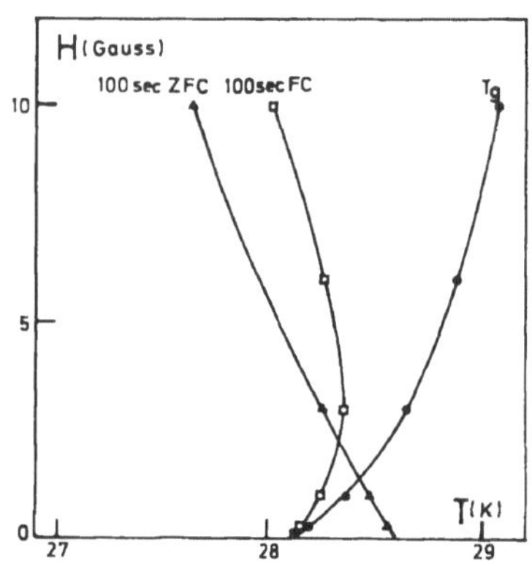

Fig. 2. 100 sec ZFC, 100 sec FC and T_g plotted in an H-T diagram. The 100 sec ZFC line represents the disappearance of time effects of the zero field cooled magnetization at an observation time of 100 sec. The 100 sec FC line corresponds to the onset of a decrease of the field cooled magnetization at an observation time of 100 sec.

The ZFC measurements were performed by cooling the sample in zero field from a temperature above T_g. The field was then applied and the change of the magnetization with time recorded at constant temperature. The time effects of the ZFC magnetization are well interpreted within a phenomenological model prescribing a wide distribution of relaxation times for the spin clusters[1-3]. In this model it is assumed that the spectrum exhibits a reasonably well defined maximum value τ_{max}. As the temperature is lowered the spectrum evolves towards longer relaxation times. In the ZFC experiment the increase of the magnetization with time is governed by a polarization of cluster moments with relaxation times shorter than the observation time and the relaxation effects disappear when the observation time equals τ_{max}. In Fig. 2 we have plotted the 100 sec ZFC line in an H-T diagram which corresponds to the disappearance of relaxation at an observation time of 100 sec and is thus equivalent to the line at which τ_{max} equals 100 sec. Fig. 2 also shows how T_g varies with applied field. The ZFC and T_g lines cross at H = 1.4 Gauss and T_g = 28.44 K. At temperatures above T_g (at H = 1.4 Gauss) the equilibrium value of the ZFC magnetization, reached in a shorter time than 100 sec, displays a C/T dependence. For $T < T_g$ the equilibrium value of the ZFC magnetization, reached after a longer time than 100 sec, gradually decreases with decreasing temperature from the maximum value C/T_g at T_g. For every field there is a charactaristic time, where the ZFC magnetization attains the maximum value C/T_g at T_g. It was previously proposed by Lundgren et al.[3] that this characteristic time is coupled to a critical value of the cluster relaxation times, τ_{crit}, where intercluster interaction becomes strong enough to cause cluster recombinations. These cluster recombinations give a reduced net magnetization of the sample causing the deviation of the equilibrium magnetization from a C/T dependence.[3] τ_{crit} was estimated to be 10^5 sec at zero field, found to be $5 \cdot 10^2$ sec at 1 Gauss and, as is shown in Fig. 2, 100 sec at 1.4 Gauss.

In the ZFC experiment the total change of the magnetization will go from zero to $\sim C/T$ $(T > T_g)$ at equilibrium. Looking at the FC magnetization we also expect an increase of the magnetization with time at least at temperatures and fields where the observation time is shorter than τ_{crit}. However, the magnitude of this change being considerably smaller $(\sim(\Delta T/T)C/T)$ than in the ZFC experiment. We are certainly able to observe this effect at fields smaller than 3 Gauss (see Fig. 1). At higher fields this increase is masked by a decrease as described experimentally above. This decrease can be assumed to arise from rearrangements of the spins within the large recombined spin clusters mentioned above.

These experiments show that the FC magnetization is not a true equilibrium state in the thermodynamic sense. The equilibrium lying somewhat below the FC magnetization curve. It is therefore of importance to study the approach to equilibrium both from below (ZFC experiments) and from above (FC experiments) of different types of spin glass systems to obtain information about the true equilibrium state for spin glasses.

REFERENCES

1. L. Lundgren, P. Svedlindh, and O. Beckman, Measurement of complex susceptibility on a metallic spin glass with broad relaxation spectrum, J. M. M. M. 25:33(1981).
2. L. Lundgren, P. Svedlindh, and O. Beckman, Low-field susceptibility measurements on a Au-Fe spin glass, J. Phys. F 12:2663(1982).
3. L. Lundgren, P. Svedlindh, and O. Beckman, Experimental indications for a critical relaxation time in spin glasses, Phys. Rev. B 26:3990(1982).

SPIN GLASSES: RECENT THEORETICAL DEVELOPMENTS

A. P. Young

Department of Mathematics
Imperial College
London SW7 2BZ
U.K.

INTRODUCTION

The considerable recent interest in phase transitions in random magnetic systems is reflected in there being several lectures on this topic presented here (Aharony, Coles, Cowley and myself). From a theoretical point of view, probably the simplest model which can describe the properties of most of these systems has the Hamiltonian

$$H = -\sum_{\langle ij\rangle} J_{ij}\, \underline{S}_i \cdot \underline{S}_j - \sum_i \underline{h}_i \cdot \underline{S}_i \tag{1}$$

where $\underline{S}_i$ is an m-component classical spin on site i of a d-dimensional lattice, J_{ij} is a pair interaction and $\underline{h}_i$ is a local field. If $\underline{h}_i$ is independent of site i (a uniform field) and J_{ij} are random, the model can represent a system with two types of magnetic atoms A and B. If one species in non-magnetic then ordering disappears when the concentration of this species exceeds the percolation concentration (discussed by Aharony). This type of behaviour is fairly well understood. Another very relevant case is where the J_{ij} are not random but the local fields are disordered, and on average, point equally in all directions. Aharony and Cowley will discuss this more controversial topic. Spin glasses are characterised by extreme disorder in the interactions J_{ij} such that both signs occur with significant probability and in a random way. Professor Coles will describe systems in nature which have this property.

A simple looking spin glass model has been proposed by Edwards and Anderson [1] (EA). They consider a spin on each lattice site, as in eq. (1), and assume each interaction is an independent random variable with a probability distribution $P(J_{ij})$, and frequently a symmetric distribution such as a Gaussian

$$P(J_{ij}) = (2\pi J)^{-\frac{1}{2}} \exp[-J_{ij}^2/2J^2] \tag{2}$$

or a binary distribution

$$P(J_{ij}) = \tfrac{1}{2}[\delta(J_{ij} - J) + \delta(J_{ij} + J)] \tag{3}$$

is used. It will, however, be necessary to include an asymmetric distribution to describe the transition between spin glass and ferromagnetic behaviour. Frequently Ising spins (m=1) where $S_i=\pm 1$, are taken to further simplify the problem.

To get a feeling for the effects that a random sign in the interaction may have, consider the single square, with Ising spins at the corners shown in fig. 1. One of the interactions is -J the others +J. Suppose we fix spins A, B and C to be ↑. Then spin D receives conflicting instructions. From A it wants to be ↓ but from C it wants to be ↑. This is known as 'frustration' and always occurs when the product of the interactions around a loop is negative. Professor Toulouse will discuss this concept more in his lectures. From this trivial example we can learn two things which are important for a large system. First of all it is highly non trivial to determine the ground state when there is a frustration since it is impossible to minimise separately the energy of each interaction. Secondly there are many ways of making the best possible compromise. This means that there will be many states of energy equal to, or very close to, the ground state energy but which have very different spin configurations.

Professor Coles will discuss the experimental situation in detail in his lectures. Here I just want to mention a couple of basic experimental facts to provide motivation for the theory. The magnetic susceptibility, χ, has a rather abrupt change in behaviour

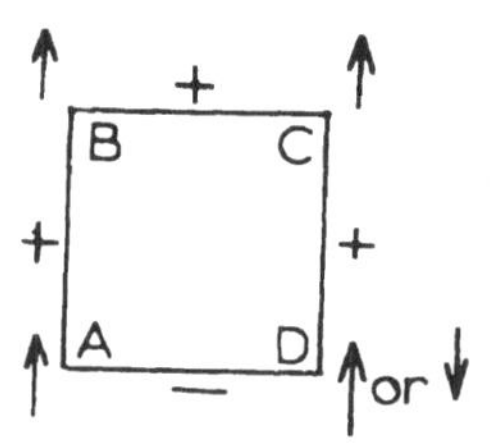

Figure 1. A single square with spins at the corners and one -ve interaction (between C and D), which illustrates the concept of frustration.

at the 'freezing temperature' T_f and, below T_f, there are very slow decay processes and irreversibility [2]. Also the non-linear susceptibility, $\chi_{n\ell}$, which is defined by

$$m = \chi h - \chi_{n\ell} h^3 + \ldots , \quad (4)$$

where m is the magnetisation per spin, increases very dramatically [3] near T_f and possibly diverges, although the experiments cannot prove conclusively that a real divergence occurs.

Historically there have been two views of the freezing observed in χ. The oldest is the hypothesis [4] that relaxation times smoothly increase as the temperature, T, is lowered and that T_f is simply the temperature where the relaxation time equals the measuring time. This approach predicts that the freezing temperature varies with the logarithm of the frequency, ν, i.e.

$$\frac{\nu}{T_f}\frac{dT_f}{d\nu} = \frac{1}{\ell n(\nu_o/\nu)} \quad (5)$$

where ν_0 is a constant. However, it is found that, at least for some systems[5], very little frequency dependence is observed and ν/ν_0 is much smaller than any reasonable theoretical estimate. Furthermore this gradual freezing hypothesis has no explanation for the rapid growth of $\chi_{n\ell}$ at T_f.

The alternative picture, due to Edwards and Anderson [1], is that a sharp phase transition occurs at T_f to a phase where the spins are locked in apparently random directions but each spin stays for infinite time with a definite preferred orientation. This effect certainly occurs in a mean field theory (MFT) as we shall discuss in some detail below. The problem is that several subsequent calculations [6] predict that the lower critical dimension (LCD) for the spin glass problem is four. The LCD is that dimension below which fluctuations, neglected in MFT, destroy the ordering that MFT predicts. Consequently no transition should occur in d=3.

We see that problems arise in both of the theoretical approaches to the spin glass problem. In these lectures I propose to cover two topics. First of all I shall describe the MFT. This was in a confused state for a long time but the difficulties with it have recently been resolved. Secondly I shall discuss relevant work on both the statics and dynamics of models with short range interactions with a view to understanding experiments. Unfortunately there is much less progress to report on this than on the MFT.

MEAN FIELD THEORY

The MFT of ferromagnetism has been very successful in describing the behaviour of many systems, except in a small temperature range close to the critical point. It is natural to strive for an analogous approximation in spin glasses. Unfortunately it is not immediately obvious what this approximation should be. Sherrington and Kirkpatrick [7] (referred to as SK) proposed that since MFT for ferromagnetism is exact in the infinite range limit (i.e. each spin interacts equally with every other spin) one should take MFT for spin glasses to be the exact solution of an infinite range model where the distribution $P(J_{ij})$ is the same for all pairs of spins, i and j, no matter how far apart they are. It is important to realise that the dimensionality of the space in which the spins are positioned plays no role in this model. In order to have a non trivial thermodynamic limit both the mean and the variance of the distribution have to be scaled by N^{-1}, where N is the number of spins. We shall only consider Ising spins so the model has a Hamiltonian, H, given by

$$H = -\sum_{<ij>} J_{ij} S_i S_j - h \sum_i S_i \qquad (6)$$

where $S_i = \pm 1$ and

$$P(J_{ij}) = (N/2\pi J)^{\frac{1}{2}} \exp[-NJ_{ij}{}^2/2J^2] \qquad (7)$$

for all i and j. In eq. (7) the mean of the distribution has been set equal to zero. The free energy, F, is given by

$$F = -T < \ell n [\sum_{\{S_i = \pm 1\}} \exp(-H/T] >_J \qquad (8)$$

where $< ... >_J$ denotes an average over many samples with different interactions taken from the same distribution, eq. (7). This average is often carried out by the 'replica trick' where $<\ell n Z>_J$ (Z is the partition function) is written as

$$<\ell n Z>_J = \lim_{n \to 0} \frac{1}{n} (<Z^n>_J - 1) \qquad (9)$$

One evaluates $<Z^n>_J$ for integer n by performing the J_{ij} average before the spin trace and one then takes the limit $n \to 0$. The spin glass order parameter $q^{\alpha\beta}$ is defined to be

$$q^{\alpha\beta} = <S_i^{\alpha} S_i^{\beta}> \qquad (\alpha \neq \beta) \qquad (10)$$

where $< ... >$ indicates a statistical mechanics average over the non-disordered replica Hamiltonian H_{eff}, defined by $\exp[-H_{eff}/T] = \langle \exp -\sum_{\alpha=1} H_\alpha/T \rangle_J$, and $\alpha(=1 \ldots n)$ and β specify replicas. One can also show that [7,8]

$$q = \langle \langle S_i \rangle_T^2 \rangle_J = q^{\alpha\beta} \tag{11}$$

for any $\alpha \neq \beta$, where $\langle \ldots \rangle_T$ indicates a statistical mechanics average for the disordered Hamiltonian of eq. (6). SK assumed that all the $q^{\alpha\beta}$ $(\alpha \neq \beta)$ are equal to q and obtained a self consistent expression for this order parameter. They found $q \to 1$ as $T \to 0$ and (with $h = 0$) $q \propto (T-T_c)$ as $T \to T_c = J$, the transition temperature.

Subsequently Almeida and Thouless [9] (AT) showed that the replica symmetric solution of SK is unstable if $h < h_c(T)$ where $h_c(T) \propto (T_c-T)^{3/2}$ for $T \to T_c$ and $h_c(T) \to \infty$ as $T \to 0$. Below this AT line one has to 'break replica symmetry', i.e. $q^{\alpha\beta}$ depends on α and β. Parisi [10] has proposed a very interesting such scheme involving an infinity of order parameters which can be represented by a function q(x) with $0 \leq x \leq 1$. Sums over replicas go over into integrals over x, i.e.

$$\lim_{n \to 0} \frac{1}{n(n-1)} \sum_{\alpha \neq \beta} q^{\alpha\beta} = \int_0^1 q(x)\, dx \tag{12}$$

For $h \to 0$, $q(x = 0) = 0$, $\int_0^1 q(x)dx = 1 - T/T_c$ and for all h, q (x) is monotonic so q (1) is the largest value. Fig. 2 shows q (1), $\int_0^1 q(x)dx$

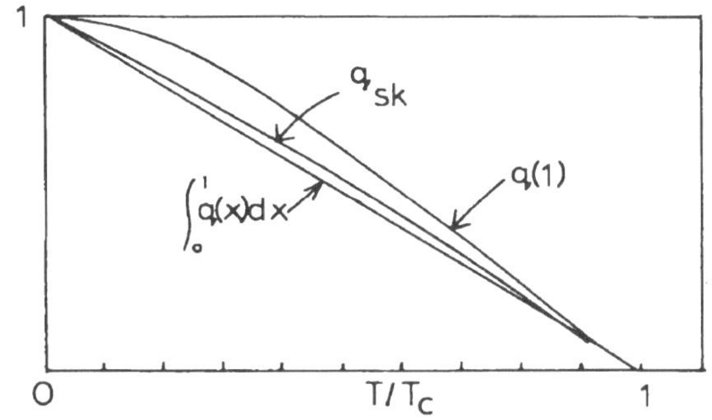

Fig. 2. Plots $\int_0^1 q(x)\, dx$ and $q(x=1)$ from Parisi's theory against T/T_c for $h = 0$. Also shown is q_{SK}, the prediction of the Sk theory.

and q_{SK} (the SK prediction). The obvious questions to ask are (i) what is the significance of the variable 'x', and (ii) with a suitable interpretation of x, is the Parisi theory the exact solution of the model? A partial answer to the first question is obtained from the following argument [8]. The replica Hamiltonian H_{eff}, is invariant under permutation of the replicas whereas Parisi's solution for $q^{\alpha\beta}$ breaks this symmetry. There must therefore be other distinct but equivalent solutions obtained from Parisi's by permuting replicas and we should include all of then in a statistical mechanics average. This restores replica symmetry. Hence q in eq. (11) is $q^{\alpha\beta}$ for any $\alpha \neq \beta$ but averaged over distinct solutions, which is equivalent to taking one solution and averaging over replicas, i.e.

$$q = \lim_{n \to 0} \frac{1}{n(n-1)} \sum_{\alpha \neq \beta} q^{\alpha\beta} , \tag{13}$$

where the $q^{\alpha\beta}$ refer to one solution, and this is just $\int_0^1 q(x)\,dx$ by eq. (12). To understand the significance of q(x) for a particular x it is necessary to discuss a different approach, proposed by Thouless et al. [11] (TAP).

They considered a single sample and derived equations [12] for the magnetisations m_i which read

$$m_i = \tanh\left[\frac{1}{T}\left\{\sum_j J_{ij} m_j + h - m_i \sum_j \frac{J_{ij}^2}{T}(1 - m_j^2)\right\}\right] . \tag{14}$$

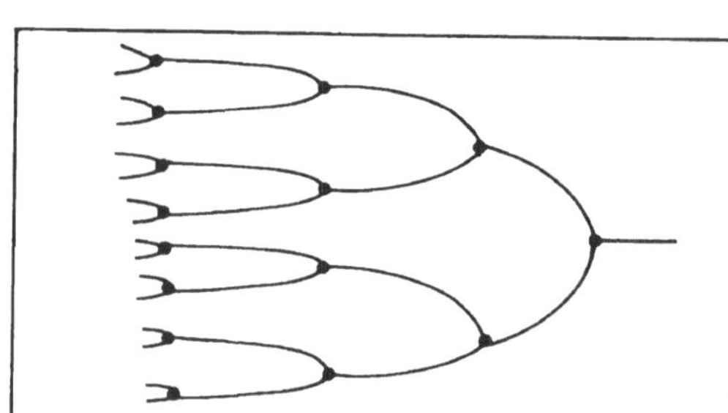

Fig. 3. Schematic sketch indicating successive bifurcations of TAP solutions as the temperature is lowered.

One then has to solve these equations for a given $\{J_{ij}\}$ and carry out the bond average. In fact the problem is even more complicated because there are many [13] solutions of these equations below the AT line. To be precise the number of solutions, N_s, is given by

$$N_s \propto \exp[N\alpha(T,h)] \tag{15}$$

where $\alpha(T,h) = 0$ for $h > h_c(T)$ (only one TAP solution) and $\alpha(0,0) \approx 0.2$. As the temperature is reduced at fixed h the number of solutions increases. Thus solutions keep 'bifurcating' as the temperature is lowered [14], see fig. 3. In a sense, therefore, everywhere below the AT line is a critical point. For the TAP solutions to exist as separate entities at finite T they must be separated by infinitely high barriers when $N \to \infty$, otherwise thermal fluctuations would mix solutions together into one solution. We may, therefore, call each TAP solution a stable 'phase'.

Statistical mechanics averages must, therefore, involve a weighted average over phases. Denoting a particular phase by 's' then the corresponding weight is given by [8]

$$P(s) = \exp[-F_s/T] \,/\, Z \tag{16}$$

where F_s is the free energy of the TAP solution and $Z = \sum_s \exp(-F_s/T)$. $m_i = \langle S_i \rangle_T$, is given by $m_i = \sum_s m_i{}^s P(s)$ where $m_i{}^s$ is the magnetisation on site i for phase 's'. The statistical mechanics order parameter, q, eq. (11) is, therefore, given by

$$q = \langle \sum_{s,s'} P(s)\, P(s')\, q^{ss'} \rangle_J \tag{17}$$

where

$$q^{ss'} = N^{-1} \sum_i m_i{}^s\, m_i{}^{s'} \tag{18}$$

is the overlap between solutions s and s'. One can also define the Edwards-Anderson order parameter as the value of $(m_i{}^s)^2$ in one phase averaged over phases, i.e.

$$q_{EA} = \sum_s P(s)\, q^{ss} \quad . \tag{19}$$

Below the AT line, where many phases occur one has $q_{EA} > q$.

We can now obtain [15] a direct interpretation of Parisi variable x by comparing eqs. (12), (13) and (17). This gives

$$\frac{dx}{d\hat{q}} = < \sum_{s,s'} P(s)\, P(s')\, \delta(q^{ss'} - \hat{q})> , \tag{20}$$

which is just the probability that correctly weighted solutions have overlap given by $\hat{q}$. In other words the inverse function x(q) is more significant than q(x). We shall call eqs. (13) and (20) the replica interpretation of x. One can also show that

$$q_{EA} = q(1) . \tag{21}$$

If one follows the dynamics of the system one can only obtain statistical mechanics averages over times longer than the time to hop over barriers between phases. Since these barriers diverge for $N\to\infty$ one expects that relaxation times also diverge in this limit, i.e. the SK model is non-ergodic. Direct numerical evidence for this has been obtained [16]. Sompolinsky [17] has given a different interpretation of Parisi's x in terms of a sequence of these time scales for crossing barriers. This picture agrees with eq. (21) for q_{EA} but predicts that $q = q(0)$ (=0 for $h\to 0$) instead of eqs. (13) and (12). Unlike eq. (13), the prediction $q = q(0)$ leads to a violation of linear response theory results such as $T\chi_{ii} = 1-q$, where χ_{ii} is the local susceptibility evaluated by statistical mechanics.

Recently numerical simulations [18] have been undertaken to test out whether Parisi's theory is correct with a suitable interpretation of x. Results for q and $q^{(2)} = \langle\langle S_i S_j\rangle_T^{\,2}\rangle_J$ $(= \int_0^1 q^2(x)\,dx$ in the

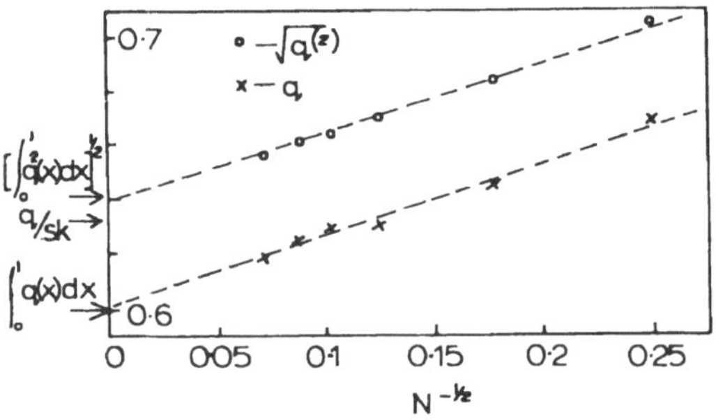

Fig. 4. Numerical data from ref. 18 for q and $\sqrt{q(2)}$ at T = 0.4, h = 0, plotted against $N^{-\frac{1}{2}}$ for various sizes. The results appear to converge well to the predictions of the Parisi theory, $\int q(x)$ and $[\int q^2(x)dx]^{\frac{1}{2}}$, respectively.

replica interpretation of x) at $T = 0.4$ and $h = 0$ are shown in fig. 4 for different sizes and appear to be converging to Parisi's results with the replica interpretation of x. The numerical data does not appear to be consistent with Sompolinsky's dynamical interpretation of x. In his theory Sompolinsky uses equations which are only valid for $N = \infty$ to describe fluctuations over barriers (which only occur for finite N). This may account for the differences between his results and those of the replica interpretation.

To summarise the SK model is characterised by many phase with infinite barriers between them. Statistical mechanics involves a weighted average over (in principle) all the phases and gives different results from a time average which involves only one phase. The Parisi theory with the replica interpretation of x appears to be the exact solution. It would be very desirable to rederive Parisi's equations directly from the TAP formalism without replicas. We have only discussed the SK model with Ising spins. A similar picture holds for vector spins but, in addition, there are interesting effects associated with transverse ordering,[19] which we do not have space to discuss.

FLUCTUATIONS

The MFT presented above is rather complicated in the non-trivial phase below the AT line but is simple above that line where the SK solution is correct. In particular the paramagnetic phase, $T > T_c$ $h = 0$, is fairly trivial, so we can set about including fluctuation effects, which appear when the range of interactions becomes finite, through the renormalisation group[20] (RG) approach. Since the order parameter is $q^{\alpha\beta} = \langle S^\alpha S^\beta \rangle$ (for Ising spins) it is necessary to obtain a Ginsburg-Landau Wilson (GLW) effective Hamiltonian in terms of composite fields $S_i^\alpha S_i^\beta$. This is achieved by writing

$$Z = \int_{-\infty}^{\infty} \prod_i \prod_{\alpha<\beta} [dQ_i^{\alpha\beta}\, \delta(Q_i^{\alpha\beta} - S_i^\alpha S_i^\beta)] \sum_{\{S_i^\alpha = \pm 1\}} \exp[-H_{eff}/T] \qquad (22)$$

The trace over the spins can be carried out to give the GLW Hamiltonian in terms of continuous variables $Q^{\alpha\beta}$ as[21]

$$H_{GLW} = \int d^d x \{\tfrac{1}{2} [r + \nabla^2] \, \mathrm{Tr}\, Q^2(x) + w \mathrm{Tr} Q^3 + \ldots\} \qquad (23)$$

where a continum approximation has also been made. In ferromagnetic systems MFT predicts critical exponents correctly if d is greater than the upper critical dimension (UCD), which is equal to 4 in that case. The spin glass problem is different in that the lowest order interaction term in eq. (23) is cubic in the fields Q, whereas for ferromagnets one has a quartic interaction. The same dimensional arguments which give an UCD of 4 with quartic coupling predict an UCD of 6 if there is a cubic term. One can therefore obtain critical exponents in 6-ε dimensions as an expansion in ε. For instance, the exponent β, which describes the growth of ordering below T_c, i.e. $q = \langle Q^{\alpha\beta} \rangle \sim (T_c - T)^{\beta}$, is given by [21]

$$\beta = 1 + \left(\frac{m+1}{2m-1}\right) \frac{\varepsilon}{4} + O(\varepsilon^2) \qquad (24)$$

where we have generalised H_{GLW} for m-component vector spins. β is predicted to be greater than its mean field value, which is rather unusual. Similarly the exponent η, which describes the decay power law correlations at the critical point, is negative, which is also rather strange. Even more bizarre values for exponents will appear below.

Professor Coles describes in his lectures how ferromagnetism, spin glass behaviour and paramagnetism can all occur as the temperature and concentration of one of the atomic species are varied. In the Edwards-Anderson model concentration variations can be mimicked by varying the ratio of the mean to the variance of the distribution. For the infinite range (SK) limit the phase diagram is shown in fig. 5. A multicritical point, X, where spin glass, ferromagnetic and paramagnetic phases come together is found. Experiments also find such a multicritical point. It is clearly, then of interest to evaluate corrections to critical exponents at X. .

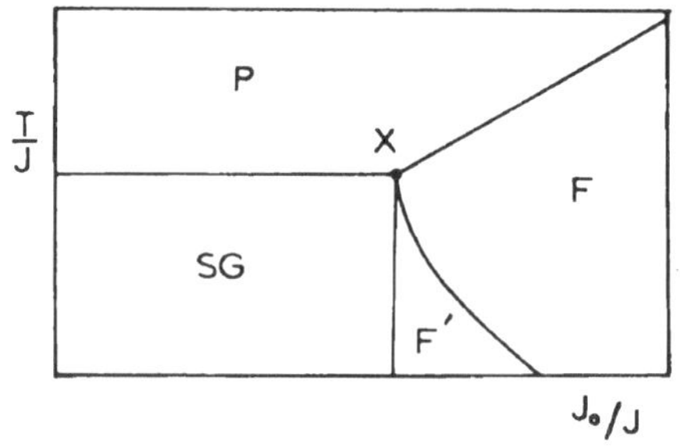

Fig. 5. Sketch the phase boundary of the Sk model in terms of variance J/N, and mean, J_0/N, of the exchange distribution. P, F and SG denote paramagnetic ferromagnetic and spin glass phases respectively. F' is a ferromagnetic phase with replica symmetry breaking (i.e. is non-ergodic).

With $h = 0$ there should be two 'relevant' exponents, y_1 and $y_2 (< y_1)$ $y_1 = \nu^{-1}$ (ν describes the divergence of the correlation length as X is approached from the paramagnetic phase), and the ratio y_2/y_1 gives the shape of the critical lines close to X. Unfortunately for $m = 2$ and 3 both y_1 and y_2 are found to be complex [22]. Taken literally this would predict a spiralling in of the phase boundaries towards X, which is surely unphysical. This result is most perplexing since the calculation is performed in the paramagnetic phase, where everything appears to be understood and there is no problem with replica symmetry breaking. It appears that the ε expansion does not make any sense for the multicritical point X, and possibly also for the ordinary spin glass transition. This may mean that the behaviour for $d < 6$ is qualitatively different from that for $d > 6$ (perhaps a Kosterlitz-Thouless [23] type phase with an infinite correlation length but $q = 0$?).

In fact this question may be rather academic. Another important concept is that of the lower critical dimension (LCD) below which no transition occurs. To be precise we mean that correlations are of finite range, except possibly at $T = 0$, for d below the LCD. For ferromagnetism the LCD is 1 for Ising spins and 2 for vector spins with continuous rotational symmetry. The first calculation of the LCD for spin glasses was done by Fisch and Harris [24]. They calculated the spin glass susceptibility χ_{SG}, defined by

$$\chi_{SG} = N^{-1} \sum_{i,j} \langle\langle S_i S_j \rangle_T^2 \rangle_J \tag{25}$$

by means of a high temperature series expansion in powers of $\tanh^2(J/T)$ for an Ising model on a d-dimensional simple cubic lattice with nearest neighbour interactions having the '$\pm J$' distribution of eq. (3). χ_{SG} is the response function which diverges at a spin glass transition just as the uniform susceptibility, χ, diverges at a ferromagnetic critical point. For the model considered by Fisch and Harris, the non-linear susceptibility, χ_{SG}, defined by eq. (4), which is observed experimentally [3] to become very large, is given by $\chi_{n\ell} = \chi_{SG}/(3T^3)$. Fisch and Harris found a divergence with a mean field exponent, $\gamma = 1$, for $d > 6$ as expected, non mean field exponents for $4 < d < 6$ and no divergence at all for $d < 4$, implying the LCD is 4. Actually the series is rather irregular for $d \lesssim 4$ and different methods of analysing it give different results, some even predicting a transition in $d = 3$ (R. Palmer, private communication), so the situation is not completely clearcut.

If the system is sufficiently small, and for Ising spins, one can perform the statistical mechanics average, $\langle \ldots \rangle_T$ exactly. By an ingeneous transfer matrix technique Morgenstern and Binder [25] (MB) have been able to carry this out for samples as large as 18×18 in $d = 2$ and $4 \times 4 \times 10$ in $d = 3$. MB calculated the correlation function

$$\Gamma(R_{ij}) = \langle\langle S_i S_j \rangle_T^2 \rangle_J , \tag{26}$$

where R_{ij} is the distance between sites i and j. They found that $\Gamma(R_{ij}) \propto \exp(-R_{ij}/\xi)$ where the correlation length ξ increases smoothly as T decreases and apparently diverges only as $T \to 0$. Subsequent Monte-Carlo simulations [26] on much larger lattices, up to 128^2, have confirmed qualitatively this prediction for $d = 2$, although MB somewhat underestimate ξ because of finite size effects. Since the linear dimensions used by MB in their $d = 3$ calculations are very small one expects that finite size effects are even more important here than in $d = 2$. Once again, then, the $d = 3$ situation is not completely clearcut but MB's results are certainly consistent with an LCD equal to 4.

DYNAMICS

If the theoretical predictions that the LCD is 4 are correct one presumably has to turn to the older picture of 'gradual freezing'. It is therefore important to understand dynamics as well as statistical mechanics. One particularly useful tool for studying dynamics is Monte-Carlo (MC) simulations. We shall only discuss simulations on Ising spins, which are the most convenient for numerical work. The MC method flips spins with a certain probability designed to bring the system to equilibrium. Once in equilibrium both static quantities and time (in units of MC steps per spin) dependent correlation functions can be evaluated. This stochastic dynamics represents coupling of the spins to other degrees of freedom (such as phonons) which act as a heat bath. For vector spins there would be additional dynamics, coming from spin precession, which is not included in the MC simulation. The most natural dynamical correlation function to study is the auto-correlation function q(t), which, for Ising spins, is given by

$$q(t) = N^{-1} \sum_{i=1}^{N} \langle\langle S_i(t_0)\, S_i(t_0 + t) \rangle_T \rangle_J , \tag{27}$$

where the average $\langle \ldots \rangle_T$ is now performed by making several runs of the MC simulation, and t_0 is a sufficiently large number of MC steps that the system has settled down to equilibrium. The easiest model to study is the nearest neighbour EA model with either a Gaussian or '$\pm J$' distribution on a square lattice. Early work [27] showed clearly a rapid increase in relaxation times at lower temperatures but was not sufficiently accurate to give a quantitative description. In order to be more precise it is useful to have a single parameter which characterises the decay of q(t) independent of its form (which changes from roughly exponential to roughly logarithmic as the temperature is reduced). Since $q(t) \to 0$ as $t \to \infty$ for $h = 0$ (if we are above any possible transition) then, in an earlier NATO school, Kirkpatrick [28] suggested calculating

$$\tau = \int_0^\infty q(t)\, dt , \tag{28}$$

τ can be thought of as an average relaxation time [26]. It is found the $\ln \tau$ varies with T more smoothly than τ itself, suggesting that relaxation is by thermal activation over barriers. Defining a characteristic barrier height, ΔE, by $\Delta E = T \ln \tau$ then some recent fairly precise, results [26] for a $\pm$ J symmetric distribution on 64^2 and 128^2 lattices are shown in fig. 6. It is not possible to go to lower temperatures because the system cannot be equilibrated within available computer time. Over the temperature range studied one has $\Delta E = a + b/T$ which shows no sign of divergence (transition) except at $T = 0$. The growth of ΔE at smaller T is presumably related to the increase in the range of spatial correlations, ξ, which also apparently diverges [25,26] as $T \to 0$. A temperature dependent ΔE causes relaxation times to increase more rapidly than a simple Arrhenius law ($\Delta E = $ const.) but is insufficient [26] to explain the virtual independence of freezing temperature with frequency observed [5] in certain materials at least. The growth of ξ and χ_{SG} is also not sufficiently rapid [26] to explain the experimental data [3] on $\chi_{n\ell}$. There are not yet such detailed simulations on a 3-dimensional model but some preliminary calculations [29] give rather similar results to $d = 2$.

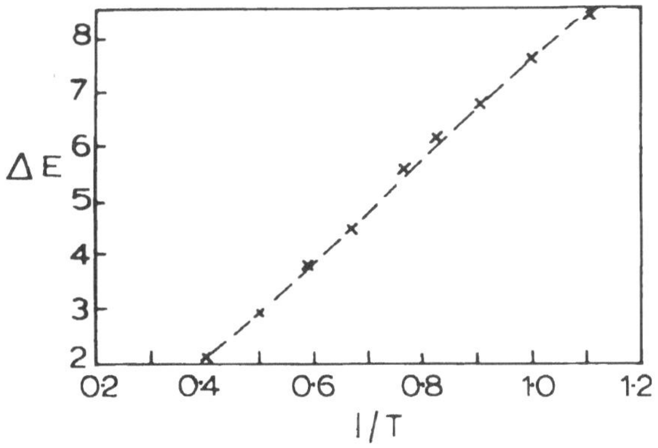

Fig. 6. Shows a 'characteristic barrier height', ΔE, for a 128 x 128 $d = 2$ Ising spin glass model with '$\pm$J' bands. Data is from ref. 26.

CONCLUSIONS

The MFT of spin glasses is now rather well understood. However, the LCD is probably equal to 4 so MFT should be irrelevant. Certainly short range models in $d=2$ show a smooth increase in correlation length and energy barriers as T is reduced, and the same may well be true in $d=3$ though sufficiently precise numerical studies remain to be done. It if therefore a puzzle why many experiments show such a sharp change in behaviour at T_f and can, in fact, be described to a reasonable degree,[3,30] by predictions of the SK model.

In fact the most precise experiments are on metallic systems with RKKY interactions. It is usually assumed that $\cos(2k_F R)/R^3$ falls off sufficiently fast that it is effectively short ranged but there is no rigorous demonstration of this. It would be interesting to see if careful measurements of say $\chi_{n\ell}$ on an insulating spin glass with known short range interactions, such as $Eu_x Sr_{1-x} S$[31], would agree better with theory.

REFERENCES

1. S. F. Edwards and P. W. Anderson, J. Phys. F 5, 965 (1975).
2. C. N. Guy, J. Phys. F 8, 1309 (1978); Solid State Comm. 35, 113 (1980); A. Malozemoff and Y. Imry, Phys. Rev. B 24, 489 (1981).
3. R. Omari, J. J. Prejean and J. Souletie (unpublished); P. Monod and H. Bouchiat, J. de Physique Lett., 43, 45 (1982); B. Barbara, A. P. Malozemoff and Y. Imry, Phys. Rev. Lett. 47, 1852 (1981).
4. J. L. Tholence and R, Tournier, J. de Physique, 35, C4, 229 (1974); E. P. Wohlfarth, Physica 86-88B, 852 (1977).
5. E. D. Dahlberg, M. Hardiman, R. Orbach and J. Souletie, Phys. Rev. Lett. 42, 401 (1979); C. A. M. Mulder, A. J. van Duyne Duynevelt and J. A. Mydosh, Phys. Rev. B 25, 515 (1982).
6. R. Fisch and A. B. Harris, Phys. Rev. Lett. 38, 785 (1977); I. Morgenstern and K. Binder, Phys. Rev. B 22, 288 (1980); A. J. Bray and M. A. moore, J. Phys. C 12, 79 (1979).
7. D. Sherrington and S. Kirkpatrick, Phys. Rev. Lett. 35, 1792 (1975).
8. C. de Dominicis and A. P. Young, (to be published in J. Phys. A).
9. J. R. L. de Almeida and D. J. Thouless, J. Phys. A 11, 983 (1978).
10. G. Parisi, J. Phys. A 13, 1101 (1980); Phys. Rep. 67, 25 (1980) and references therein.
11. D. J. Thouless, P. W. Anderson and R. J. Palmer, Phil. Mag., 35, 593 (1977).
12. A more rigorous derivation of the equations is given by C. de Dominicis, Phys. Rep. 67, 37 (1980).
13. A. J. Bray and M. A. Moore, J. Phys. C 13, L469 (1980); C. de Dominicis, M. Gabay, T. Garel and H. Orland, J. de Physique

41, 923 (1980); F. Tanaka and S. F. Edwards, J. Phys. F 10, 2471 (1980).
14. U. Krey, J. Mag. Mag. Mater. 6, 27 (1977).
15. G. Parisi (preprint); A. Houghton, S. Jain and A. P. Young (to be published in J. Phys. C).
16. N. D. Mackenzie and A. P. Young, Phys. Rev. Lett. 49, 301 (1982).
17. H. Sompolinsky, Phys. Rev. Lett., 47, 935 (1981).
18. N. D. Mackenzie and A. P. Young (to be published).
19. M. Gabay and G. Toulouse, Phys. Rev. Lett. 42, 309 (1981); Cragg, D. M. and Sherrington, D., Phys. Rev. Lett. 16, 1190 (1982).
20. See the articles in Vol. 6 of 'Phase Transitions and Critical Phenomena', eds. C. Domb and M. S. Green, (Academic Press, New York, 1976).
21. A. B. Harris, T. Lubensky and J. H. Chen, Phys. Rev. Lett., 36, 415 (1976).
22. J. H. Chen and T. C. Lubensky, Phys. Rev. B 16, 2106 (1977).
23. J. M. Kosterlitz and D. J. Thouless, J. Phys. C 6, 1181 (1973).
24. R. Fisch and A. B. Harris, Phys. Rev. Lett. 38, 785 (1977).
25. I. Morgenstern and K. Binder, Phys. Rev. Lett. 43, 1615 (1979); Phys. Rev. B 22, 288 (1980); Z. Phys. B 39, 227 (1980).
26. A. P. Young (to be published in Phys. Rev. Lett.).
27. K. Binder and D. Stauffer, Z. Phys. B 34, 97 (1979); S. Kirkpatrick, Phys. Rev. B 16, 4630 (1977).
28. S. Kirkpatrick in 'Ordering in Strongly Fluctuating Condensed Matter Systems", ed. T. Riste, Plenum, p. 459 (1980).
29. A. P. Young (unpublished).
30. R. V. Chamberlin, M. Hardiman, L. A. Turkevich and R. Orbach, Phys. Rev. B 25, 6720 (1982); Y. Yeshurun, L. J. P. Ketelsen and M. B. Salamon, Phys. Rev. B 26, 1491 (1982); N. Bontemps, J. Rajchenbach and R. Orbach, J. de Physique Lettres, 44, L47 (1983).
31. J. Ferre, J. Rajchenbach and H. Maletta, J. Appl. Phys. 52 (3), 1697 (1981).

CRITICAL AND MULTICRITICAL POINTS, LINES AND SURFACES OF SPIN GLASSES IN THE PRESENCE OF MAGNETIC FIELDS AND ANISOTROPY

David Sherrington

Physics Department
Imperial College
London SW7 2BZ, UK

As Dr. Young has pointed out in his lectures[1] the mean field theory of a simple model Ising spin glass[2] is much more subtle than that of a conventional magnet while normal renormalization group theory applied to short-ranged models[3] yields perplexing and apparently unphysical[4,5] results, even in the paramagnetic phase. I shall not attempt here to improve upon the RG problem but rather to develop further the parameter space of potentially interesting mean field problems in the hope of demonstrating a great richness of phases, transitions and crossovers. Thus, I shall restrict my comments to infinite-ranged models studied within replica theory[6] but with the full panoply of replica-symmetry (RS) breaking[7].

Explicitly, I shall concentrate on a model of a classical m-vector spin glass in the presence of local uniaxial anisotropy and a magnetic field, as characterized by the Hamiltonian

$$H = -\sum_{(ij)} J_{ij}\underline{S}_i\cdot\underline{S}_j - D\sum_i (S_{i1})^2 - \sum_i \underline{H}\cdot\underline{S}_i; \quad |\underline{S}_i|^2 = m \tag{1}$$

where the J_{ij} are randomly distributed with probability

$$P(J_{ij}) = (N/2\pi J^2)^{\frac{1}{2}} \exp\left(-N(J_{ij} - J_0/N)^2/2J^2\right) \tag{2}$$

For details of the analysis the reader is referred elsewhere, eg. Ref.8, but briefly the method of analysis proceeds as follows: (i) the replica method[6] is used to average the free energy, yielding an effective pure model to be analysed in the limit $n\to 0$. (ii) mean field theory or steepest descents yields a set of order parameters of which the relevant ones are

$\underline{m}^{\alpha} = \langle \underline{S}^{\alpha} \rangle$, $q^{\alpha\beta}_{T,L} = \langle S^{\alpha}_{T,L} S^{\beta}_{T,L} \rangle$ where α,β are replica labels and

T, L denote transverse and longitudinal (with respect to the anisotropy axis 1).
(iii) the replica symmetric approximation ($m^{\alpha}=m$, $q^{\alpha\beta}=q; \alpha\neq\beta$) gives a measure of a sub-set of phase transition locations
(iv) stability analysis of a generalised fluctuation matrix in replica-space, $\lim_{n\to 0} \partial^2 nF(n)/\partial X^{\alpha\beta}\partial X^{\alpha\delta}$ where F(n) is the averaged replicated free energy function and $X^{\alpha\beta}$ is a generic order parameter, essentially identifies all the true phase transitions
(v) an extension of Parisi's analysis[7] gives properties in phases where replica-symmetry is broken and identifies crossovers in the anomalous behaviour which is characteristic of non-ergodicity.

Let us initially take $J_0=0$. If D and H are also both zero we have an isotropic spin glass with a transition at T=J. The phase structure for $D\neq 0$, H in the 1 direction, is indicated in Fig.1. For H=0 the appropriate transition lines are (a), (a′), (b), (b′) and the phases are paramagnetic (PARA; m=q=o), longitudinal spin glass (LSG $m, q_T=0, q_L\neq 0$), transverse spin glass (TSG; $m, q_L=0, q_T\neq 0$) frozen isotropically throughout the (m-1) dimensional plane orthogonal to 1, and mixed (longitudinal and transverse) spin glass (MSG; $m=0, q_L\neq 0$, $q_T\neq 0$). In all the spin glass phases RS is broken and Parisi analysis yields continuously growing non-ergodicity beneath phase lines. The non-ergodicity is most conveniently characterized by the anomalous response Δ^{ℓ} which measures the difference between field-cooled (FC) and zero-field-cooled (ZFC) local susceptibilities and is determined by

$$\Delta^{\ell}_{T,L} \equiv T(\chi^{\ell}_{T,L}(FC)-\chi^{\ell}_{T,L}(ZFC)) = \int_0^1 (q_{T,L}(x)-q^{max}_{T,L})dx, \qquad (3)$$

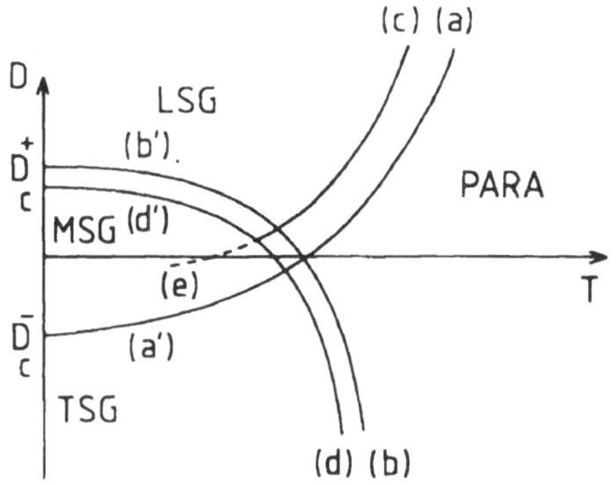

Fig.1: Schematic phase diagram of a classical spin glass with local uniaxial anisotropy. (a),(a′),(b),(b′) correspond to zero magnetic field; (d),(d′),(c),(e) describe modification due to a coaxial field. Solid lines designate phase transitions, the dashed line a crossover.

where the q(x) are Parisi order functions defined in the interval (0,1) and related to the extent of correlation between different solutions of the unaveraged system[1].

The application of a longitudinal field has rather different effects longitudinally and transversely. The temperature for onset of non-zero q_T is simply reduced as indicated by curve (d), by an amount scaling as H^2 for small H. A finite q_L is induced at all temperatures but for $D>D_c(H)$, a critical value, there remains a sharp longitudinal RS-breaking transition marking the onset of non-zero Δ_L^ℓ; this is indicated by curve (c), with the temperature reduction relative to (a) scaling as $H^{2/3}$ for small H. $D_c(H)$ corresponds to the point at which curves (c),(d) intersect; in (D,T,H) this intersection defines a multicritical line. The continuation of (c) beyond (d), that is (e), has a more subtle significance designating a crossover in Δ_L^ℓ. For $H\neq 0$ not only is Δ_T^ℓ everywhere non-zero to the left of (d) but so is Δ_L^ℓ, due to field induced fluctuation coupling. Thus (e) cannot indicate the onset of a longitudinal anomaly as (c) does. Rather, it characterizes a crossover from a region of slow growth of Δ_L^ℓ as the temperature is reduced beneath (d) to a more rapid growth as observed beneath (c); just below (d) Δ_L^ℓ grows essentially as $O(\tau^3)$, where τ is the reduced temperature relative to (d), while beyond (e) this changes to $O(\tau^2, \tau H^{2/3})$. Thus, in the region $D<D_c(H)$, which includes the case of an isotropic spin glass in a field[9,10], there is only one true phase transition (to a transverse spin-glass in the presence of field-induced magnetization) but with a possible crossover in the anomalous part of the longitudinal response at a lower temperature (for D great enough). The transverse spin glass ordering is reminiscient of a spin-flop transition but differs from it in that (at least in the present model) there is no critical field beyond which it is suppressed.

The inclusion of non-zero J_o permits the occurrence of spontaneous ferromagnetism when J_o is large enough. This is illustrated for $D_c^+ > D > 0$ in Fig. 2, which also indicates the features

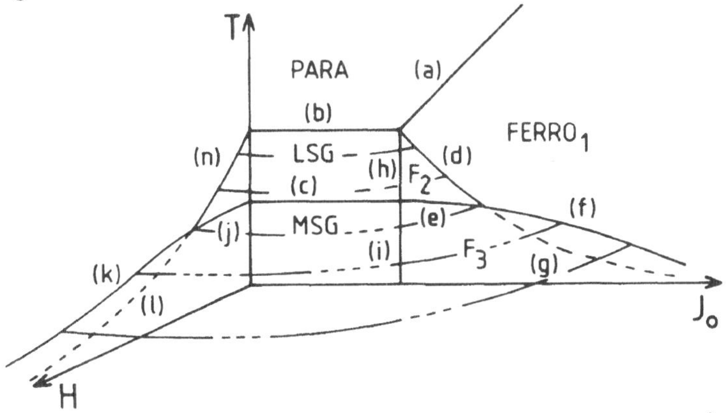

Fig.2: Schematic phase diagram for D in $(0,D_c^+)$.

discussed above in a (T,H) plane (thus (i),(j) designate the onset of transverse ordering, (n) a true longitudinal spin-glass transition and (l) a longitudinal anomaly crossover). There are now three different ferromagnetic phases, the simplest ($FERRO_1$) having spins frozen collinearly and behaving ergodically, a second (F_2) again collinear but non-ergodic, and the third (F_3) having randomly canted moments (collinear ferromagnet plus transverse spin-glass) and non-ergodic but with a weak-to-strong crossover in the longitudinal anomaly along curve (g). Note that although curve (a) (PARA to $FERRO_1$) is an isolated line (for H=0 only), the other transitions represent surfaces ((n) to (d), (l) to (g), (j) to (e), (k) to (f)). For $D > D_c^+$ the regions MSG and F_3 are suppressed. For D=0 the regions LSG and F_2 are suppressed.

For a transverse field the situation is more complicated since there are now two special directions but it is straightforward to extend the qualitative argument. Let us consider explicitly only the case where the transverse magnetization is spontaneous and therefore unavoidable, for $D < 0$, $J_0 > 0$. In this case the (J_0,T) plane of Fig.2 is modified in $0 > D > D_c^-$ with LSG replaced by TSG, $FERRO_1$ by a collinear basal-plane ergodic ferromagnet, F_2 by a non-ergodic ferromagnet with moments randomly canted throughout the (m-1)-dimensional basal plane, F_3 by a non-ergodic ferromagnet with overall magnetization in the basal plane but moments canted throughout the whole m-dimensional space. Within the new phases F_2 and F_3 are anomaly crossover in the direction of magnetization. The phase-line between $FERRO_1$ and F_2 now has a shape more like (e) (f) but starting at the (PARA-TSG-$FERRO_1$ -F_2)multicritical point, while (d) now indicates a cross-over.

Let us turn to the experimental situation. Equ.(1) with D=0, m=0 corresponds to a Heisenberg model, which is the most common experimental situation (albeit more normally with site disorder and oscillatory exchange). Transverse spin glass freezing below the critical concentration[11] for ferromagnetism but in an applied field and above the critical concentration in the spontaneous field[12] have both been observed in Mössbauer experiments on <u>Au</u> Fe. The onset of weak followed by crossover to strong longitudinal anomalous response has also been observed in Faraday rotation on $Cd_{.45}Mn_{.55}Te$[13]. There are many other observations of the $H^{2/3}$ shift in the onset of strong irreversibility in spin glasses as well as some of the corresponding behaviour in ferromagnets of type F_3 . $D \neq 0$ is appropriate to the dilute hexagonal alloys[14] <u>Zn</u> Mn ($D > 0$) and <u>Cd</u> Mn ($D < 0$), while the related <u>Mg</u> Mn has $D \sim 0$. In these systems the effective $|D|$ decreases with concentration and for the concentrations available <u>Zn</u> Mn has $D > D_c^+$, <u>Cd</u> Mn has $D < D_c^-$. They exhibit the single transitions of the type predicted above. Unfortunately the region $D_c^+ > D > D_c^-$ has yet to receive experimental investigation.

Here we have considered only the simplest classical spin model with uniform uniaxial anisotropy and uniform field and only within mean field theory (albeit a sophisticated version). Already we have seen an interesting range of transitions and crossovers, some of the former accompanied by the onset of a simple thermodynamic order parameter but most involving more complex non-ergodic effects. Within mean field theory extensions for classical vector spins are straightforward, for quantum spins more complicated due to lack of commutation, and for Potts[15] or generalised ($S > \frac{1}{2}$) Ising models complicated by first order transitions and inadequacy of standard Parisi formulation. Short ranged models beyond mean field theory are uncharted territory for which even simpler models remain a mystery[1]. However, as Dr. Young has mentioned[1], true thermodynamics may be irrelevant to real experiment. It is interesting to note that on the basis of experiments to date mean field theory appears to be in conceptual accord with experiment, including critical exponents within spin-glass phases while in paramagnetic phases critical exponents differ from those of mean field theory[16]. It is thus tempting to suggest that non-ergodicity quenches the critical fluctuations which modify exponents at normal transitions.

REFERENCES

1. A.P. Young, these lectures (1983).
2. D. Sherrington and S. Kirkpatrick, Phys. Rev. Lett. 35, 1792 (1975).
3. A.B. Harris, T.C. Lubensky and J.H. Chen, Phys. Rev. Lett. 36, 415 (1976).
4. J.H. Chen and T.C. Lubensky, Phys. Rev. B16, 2106 (1976).
5. A.J. Bray and S.A. Roberts, J. Phys. C13, 5405 (1980).
6. S.F. Edwards and P.W. Anderson, J. Phys. F5, 965 (1975).
7. G. Parisi, J. Phys. A13, 1101 (1980).
8. D. Elderfield and D. Sherrington, J. Phys. C. (to be published, 1983).
9. M. Gabay and G. Toulouse, Phys. Rev. Lett. 47, 201 (1981).
10. D.M. Cragg, D. Sherrington and M. Gabay, Phys. Rev. Lett. 49, 158 (1982).
11. W. Marschmann, J. Lauer and W. Keune, J. Mag. Mag. Mat. 31-34, (1983).
12. J. Lauer and W. Keune, Phys. Rev. Lett. 48, 1850 (1982).
13. H. Kett, W. Gebhardt and V. Krey, J. Mag. Mag. Mat. 25, 215 (1981).
14. H. Albrecht, F.T. Hedgecrock and P. Monod, Phys. Rev. Lett. 48, 819 (1982).
15. D. Elderfield and D. Sherrington, J. Phys. C. (to be published, 1983).
16. R. Tournier, comment at meeting on "Spin Glasses" (Orsay, 1983).

SOME RENORMALIZATION GROUP RESULTS ON THE QUENCHED GAUGE FIELD MODEL OF THE SPIN GLASS

R.K. Ritala

Research Institute for Theoretical Physics
University of Helsinki
Siltavuorenpenger 20C
SF - 00170 Helsinki, Finland

INTRODUCTION

We study the effects of the dense frustration to the ferromagnetic transition by Callan-Symanzik -renormalization group equations. A qualitative agreement with previous $O(\varepsilon)$ -results[1] is found by deriving a Ward -identity. Our result is, however, independent of the order in ε and thus more general. The dense frustration opens an unstable manifold in parameter space for the renormalization transformation at the ferromagnetic fixed point below the critical dimension $d_c = 4$. Thus no transition is found.

We discuss also the possibility that the properties of the spin glass could be explained by a very large but finite correlation length, signalizing the destroyed ferromagnetic transition. Recently Young[2] has put forward an idea of somewhat similar nature. He argues that long correlations should have a decisive role in experimental systems (ie. d = 3). This could be an explanation for surprisingly good agreement of the mean field theory and experimental results, which seems contradictory to the general belief that the upper and lower critical dimensions of the spin glass transition are 6 and 4, respectively. Our conclusion is negative, but does not rule out Young's suggestion, because the large correlation lengths are of different origin.

THE MODEL; TRICRITICALITY

We consider the following XY -spin glass model defined on the continuum for a complex field $\phi = \phi_x + i\phi_y$:

$$H = \int d^d x\Big(\alpha|\phi(x)|^2 + \frac{\beta}{4!}|\phi(x)|^4 + |(\nabla - ig\vec{A}(x))\phi(x)|^2\Big) \quad (1a)$$

$$P\{\vec{A}(x)\} = N \exp\Big(-\frac{1}{2}\int d^d x\big(a|\vec{A}(x)|^2 + |\nabla x\vec{A}(x)|^2 \big)\Big) \quad (1b)$$

which can be derived from a general anisotropic lattice spin glass model[3]. Hertz[1] introduced this Hamiltonian heuristically.

The quenched random field operator $\vec{A}$ is relevant only if the probability distribution $P\{\vec{A}(x)\}$ is invariant under local (gauge) U(1) -transformation as is the Hamiltonian H. Thus we choose to study a = 0, the densely frustrated case. The eventual transition at the critical renormalized temperature would hence be tricritical.

We have left out the random temperature contribution, which is relevant for the renormalization group and arises naturally in H. This effect does not change qualitatively our results. The properties of the pure random temperature model are discussed in literature[4].

THE WARD -IDENTITY AND THE CALLAN-SYMANZIK -EQUATIONS[5]

Let us suppose that the U(1) -gauge symmetry is not broken spontaneously above the ferromagnetic critical temperature. Then the following identity for the 1-particle-irreducible (or vertex-) functions can be derived:

$$\frac{\partial}{\partial k^2}\Gamma_R^{(2,0)}(k;\alpha_R,g_R,\beta_R)\Big|_{k=0} = \frac{1}{g}\frac{\partial}{\partial p_\mu}\Gamma_{R,\mu}^{(2,1)}(p,q;\alpha_R,g_R,\beta_R)\Big|_{\substack{k=0\\p=0}} \quad (2)$$

where the upper indices give the number of external thermodynamic fields and quenched random fields, respectively. The subindex R denotes renormalized quantities and the second lower index in $\Gamma^{(2,1)}$ gives the component of the external $\vec{A}$ -field.

Combining eq. (2) with renormalization conditions we arrive at the relationship

$$g = g_R \quad (3)$$

which thus holds for any order in ε. The coupling constant of the quenched field-magnetization -interaction is unchanged when the theory is renormalized.

The Callan-Symanzik -equations for dimensionless quantities (denoted with tildas) now reads as

$$\left\{ \alpha_R \frac{\partial}{\partial \alpha_R} + f_\beta(\tilde{\beta}_R,\tilde{g}_R) \frac{\partial}{\partial \tilde{\beta}_R} - \frac{\varepsilon}{4} \tilde{g}_R \frac{\partial}{\partial \tilde{g}_R} - \frac{N}{2} \eta(\tilde{\beta}_R,\tilde{g}_R) \right\} \Gamma_R^{(N)} = 0 \qquad (4)$$

with

$$f_\beta = - \frac{\varepsilon}{2} \left(\frac{\partial \log \tilde{\beta}}{\partial \tilde{\beta}_R} \right)^{-1}_{\tilde{g}} \quad ; \quad \eta = \alpha_R \frac{\partial}{\partial \alpha_R} \log Z_\phi$$

We have neglected the inhomogeneous part from eq. (4) as can be done near the critical point.

The fixed point is $\tilde{g}_R = 0$, $\tilde{\beta}_R = \beta_0$; the ferromagnetic fixed point. The stability analysis shows, however, that this point is hyperbolic: starting the scaling from nonzero $\tilde{g}_R$ the system first approaches the fixed point but in the end diverges. Thus no phase transition exists.

The only result depending on the order of ε is the value of β_0, the instability is exact. In the previous $O(\varepsilon)$ -results the eq. (3) does not hold, because the contribution from one graph was omitted.

THE CORRELATION LENGTH IN THE FRUSTRATED SYSTEM

The fixed point of eq. (3) has two unstable manifolds: the usual one parametrized by temperature and the frustration manifold. We now fix the temperature to the critical value and use $\tilde{g}_R$ as an effective temperature. Especially interesting is the critical exponent of the correlation length.

In the vicinity of the fixed point the eq. (4) has a scaling like solution. By fixing $\tilde{\beta}_R$ at the fixed point value and scaling by a factor $\rho = (\alpha \tilde{g}_R{}^{4/\varepsilon})^{1/2}$ we find

$$\Gamma_R^{(N)}(k_i;\alpha_R,\tilde{g}_R,\beta_0) = \alpha_R^{(N+d-Nd/2)/2} \, \tilde{g}_R^{\,2(N+d-Nd/2-N\eta(\beta_0,0))/\varepsilon} \times F^{(N)}\left(\frac{k_i \tilde{g}_R{}^{2/\varepsilon}}{\alpha^{1/2}} \right) \qquad (5)$$

where $F^{(N)}$ is an arbitrary factor. Thus the correlation length is given by

$$\xi \sim \tilde{g}_R^{-2/\varepsilon} \qquad \text{at } T = T_c \text{(ferromagnetic)} \qquad (6)$$

This critical exponent is too large for the spin glass properties to be explained by a broadened ferromagnetic transition. The result (6) is again a direct consequence of the Ward -identity (2).

The details of these calculations will be presented elsewhere[3]. The author is grateful to professors J.A. Hertz and S. Stenholm for valuable discussions.

REFERENCES

1. J.A. Hertz, Phys. Rev. B18, 4875 (1978)
2. A.P. Young, Phys. Rev. Lett. 50, 917 (1983)
3. R.K. Ritala, J.A. Hertz, to be published
4. for example T.C. Lubensky, Phys. Rev. B11, 3573 (1975)
5. see D.J. Amit, "Field Theory, the Renormalization Group and Critical Phenomena", McGraw-Hill, New York (1978)

FRUSTRATION, LANDAU LEVELS AND SUPERCONDUCTING DIAMAGNETISM ON NETWORKS

Gérard Toulouse

Laboratoire de Physique de l'ENS
24 rue Lhomond
75231 Paris, France

INTRODUCTION

The concept of frustration[1] arises naturally in a school devoted to multicritical phenomena because physical systems with competing interactions (or constraints) tend to have a rich phase diagram with many phases, transition lines and multicritical points.

The reason is simply that, in such systems, any ground state or phase results from an unsatisfactory compromise between the interactions, therefore it has little energetic stability and a small modification of external parameters may lead to a qualitatively different state. As an example[2], one may refer to the complex phase diagram observed in CeSb.

In disordered frustrated systems, the number of different possible states may become so large indeed that an infinitesimal variation of an external parameter (the magnetic field, say) leads to a new equilibrium state, far out in phase space. In this limit, the phenomena of ergodicity breaking, "violation" of linear response theory, characteristic of spin glasses[3], are obtained.

From another viewpoint, the percolation multicritical point, which mixes thermal and geometrical critical phenomena, has a natural counterpart : the frustration multicritical point, when frustration disorder is considered instead of dilution disorder[4]. Many other problems involving aspects of frustration could be mentioned : fully frustrated lattices, gauge theories, amorphous structures, hydrodynamic instabilities and patterns, defects due to boundary conditions, optimization problems under constraints.

However, only one specific aspect of frustration will be discussed in these lectures, namely the frustration induced by a magnetic field applied on a two-dimensional network and its effect on the spectrum of electronic levels and on the superconducting diamagnetic properties. Periodic, fractal or disordered structures will be considered. In these problems, continuous variation of the applied magnetic field allows for a fine tuning of the frustration, that is not easy to achieve otherwise.

I. SUPERCONDUCTING TRANSITION NEAR THE PERCOLATION THRESHOLD : A MULTICRITICAL POINT

After many years of careful study, the Tel-Aviv group of G. Deutscher[5] has been able to prepare random mixtures of superconducting and insulating elements, such as InGe films, which are well described by a percolation model. The diamagnetic properties of such mixtures and their variation across the percolation threshold (for the superconducting component) raise an interesting problem.

The idea is that superconducting diamagnetism, being essentially due to the loops in the percolation clusters, will provide information on the geometry (essentially, the multiple-connectedness) of those clusters. Superconducting fluctuations being generally small, they are altogether neglected in the present theoretical analyses[6,7]. Mean field theory is used to describe the thermal (superconducting) critical phenomena, the emphasis being on the geometrical (percolation) critical aspects.

The experimental data[5] bear on the upper critical field H_{c2} measured close to T_0, the superconducting transition temperature, on the metallic side of the percolation threshold. On the insulating side of the same threshold, a quantity of interest is the diamagnetic susceptibility χ, for which the present data come from numerical simulation only[8]. Thus one defines two critical exponents, k and b,

$$\left.\frac{dH_{c2}}{dT}\right|_{T_0} \sim \frac{1}{(p-p_c)^k} \qquad , p > p_c,$$

$$-\left.\frac{d\chi}{dT}\right|_{T_0} \sim \frac{1}{(p_c-p)^b} \qquad , p < p_c,$$

where p is the concentration of the superconducting component and p_c the percolation threshold. Eventual dependence of T_0 on p is due to fluctuations and ignored at the mean field level. The two quantities above measure the response of the film to a small perpendicular magnetic field.

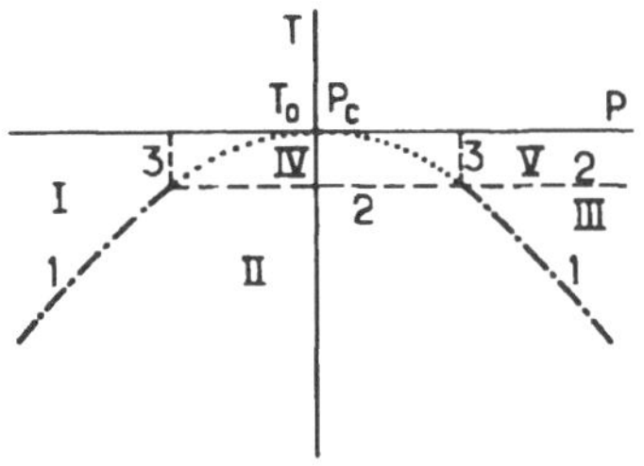

Fig. 1. Temperature-concentration phase diagram around the transition temperature ($T = T_0$) and the percolation threshold ($p = p_c$) for a random superconductor-insulator mixture of size L×L. Definitions of the boundaries 1, 2, 3 separating regions I to V, are given in the text.

The published experimental estimate[5] for k is : $k = 0.6 \pm 0.05$. The numerical estimate[8] for b is : $b = 1.55 \pm 0.04$.

The scaling theory developed in Ref.7 leads to a scaling relation between k and b : $b + 2k = 2\nu$, where ν is the percolation length exponent ($\nu = 4/3$ in two dimensions). Furthermore, explicit expressions for exponents k and b are obtained in terms of exponents ν, t and β, where t is the conductivity exponent and β is the infinite cluster density exponent :

$$k = \frac{t-\beta}{2} \quad ,$$

$$b = 2\nu - t + \beta .$$

Using presently accepted values for t and β in two dimensions, one obtains : $k \simeq 0.57$ and $b \simeq 1.53$, in good accord with the existing data.

The scaling aspect of the analysis is based on the recognition that there are three lengths of importance in the problem :
i) the characteristic length of the sample, L,
ii) the superconducting coherence length, ξ_s,
iii) the percolation correlation length, ξ_p.
Boundaries 1 of the phase diagram shown in Fig. 1 correspond to $\xi_s \sim \xi_p$. Boundaries 2 correspond to $\xi_s \sim L$. Boundaries 3 correspond

to $\xi_p \sim L$. They divide the phase diagram into 5 regions, denoted I to V, and the analysis amounts to matching the behaviours of H_{c2} and χ across the boundaries.

In fact, this analysis may be oversimplified. Since this is a matter of present controversy, we will not go into any details. Another theory[6] predicts different expressions for k and b :

$$k = t - \beta \quad , \quad b = 2\nu - t \quad ,$$

which do not agree well with the data, but the latter are presently too scanty to be really conclusive.

However, both theories do agree on two important conclusions :
i) no new exponent is involved in these superconducting diamagnetic properties, contrary to what one might have expected intuitively a priori ; the diamagnetic currents do not appear to provide additional information on the geometry, in comparison with a normal conductivity measurement,
ii) the infinite percolation cluster at threshold behaves essentially as a fractal, self-similar structure, the relations between its diamagnetic susceptibility and its normal conductivity are the same which hold for a regular (non disordered) fractal structure, such as a Sierpinski gasket (see below) ; contrary to early expectations, the distinction between backbone cluster and dead ends does not appear to be relevant either.

In conclusion, this multicritical point offers an interesting theoretical challenge and clearly calls for more experimental and numerical data. Extension to higher fields and to three-dimensional samples should also be pursued.

II. LANDAU LEVELS ON PERIODIC STRUCTURES

The theory involved in the computation of the upper critical field of a superconductor (neglecting fluctuations) consists in solving the linearized Ginzburg-Landau equation, which is essentially equivalent to the Schrödinger equation. The eigenvalues of the Schrödinger equation, for free electrons in a magnetic field, are called Landau levels. We shall be interested in the spectrum of eigenvalues, as modified by a potential or by the geometry of the underlying space. For superconducting applications, one is only interested in the lowest eigenvalue (edge of the spectrum) ; besides, in the preceding Section, only the low-field behaviour was considered. Therefore, we are now setting the problem into a larger framework, by considering the whole spectrum of eigenvalues.

Fig. 2. Spectrum of a tight-binding model on a square lattice in a perpendicular magnetic field. The eigenvalue energy is the vertical variable, in arbitrary units, and ϕ, the reduced magnetic flux through one lattice cell, is the horizontal variable, ranging from 0 to 1. From D.R. Hofstadter[9].

Consideration of Fig.2 provides a good start. This is the Landau level spectrum for a tight-binding problem on a square lattice, as determined numerically by D.R. Hofstadter[9]. The abscissa ϕ is the magnetic flux through one lattice cell $\Phi = Ha^2$, in units of the flux quantum $\Phi_0 = hc/e$.

Worthy of note are the following features ;
i) the periodicity in ϕ along the horizontal axis, with period 1, due to gauge invariance,
ii) the linear behaviour of the electron (and hole) Landau levels, when ϕ is small, which is analogous to the structure of Landau levels for electrons in free space,
iii) the existence of gaps which keep their identity, when ϕ increases, trough a succession of closures and openings ; to each gap can be attributed a gap index p ; a p-gap has period $\frac{2}{|p|}$.

The variation of the lowest eigenvalue as a function of ϕ determines the effect of the frustration, induced here by the magnetic field, on the transition temperature (or ground state energy) of a magnetic or superconducting transition, treated in mean field theory[10]. When the flux through one lattice cell is half a flux quantum, $\phi = 1/2$, the lattice is effectively fully frustrated. Remark that the maximum reduction of the mean field transition temperature obtains for values of ϕ different from 1/2, contrary to intuitive expectations (for similar findings on the effects of frustration, see Ref.11).

The linear variation of the spectrum edge, in low field, implies that the upper critical field H_{c2} varies linearly below T_0, for a periodic network as it does for a bulk continuous material. However, we shall see below that this property does not hold for a self-similar network.

The closures of the gaps in Fig. 2 are interesting to observe. Note for instance that, for $\phi = 1/2$, the spectrum comprises two bands which touch each other. This means that at this energy the density of states has an isolated zero with a non-analytic behaviour. Other such touchings of bands can be observed on the central horizontal axis. The case of a triangular lattice has been studied also[12]. In this geometry, a gap closure due to touching of two bands is observed when the reduced flux through an elementary triangle is $\phi = 1/12$. I do not know whether one can make predictions for such band touchings, based on geometrical considerations alone. It has been noticed[13] that, in the absence of a magnetic field, such isolated zeroes occur for the honeycomb and diamond lattices, both of which have shortest cycles made up of six bonds.

Notwithstanding these occasional closures, the gaps can be followed throughout the spectrum, as noted above. Five years ago, G.H. Wannier[14] remarked that the integrated density of states below any gap was a linear function of ϕ, with an integer (positive or negative) linear coefficient. The analysis of Wannier was based on a nesting principle enunciated by Hofstadter[9]. Wannier's integer coefficient is equal to the gap index and it shows up also as the quantum number in the quantized Hall conductance[15,16]. This is an indication that Wannier's result is a property of great generality and that there should exist a general proof.

Indeed, it is a direct consequence of gauge invariance that in a gap of the density of states for non interacting electrons in a two-dimensional potential, the integrated density of states is a linear function of magnetic field with quantized coefficient. The proof is as follows.

Consider a two-dimensional sample of area S containing N non interacting electrons (assumed spinless without any loss of generality) in a magnetic field H. The Fermi level is assumed to lie in a gap of the bulk density of states and we are interested in

$\left.\frac{dN}{dH}\right|_{\mu}$.

Consider now the effect of a small change of magnetic field δH such that $S.\delta H = \Phi_o$. Note that δH can be arbitrarily small for large enough samples. Since the Fermi level lies in a gap of the bulk, the number of states below it can only vary by transit of states via the edges. The edge levels are transversely localized - they can run along the edge but they cannot penetrate inside the sample. The topology of the edge is that of a ring and, by gauge invariance, a change of magnetic flux through an annulus by one flux quantum cannot modify the spectrum of the electron states on the annulus. Thus the spectrum of the edge levels is globally unchanged when H receives an increment δH and the only change is that a number δN (integer positive, negative or zero) of levels has crossed downwards the Fermi level μ. As a consequence :

$$\left.\frac{dN}{dH}\right|_{\mu} = \frac{\delta N}{\delta H} = \frac{S}{\Phi_o} \cdot \delta N \qquad \text{, q.e.d.}$$

For the particular case of the two-dimensional square lattice of Wannier-Hofstadter, one may introduce the notation $W = \frac{N}{S} a^2$ for the integrated density of states per site (a is the lattice parameter) and one obtains :

$$\left.\frac{dW}{d\phi}\right|_{\mu} = \text{integer.}$$

The relationship with the quantized Hall conductance is contained in the Středa-Widom[17] formula :

$$\sigma_H = ec \left.\frac{dn}{dH}\right|_{\mu}$$

which applies when the Fermi level lies within a gap of the density of states of the bulk ; n is the electronic density per unit area. The core of the argument is that the Hall conductance, being a ratio of current over voltage, is proportional to $\left.\frac{dm}{d\mu}\right|_H$, where m is the magnetization per unit area - because the magnetization of an edge state is proportional to the current flowing along the edge. The equality of crossed second derivatives of the thermodynamic potential at fixed H and μ leads to the formula for σ_H written above.

Consideration of Fig. 2 shows that a periodic potential would lead to wild oscillations of the Hall conductance between quantized positive and negative values. Such strange effects have not been observed yet experimentally because a lattice parameter comparable in length with the cyclotron radius is required. In available magnetic fields, this length is at least of order 100 Å, a distance much larger than usual interatomic distances. With the advent of new microelectronic techniques or perhaps in some other chemical ways with new materials, these effects will probably become observable. Anyway, this discussion has shown that the solution of the Schrödinger equation, with a periodic potential in a magnetic field, involves a lot of non trivial physics.

III. LANDAU LEVELS ON STRIPS AND THE ROLE OF EDGE STATES

In order to observe the genesis of the asymptotic spectrum of Fig. 2, and to get a more detailed understanding of the interplay between bulk and edge levels, we have investigated the spectrum of two-dimensional regular strips which are infinite in extent in one direction and finite in the other[16] (Fig. 3). Two sets of boundary conditions have been considered. One pertains for a tight-binding problem. The other applies if we regard the strips as networks of superconducting wires connected at the nodes. The tight-binding boundary conditions turn out to be "repulsive", and the superconductiong network boundary conditions "attractive", in the sense that for the latter there are edge states with energies below the lowest bulk level and above the highest. These bound states, in the superconductor context[18], define an upper critical field H_{c3}, due to nucleation of superconductivity at the edges. More important, from our viewpoint here, are the edge levels which fill the gaps. This is shown in Fig. 4 which represents a graph of the spectrum for a strip of width N = 13, with tight-binding boundary conditions.

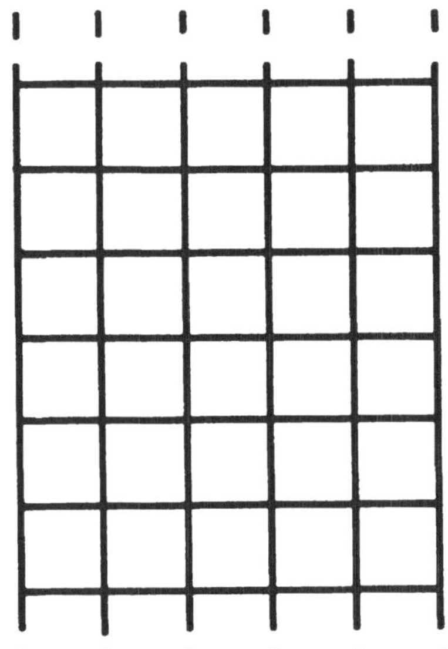

Fig. 3. A regular two-dimensional strip of width N = 6.

Fig. 4. Spectrum of a tight-binding model on a regular strip of width N = 13 in a magnetic field. Same variables as in Fig. 2. The numerical mesh is $\Delta\phi = 5\times10^{-3}$ and $\Delta k = 5\pi\times10^{-2}$.

The crucial fact is that for strips there is continuity of the electronic levels as a function of field (Fig. 4) whereas for the infinite lattice this continuity does not exist (Fig. 2). The lack of continuity versus field for the electronic levels, with or without a periodic potential, is a the root of all the surprising features associated with electron diamagnetism and Landau levels. The continuity of the states is restored for bounded systems because the bulk levels transit via the edges.

In order to get a more detailed vision, we have plotted the eigenvalue energy as a function of k, the wave vector in the infinite direction of the strip, and of ϕ, the reduced magnetic flux. For a strip of width N, there are N eigenvalue sheets stacked above one another. Fig. 5 is an overview of the uppermost sheet for a strip of width N = 7. This energy sheet comprises rather wrinkled portions but it contains also steep flat slopes which correspond to edge states. In these flat slopes, the eigenvalue energy, instead of being a function of two independent variables k and ϕ, depends only on the composite variable $k - \pi N\phi$. This is a manifestation of the special gauge invariance property which the edge states possess, because they are transversely localized[16]. When inserted into the formula for σ_H, this property of the energy leads to quantization in the gaps of the density of bulk states.

IV. SPECTRUM OF THE SCHRÖDINGER EQUATION ON REGULAR FRACTAL STRUCTURES

We have mentioned, in Section I, that the infinite cluster at percolation threshold is a self-similar structure. Of course, it is a quite irregular, disordered structure. In a bond percolation model, for instance, the coordination number on the cluster varies from site to site. However, there are indications that in many physical properties the fractal character dominates and the disorder is not relevant. Anyway, the study of regular fractal structures appears as a promising way of approach towards the physics of disordered and amorphous systems. Last but not least, one can obtain exact results for these regular structures.

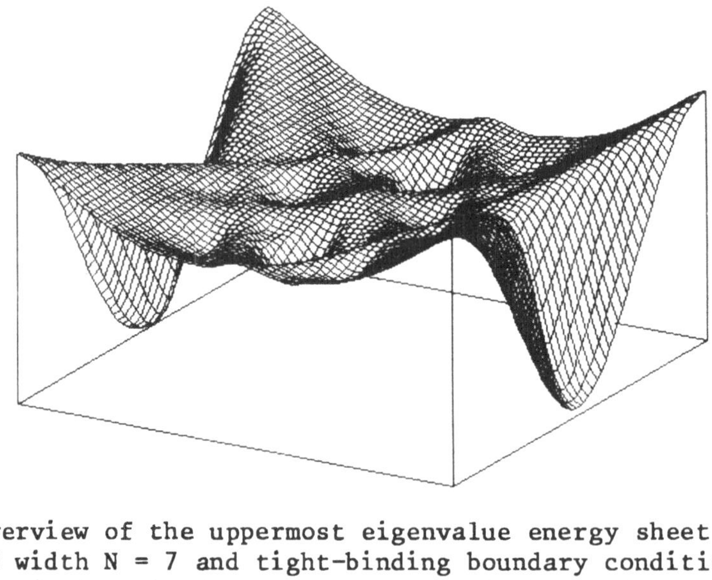

Fig. 5. Overview of the uppermost eigenvalue energy sheet, for strips of width N = 7 and tight-binding boundary conditions. The vertical variable is the energy and the horizontal variables are the momentum k (front right axis) and the reduced magnetic flux ϕ(front left axis).

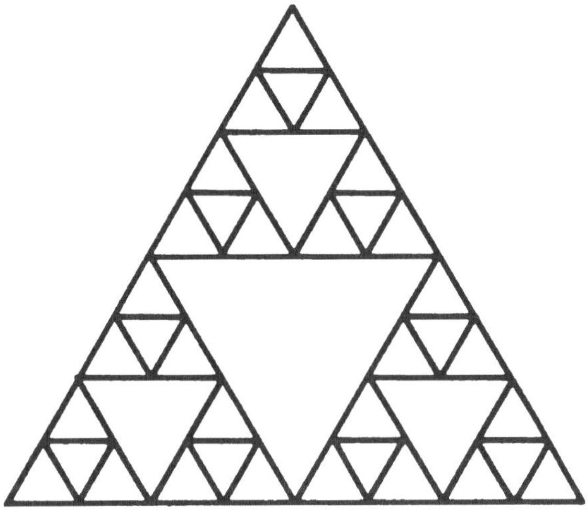

Fig. 6. Two-dimensional Sierpinski gasket, after four stages of iteration starting from an elementary triangle.

Fig. 6 shows a portion of a Sierpinski gasket in two dimensions, a hierarchical structure made of triangular units. The iteration process is to be repeated ad infinitum and the resulting structure is clearly self-similar. One can define Sierpinski gaskets in higher dimensions but since our main interest here is in the response to an applied magnetic field we shall remain in dimension two.

Rammal[19] has succeeded in obtaining a complete description of the spectrum and of the eigenmodes of the harmonic vibrations problem (a variant of the tight-binding problem). The spectrum comprises two tangled components:a discrete spectrum, corresponing to local, highly degenerate, modes and a pure point spectrum, whose support is a Cantor set of zero measure, corresponding to hierarchical states. In two dimensions, the number of local modes is twice the number of extended modes.

These results are startling. They show that the geometry of the gaskets, though regular, is able by itself to localize an important fraction of the states, in the absence of any disorder[20]. In the case of a periodic lattice without field, all states are extended, and the spectrum of Fig. 2 is the result of a battle between the potential which tends to form bands and the magnetic field which tends to produce gaps and degenerate levels. In the case of a Sierpinski gasket, the zero-field spectrum contains an already infinite number of gaps and degenerate states. What will then be the effect of a magnetic field ?

No complete solution yet is available for the asymptotic spectrum in the presence of a field but some important clues have been obtained from the study of the iteration properties of the spectrum for finite gaskets[21]. It appears that the magnetic field, in first order, lifts the degeneracy of the local states. Of course, the periodicity of the spectrum with respect to ϕ, the reduced magnetic flux through the basic triangles, still exists. We refer the reader to Ref. 21 for the symmetry and nesting properties of the spectrum which have already been discovered.

Two important sets of questions remain unsolved.

The first concerns the gaps. Can one define a gap index ? Is there a quantized linear law for the integrated density of states below a gap, as in the case of a periodic lattice ? Is the effect of the magnetic field qualitatively different for the two components of the zero-field spectrum ?

The second brings us back to the multicritical point of Section I. What is the behaviour of the lowest eigenvalue, the edge of the spectrum, in small fields ? Two different guesses have been presented[6, 21]. Both involve an exponent which is related to the fractal and spectral dimensionalities of the gasket[22] and both predict an initial horizontal slope instead of the linear variation, observable in Fig. 2 for a periodic lattice. This is a well posed problem. When we know the solution of this question for a Sierpinski gasket, we will know likewise how the upper critical field for a superconducting transition varies in the vicinity of a percolation threshold.

ACKNOWLEDGEMENTS

The author expresses his debt to his collaborators B.I. Halperin, M.T. Jaekel, T.C. Lubensky, R. Rammal. Inspiring discussions with S. Alexander are also gratefully acknowledged.

REFERENCES

1. G. Toulouse, Comm. on Phys. 2 : 115 (1977); see also in Springer Lecture Notes in Physics, Vols 115 (1980) and 149 (1981).
2. J. Rossat-Mignod, P. Burlet, J. Villain, H. Bartholin, Wang-Tcheng-Si, D. Florence, O. Vogt, Phys. Rev. B 16 : 440 (1977); see the lectures of P. Bak, R. Currat, M.E. Fischer, Y. Shapira, in this volume.
3. See the lectures of A.P. Young in this volume; also G. Toulouse, in "Anderson Localization", Springer Series in Solid State Sciences, Vol. 39 (1982).
4. See the lectures of A. Aharony, B. Coles, A.P. Young in this volume; also B. Derrida, J. Vannimenus, Phys. Rev. B (1983).

5. G. Deutscher in Springer Lecture Notes in Physics, Vol. 149 (1981); G. Deutscher, I. Grave, S. Alexander, Phys. Rev. Letters 48 : 1497 (1982).
6. S. Alexander, Phys. Rev. B 27 : 1541 (1983); S. Alexander, E. Halevi, J. Physique (1983).
7. R. Rammal, T.C. Lubensky, G. Toulouse, J. Physique Lettres 44 : L 65 (1983); G. Toulouse, R. Rammal, Helvetica Physica Acta (1983).
8. R. Rammal, J.C. Angles d'Auriac, to be published.
9. D.R. Hofstadter, Phys. Rev. B 14 : 2239 (1976).
10. S. Teitel, C. Jayaprakash, Phys. Rev. B 27 : 598 (1983).
11. J. Zittartz, J. Magn. Magn. Mat. 31-34: 1105 (1983).
12. F.H. Claro, G.H. Wannier, Phys. Rev. B 19 : 6068 (1979).
13. N. Pottier, Thèse, Université Paris 6 (1976).
14. G.H. Wannier, Phys. Stat. Solidi (b) 88 : 757 (1978).
15. D.J. Thouless, M. Kohmoto, M.P. Nightingale, M. den Nijs, Phys. Rev. Letters 49 : 405 (1982); P. Středa, J. Phys. C 15 : L 1299 (1982).
16. R. Rammal, G. Toulouse, M.T. Jaekel, B.I. Halperin, Phys. Rev. B (1983); R. Rammal, T.C. Lubensky, G. Toulouse, Phys. Rev. B 27 : 2820 (1983).
17. P. Středa, J. Phys. C 15 : L 717 (1982); A. Widom, Phys. Letters A 90 : 474 (1982).
18. D. Saint-James, P.G. de Gennes, Phys. Letters 7 : 306 (1963).
19. R. Rammal, to be published; E. Domany, S. Alexander, D. Bensimon, L.P. Kadanoff, to be published.
20. For a discussion of tangled spectra of localized and extended states in quantum percolation, see S. Kirkpatrick, "Ill-condensed Matter", North-Holland (1979).
21. R. Rammal, G. Toulouse, Phys. Rev. Letters 49 : 1194 (1982).
22. S. Alexander, R. Orbach, J. Physique Lettres 43 : L 625 (1982); R. Rammal, G. Toulouse, J. Physique Lettres 44 : L 13 (1983).

INTRODUCTION TO CHAOS:

PHENOMENON, STRUCTURE OF SPECTRA AND DIFFUSION

Siegfried Grossmann

Fachbereich Physik, Philipps-Universität
Renthof 6, D-3550 Marburg, F.R.G.

Stefan Thomae

Institut für Festkörperforschung der KFA
D-5170 Jülich, F.R.G.

1. INTRODUCTION

In these lectures we present a brief introduction into a dynamical quality which is often called deterministic chaos. It is observed in many physical and other systems in which three or more order parameters or macroscopic variables $x_\alpha(t)$ competitively determine the time behavior. The equations of motion

$$\dot{x}_\alpha = f_\alpha(x,a) \quad , \quad \alpha = 1,2,\dots,d \tag{1}$$

are coupled ordinary differential equations (o.d.e.s), depending on an externally controlled parameter a. If there is only one variable $x_1(t)$ the stationary solution is time independent (or goes to infinity). If there are two variables $x_1(t)$, $x_2(t)$ limit cycle behavior is possible besides the time independent steady state (or x_α to infinity). The periodic (although in general not sinusoidal) oscillation is selfgenerated and not externally imposed.

If there are three variables $x_\alpha(t)$, $\alpha = 1,2,3$, besides time independent steady state and limit cycle a third form of dynamical behavior is possible: permanent time dependent, nonperiodic, bounded motion, sensitively depending on the initial state, unpredictable in the long run, apparently pseudo-stochastic, irregular, chaotic [1].

In sect.2 a survey will be given on the rich variety of systems featuring this behavior. The observation [2] that the third variable may be replaced by a periodic external driving led to considerable progress in realizing such systems experimentally. So, a complex order parameter system with not too particular nonlinear coupling is supposed to become chaotic if coupled to an external sinusoidal signal. This fact was known already to Rayleigh [3] , who introduced the parametric oscillator (described by the Mathieu equation) and the driven unharmonic oscillator (represented by the Duffing equation).

In many cases the theoretical treatment of dissipative systems can even be restricted to one variable x_τ, if the time is a discrete variable $\tau = 0,1,2,\ldots$ The equation of motion then is a 1-dimensional discrete map

$$x_{\tau+1} = f_\alpha(x_\tau) \quad , \quad \tau = 0,1,2,\ldots \tag{2}$$

Conservative systems are described either by including $x_{\tau-1}$ or by considering 2-dimensional area preserving maps. The same holds true for the description of solid lattice systems. In that case τ denotes the lattice site number.

Sect.3 is devoted to this reduction to discrete maps and the time development generated by them. We discuss various aspects of chaotic motion in the language of discrete nonlinear 1-dimensional dynamics. - The structure of spectra and correlation functions is considered in sect.4. In particular the simultaneous appearance of broad bands and sharp peaks along an inverse cascade, socalled periodic chaos [4] is described.

The inverse cascade can be generated by a renormalization group transformation for discrete maps [5,6]. As will be shown in sect.5, this is particularly simple for piecewise linear maps.

Delocalized pseudo-Brownian diffusive motion is generated by periodic discrete maps ([7] - [9] for dissipative systems, [10] for conservative ones). We treat this subject in sect.6. Interesting aspects are the universal features of the onset of diffusion and its critical corrections, the particular critical behavior at thresholds of new channels within the diffusive regime, the symmetry breaking states as well as bands of non-chaotic running modes.

In the concluding sect.7 the anomalous diffusion induced by a super-structure and diffusion on a fractal are treated within the frame of discrete nonlinear dynamics eq. (2).

2. EXPERIMENTAL OBSERVATION OF SELFGENERATED NOISE

In an increasingly larger variety of systems the macroscopically observable amplitudes $x_\alpha(t)$ remain finite with a permanent nonperiodic bounded time dependence although the boundary conditions are time independent. A pioneering experiment was performed by G. Ahlers [11] , who observed this behavior in the effective heatflow through a Rayleigh-Bénard fluid layer heated from below. The external parameter a in this experiment is the Rayleigh-number $Ra \propto \Delta T$. $\Delta T/H$ is the temperature gradient across the layer of thickness H. The relevant variables x_α are at least the Nusselt number Nu(t), the magnitude of the fluid's convective motion, which is spatially ordered in a regular roll pattern, and the magnitude of the temperature field with a corresponding spatial structure.

If one truncates the equations of motion to these three amplitudes denoted by z(t), x(t), y(t), one gets from the Navier-Stokes equation and the heat equation a three-component set of coupled o.d.e.s (1), the famous Lorenz equations [1]

$$\dot{x} = -\sigma x + \sigma y, \quad \dot{y} = -y + ax - xz, \quad \dot{z} = -bz + xy. \tag{3}$$

σ is the Prandtl number usually chosen as 10, $b \cong 8/3$, $a = \Delta T/\Delta T_0$ the external control parameter. ΔT_0 is the temperature difference at the onset of convection. Although (3) is expected to be a sufficient approximation to the fluid equations for $a \cong 1$, only, they have become the prototype of chaos-producing equations. Chaos starts at $a_T = \sigma(\sigma+b+3)/(\sigma-1-b) \cong 24.74$ when (3) represents the real fluid hardly anymore.

Ahlers originally used a fluid layer with aspect ratio Γ = length/2 $\times$ hight = 5.27. The value of Γ affects in particular the fluid's behavior in the transition regime from the laminar, regular state to the chaotic motion. If Γ is small or competing additional modes are suppressed by other experimental measures (e.g. by a magnetic field in Hg convection [12]), the transition is characterized by the appearance of a τ_0-periodic motion, possibly locking of two competing oscillations, followed by a slowing down through period doublings $2\tau_0, 4\tau_0, \ldots, 2^n\tau_0, \ldots$ showing up as successive bifurcations of the amplitude. In the τ_0-periodic state the sequence of maxima is $z_0 = z_1 = \ldots$, in the $2\tau_0$ state it is $z_0, z_1, z_2 = z_0, z_3 = z_1, \ldots$ with $|z_0-z_1|$ increasing from 0 at the bifurcation point a_1 with increasing $a \propto \Delta T$. At a_2 in addition z_2 starts to differ from z_0, so $z_0, z_1, z_2, z_3, z_4 = z_0, \ldots$ constitutes a period $4\tau_0$ state, and so on.

The sequence of intervals $L_n = a_{n+1} - a_n$ with stable period $2^n\tau_0$ scales geometrically in length (Fig. 1, l.h.s.). Asymptotically with $n\to\infty$ one expects from theory

$$L_{n+1} = L_n/\delta. \tag{4}$$

For dissipative systems with a quadratic maximum (common case) it is always $\delta \cong 4.67$ [4-6] . This universality of δ has been realized by Feigenbaum [5] for the discrete dynamics and has since then been observed in many systems. In discrete dynamics with discontinuous (finite) slope at the maximum it is $\delta = 2$ (see sect.5). If the order z of the maximum varies from $z = 1$ to $z = 2$ (regular, quadratic), $\delta(z)$ varies from its "classical" integer value 2 to its "critical" one 4.669 201 609... In conservative 2-dim.discrete regular maps it is $\delta = 8.721\ 097\ 200...$ [13] .

The magnitude of bifurcation of the physical variables also shrinks with increasing n. It scales down by a factor $\alpha = 2.502\ 907\ 875...$ [5] at corresponding states in L_{n+1} and L_n. ($\alpha_1 = 4.018\ 076\ 704...$ and $\alpha_2 = 16.363\ 896\ 879...$ are the rescaling factors for 2-dim. conservative maps [13].) This results in a successive decrease of the peak heights of the leading frequencies $(2\pi/\tau_0)\ 2^{-n}$. It is roughly $I_{n+1}/I_n \cong (1/2\alpha)^2$ corresponding to $\cong 14$ dB (cf. end of sect.4).

Both, the intensity drop as well as the interval shrinking along the cascade of subharmonic generation have been observed in many open, far from equilibrium situations. A representative, though presumably not complete survey is given in table 1. There are furthermore suggestions of additional experimental realizations of period doubling and selfgenerated noise, see table 2.

The transition region of period doubling is of finite extension due to the geometrically decreasing length of stable period 2^n intervals. $\sum_{n=0}^{\infty} L_n = L_0\ (1+\delta^{-1}+...) = L_0/(1-\delta^{-1})$. At about $L_0/(\delta-1)$ or 27% beyond the first interval of the cascade is the transition point a_∞ behind which selfgenerated noise or chaos occurs. This happens gradually, accompanied by a reduction of line structures in the spectra and interupted by windows of stable periodic states. In the following we shall concentrate on this parameter range (see fig.1, r.h.s.).

We conclude this section with a summary of the phenomenon in table 3 and a typical example (fig.2). One has to expect such a behavior whenever a few degrees of freedom (or order parameters) interact nonlinearly in an open system, entirely static or harmonically driven. It does not matter whether it is a classical or quantum mechanical system as e.g. He II. It is one of only a few possible routes to chaos.

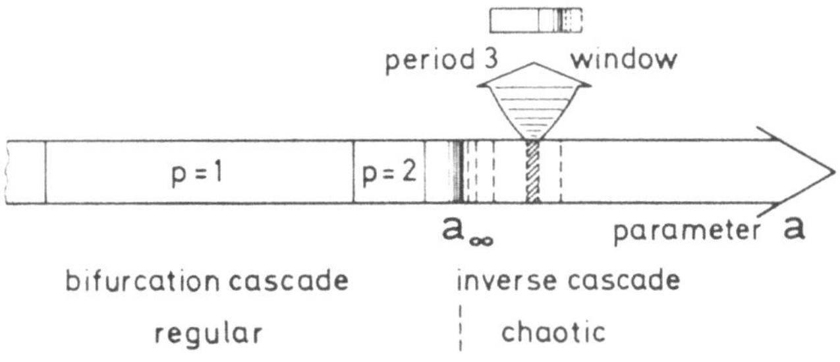

Fig. 1: Bifurcation sequence (schematic). a_∞ strange attractor, onset of chaos. L.h.s. direct cascade $p=2^n$, r.h.s. inverse cascade with example of regular window.

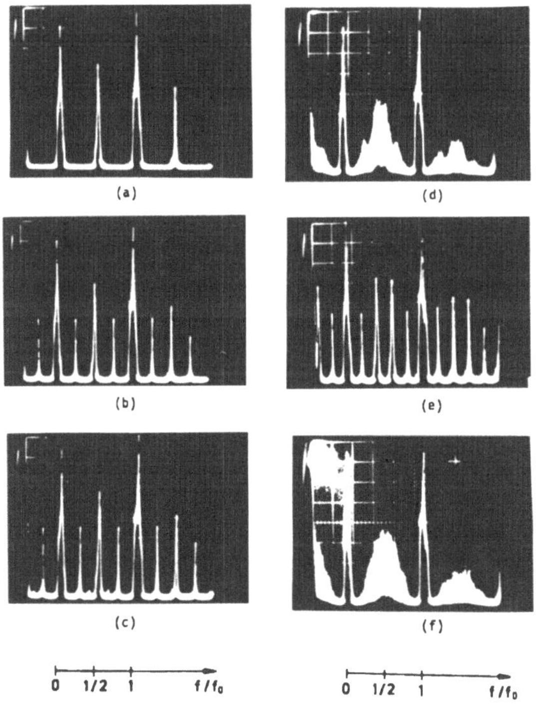

Fig. 2: Spectra of a nonlinear electric oscillator which is harmonically driven with increasing voltage (W. Lauterborn, W. Meyer-Ilse, Göttingen 1982)

Table 1: Various experiments showing subharmonic generation and self-generated noise [11,14-33] .

Rayleigh-Bénard heat flow in fluid layers, $a \propto \Delta T$.
G.Ahlers 1974, A.Libchaber et al. 1980,
J.P.Gollub et al. 1980, M.Giglio et al. 1981.

Momentum flow in Taylor's concentric rotating cylinders, $a \propto \Omega$. H.L.Swinney et al., 1975, 1979, Yu.N.Belyaev et al.1979.

"Chemical turbulence": concentration fluctuations in stirred open tank reactors, $a \propto$ flowrate. O.E.Rössler 1976, L.F.Olsen, H.Degn 1977, R.A.Schmitz et al. 1977, J.C.Roux et al. 1980/81.

Nonlinear RCL (varactor diode) circuits, $a \propto$ driving voltage. P.S.Linsay 1981, J.Testa et al. 1982, W.Lauterborn et al. 1982.

Acoustic cavitation noise, driven piezo-electrically, $a \propto$ driving voltage. W.Lauterborn, E.Cramer 1981.

"Optical turbulence" in lasers. H.Haken 1975, H.M.Gibbs et al. 1981, T.Arecchi et al. 1982.

Faraday's shallow water surface waves, R.Keolian et al. 1981.

Quantum vortex line generation by 1st sound in superfluid He II. C.W.Smith et al. 1982.

"NMR turbulence", E.Brun et al. 1982.

Table 2: Candidates to show nonlinearity noise and period doubling bifurcations [2,34-37] .

"Solid state turbulence": charge density waves, superionic conductors, B.A.Huberman et al. 1979.

Noisy Josephson junctions, SUPARAMPS, B.A.Huberman et al. 1980
R.Y.Chiao et al. 1976; N.F. Pedersen et al. Phys.Lett 90A,150(1982)

Pinned dislocation lines, C.Herring et al. 1980.

Nonlinearly coupled plasma waves, D.A.Russel et al. 1981.

Table 3: Change in dynamical quality in some physical observable A under increasing external stress; one possible route among others.

Temporal behavior A(t)	Power spectrum $\tilde{A}(\nu)$
• stationary, A = const	-
•• periodic quasiperiodic	discrete line(s)
∴ period doubling	subharmonic generation discrete lines series
∷ permanently t-dependent irregular, nonperiodic pseudo-stochastic sensitive to disturbances correlations decay	broad band spectra self-generated noise

3. DISCRETE NONLINEAR 1-DIMENSIONAL DYNAMICS

How can we explain the regular structures in the spectra as well as the broad band noise, the regular periods together with the chaotic aspects on slight changes of the static stress parameter of so many systems? This certainly cannot be achieved by a linear theory. Then a transformation to the principal axes always leads to a set of eigenvalues, implying either a steady state, a periodic state, or at most a quasiperiodic one, either stable or unstable (i.e. increasing without limit). A linear theory cannot cope with phenomena as described in sect.2.

Therefore, whatever theoretical description is used, it must retain at least the essentials of the nonlinear aspects. This, in general, makes an analytical treatment impossible. One either falls back upon numerical methods or reduces the theory so drastically, albeit carefully preserving the basic nonlinearity, that analytic treatment becomes possible. This motivates the projection onto discrete time dynamics of only one variable for dissipative systems or two variables for conservative systems.

A stroboscopical description of an observable at discrete times t_o, t_1, $t_2, \ldots t_\tau, \ldots$ provides discrete trajectories x_o, $x_1, \ldots, x_\tau \ldots = \{x_\tau\}_{\tau=o}^{\infty}$. The astonishing recent discovery is that these trajectories may still reflect the macroscopic, deterministic equations of motion of the system of interest, not only in steady stable situations, $x_\tau \to x^* = \text{const}$ or in periodic states $x_{\tau+p} = x_\tau$, $\tau = 0,1,\ldots,p-1$, but also for pseudo-stochastic motions. Their typical feature is a strong enhancement of initial uncertainty (e.g. due to finite precision of measurement). Hence long-time predictability for these states is lost, though short-time predictions by means of (2) are still possible. Thus the dynamical law $f_a(x)$ may produce trajectories which look chaotic. They are nonperiodic, permanently τ-dependent, erratically varying in a bounded range (henceforth normalized to [0,1] or [-1, +1]), filling it eventually densely (although in general not equally dense). Nearby initial points x_0, $x_0+\varepsilon_0$ may create trajectories which quickly diverge in distance, reflecting sensitivity to disturbances. Correlations decay with increasing time between the events.

In mathematical terms the system behaves *ergodic* and *mixing* and can be described "as if" it were stochastic. The basic quantities are the invariant density $\rho^*(x)$, the stationary correlation function c_τ, the spectrum S_ω. They all depend on the map (2), i.e. the dynamical law $f_a(x)$, where a indicates the map's change with varying external parameter.

The invariant density is defined by

$$\rho^*(x)dx = \lim_{T\to\infty} (1/T) \sum_{\tau=0}^{T-1} \chi_{[x,x+dx]} (x_\tau). \qquad (5)$$

It is the relative number of visits to the subinterval [x,x+dx] in the course of time. Ergodicity means that $\rho^*(x)$ does *not* depend on x_0 for almost every (a.e.) initial value x_0 and that it induces an absolutely continuous measure by

$$m(I) = \int_I \rho^*(x')dx', \quad I \subseteq I_{tot}.$$

Its invariance under time translations means

$$m(I) = m(f^{-1}(I)).$$

f^{-1} denotes the inverse map, I is any measurable subinterval of I_{tot}, the total domain of definition of the map.

The infinitesimal generator of the time development, H, which defines the master-equation

$$\rho_{\tau+1}(x) = H\rho_\tau, \quad \rho_{\tau=0} = \rho_0 \qquad (6)$$

for any initial distribution $\rho_0(x)$ is determined by $f_a(x)$ via

$$(H\rho)(x) = \int_{I_{tot}} \delta(x-f_a(y)) \; \rho(y)\, dy. \qquad (7)$$

Starting with $\rho_0 = \delta(x-x_0)$ one recovers the deterministic law (2), and (6) creates the trajectory beginning with x_0. If f is also mixing a smooth initial distribution approaches the invariant density ρ^*, which solves the socalled Frobenius-Perron equation

$$\rho^*(x) = \int \delta(x-f_a(y))\rho^*(y)dy. \tag{8}$$

Ergodicity implies the equivalence of the time means with the $\rho^*(x)$-mean for a.e. x_0. While time means are frequently used in numerical computations, analytical statements are based on ensemble means with ρ^*.

A measure for the sensitivity of the motion to disturbances is the divergence rate of trajectories expressed by the Lyapunov exponent

$$\lambda = \langle |\ell n\, f_a'(x)| \rangle \quad . \tag{9}$$

Two trajectories initially a distance ε_0 apart have a mean distance

$$\varepsilon_\tau \cong e^{\lambda\tau}\varepsilon_o$$

after τ steps. λ is negative in the stable state: $\lambda < 0$. Marginal stability (for instance at a bifurcation point) is characterized by $\lambda = 0$. $\lambda > 0$ indicates instability. Then, at least in the mean, trajectories starting near a given trajectory diverge from it. A reliable prediction can be given over

$$\tau_{pre} \cong \lambda^{-1}\ell n(\varepsilon_{max}/\varepsilon_{min}) \tag{10}$$

time steps if ε_{max} is the tolerable x-error and ε_{min} the accuracy to which the initial state can be fixed. To widen the range of forecasting by increasing the precision of measurement is difficult, since τ_{pre} depends from ε_{min} only logarithmically. Since $\ell n(\varepsilon_{max}/\varepsilon_{min})$ usually is smaller than about 7, the intrinsic dynamical divergence time λ^{-1} essentially gives the range of prediction.

For systems with d variables x_α, $\alpha = 1,2,\ldots,$ d there are d Lyapunov exponents λ_α. $e^{\lambda_1} \geq e^{\lambda_2} \geq \ldots \geq e^{\lambda_d}$ are the mean contraction (dilation) rates with $e^{\lambda_\alpha} = \lim\, [j_\alpha(T)]^{1/T}$, $j_\alpha(T)$ eigenvalue of $J(x_{T-1})\; J(x_{T-2})\ldots\; J(x_0)$, $J(x)$ the Jacobian $(\partial f_\alpha/\partial x_{\alpha'})$. Chaotic behavior shows up if at least one λ_α is positive. The total sum $\sum_{\alpha=1}^{d} \lambda_\alpha$ is a measure for the change in the d-dimensional volume per unit time. Many systems of physical interest show volume contraction with simultaneous dilation along some of the d directions. Since the motion in the phase space is bounded, this stretching leads necessarily to a folding of trajectories. Because of the volume shrinking they eventually fill a manifold with a dimension considerably lower than d. For regular motions this dimension is integer, for chaotic motions the notion of fractional dimension is more appropriate.

Two such non-integer dimensions have proven to be useful, although a lot of others can be introduced (see e.g. [38]).

i) The information dimension is defined by

$$I(\varepsilon) = \sum_{i=1}^{N(\varepsilon)} p_i \ \ln(1/p_i) \cong \ln(1/\varepsilon)^{D_I}, \quad \varepsilon \to 0. \qquad (20)$$

It tells us how the information obtained in a measurement with uncertainty ε increases with increasing precision. p_i is the fraction of time which the trajectory spends in a subcube of extension ε . There are $N(\varepsilon)$ such cubes necessary to cover the attractor which the trajectory approaches after the transient stage. $I(\varepsilon)$ is the amount of information about this attractor weighted with the frequencies with which the orbits visit different parts of it. According to [39], [38] D_I is related to the Lyapunov exponents.

$$D_I = k + (\lambda_1 + \lambda_2 + \ldots + \lambda_k)/|\lambda_{k+1}| \ . \qquad (21)$$

k is the largest integer for which the partial sum in the numerator is positive; $D_I = 0$ if $\lambda_1 < 0$, $D_I = d$ if the total λ-sum is positive.

ii) The Hausdorff or fractal [40] dimension D_H of a set S in a d-dimensional Euklidean space tells how the number of covering ε-cubes has to increase asymptotically with decreasing ε if the set is to have a finite measure. More precisely, let $r \geqq 0$ be any real number and

$$V_r = \lim_{\varepsilon \to 0} \ \inf_{\varepsilon_i \leqq \varepsilon} \ \sum_i (\varepsilon_i)^r \qquad (22a)$$

be the minimal volume of an r-dimensional covering. Imagine for example the set of interest is a 2-dimensional surface. Then, if $r = 3$ one gets $V_r = 0$ but if $r = 1$ one finds $V_r = \infty$. The Hausdorff dimension D_H is defined by the inequalities

$$V_r = \infty \text{ for all } r < D_H, \quad V_r = 0 \text{ for all } r > D_H. \qquad (22b)$$

V_{D_H} may be either 0 or ∞, but often is finite. In this case one has

$$N(\varepsilon) \cdot \varepsilon^{D_H} = V_{D_H} \ \text{ finite, i.e. } N(\varepsilon) \propto \varepsilon^{-D_H} \qquad (22c)$$

if all ε-cubes are equal.

The classical Cantor set S_C may serve as an example: S_C is the point set obtained by deleting the center third of [0,1], from the remaining subintervals again the center thirds, etc. The limit $\varepsilon \to 0$ is conveniently realized by $\varepsilon_n = 1/3^n$, $n \to \infty$. The number of subintervals increases as 2^n. So, $2^n/3^{D_H n}$ is finite for $n \to \infty$, if and only if

$$2/3^{D_H} = 1 \ , \quad D_H = \ln 2/\ln 3 \cong 0.631.$$

The Cantor set construction may be generalized as follows [40]: Replace each interval by ℓ subintervals obtained by reduction with factors k_i, $i = 1,2,\ldots, \ell$ and repeat this procedure ad infinitum. Then D_H is determined by

$$\sum_{i=1}^{\ell} k_i^{D_H} = 1. \qquad (23)$$

Other important tools for the description of the f_a-dynamics are stationary correlation functions.

i) The temporal correlation function

$$c_\tau = \langle \delta x \; \delta x_\tau \rangle = \lim_{T\to\infty} \frac{1}{T} \sum_{t=0}^{T-1} \delta x_t \; \delta x_{t+\tau} \qquad (24)$$

has distinct characteristic features in the different dynamical states. It is periodic if the trajectories merge into a periodic attractor, but it decays to zero with $\tau \to \infty$ in chaotic states. A mixed behavior does also occur: the initial variance $\Gamma_0 = \langle(\delta x)^2\rangle$ decays to a permanently oscillating function. We called this "periodic chaos" [4] ; in these states $c_{p\tau} - c_{p\infty}$ is simply decaying (p denotes the period).

The Fourier transform

$$S_\omega = \frac{1}{T} \sum_{\tau=0}^{T-1} e^{i\omega\tau} c_\tau, \; \omega = 2\pi\nu/_T, \; \nu = 0,1,\ldots,T-1 \qquad (25)$$

is identical with the power spectrum of the trajectory's Fourier amplitude, averaged over x_0 according to the stationary invariant density ρ^*. S_ω is a quantity directly accessible by measurement.

S_ω has only δ-peaks before chaos sets in. A broad band structure shows up in the ergodic, mixing states reflecting the correlation decay. The gross features of these broadband spectra will be traced back to some gross features of $f_a(x)$. In the states of periodic chaos one finds sharp lines on top of the broad background.

ii) Another characteristic correlation function is the analogue of the static spatial correlation in statistical physics. As introduced in [41] one considers

$$C(\ell) = \int_{y\leqq\ell} d^d y c(y) = \langle \text{number of pairs with distance} \leqq \ell \rangle$$

$$\text{with } c(y) = (1/T^2) \sum_{\tau_1\tau_2=0}^{T-1} \delta(\vec{x}_{\tau_1} - \vec{x}_{\tau_2} - \vec{y}). \qquad (26)$$

The correlation integral $C(\ell)$ defines another critical exponent on chaotic attractors.

$$C(\ell) \propto \ell^\nu \quad \text{for } \ell \to 0. \qquad (26')$$

Schwarz' inequality implies [41] $\nu \leqq D_H$.

Let us finish this brief introduction into discrete dynamics by mentioning that recently the development of a linear response theory [42] has started.

How does the map $f_a(x)$ which generates the trajectories, spectra, correlations, prediction range etc. look like? One would wish, of course, to deduce it from the correct equations of motion. This is not easily possible, since in general $f_a(x)$ is not uniquely invertible, whereas o.d.e.s or p.d.e.s of physical systems are. A usual route to get at least an idea, which map $f_a(x)$ should approximately represent a given system, is to consider a mode expansion of the original equations, carefully retaining their nonlinearity. For example, the Lorenz equations (3) are a three-mode approximation of the fluid equations

$$\partial_t v_i = -(\vec{v}\cdot \text{grad})\, v_i - \text{grad}_i p/\rho_0 + \nu\Delta v_i, \qquad (27a)$$

$$\partial_t T = -(\vec{v}\cdot \text{grad})T + \lambda\Delta T. \qquad (27b)$$

Other systems are described from the very start by a coupled set of o.d.e.s (1). If the phase space dimension d is rather small (but $d \geqq 3$) one may use a Poincaré plot or a Lorenz plot [1] to find $f_a(x)$ from the numerical solution of (1). The Lorenz plot is defined by the sequence of maxima $x_0, x_1, x_2, \ldots$ of one of the variables or, alternatively, by the series of time intervals $\Delta t_1, \Delta t_2, \Delta t_3, \ldots$ between successive events of equal amplitudes, or by any other sequence of values of an appropriately chosen property. That these particular stroboscopic data assemblings lead to a deterministic map $f_a(x)$ can be explained for the three equ.s (3). Let z_0 be a maximum, so $\dot{z}_0 = 0$. This fixes $x_0 y_0$ at t_0. Since it turns out that the solutions of (3) for suitable values of the parameters are attracted to a $D_H \cong 2.06$ dimensional limit set, those two initial conditions determine the solution and in particular the next maximum z_1, or Δt_1, etc. with a similar precision as D_H approximates 2.

The (2-dimensional) Poincaré plot maps successive crossings of the solution through a conveniently chosen surface section. The very existence of a low (d) dimensional map signals that the physical system of interest has only a few effective degrees of freedom which are *not* disturbed by thermal noise. If noise were effective, one would not have a law $f_a(x_\tau)$, since even nearly equal x_τ at different times will have quite different fates, leading to a *distribution* of $x_{\tau+1}$'s instead of a well determined value.

No analytical procedure is known to derive f_a from the set of o.d.e.s. Numerical methods are also suited to data analysis, see [43]. The appropriate dimension d may be found from a single measured quantity A(t): At first a d-dimensional phase space is defined by $A(t+\alpha\Delta t) \equiv x_\alpha(t)$. Sometimes $d^\alpha A/dt^\alpha$ is considered, instead. It seems to be qualitatively independent of the delay time Δt. The optimal dimension d is then found for instance by considering the conditional probability to find $A(t+(d+1)\,\Delta t)$ provided the $x_\alpha(t)$ for $1 \leqq \alpha \leqq d$ are given. The smallest d for which this is sharply peaked is a good choice.

Clearly, since $f_a(x)$ is derived by an approximation, what is left of the original dynamics must be contained in typical structural properties of $f_a(x)$. One of them is that $f_a(x)$ *cannot* be monotonous. This would contradict the boundedness of the physical observables. $x_0, x_1, x_2, \ldots$ can neither increase nor decrease indefinitely. $f_a(x)$ must therefore have at least one maximum and minimum or equivalently have discontinuities. Its elementary parts are pieces of low order curves. The parameter a can affect the height of the maxima (minima, jumps) as well as the number of pieces or details in their form.

Most phenomena observed so far seem to be related to the height (width) of the maximum, thus the most important kind of parameter dependence. Hence, the "generic" form of $f_a(x)$ is

$$f_a(x) = 1-\mu|x|^z, \quad \mu = 2a, \quad x \in [-1,+1], \; 0<\mu\leqq 2. \tag{28}$$

μ is the "height" of the map.

The order z of the maximum will be 2 if the physics is expected to be analytic in x. For z = 1 the slope has a discontinuity, so representing some kind of a bivalent situation.

Modifications of (28) are asymmetries as for instance a shift of the maximum, deformations of the flanks, several pieces like (28), discontinuous maps like $f(x) = ax \pmod 1$, $x\in[o,1]$, $a \geqq 1$. This leads to modifications or superpositions of the elementary phenomena. For instance, several attractors can appear simultaneously if the condition of negative Schwarzian derivative S_f of the maps [44, 45] is not met [46]. S_f is defined by

$$S_f = f'''/f' - (3/2)(f''/f')^2, \quad f(x)\in C^3. \tag{29}$$

$S_f < 0$ (with exception of the extremum $f'=0$) means that $1/\sqrt{|f'|}$ is convex. For the generic form (28) we find

$$S_f = (1-z^2)/2x^2 < 0 \quad , \quad \text{if } z > 1, \text{ all } \mu . \tag{30}$$

As has been discussed by Mayer-Kress and Haken [46], the bifurcation behavior as a function of a is changed if $S_f > 0$; there is coexistence of point attractors and periodic or chaotic attractors, and all of these attractor types may be destroyed by arbitrary small amounts of external noise. A direct transition from a fixed point to a chaotic attractor is possible, so providing a model for the observation in Rayleigh-Bénard fluid layers with not too small aspect ratio Γ [11] . An example [47] of such a map is

$$f_{a,b}(x) = 4a\, x(1-x) \{1 + 8\, bx\, [1-3x +2(2-x)x^2]\}/(1+b).$$

It is still symmetric with respect to $x_c = 1/2$, is convex down if $-5/8 < b < 0$, but has $S_f > 0$ for increasing b.

We conclude this introductory discussion of the basic concepts of discrete dynamics by mentioning the effects of external noise, not to be mixed up with the selfgenerated pseudo-stochastic nonlinearity noise of the map $f_\alpha(x)$ itself. It may be coupled to the f-dynamics parametrically (e.g. noisy α) or additively:

$$x_{\tau+1} = f_\alpha(x_\tau) + \sigma\xi_\tau. \qquad (31)$$

The ξ_τ are independent Gaussian random variables with unit variance. The external noise adds to the selfgenerated nonlinearity pseudo-noise. It sometimes destroys attractors, it sometimes induces scarce but random transitions between different stable attractors. For increasing noise level σ the structure of the spectrum is washed out. In particular the largest observable order of subharmonics or of periodic windows decreases. All these items have been explored and discussed in [48-50].

4. STRUCTURE OF SPECTRA

Let us now discuss the properties of nonlinearity noise, using the concepts introduced before.

A particularly simple map is the hat (or tent) map, see e.q.(28) with z = 1. If $\mu < 1$ there is one stable fixed point $x^* = (1+\mu)^{-1}$; this is unstable if $1 < \mu \leqq 2$. In the unstable regime the Lyapunov exponent is $\lambda = \ln\mu > 0$, so the trajectories are sensitive to the initial condition. There is no stable higher period p. These would be solutions of

$$x^* = f^{[p]}(x^*) \;, \quad f^{[p]} = f(f^{[p-1]}(x)), \; f^{[1]} = f. \qquad (31)$$

The higher iterates $f^{[p]}$ are broken linear transformations too, with slope $\mu^p > 1$ if $\mu > 1$.

Although there is no stable period, the directly calculated trajectories seem to show some periodicity; its value p depends on μ. This phenomenon can be explained as follows. We consider first the tent map for $\mu = 2$. For a.e. x_0 the trajectories fill the entire interval [-1, +1] densely, apparently everywhere with equal density. The invariant density $\rho^*(x) = 1/2$ can be determined from the Frobenius-Perron equation (8). The argument is: ρ^* = const *is* a solution; due to ergodicity it is the only solution [51,52,4] ; by normalization find const = 0.5.

Of course, there are also initial values x_0 which lead to orbits of period p; *all* p are possible, since in particular period p = 3 is possible. More precisely: whenever a map $f_\alpha(x)$ allows to construct a piece of a trajectory with two successive increasing (or decreasing) steps followed by a jump below (above) both pre-images

$$x_0 < f(x_0) < f^{[2]}(x_0) \;, \quad f^{[3]}(x_0) \leq x_0$$

the famous result of Li and Yorke ([53], "Period 3 implies chaos") applies: all periods p are present. This famous result in turn is the special case of the magnificent and fundamental theorem of Šarkovskii [54].

For the $\mu = 2$ tent map precisely all rational x_0 lead to periodic orbits. Hence the Lebesgue measure of all p-periodic initial points is 0. Furthermore they are all unstable ($|f'| = 2$) and cannot be detected among the measure-1 set of nonperiodic orbits if arbitrarily small external noise is present. But note, on the computer x_0 always is rational, so *only* the periodic orbits are produced and not the typical ones, which are chaotic.

The informational dimension is $D_I = 1$, the Hausdorff dimension $D_H = 1$, too.

The correlation functions can be calculated exactly. Temporal correlations decay immediately [4].

$$c_\tau = c_0 \delta_{\tau,0} \quad , \quad c_0 = \langle (\delta x)^2 \rangle = 1/3. \tag{32}$$

Hence the spectrum is white, S_ω = const. = c_0/T. The pair distribution function is independent of the pair's extension, $c(y) = 1/2$, so the correlation exponent is $\nu = 1$, $C(\ell) = \ell/2$.

It might be useful to explain the temporal correlation decay in this simple example. Since $\langle x \rangle = 0$ it is $c_\tau = \int dx_0 \rho^*(x_0)\, x_0 x_\tau(x_0)$. The main reason that the correlation decays at all, is its representation by an integral over a smooth ρ^*; that it occurs so fast, namely already for $\tau = 1$, is due to the particular properties of the map: (i) ρ^* is symmetric, as well as ii) $x_\tau(x_0) = x_\tau(-x_0)$ for $\tau \geqq 1$ is symmetric, while x_0 itself is antisymmetric.

The following warning is in order: The symmetry of the invariant density is *not* a necessary consequence of the symmetry of the respective map or vice versa. Skew hat maps are an example. Figure 3 shows c_τ and the cos-transform for hat maps with skewness $2\alpha-1$. The invariant densities turn out to be $\rho^* = 1$ for every α. The correlation functions $c_\tau = c_0(2\alpha-1)^\tau$ decay monotonously or oscillatory depending on α above or below 1/2 after roughly $-1/\ln|2\alpha-1|$ steps. This shows that there are deviations from pure Brownian-like motion, although one still would call the trajectories chaotic. Similar observations have been described also for the parabola map if $\mu < 2$ [55].

If μ is decreased, the attractor is the subinterval $[1-\mu, 1]$. Near $\mu \cong \sqrt{2}$ the trajectories show an oscillatory component of period 2. To understand it we consider the second iterate $f_a^{[2]}(x)$, see fig.4, for the parabola. Two subintervals (determined by the position of the unstable fixed point) show up, in which $f_a^{[2]}$ looks similar (up to ± 1) to f_a itself at $a = 1$. Precisely for $\tilde{a}_1 = 0.919643...$ $f_a^{[2]}$ maps the entire subinterval $\tilde{I}_1$ on $\tilde{I}_1$ as well as

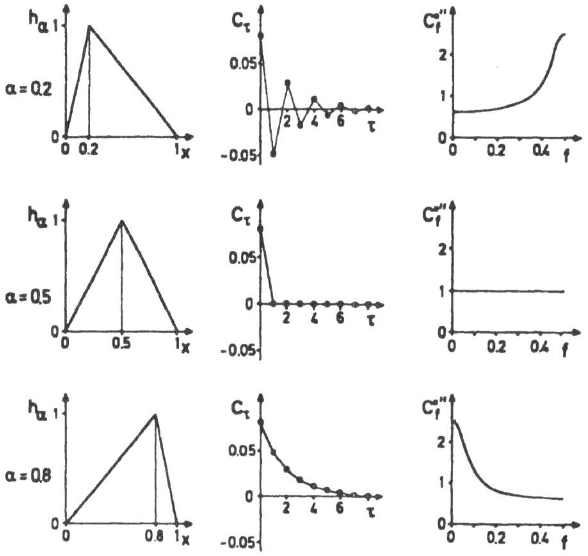

Fig.3: Correlation decay of several hat maps $h_\alpha(x)$ with maximum at $\alpha \in (0,1)$, $\rho^*_\alpha(x) = 1$, $c_0 = 1/12$ for all α

$\tilde{I}_2$ on $\tilde{I}_2$. Although parts of a 4th order parabola the sub-maps are ergodic and mixing [45], hence correlations decay. But $f_{\tilde{a}1}$ itself maps $\tilde{I}_1$ to $\tilde{I}_2$ and vice versa. The motion can thus be described as clearly periodic with superimposed small-scale chaos. The spectrum contains a sharp peak at $\omega/2\pi = 1/2$ but also broad band noise. This latter is reduced in level (since the amplitude only varies in $\tilde{I}_{1,2}$ instead of the full interval [0,1] and has some gross structure which reflects the skewness.

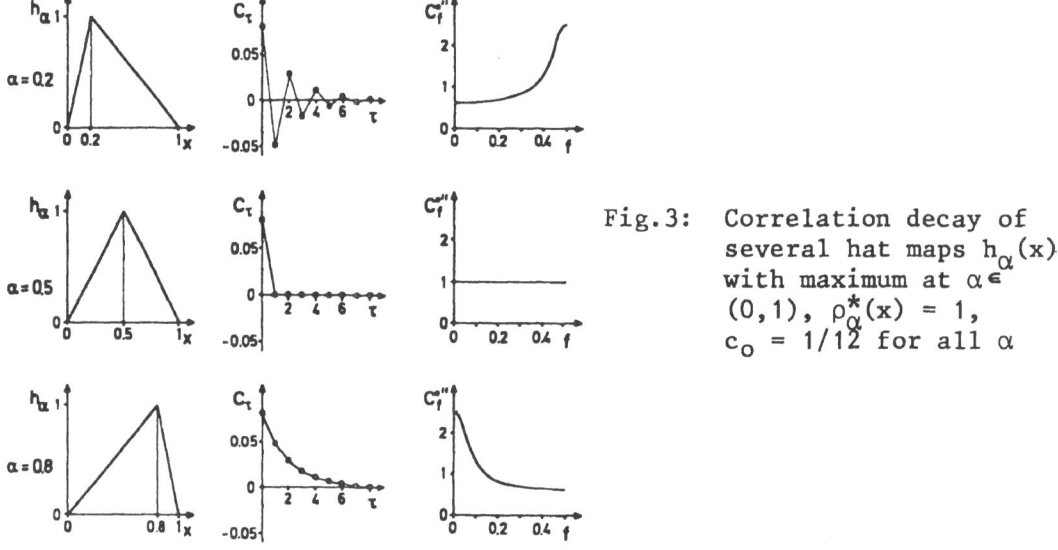

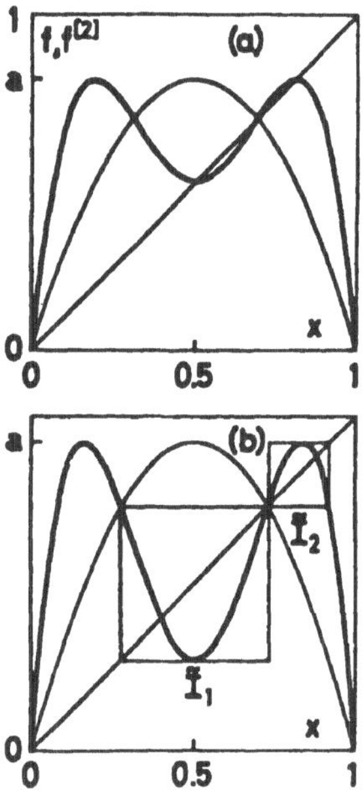

Fig.4: Iterated map $f_a^{[2]}$ together with f_a (thin line), parabola with $a = 0.919... = \tilde{a}_1$, period 2 chaos

If a is slightly above $\tilde{a}_1$ there occur occasional jumps from $\tilde{I}_1$ to $\tilde{I}_2$ and vice versa under $f_{\tilde{a}_1}^{[2]}$. The periodic component thus decays. The decay time is $\propto \sqrt{a-\tilde{a}_1}$ [56]. The reason is that the fraction of $\tilde{I}_{1,2}$ under the overshooting tip of $f_a^{[2]}$ decreases as this square root. The situation is basic for transfer by deterministic chaotic motion between different intervals, leading to diffusion-like hopping.

If a is slightly below $\tilde{a}_1$, the attractors of $f_a^{[2]}$ are even smaller than $\tilde{I}_{1,2}$. Thus there are two separate bands which are filled by the trajectories. Therefore $\tilde{a}_1$ is called a band splitting (merging) point.

There are more such band splitting points $\tilde{a}_n$. Slightly above $\tilde{a}_n$ there are 2^{n-1} bands, slightly below 2^n. They are periodically visited by the trajectory. Thus, each band splitting results in a period doubling, i.e. a new subharmonic in the spectrum. Simultaneously the decreasing width of the bands reduces the chaos level. The $\{\tilde{a}_n\}$ constitute the inverse cascade of states of periodic chaos [4]. The intervals $\tilde{a}_n-\tilde{a}_{n+1}$ shrink geometrically with the same δ, the band widths with the same α as along the period doubling forward cascade, denotes by $\{a_n\}$. Both converge to the same limit, $a_n \rightarrow a_\infty \leftarrow \tilde{a}_n$.

The spectra along the inverse cascade are displayed in fig.5. The mean chaos level scales approximately as [57,58]

$$S_{noise,n+1}/S_{noise,n} \cong (\alpha^{-2} + \alpha^{-4})/2, \tag{33}$$

the spectral power in successive subharmonics as ([59,58], for a survey

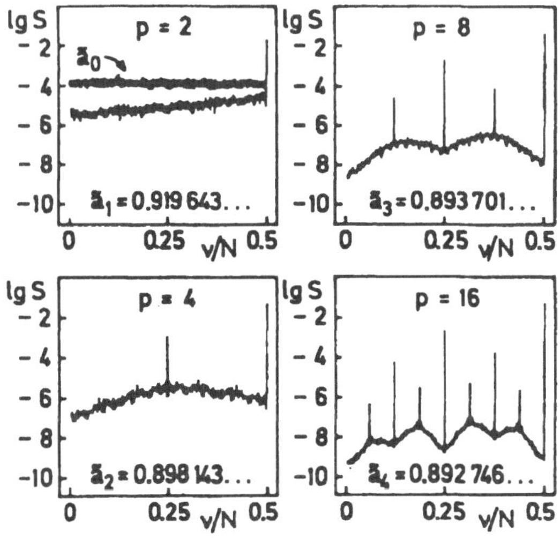

Fig.5: Spectra of periodic chaos along the inverse cascade

see [60])

$$S_{n+1}/S_n \cong [(\alpha - 1)/2\alpha^2]^2. \qquad (34)$$

If there is a parabolic maximum, $z = 2$, the corresponding $\alpha = 2.502$ yields a chaos decrease by 10.3 dB and peak high decrease by 18.4 dB. More details about the structure of the spectra, in particular for higher periods, are given in [61].

5. RENORMALIZATION GROUP

Direct and inverse subharmonic generation can be treated by a renormalization transformation (RGT), introduced by Feigenbaum [5] and by Coullet and Tresser [6]. It is based on the idea that $f^{[2]}$ in a subinterval resembles f in the original interval if the parameters are properly renormalized. For a parabolic maximum this is shown in fig.6. The analytic definition of the RGT of an arbitrary map is

$$(Tf)(x) = -(1/q)f^{[2]}(-qx). \qquad (35)$$

q scales the reduced interval to the original interval length. The reduced interval is fixed by demanding $Tf(o) = 1$. The sign takes care of the upside-down property of $f^{[2]}$.

T is a functional transformation in general. It becomes particularly simple for the broken linear transformation (28) with $z = 1$. We shall take this map for an easy exercise. T transforms a period p state to a p/2 state, describing period-halfing or going *backwards* along the cascade. The RGT's fixed point is reached by successively applying T^{-1} on a map with a certain property kept fixed. For instance $f^{[2]}(x_c) = f^{[3]}(x_c)$ defines the basic cascade with periods $p = 1 \times 2^n$; $f^{[4]}(x_c) = f^{[7]}(x_c)$ the period-3 cascade

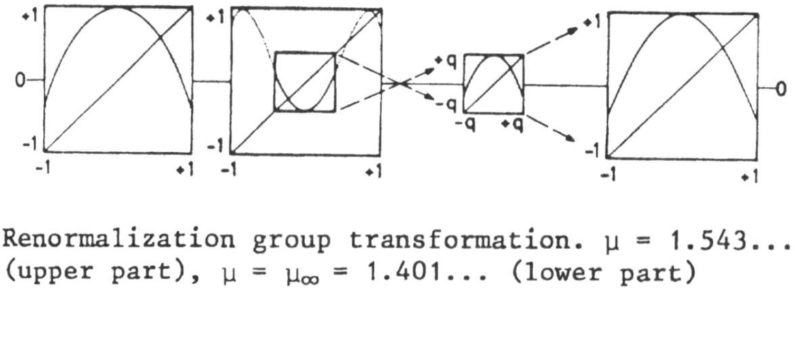

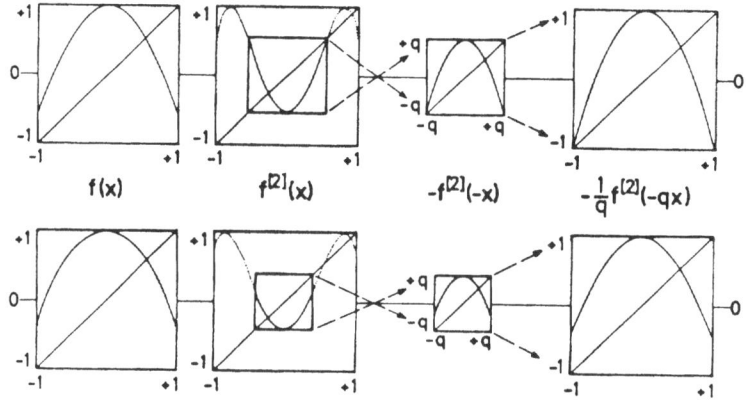

Fig.6: Renormalization group transformation. $\mu = 1.543... = \tilde{\mu}_1$ (upper part), $\mu = \mu_\infty = 1.401...$ (lower part)

with $p = 3 * 2^n$, etc. x_c denotes the phase point mapped on the maximum.

For the tent map the RGT degenerates to a parameter transformation instead of a functional transformation.

$$f_\mu^{[2]}(x) = 1-\mu|1-\mu|x|| = 1-\mu+\mu^2|x|, \text{ x small.}$$

$$(Tf_\mu)(x) = (\mu-1)/q \quad -\mu^2|x| = 1-\mu'|x|. \qquad (36')$$

$$\text{RGT: } q(\mu) = \mu-1 \equiv 1/\alpha(\mu) \; ; \; \mu' \equiv T\mu = \mu^2 \qquad (36'')$$

Remembering that T^{-1} creates period doubling let us start with period 1 chaos at $\tilde{\mu}_0 = 2$. The first band splitting (period 2 chaos) happens at $\tilde{\mu}_1 = T^{-1}2 = \sqrt{2}$. The reduction factor is $\alpha_1 = (\sqrt{2}-1)^{-1} \cong 2.41$ for the central band and $\alpha_2 = \sqrt{2}\alpha_1 \cong 3.41$. Period 4 band splitting happens at $\tilde{\mu}_2 = T^{-1}\tilde{\mu}_1 = \sqrt{\sqrt{2}}$, and the general inverse cascade is

$$\tilde{\mu}_n = 2^{2^{-n}} \to 1. \qquad (37)$$

The limit $\mu_\infty = 1$ is the boundary of the chaotic regime. The direct cascade of subharmonics is missing for the hat map, there is *only* the inverse cascade of periodic noisy states.

$\mu_\infty=1$ is the fixed point solution of $T\mu^* = \mu^*$. Linearizing one gets precisely one eigenvalue. $\mu = \mu_\infty + \varepsilon$: $L_{\mu^*}\varepsilon = \varepsilon'$ or $\varepsilon' = 2\varepsilon$. So the distance between successive bandsplitting states decreases asymptotically by 1/2, which means

$$\delta = 2. \qquad (38)$$

This can, of course, also be derived from

$$\delta = (\tilde{\mu}_n - \tilde{\mu}_{n+1}) \; / \; (\tilde{\mu}_{n-1} - \tilde{\mu}_n) \text{ using eq.(37).}$$

The scaling factors $\alpha_n = (2^{2^{-n}}-1)^{-1}$ diverge when approaching the fixed point. This is in contrast to maps with smooth maximum. In particular for the 2nd order maximum of a parabola map $\alpha_n \to \alpha = 2.502...$ is finite. A formal consequence of $\alpha_n \cong 2^n/\ln 2 \to \infty$ is the singular nature of $f^* = 1 - |x|$; this limiting fixed point map does not belong to the domain of the functional transformation T.

The physical consequence is the very rapidly decreasing chaos level along the inverse cascade as well as the fast submersion of peaks of higher subharmonics in an external noise; roughly it is $S_n \propto (\ln\sqrt{2})^n/2^{n^2}$. The bands become rather thin threads near $\mu_\infty = 1$.

The attractor which is approached for $n\to\infty$ is even more strange as usual. The number of bands increases with $n\to\infty$ indefinitely, their total width shrinks to zero. The limiting attractor has the cardinality of the continuum, but measure zero. Its Hausdorff dimension is $D_H = 0$, while for parabolic maximum it is $D_H \cong 0.53$ (see [62]). This can be understood from the general expression (23). For 2nd-order maxima the reduction factors of the two successively created new bands are α and approximately α^2. Then D_H follows from (23).

$$1/\alpha^{D_H} + 1/\alpha^{2D_H} = 1, \text{ or } \alpha^{D_H} = (1+\sqrt{5})/2, \quad (39)$$

which is the golden mean, the "most irrational number". For the hat map one finds

$$D_H(n) = \ln\,(1+2^{-D_H/2}) \;/\; \ln\alpha_n \quad (40)$$

The hat map has been studied extensively also by H.Mori et al. [63-65]. In particular the eigenvalues of the Frobenius-Perron operator (7) and the eigenfunction expansion of the temporal correlation functions have been elucidated, and the spectra along the inverse cascade.

6.SELFGENERATED DIFFUSION

Up to now we have considered maps for which the phase was limited to a certain interval I_{tot}. If the maximum and a δ-neighbourhood are larger than the largest x_0, the trajectories may leave the initial interval. (This happens e.g. under $f^{[2^n]}$ above band merging values $\tilde{a}_n$.) Having been mapped out of the initial interval I_{tot} the future fate of the phase x_τ depends on the definition of the map in the neighbourhood of I_{tot}. In some physical systems one meets a periodic extension of the map onto the whole real axis.

$$X_{\tau+1} = F_a\,(X_\tau) = X_\tau + f_a\,(X_\tau), \; -\infty < X_\tau < \infty. \quad (41)$$

Examples are Josephson junctions (in which the current is $I = I_c \sin\phi$ and depends periodically on the phase difference ϕ) or solids with periodic lattice. $f_a(X+1) = f_a(X)$ is assumed to be 1-periodic. It may be sinusoidal or broken linear (see fig.7) for instance. The dynamics generated by (41) is translational invariant, i.e. $\{X_\tau\}$ and $\{X_\tau+M\}$ are equivalent, M integer.

The basic interval (cell) has length ℓ. Different cells are denoted by N = 0, +1, +2,..., -1, -2,... X_τ varies at random over the whole axis if f_a is a chaotic map of sufficient amplitude. It

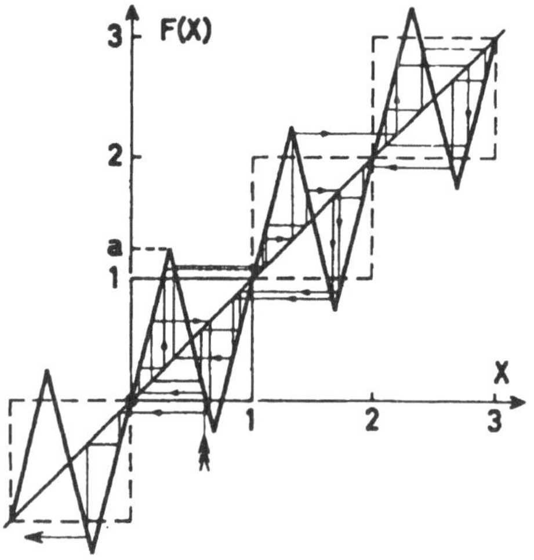

Fig.7: Piecewise linear periodic map with pseudo-stochastically hopping trajectory

can be decomposed into two pseudo-random variables

$$X_\tau = N_\tau + x_\tau , \quad 0 \leqq x_\tau < 1. \tag{42}$$

The fractional part $x_\tau(x_0)$ denotes the position within the box $N_\tau(x_0)$.

Whenever x_τ is mapped into the subinterval $I_{\tilde{\delta}} \subset [0,1]$ in which $f_a(x_\tau) \geqq 1$ the phase jumps by $\Delta(x_\tau)$, $\Delta(x) = [F_a(x)]$ is the integer part of the map. It describes the magnitude of the jump, and its positive or negative direction.

In general there is a mean drift and a diffusive broadening of trajectories starting somewhere in the first cell, N = 0. The drift velocity is

$$v = \langle \Delta \rangle = \lim_{t \to \infty} (1/t) \sum_{\tau=0}^{t-1} (X_{\tau+1} - X_\tau). \tag{43}$$

The variance can be proven to increase $\propto$ t. For the diffusion coefficient

$$\langle (X_t - \langle X_t \rangle)^2 \rangle / t = 2 D \tag{44}$$

one can derive a correlation function formula [8,66].

$$2D = \lim_{t \to \infty} (1/t) \sum_{\tau=0}^{t-1} \sum_{\lambda=0}^{t-1} \langle \delta\Delta(x_\tau) \quad \delta\Delta(x_\lambda) \rangle , \tag{45'}$$

$$= \langle (\delta\Delta)^2 \rangle \, (1 + 2 \lim_{t \to \infty} (1/t) \sum_{\tau=0}^{t-1} \sum_{\lambda=1}^{\tau} C_\lambda^0). \tag{45''}$$

C_λ^0 is the normalized correlation of the jump function $\Delta(x)$. The averages are either time or ensemble means (by ergodicity). The statistical ensemble is entirely determined by the fractional part of the map F, namely $g(x) = F(x) - \Delta(x)$, $0 \leqq g < 1$. It is independent of the cell number N due to tranlational invariance.

If the length $\tilde{\delta}$ of the jump region $I_{\tilde{\delta}}$ is small, successive jumps are uncorrelated, $C_\lambda^0 = 0$.

$$2D = \langle (\delta\Delta)^2 \rangle \cong \tilde{\delta} . \rho^*(x \in I_{\tilde{\delta}}) \tag{46}$$

describes the onset of diffusion. Since $\tilde{\delta} \propto (a - a_c)^{1/z}$ one immediately finds a universal power law which describes the onset of diffusion at $a_c = 1$ [7,8]. z is the order of the maximum of g(x) (or F(x)), as usual.

$$D \propto (a - a_c)^{1/z}. \tag{47}$$

But if $g_a(x)$ creates an invariant density which has peculiarities for $a \to a_c$ one gets critical corrections of power law or logarithmic type [66]. We show another, similar phenomenon in fig.8. It displays the behavior at the opening of a "new channel", namely, the possibility to jump to the next nearest cell. As function of $\varepsilon = |a - 2| \ll 1$ the diffusion coefficient varies as

$$D_- = (1/8) \, [1 + \varepsilon \ln 4\varepsilon / \ln 2] , \quad D_+ = (1/8)[1 - \varepsilon \ln \varepsilon / \ln 2]. \tag{48}$$

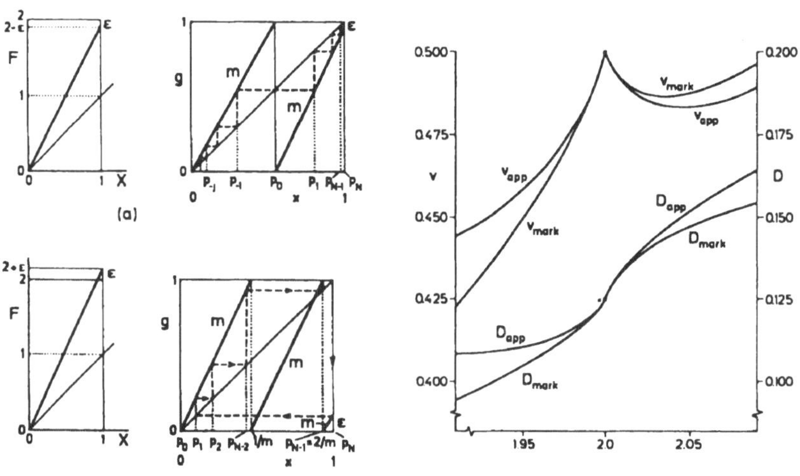

Fig.8: Critical behavior of drift and diffusion coefficient at the onset of jumps to the next nearest cell for the broken linear maps with height $2 - \varepsilon$ or $2 + \varepsilon$.

A new phenomenon appears for periodic maps with smooth maximum, i.e. if $|F'(x)| < 1$ near the top. Then there may be stable fixed points or stable periodic attractors of the reduced, elementary g-statistics (fig.9). That leads to entirely regular X_τ-trajectories with a well defined sequence of jumps between cells. If there is a surplus of jumps into one direction, one gets "running modes"; one may also find "localized modes", where for instance the phase jumps to and fro two cells, a distance $[a]$ apart.

The regular modes always occur in certain intervals [8,9], since the g-attractors stabilize in open intervals of the parameter a. There are five such regular "bands" connected with nearest neighbour jumps of period 1 or 2 attractors of the sinusoidal map (fig.10)

$$F_\mu(X) = X + \mu \sin 2\pi X. \quad (49)$$

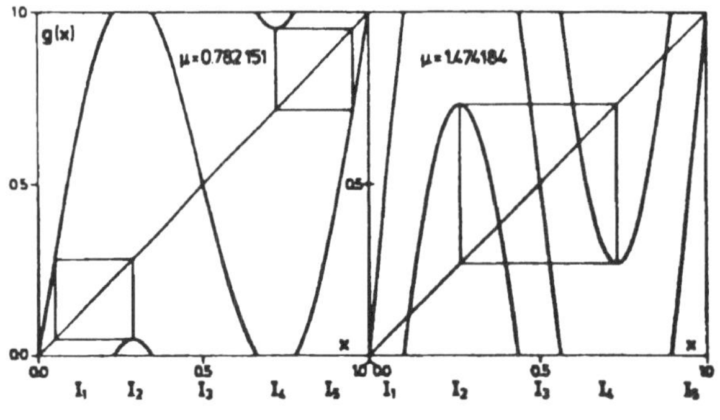

Fig.9: Reduced sinusoidal map for some parameters μ (see eq. (49)) having stable period 2.

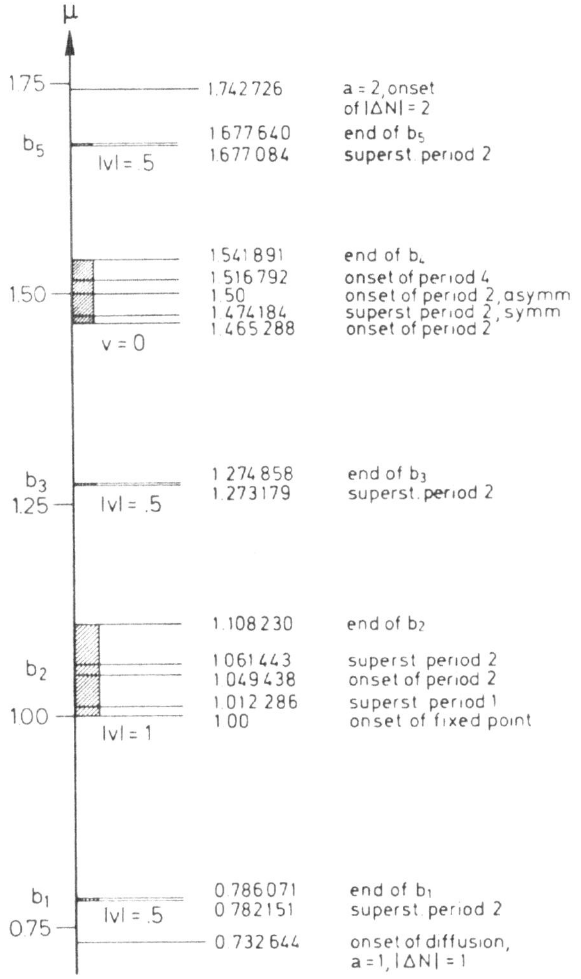

Fig.10: Location of nondiffusive regular bands with nearest neighbour jumps. Each band consists itself of period doubling cascades, inverse cascades, windows, governed by the universal numbers δ and α. These bands repeat for $\mu > 1.74...$ and so on. Further bands with different periods also exist.

Fig.9 shows as examples the stable g-map cycles in the running states of band b_1 and the localized band b_4. Just before the onset of running mode bands there is critical enhancement of diffusion, before the threshold to the localized band there is critical reduction of D. This has been derived by Schell et al. [9] for the sinusoidal map. In general one has to expect again a powerlaw dependence on the distance from the band edges with an exponent 1/z. This is typical for intermittency effects [67-70].

7. ANOMALOUS DIFFUSION

Though diffusion with $\langle(x_t-\langle x_t\rangle)^2\rangle \propto t$ is most commonly found in physical systems, there are also examples for a different temporal behavior of the variance, e.g. the spreading of smoke plumes in turbulent flows, i.e. the transport of passive scalars in turbulent velocity fields [71]. Such anomalous diffusion processes can also be generated by 1-dimensional discrete maps. We will briefly describe two ways how the maps in question can be constructed.

The essence of the diffusion mechanism considered in sect.6 may be summarized as follows: The fractional part g(x) creates the x_τ-microdynamics with a well defined invariant density which in turn determines the probabilities for jumping to other cells. If the microdynamics is sufficiently chaotic, successive jumps are not correlated and the ensueing N_τ-diffusion is normal, i.e. $\langle (N_t - \langle N_t \rangle)^2 \rangle \propto t$. Since all cells have the same size, X_τ changes linearly with N_τ. Hence X_τ-diffusion is normal as well.

Anomalous diffusion may be obtained by a conjugation transformation [4] of the map F_a, which is defined by a continuous invertible function h(X). Introducing a new dynamical variable $Y_\tau = h(X_\tau)$ we have

$$Y_{\tau+1} = G_a(Y_\tau), \text{ with } G_a = h \circ F_a \circ h^{-1}. \tag{50}$$

A suitable choice of h leads to anomalous Y_τ-diffusion. As an example we consider $h_\alpha(X) = \text{sign}(X)|X|^\alpha$, $\alpha > 0$. For sufficiently large t one finds

$$\langle (Y_t - \langle Y_t \rangle)^2 \rangle \propto t^\alpha \tag{51}$$

provided, of course, that X_τ shows normal diffusion. A detailed discussion of the characteristics of such maps will be given in [72].

Unfortunately, all these maps have a flaw in common, their global inhomogeneity on the Y-scale. The analogy to turbulent flows, on the other hand, suggests a different approach. Frequently the flows are statistically homogeneous and the observed anomalous diffusion seems to be rather a consequence of the superposition of many scales of

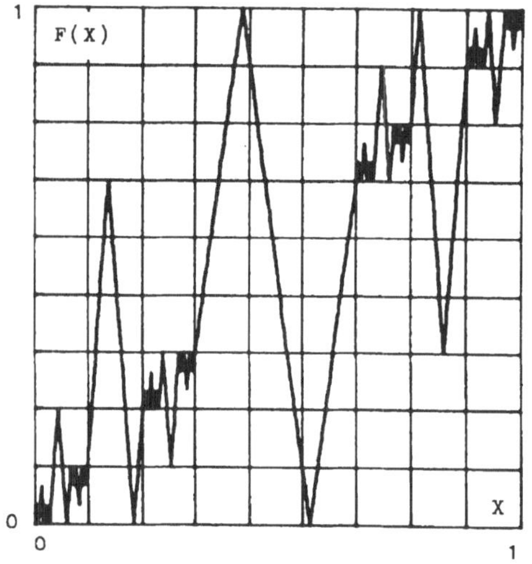

Fig.11: A map with scales of motion ranging from 3^0 down to 3^{-2}.

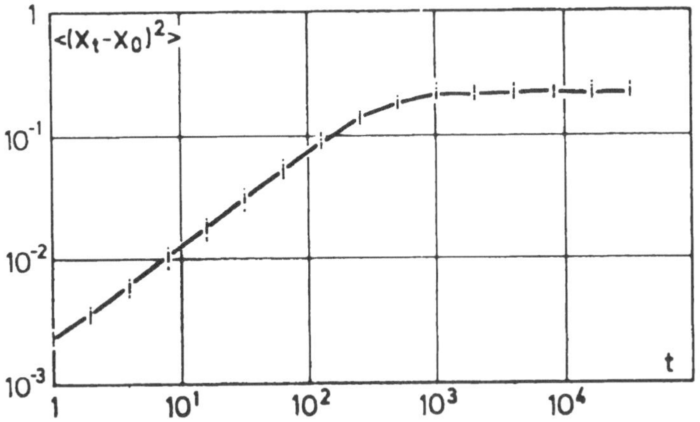

Fig.12: Numerically determined variance vs. time. The data points represent 20 runs $t = 2^0, 2^1, \ldots, 2^{15}$. For each value of t averaging was performed over a sample of 50,000 values.

motion [73]. Fig.11 shows a map which mimics this dynamical structure. The complete motion of the phase X_T may be understood as a hopping motion between the left third [0,1/3] and the right third [2/3,1] (typical scale 1) onto which a smaller scale hopping motion between the left and the right thirds of the thirds ([0,1/9] $\rightleftarrows$ [2/9,3/9], [6/9,7/9] $\rightleftarrows$ [8/9,1]) (typical scale 3^{-1}) is superimposed, which in turn is modulated on an even smaller scale (3^{-2}) and so on until finally a smallest scale (3^{-n}) is reached. Because of its self-similar structure this map may be considered a finite approximation to a fractal in the sense of Mandelbrot [40]. Fig.12 shows the result of a numerical simulation of the diffusion process within the interval [0,1] for a map with scales of motion ranging from 1 down to 3^{-10}. For not too long times one finds indeed a behavior $\langle (X_t - X_0)^2 \rangle \propto t^{\alpha}$ where α is approximately 0.75. Since the phase is confined to [0,1], saturation must occur for long times. Details of this type of discrete dynamics with many scales of motion will be published elsewhere [74].

References

1. E.N.Lorenz, J.Atm.Sci. 20,130 (1963)
2. B.A.Huberman, J.P.Crutchfield, Phys.Rev.Lett. 43,1743 (1979)
3. Lord Rayleigh, Phil.Mag. 15,229 (1883); 16,50 (1883);24,145(1887)
4. S.Grossmann, S.Thomae, Z.Naturforsch. 32a,1353 (1977)
5. M.J.Feigenbaum, J.Stat.Phys. 19,25 (1978)
6. P.Coullet, C.Tresser, J.Physique (Paris), Colloque 39,C5-25(1978)
7. T.Geisel, J.Nierwetberg, Phys.Rev.Lett. 48,7 (1982)
8. S.Grossmann, H.Fujisaka, Phys.Rev. A26,1779 (1982)

9. M.Schell, S.Fraser, R.Kapral, Phys.Rev. A26,504 (1982)
10. B.V.Chirikov, Phys.Rep. 52,263 (1979)
11. G.Ahlers, Phys.Rev.Lett. 33,1185 (1974)
12. A.Libchaber, C.Laroche, S.Fauve, J.Physique (Paris), Lettre 43, L-211 (1982)
13. R.Helleman, in: Fundamental Problems in Statistical Mechanics, Ed. E.G.D.Cohen, North Holland,Amsterdam etc. 1980, p165
P.Collet, J.-P.Eckmann, H.Koch, Physica 3D,457 (1981),J.M.Greene, R.S.Mc Kay, F.Vivaldi, M.J.Feigenbaum, Physica 3D,468 (1981)
14. A.Libchaber, J.Maurer, J.Physique(Paris), Colloque, C3-51(1980)
15. J.P.Gollub, S.V.Benson, J.Fluid Mech. 100,449 (1980)
16. M.Giglio, S.Musatti, U.Perini, Phys.Rev.Lett. 47,243 (1981)
17. J.P.Gollub, H.Swinney, Phys.Rev.Lett. 35,927 (1975)
18. R.P.Fenstermacher, H.L.Swinney, J.P.Gollub, J.Fluid Mech. 94, 103 (1979)
19. Yu.N.Belyaev, A.A.Monakhov, S.A.Shcherbakov, I.M.Yavorskaya, JETP-Lett. 29,295 (1979)
20. O.E.Rössler, Z.Naturforsch. 31a,1168 (1976); 1664 (1976)
O.E.Rössler, K.Wegmann, Nature 271,89 (1978)
21. L.F.Olsen, H.Degn, Nature 267,177 (1977)
22. R.A.Schmitz, K.R.Graziani, J.L.Hudson, J.Chem.Phys. 67,3040(1977)
23. J.C.Roux, A.Rossi, S.Bachelard, C.Vidal, Phys.Letters 77A,391 (1980); J.S.Turner, J.C.Roux, W.D.McCormick, H.L.Swinney, Phys.Letters 85A,9 (1981)
24. P.S.Linsay, Phys.Rev.Lett. 47,1349 (1981)
25. J.Testa, J.Pérez, C.Jeffries, Phys.Rev.Lett. 48,714 (1982)
26. W.Lauterborn, W.Meyer-Ilse, Diplomarbeit, Göttingen, 1982
27. W.Lauterborn, E.Cramer, Phys.Rev.Lett. 47,1445 (1981)
28. H.Haken, Phys.Letters 53A,77 (1975); 62A,133 (1977)
29. H.M.Gibbs, F.A.Hopf, D.L.Kaplan, R.L.Shoemaker, Phys.Rev.Lett. 46,474 (1981)
30. F.T.Arecchi, R.Meucci, G.Puccioni, J.Tredicce, Phys.Rev.Lett. 49,1217 (1982)
31. R.Keolian, I.Rudnick, L.A.Turkevich, S.J.Putterman, J.A.Rudnick, Phys.Rev.Lett. 47,1133 (1981)
32. C.W.Smith, M.J.Tejwani, D.A.Farris, Phys.Rev.Lett. 48,492(1982)
33. D.Meier, R.Holzner, B.Derighetti, E.Brun, in:Evolution of Chaos and Order, Ed.H.Haken, Synergetics Symposium 1982, Springer: Berlin etc. 1982
34. B.A.Huberman, J.P. Crutchfield, N.H.Packard, Appl.Phys.Lett. 37,750 (1980)
35. R.Y.Chiao, P.T.Parrish, J.Appl.Phys. 47,2639 (1976)
36. C.Herring, B.A.Huberman, Appl.Phys.Lett. 36,975 (1980)
37. D.A.Russell, E.Ott, Phys.Fluids 24,1976 (1981)
38. J.D.Farmer, E.Ott, J.A.Yorke, Physica D, 1982/83
39. J.Kaplan, J.Yorke, in: Funct.Diff.Equations etc., Eds.H.O.Peitgen, H.O.Walter, Springer: Berlin etc., 1978, p.228
40. B.Mandelbrot, Fractals:Form,Chance, and Dimension,Freeman: San Francisco, 1977

41. P.Grassberger, I.Procaccia, Measuring the Strangeness of Attractors, Preprint 1982
42. J.Heldstab, H.Thomas, T.Geisel, G.Radons, Z.Phys.B,1983 to appear
43. N.H.Packard, J.P.Crutchfield, J.D.Farmer, R.S.Shaw, Phys.Rev. Lett. 45,712 (1980); H.Froehling, J.P.Crutchfield, D.Farmer, N.H.Packard, R.Shaw, Physica D, 605 (1981)
44. D.Singer, SIAM J.Appl.Math. 35,260 (1978)
45. M.Misiurewicz, Absolutely continuous measures for certain maps of an interval, Publ.Math. IHES, 1980
46. G.Mayer-Kress, H.Haken, Physica D, 1983, to appear
47. G.Mayer-Kress, H.Haken, in:Evolution of Order and Chaos, Int. Symp.on Synergetics, Elmau,Ed.H.Haken,Springer:Berlin,etc.1982
48. J.P.Crutchfield, B.A.Huberman, Phys.Lett. 77A,407 (1980)
49. G.Mayer-Kress, H.Haken, J.Stat.Phys. 26,149 (1982)
50. J.P.Crutchfield, J.D.Farmer, B.A.Huberman,Phys.Rep. 92,45(1982)
51. A.Lasota, J.A.Yorke, Trans.Am.Math.Soc. 186,481 (1973)
52. T.Y.Li, J.A.Yorke, Trans.Am.Math.Soc. 235,183 (1978)
53. T.Y.Li, J.A.Yorke, Amer.Math.Monthly 82,985 (1975)
54. A.N.Šarkovskii, Ukr.Mat.Z. 16,61 (1964)
55. J.J.Kozak, M.K.Musho, M.D.Hatlee,Phys.Rev.Lett. 49,1801 (1982)
56. S.J.Shenker, L.P.Kadanoff, J.Phys.A:Math.Gen.14,L 23 (1981)
57. B.A.Huberman, A.B.Zisook, Phys.Rev.Lett. 46,626 (1981)
58. S.Thomae, S.Grossmann, Phys.Lett. 83A,181 (1981)
59. M.J.Feigenbaum, Phys.Letters 74A,375 (1979)
60. M.J.Feigenbaum, in:Nonlinear Phenomena in Chemical Dynamics, Eds.C.Vidal, A.Pacault, Proc.Int.Conf.Bordeaux,Sept.7-11,1981, Springer:Berlin etc., 1981, p.95
61. S.Thomae, S.Grossmann, J.Stat.Phys. 26,485 (1981)
62. P.Grassberger, J.Stat.Phys. 26,173(1981)
63. H.Mori, B.-C.So, T.Ose, Progr.Theor.Phys. 66,1266 (1981)
64. H.Mori, in:Nonlinear Phenomena in Chemical Dynamics, Eds.C.Vidal, A.Pacault,Proc.Int.Conf.Bordeaux,Sept.7-11,1981, Springer:Berlin etc., 1981, p.88
65. H.Shigematsu, H.Mori, T.Yoshida, H.Okamoto,two preprints Kyushu Univ., 1982
66. H.Fujisaka, S.Grossmann, Z.Phys. B48,261 (1982)
67. Y.Pomeau, P.Manneville,Phys.Lett.75A,1(1979);Comm.Math.Phys. 74,189 (1980)
68. G.Mayer-Kress, H.Haken, Phys.Lett. 82A,151 (1981)
69. J.-P.Eckmann, L.Thomas, P.Wittwer,J.Phys.A:Math.Gen.14,3153(1981)
70. J.E.Hirsch, B.A.Huberman,D.J.Scalapino,Phys.Rev. A25,519 (1982)
71. A.S.Monin,A.M.Yaglom:Statistical fluid mechanics, Vol.I,II, MIT Press: Cambridge, Mass.,London, 1971
72. H.Fujisaka et al., to be published
73. L.F.Richardson:Weather prediction by numerical process, Cambridge University Press 1922 (reprinted 1965)
74. S.Grossmann et al., to be published

Mathematical aspects are treated in:

75. P. Collet, J.-P. Eckmann, Iterated Maps on the Interval as Dynamical Systems, Birkhäuser, Basel, 1980

76. I. Gumowski, C. Mira, Recurrences and Discrete Dynamic Systems, Springer, Berlin,1980
77. G. Targonski, Topics in Iteration Theory, Vandenhock and Ruprecht, Göttingen, 1981

MULTICRITICAL BEHAVIOUR OF A NONEQUILIBRIUM SYSTEM

Tormod Riste and Kaare Otnes

Institute for Energy Technology

N-2007 Kjeller, Norway

INTRODUCTION

In the first part of this school we only dealt with multicriticality in equilibrium systems. Later we discussed nonequilibrium systems and chaos. This talk will address the problem: does multicriticality exist also in nonequilibrium systems and, if it does, what similarities/differences do we find?

Most of our knowledge of phase transitions in nonequilibrium systems comes from experimental and theoretical studies on lasers. At the last Geilo school Degiorgio[1] reviewed extensively what was known about simple criticality in one-mode lasers. Since then Agarwal and Dattagupta[2] have shown theoretically that multicriticality is to be expected for a two-mode laser. This follows naturally from the fact that the system has two coupled order parameters. Antoranz et al.[3] have shown that in a two-mode laser with a saturable absorber the first transition is to time-dependent states. Hence a multicritical point in a nonequilibrium system may involve, not only a change of behaviour of the phase boundary, but also a change of the attractor character. This problem was recently dealt with in a theoretical paper by Brand et al.[4].

The ability of two-mode lasers with saturable absorbers to show multicriticality, multistability and breaking of time symmetry is shared by other nonequilibrium systems with two diffusive agents: Rayleigh-Bénard convection in: two-component liquids, charged liquids in a vertical magnetic field, nematic liquid crystals in a vertical magnetic field. Our work is on the latter system.

CONVECTION IN A NEMATIC LIQUID CRYSTAL

In a nematic liquid crystal, as in any other liquid, steady convection sets in when the vertical temperature gradient exceeds a well-defined threshold value. Special features of liquid crystals are low threshold gradients and the susceptibility to influence by a magnetic field, both of which may be ascribed to the anisotropy of the constituent molecules. We have earlier made use of the effect of a magnetic field to demonstrate some phase-transition characteristics of the thermoconvective instability[5]: the mean-field character of the order parameter (i.e. the convective velocity), order parameter relaxation and data collapse. The convective threshold is just one of several instabilities that may occur in a liquid crystal, as shown in an extensive review by the Orsay group[6] in 1978. In 1977 Lekkerkerker[7] predicted that in a vertically aligned nematic, heated from below, the steady flow regime should be preceded by one of oscillatory flow. This problem was subsequently treated by several authors[8-10].

It is customary in driven (nonequilibrium, open) systems to consider the behaviour of the order parameter, or any other observable, as a function of a dimensionless control parameter R, which for a convective system is given by

$$R = \frac{\beta g l^3 \Delta T}{\nu \kappa} \tag{1}$$

β: volume thermal expansion
g: gravitational acceleration
l: vertical layer thickness of the liquid
ΔT: vertical temperature difference
ν: kinematic viscosity
κ: heat diffusivity

R, the Rayleigh number, is thus conveniently varied through the vertical temperature difference. For steady convection the threshold is at $R_c = 1708$, which determines the phase transition at ΔT_c.

Formula (1) is strictly valid only for shallow vessels with infinitely wide top and bottom plates. Estimates for ΔT_c^s, the threshold to the steady regime, give ~10° for a 1 cm deep layer of a nematogen liquid in its isotropic phase. According to (1) ΔT_c^s should decrease showly with H through ν and κ.

ΔT_c^o, the threshold to the oscillatory or, more generally, time dependent regime, depends on the strength of the applied vertical field through

$$\Delta T_c^o(H) = \Delta T_c^o(0)\ [1 + (H/H_c)^2] \tag{2}$$

H_c, the (Fredericks) critical field for alignment of the nematic sample is for l = 1 cm ~ 10 Gauss. $T_c^{(o)}$ is not given by (1) but by a very complex formula which, according to Dubois-Violette and Gabay[10], yields $\Delta T_c^{(o)}(0) < \frac{1}{2}\Delta T_c^{(s)}$. The latter authors predict that the oscillatory instability is an inverted bifurcation, as in a first-order transition, which possibly agrees with some experiments by Guyon et al.[9]. Previous theoretical work[7,8] had given continuous transitions at $\Delta T_c^{(o)}$ and discontinuous ones at $\Delta T_c(s)$.

The time dependence of the flow has been assumed to be as overstable oscillations[7], but may on general grounds be expected to be periodic, quasiperiodic or irregular. Guyon et al.[9] only observed induced, transient oscillations, whereas we have earlier reported [11,12] persistent, nonlinear oscillations within part of the oscillatory regime. Ideally one should want the experiments to reveal the attractor character and the nature of the bifurcations in (ΔT,H) parameter space. Such information is already available in theoretical papers by da Costa et al.[13] for thermohaline convection and magnetoconvection, and one expects qualitative aspects of their predictions to apply to our problem too. According to their calculations, the strength of the field and the shape of the sample container (i.e. the aspect ratio) may both affect the bifurcations and the attractor character in the time-dependent regime.

The time scale in the oscillatory regime is determined by the relaxation times for temperature (t_T) and orientation (t_o). For H = 0 one estimates[9] $t_T \sim 5\cdot 10^2$ sec and $t_o \sim 10^5$ sec for a vessel of linear dimension ~ 1 cm, $t_o \sim H^{-2}$ for H >100 Gauss, to become equal to t_T for H ~ 1 kG when the oscillatory regime is suppressed. As a net result the dynamics of the system is expected to get very slow when $H \rightarrow H_1$.

EXPERIMENTAL SETUP AND METHOD

The sample cell is a parallelpiped of horizontal dimensions 50 and 4 mm and height 25 mm. With this geometry one expects two convection rolls with axes parallel to the shorter edge, but this has so far not been checked. The sample is fully deuterated para-azoxyanisole, (PAA). The nematic range is given as 118 < T < 135°C, but pure specimen, as ours, may be supercooled to ~ 100°C. Hence we could perform our experiment over a temperature range where the nematic order parameter is almost constant.

The top, bottom and side walls are made of Al, Cu and stainless steel , respectively. A vertical temperature gradient is obtained by setting the difference of the power fed to electrical heating elements at the top and bottom of the vessel. The temperature of the bottom plate is controlled and kept constant throughout the experiment. The vertical temperature difference is monitored by thermistors and thermoelements. ΔT was found to stay constant within $\pm$ 0.01° for hours. The strength of the vertical magnetic field can be chosen in the range 0-1.2 kG.

The neutron wavelength used was 1.25Å. The scattered intensity was recorded at Q ~ 1.8$Å^{-1}$, at the first liquid diffraction peak of nematic PAA. The intensity of this peak is very sensitive to the molecular orientation, which in its turn depends on the convective flow. Hence we use the molecular orientation as an internal probe for the flow and monitor it with a neutron beam. In the absence of flow the molecules are aligned parallel to the field. It can be shown that the intensity, i.e. its deviation from that of the fully aligned, flow-free state, is proportional to the order parameter of the convective state. With a vertical magnetic field the deviation is negative, hence we use an inverted intensity scale in the figures that follow.

In the nematic state there is orientational coupling between the molecules. Hydrodynamic fluctuations and oscillations of wavelengths of the size of the convections rolls show up as real-time variations of the neutron intensity, because our neutron beam covers the whole sample. The time-averaged intensity gives the order parameter.

EXPERIMENTAL RESULTS

Fig. 1 gives the time-averaged intensity as a function of ΔT. With an applied field of 1.2 kG the oscillatory convection is suppressed, and the curve defines ΔT_c^s for the onset of steady flow for this value of H. At each gradient setting used when measuring this curve we also measured intensities for other values of H, and data for 0.6 kG in the steady flow regime are also shown.

The curves of Fig. 1 are also indicated in Fig. 2, in which we also give data for the oscillatory regime when H = 0.6 kG. The circular points have all been obtained with increasing temperature gradient. A striking feature is the drop and further increase of the intensity from A to B. Another striking feature is the multistability in the oscillatory regime. A typical return path from the highest gradients is shown by the square points. From C it passes approximately along the above mentioned 0.6 kG-curve to the lowest intensity level of the oscillatory regime. It enters this regime and rejoins the circular points at the lowest gradients of the regime after making stops at intermediate intensity levels. Triangular points seen at intermediate levels are obtained in other runs where we changed the field and the gradient in an arbitrary direction, i.e. upwards or downwards, before setting at 0.6 kG. In this figure each point represents an average taken during one hour or more. Similar data at other fields make us believe the the oscillatory regime consists of a number of levels, as indicated by the broken lines. Fig. 3 gives an example of a run,

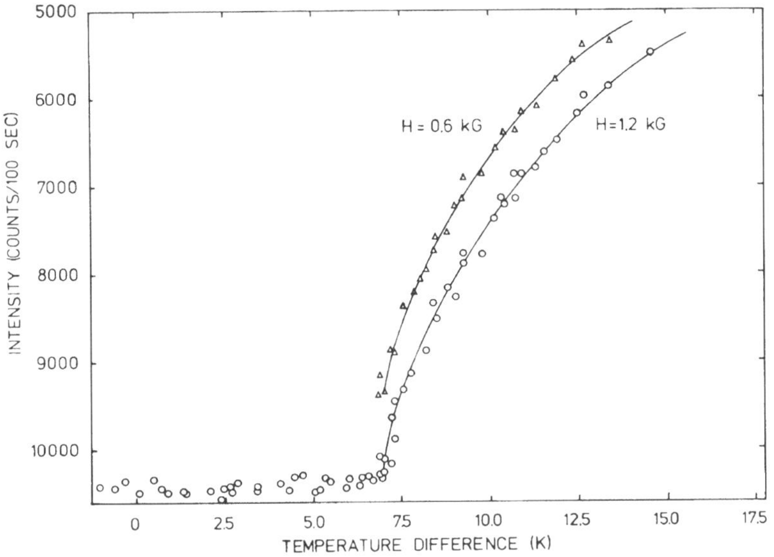

Fig. 1. Neutron intensity versus temperature difference (ΔT) between lower and upper plate, showing threshold to steady flow regime for field (H) 1.2 kG. At each ΔT data for H = 0.6 kG were measured, but only data for steady flow are given in this figure.

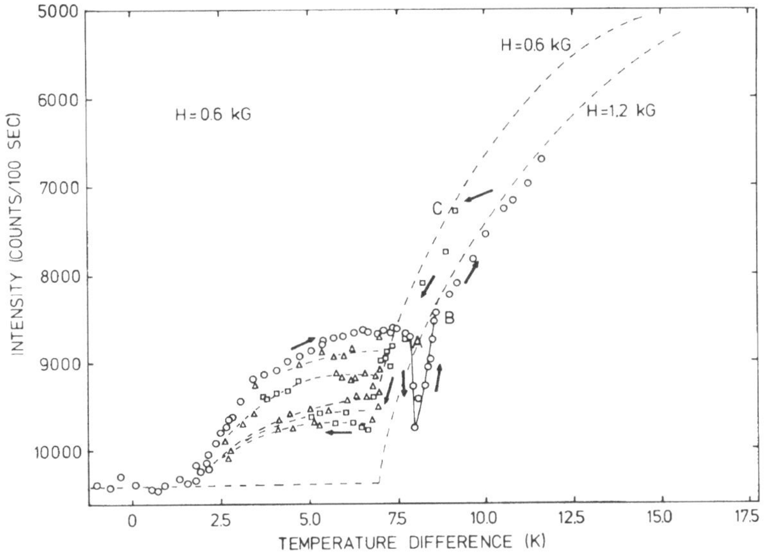

Fig. 2. Neutron intensity versus temperature difference (ΔT) for H = 0.6 kG. Broken curves are from Fig. 1. Circular and square points are for a complete cycle of increasing and decreasing ΔT, respectively. Triangular points are measured after additional variation of H. Notice intensity dip between A and B.

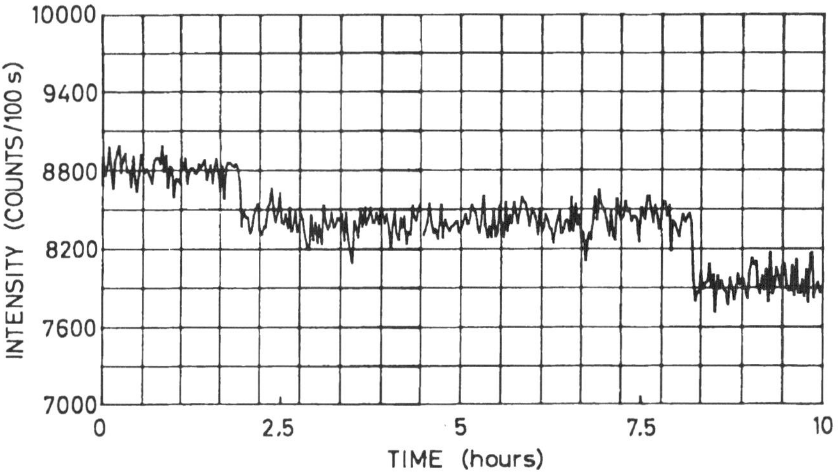

Fig. 3 Ten-hour time record of neutron intensity for H = 0.3 kG and ΔT decreasing from 3° to 2° in steps of 0.15°.

lasting ten hours, in which the intensity makes abrupt jumps between levels. The intensity anomaly between A and B is fully reproducible, and similar anomalies were observed when we let the system pass from two of the intermediate levels to the steady regime. We therefore interpret the curve at 1.2 kG as the upper stability limit of the oscillatory regime. Similarly the broken curve for 0.6 kG is the lower stability limit of the steady flow regime. The encounter with this curve on the return path at C is also marked by a small anomaly in the intensity record, but too small to be visible in the figure. The lower stability limit of the steady regime is not at C, but the lower point on the broken 0.6 kG-curve at which the system point will enter the oscillatory regime.

Fig. 4 gives the phase diagramme obtained from similar observations at different fields. Points in the upper part of the oscillatory regime are not very accurate, although intensities have been measured to a precision of 0.5 per cent.

DISCUSSION

In all of our observations ΔT_c^0, the threshold of the oscillatory regime, appears as a continuous transition. The transition between the oscillatory and the steady regimes is on the other hand discontinuous with observable stability limits. The field H_1, above which the oscillatory regime is suppressed, is from

Fig. 4 found to be 1.1$\pm$.1 kG. A preliminary value of 1.2 kG has been given by us earlier[14]. It seems safe to conclude that there is a multicritical point at H_1.

For $H > H_1$ the steady regime is entered through a continuous transition (ΔT_c^s), although the long build-up times of the flow near ΔT_c^s is not easily distinguished from a possible hysteretic behaviour. With this reservation the phase diagram of Fig. 4 represents a bicritical system. A similarity with an equilibrium system becomes apparent if one inverts the ΔT-axis and identifies it with T.

The intensity anomaly AB (see Fig. 2) that marks a transition where time symmetry is broken can be understood as a slowing-down and disappearance of time-dependent flow, followed by an appearance and speed-up of time-independent flow. This transition has been discussed by da Costa et al.[13], who expect it to depend in a sensitive manner on the stability of the states at either side of the transition.

We do not yet have complete information about the steady regime. The data shown in Fig. 1, and similar "parallel" curves for other fields, represent reproducible, lower-stability limits of the steady regime. It is possible also to obtain data, i.e. order

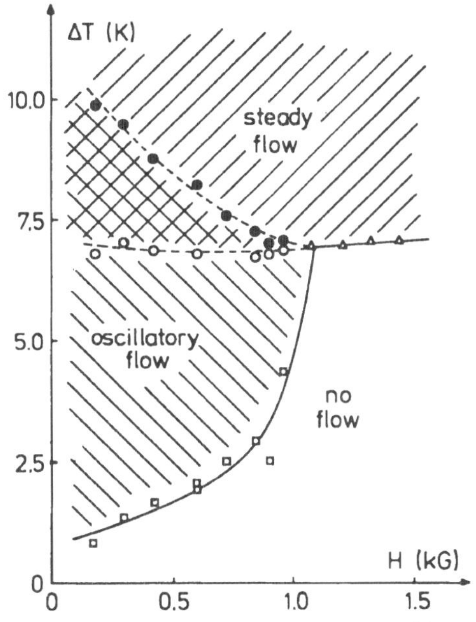

Fig. 4. Phase diagram showing confluence of first order (broken) lines and second order (full-drawn) lines at a bicritical point. First-order lines are upper and lower stability limits of oscillatory and steady flow regimes, respectively.

parameter, also to the right of these curves, but once it has attained a value at the stability limit it will stick to that curve. (See Fig. 1 of ref. 14 for data at 0.3 kG). Possibly the steady regime has some chaotic character in the sense that different spatial structures survive from the oscillatory regime.

For the oscillatory regime all our data lend support to the conclusion that there is multistability, and that the dynamics consists of motion about and between the levels. For the present sample and for the fields mostly explored so far the dynamics is too slow that we can characterize the attractor. As explained above, the time scale can be reduced through H and by reduction of the sample dimension. Some earlier experiments[12] on a cell of dimensions (25x25x3) mm^3 showed that quasiperiodic oscillations existed in certain regions of (ΔT,H)-parameters. When increasing ΔT the nonlinear character of the oscillations increased and is seen as development of singular points in the phase portrait, see Fig. 5. This is actually the behaviour predicted by da Costa et al.[13]. It is also a feature of the Lorenz attractor. Velarde[15] has treated two-component convection in the Lorenz model. This model admits multistability, corresponding to different ways and speed of rotation of the rolls. We need, however, more experiments and data processing to fully elucidate the dynamics in the time-dependent regime and the nature of the bifurcations. Fig. 2

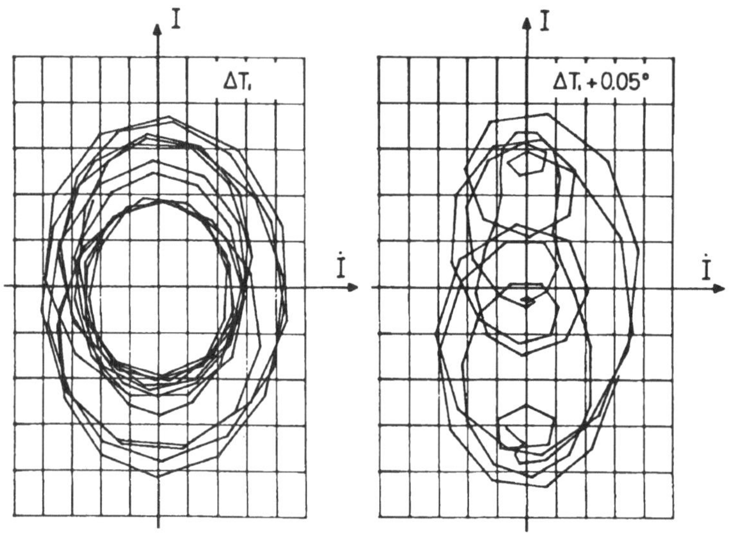

Fig. 5. Phase portrait ($\dot{I}$,I) deduced from data in oscillatory regime, showing stronger non-linearity for increasing ΔT. Data are for a smaller cell than other figures, see text.

and data at other fields indicate a higher multiplicity of levels as ΔT increases, suggesting one or more bifurcations within the oscillatory regime.

Theoretical work on nonequilibrium macroscopic systems invariably lend support to mean-field theory, implying classical exponents and negligible influence of fluctuations. The neighbourhood of the bicritical point of Fig. 4 is not sufficiently accurate that we can distinguish possible fluctuation effects. It is striking, however, that the curves for the static order parameter versus ΔT, in the vicinity of ΔT_c^o, have increasing slope as H decreases. By computer fitting our experiments give for the critical exponent β for the oscillatory regime: $0.35 \pm .01$ at H = 0.3 kG and $0.52 \pm .02$ at H = 0.6 kG. For the steady regime we get $\beta = 0.55 \pm .2$ at H = 1.2 kG.

In some earlier work[12] we have tentatively ascribed an intensity maximum near T_c^o at low fields to critical fluctuations, and gave heuristic arguments for its existence. In view of the nonclassical exponent given above for H = 0.3 kG it may be worth pursuing this point further. We do not exclude, however, that these indications of nonclassical behaviour are artifacts of the experiment, connected with multistability of the convection pattern.

To conclude, we have shown that convection in a nematic liquid crystal, subjected to a vertical magnetic field, has a multicritical phase diagram with a bicritical point. The phase lines separate regimes of no flow, time-dependent flow and steady flow. The main difference from equilibrium systems is the existence of multistability and time dependence, and the unusual behaviour of the order parameter at the boundary between the time-dependent and -independent regimes.

REFERENCES

1. V. Degiorgio in Nonlinear Phenomena at Phase Transitions and Instabilities (ed. T. Riste), Plenum Press, New York and London (1982), p. 181.
2. G.S. Agarwal and S. Dattagupta, Phys.Rev. 26A (1982) 880.
3. J.C. Antoranz, L.L. Bonilla, J. Gea, and M.G. Velarde, Phys. Rev. Lett. 49 (1982) 35.
4. H.R. Brand, P.C. Hohenberg, and V. Steinberg, Phys. Rev. 27A (1983) 591.
5. T. Riste, K. Otnes and H.B. Møller, Neutron Inelastic Scattering 1977, International Atomic Energy Agency, Vienna (1978), Vol. I, p. 511.
6. E. Dubois-Violette, G. Durand, E. Guyon, P. Manneville, P. Pieranski, Sol. St. Phys. Suppl. 14 (1978) 147.

7. H.N.W. Lekkerkerker, J. Physique Lett. (1977) L 277.
8. M.G. Velarde and I. Zuniga, J. Physique 40 (1979) 725.
9. E. Guyon, P. Pieranski and J. Salan , J. Fluid Mech. 93 (1979) 65.
10. E. Dubois-Violette and M. Gabay, J. Physique 43 (1982) 1305.
11. T. Riste and K. Otnes, ref. 1, p. 255.
12. T. Riste and K. Otnes, Proc. Yamada Conf. Sept. 1982, Physica B (in press).
13. E. Knobloch, N.O. Weiss and L.N. da Costa, J. Fluid Mech. 113 (1981) 153, and references therein.
14. K. Otnes and T. Riste, Proc. of EPS General Conference of Condensed Matter Physics, April 1983, to be published in Helvetica Physica Acta.
15. M.G. Velarde, ref. 1 p. 205.

PARTICIPANTS

Aharony, A.	Dept. of Physics, University of Tel Aviv, 69978 Tel Aviv, Israel
Andresen, A.F.	Insitute for Energy Technology, POB 40, 2007 Kjeller, Norway
Andrews, S.R.	Dept. of Physics, University of Edinburgh, Mayfield Road, Edinburgh EH9 3JZ, UK
Bak, P.	Dept. of Physics, Brookhaven National Laboratory, Upton, L.I., N.Y. 11973, USA
Beale, P.	Dept. of Theoretical Physics, Oxford University, 1 Keble Road, Oxford OX1 3NP, UK
Bedeaux, D.	Inst. for Theoretical Physics Norwegian Institute of Technology 7034 Trondheim-NTH
Belanger, D.P.	Dept. of Physics, Brookhaven National Laboratory, Upton, L.I., N.Y. 11973, USA
Bernard, L.	Institut Laue-Langevin, POB 156X, 38042 Grenoble-Cedex, France
Brookmann, D.	Lab. de Mineralogie, Université des Sciences et Techniques F-34000 Montpellier, France
Brug, J.	Section of Applied Physics, Yale University, Box 2157, New Haven, Ct. 06520, USA

Coles, B.R.	Dept. of Physics, Imperial College, London S.W.7, UK
Cowley, R.A.	Dept. of Physics, University of Edinburgh, Mayfield Road, Edinburgh EH9 3JZ, UK
Currat, R.	Institut Laue-Langevin POB 156X, 38042 Grenoble Cedex, France
Dohm, V.	Inst. f. Festkörperforschung, KFA Jülich, POB 1913, 5170 Jülich 1 W-Germany
Durlu, T.N.	Science Faculty, Ankara University, Besevler, Ankara, Tyrkey
Durand, D.	Lab. de Physique des Solides, Université Paris-Sud, Bat. 510, 91405 Orsay, France
Errandonea, G.	C.N.E.T., 196 rue de Paris, 92220 Bagneux, France
Feder, J.G.	Dept. of Physics, University of Oslo, Blindern, Oslo 3, Norway
Fisher, M.E.	Physics Dept., Cornell University, Ithaca, N.Y. 14853, USA
Fjellvåg, H.	Institute for Energy Technology, POB 40, 2007 Kjeller, Norway
Fossheim, K.	Dept. of Physics, Norwegian Institute of Technology, 7034 Trondheim-NTH, Norway
Fossum, J.O.	Dept. of Physics, Norwegian Institute of Technology, 7034 Trondheim-NTH, Norway
Galam, S.	Dept. of Physics, City College, CUNY, Convent Ave. & 138 Street, New York, N.Y. 10031, USA
Grossmann, S.	Fachbereich Physik, Universität Marburg, Renthof 5, 3550 Marburg W-Germany

Hälg, B.K.	Inst. f. Reaktortechnik, ETH Zürich CH-5303 Würenlingen, Switzerland
Hauge, E.H.	Inst. for Theoretical Physics Norwegian Institute of Technology 7034 Trondheim-NTH, Norway
Hemmer, P. Chr.	Inst. for Theoretical Physics Norwegian Institute of Technology 7034 Trondheim-NTH, Norway
Huang, C.-C.	Tate Lab. of Physics, University of Minnesota, 116 Church Street S.E. Minneapolis, Minn. 55455, USA
Høgh Jensen, M.	Physics Lab. II, H.C. Ørsted Institute, Universitetsparken 5, 2100 Copenhagen Ø, Denmark
Indekeu, J.O.	Inst. voor Theoretische Fysika Celestijenlaan 200 D, B-3030 Leuven, Belgium
Janssen, T.	Inst. for Theoretical Physics Toernooiveld, 6525 ED Nijmegen The Netherlands
Johansen, T.H.	Dept. of Physics, University of Oslo, Blindern, Oslo 3, Norway
Jug. G.	Dept. of Theoretical Physics, University of Oxford, 1 Keble Road, Oxford OX1 3NP, UK
Jøssang, T.	Dept. of Physics, University of Oslo, Blindern, Oslo 3, Norway
Kettler, P.	Hahn-Meitner-Institut, Bereich C Glienicker Str. 100, D-1000 Berlin 39, W-Germany
Kilum, D.	Physics Laboratory I, H.C. Ørsted Institute, Universitetsparken 5, 2100 Copenhagen Ø, Denmark
Knop, W.	Hahn-Meitner-Institut, Glienicker Str. 100, D-1000 Berlin 39, W-Germany

Lagerwall, S.T.	Dept. of Physics, Chalmers University of Technology, S-412 96 Gothenburg, Sweden
Lauszus, S.F.	Physics Lab. 1, University of Copenhagen, Universitetsparken 5, 2100 Copenhagen Ø, Denmark
Lebech, B.	Risø National Laboratory, 4000 Roskilde, Denmark
Lindgård, P.A.	Risø National Laboratory, 4000 Roskilde, Denmark
McCauley, J.L.	Physics Dept., University of Houston, Houston, Tx. 77034, USA
McEwen, K.	Institut Laue-Langevin, POB 156X, 38042 Grenoble Cedex, France
Montambaux, G.	Lab. de Physique des Solides, Université Paris-Sud, 91405 Orsay, France
Müller, K.A.	IBM Zürich Research Lab., Saümerstrasse 4, CH-8803 Rüschlikon, Switzerland
Navarro de Magalhaes, A.C.	Dept. de Fisica da Mat. Con., Rua Dr. Exavier Sigaud 150, 22290 URCA, Rio de Janeiro, RJ, Brasil
Newlove, S.A.	Dept. of Physics, University of Edinburgh, Mayfield Road, Edinburgh EH9 3JZ, UK
Nordblad, P.	Inst. of Technology, Univerity of Uppsala, POB 534, 851 21 Uppsala, Sweden
Olvera, M.	Caendish Laboratory, Madingley Road, Cambridge, C23 OHE, UK
Otnes, K.	Institute for Energy Technology, POB 40, 2007 Kjeller, Norway
Riste, T.	Institute for Energy Technology, POB 40, 2007 Kjeller, Norway

Ritala, R.K.	Inst.for Theoretical Physics Siltavuorenpenger 20C, SF-00170 Helsinki, Finland
Ritter, M.	Section of Applied Physics, Yale University, Box 2157, New Haven, Ct. 06520, USA
Saint-Gregoire, P.J.	GDPC Lab. du Pr Benoit, USTL, 34060 Montpellier Cedex, France
Salinas, S.R.A.	Inst. de Fisica, Universidade Sao Paulo, 05508 Sao Paulo - SP, Brasil
Sandvold, E.	Dept. of Theoretical Physics, Norwegian Institute of Technology, 7034 Trondheim-NTH, Norway
Schmittmann, B.	Dept. of Physics, University of Edinburgh, Mayfield Road, Edinburgh EH9 3ZJ, UK
Selke, W.	Inst. f.Festkörperforschung, KFA Jülich, Postfach 1913, 5170 Jülich 1, W-Germany
Shapira, Y.	Francis Bitter National Magnet Laboratory, MIT, Cambridge, Mass. 02139, USA
Sherrington, D.	Physics Dept., Imperial College, London S.W. 7, UK
Sinha, S.K.	Exxon Research & Engineering Co. Research Science Labs, POB 45, Linden, N.J. 07036, USA
Slotte, P.A.	Inst. for Theoretical Physics Norwegian Institute of Technology, 7034 Trondheim-NTH, Norway
Smith, J.	Dept. of Physics, The University, Southampton SO9 5NH, UK
Stirling, W.G.	Institut Laue-Langevin, POB 156X, 38042 Grenoble-Cedex
Styer, D.F.	Dept. of Physics, Clark Hall, Cornell University, Ithaca, N.Y. 13853, USA

Svedlindh, P.	Inst. of Technology, University of Uppsala, POB 534, 751 21 Uppsala, Sweden
Thomae, S.	IFF-Theorie III, KFA Jülich, POB 1913, 5170 Jülich 1, W-Germany
Thomas, H.	Dept. of Physics, University of Basel, Klingelbergerstrasse 82, 4056 Basel, Switzerland
Toledano, P.	Groupe de Physique Theorique Faculté des Sciences, 33 Rue Saint-Leu,80039 Amiens, France
Toulouse, G.	Ecolé Normale Supérieure, 24 rue l'Hommond, 72231 Paris Cedex 23, France
Trokiner, A.	Lab. de Physique Termique ESPCI, 10, rue Vauquelin, 75231 Paris Cedex 05, France
Vashistha, P.	Bldg. 223, Argonne National Laboratory, Argonne, Ill. 60439, USA
Vilfan, I.	J. Stefan Institute, POB 53, 6111 Ljubljana, Yugoslavia
Vlak, W.A.H.M.	Netherlands Energy Research Foundation, ECN, POB 1, 1755 ZG Petten, The Netherlands
Wolf, W.P.	Becton Center, Yale University, POB 2157, New Haven, Ct. 06520, USA
Yeomans, J.M.	Dept. of Physics, The University, Southampton SO9 5NH, UK
Young, P.A.	Dept. of Mathematics, Imperial College, London S.W. 7, UK

ORGANIZING COMMITTEE:

Jarrett, G.	Institute for Energy Technology, POB 40, 2007 Kjeller, Norway
Pynn, R.	Institut Laue-Langevin, POB 156X, 38042 Grenoble Cedex, France
Skjeltorp, A.	Institute for Energy Technology, POB 40, 2007 Kjeller, Norway

INDEX

Almeida-Thouless line, 387
Amplitude-
  mode, 202
  ratio, 1, 338
Anharmonic oscillator, 424
ANNNI model, 53, 233, 239
  high-temperature series, 55
  phase diagram, 249
  low temperature expansion, 249
  mean field theory, 249, 277-279
  in a field, 271-274
  renormalization group treatment, 279-280
Antiferromagnet
  bicritical points in 3d, 35-50
  easy axis, 36, 42
  easy plane, 41, 46
  isotropic, 41, 45
  site disordered, 341
Asymmetric clock model, 293-300
Asymptotic critical behaviour, 82, 86, 101, 104, 106
Au(Fe) spin glass, 379-381, 402

$BaMnF_4$, 179
Banana split, 258
$Ba_2NaNb_5O_{15}$, 207-211
Bicritical dynamics, 86, 97, 99
Bicritical line, 166, 168
Bicritical
  point
    degenerate, 2, 40, 45
    in 3d antiferromagnets, 35-50
    in non equilibrium system, 457
Bicritical (continued)
  point (continued)
    at structural transition 131-152
    virtual, 40, 46
Binary mixtures, 87, 126
Biphenyl, 179
Birefringence, 207-211, 330, 337
Blume-Capel model, 301

Cantor set, 247, 267, 432
Cascade
  inverse, 424, 439
  forward, 439
CeSb, 194-196, 233, 293
Chaos, 237-261, 268, 286, 423-447
  routes to, 426
  periodic, 433, 439
  small-scale, 438
Characteristic frequency, 83-84, 95
Charge density waves, 184-185
Charge transfer compounds, 240
Chiral clock model, 234, 290, 293-300
$CoBr_2.6(0.48\ D_2O,\ 0.52\ H_2O)$
  Tetracritical behaviour, 46
Coexistence curve, 1
Commensurate phases, 237-261
Competing anisotropies, 323, 356-360
Competing interactions, 54, 56, 185, 219, 271, 277, 283-288, 333-360, 409

Competing order parameters
Cone phase, 53 (*see also* modulated phase)
Conservative system, 424
Convection
  thermohaline, 453
  magneto-, 453
Correlation function, 89
  dynamic, 82
  four-point, 119
Correlation length, 92, 157, 224 336, 345, 348, 351, 407
Coupling
  reversible, 86, 95
  strain-order-parameter, 114-125
$Co_xZn_{1-x}F_2$, 342
Critical dimension (*see* upper or lower)
Critical end point, 137, 166, 276
Critical exponent, 1, 329
  crossover of, 2, 25, 36, 66
  dynamic, 84, 95, 103
  notation for, 14
Critical scattering, 336
Critical slowing down, 83
Crossover,
  dimensional, 9
  dynamic, 82, 96
  exponent, 2, 25, 36, 67, 92, 121, 133, 290, 323, 331
  Gaussian-Ising, 221
  Heisenberg-Ising, 151
  pure to random, 330
  of spin dimension, 35
  tricritical to critical, 21, 28, 90
Cr_2O_3, 45
$CsCoC\ell_3 . 2D_2O$, 28
$CsMnBr_3 2D_2O$, 43, 50
$CsMnF_3$, 46
Cubic anisotropy, 38, 129, 130, 160, 323
Cu-Mn, 365-370

DAG, 17, 24
Data collapsing, 50, 452
Defects, 137, 177, 192, 210, 211 247
Demagnetizing correction, 17, 28 43, 68, 375
Devil's staircase, 178, 237-261, 265-269, 297
Devil's top step, 234
Diffusion, 442, 445
Discommensuration, 178, 234, 285
Discrete $\emptyset^4$ model, 239, 257-261
Dissipative coupling, 81
Dissipative system, 424
Domain
  magnetic, 19
  wall, 178, 202, 289-291, 293-300
  wall energies, 234, 339
Driven system, 452
Dynamics
  near multicritical points, 81-109
  models, 84-88
$Dy_3A\ell_5O_{12}$ (see DAG)

Edge states, 416-418
Edwards-Anderson model, 322, 383
Elastic stiffness tensor, 115
Electron
  diffraction, 184
  microscopy, 179
EPR, 131, 134, 144-150
Ergodicity breaking, 390, 402, 409
$Eu_xSr_{1-x}S$, 396
Exponents (*see* critical exponents)
Extinction effects, 28

Fan phase, 53 (*see* also modulated phase)
Faraday rotation, 17
Feigenbaum convergence number, 238, 258, 426
$FeC\ell_2$, 15, 20
$Fe_xCo_{1-x}C\ell_2$, 356, 359
$Fe_{0.5}Zn_{0.5}F_2$, 330, 336, 342
Fixed point
  biconical 91, 98
  bicritical, 94
  cubic, 149, 151
  gaussian, 158, 280

Fixed point (continued)
  isotropic, 130-133, 149
      157-161
  random, 312, 324
  stable, 93, 155
  symmetric, 92
Flow instability, 454-459
Fluctuation driven transitions,
      131, 136, 155-163
Fractal
  dimension, 239, 247, 432
    of complete Devil's staircase,
      265, 268
    of percolation cluster, 316
  structure, 412, 256
  diffusion on, 424
  spectrum of Schrödinger
      equation on, 418-420
Frenkel-Kontorova model, 193
      238
Frustration, 283, 384, 405
      409-420

Gauge invariance, 414
Geometrical critical phenomena,
      410
$GdA\ell O_3$, 2, 43

Harris criterion, 310, 330
Hausdorf dimension, *see* Fractal
      dimension
Heat capacity, 77
Helicoidal phase, *see* modulated
      phase
Helium on graphite, 289
^4He superfluid, 87, 101, 122
^3He-^4He
  tricritical point, 2, 30, 84
  tricritical exponents, 23
  tricritical dynamics, 87, 103
  non-asymptotic features, 104
Hydrogen on iron, 289
Hysteresis
  effects, 15, 31, 134
  global, 178
  thermal, 207-208

Imperfections in samples, 15, 18
Incommensurate systems
  multicritical points in, 171-176
Incommensurate systems
      (continued)
  transitions in, 116, 125,
      283-288
Infinite cluster, 313
Infra-red spectroscopy, 186,227
InGe films, 410
Intercalates, 177, 239, 293
Interfacial adsorption, 301
Ising model
  1d with AFM interactions,
      239-248
  2d with AFM interactions, 343
  3d with AFM interactions, 344

Josephson junction, 266, 442

KCN, 227
$K_2Co_xFe_{1-x}F_4$, 356
Kinetic coefficient, 96, 103
  divergence of, 100
  renormalised, 98, 100
$MMnF_3$
  critical end point, 137
  cubic anisotropy, 130
  tricritical behaviour, 135
  ultrasonic experiments, 131
$K_2Mn_xFe_{1-x}F_4$, 356
$KNiF_3$, 46
Kosterlitz-Thouless transition,
      393
Krypton on graphite, 289, 290

$LaA\ell O_3$
  cubic anisotropy in, 130
  uniaxially stressed, 143-152
Landau-Khalatnikov term, 117,125
Landau levels, 409-420
  on periodic structures, 412
  on strips, 416
Lasers, 19, 451
Lifshitz condition, 174, 176,
      185
Lifshitz point, 2, 53-57, 131,
      171, 173, 233, 278
  crossover exponent for, 66
  dynamics, 85, 99
  in liquid crystals, 53, 73
  in MnP, 2, 53, 277
  in $RbCaF_3$, 53

Lifshitz point (continued)
scaling axes, 64
university class, 56
Limit cycle, 250, 423
Links and nodes model, 315
Liquid crystals
convection in, 452
critical dynamics of sound in, 126
multicritical point, 73
Liquid-gas transition, 126
Lock-in transition, 207, 285
Logarithmic correction, 15, 28, 31, 90, 330
Lorenz equations, 425
Lorentzian squared, *see* Random systems
Lower critical dimension
of Ising model in random field, 339, 343
for Lifshitz point, 75
in random systems, 321, 348
for spin glass, 385, 393, 405
Lyapunov exponent, 431, 436

Magnetic susceptibility
anomalous peak, 45
non-linear, 385, 393
transverse, 68
Magnetostriction, 67
Map
circle, 265
discrete, 265-269, 424
hat, 436
one-dimensional discrete, 424
parabola, 437
Poincaré, 266, 268
sinusoidal, 444
two-dimensional area preserving, 259, 424
Melting, 289
Memory effects, 178, 207-211
MnF_2, 2, 42, 86
MnP
anisotropic XY model for, 280-281
critical dynamics, 123
crystal structure, 59
MnP (continued)
Lifshitz point, 2, 53-70
phase diagram, 57, 63, 68
spin-wave dispersion, 61
ultrasonic attenuation, 124
$MnSi_x$, 183
$Mn_xZn_{1-x}F_2$, 342
Mode coupling, 81
Modulated phases, 50-70, 172, 177-196, 201
effect of discrete lattice, 237-261
Monolayers, 177
Monte Carlo technique, 30, 57, 233, 302
Multicritical (*see* bicritical, tricritical, etc.)
Multicritical dynamics, 81-104
Multicritical exponents, 7
Multicritical hypersurfaces, 165
Multiphase point, 233, 272
Multistability, 454

$NaNO_2$, 125, 286
Nematic-Smectic-A-Smectic-C transition, 73-78
Neutron scattering
in $CsCoC\ell_3.2D_2O$, 28
in disordered antiferromagnets, 333-360
in $FeC\ell_2$, 21
in MnP, 66
by modulated structures, 177-196
in nematic liquid crystal, 453-459
non-Lorenzian line shape, 70
in Praseodymium, 213-218
$NiC\ell_6.6H_2O$, 43
NMR, 180, 204
Noise, 425, 428
Nonasymptotic behaviour, 84, 102, 104, 107
Non-periodic motion, 423
Nowotny phase, 181
Nusselt number, 425

Order-disorder transition, 227

Parisi theory, 388
Partial differential approximant, 7,
Percolation, 313-319, 383, 409, 418
Period doubling, 425
Phase diagrams
  in low-anisotropy antiferromagnets, 42
  in MnP, 57, 68
Phason, 178, 202, 251
  gapless, 204, 285
Pinning, 178, 179, 196, 238
  energy, 254
  random, 202
  of vortices, 307-308
Polyacetylene, 185, 247
Potts model, 138, 289-290, 301, 314
Prandtl number, 425
Praseodymium, 213-218

Quantized Hall effect, 414
Quartz coefficient, 134, 136
Quartz, 186

Raman scattering, 227
Random anisotropies, 323
Random exchange, 329
Random fields, 320-322, 329, 333-360, 406
Random staggered field, 340
Random system, 309-325, 330
  structure factor of, 312, 340, 344-348
Rayleigh-Bénard convection, 269, 425, 451
Rayleigh number, 425, 452
$RbCaF_2$
  cubic anisotropy in, 130
  Lifshitz point in, 53
  tricritical behaviour of, 134
Rb_2CoF_4, 123
$Rb_2Co_xMg_{1-x}F_4$, 343
$RbMnF_3$, 45
Rn_2ZnBr_4, 180
$Rb_2ZnC\ell_4$, 201-206
Relaxational models, 84, 95, 98
Relaxational effects, 207-211, 380, 395

Renormalisation Group, 1, 81
  applied to spin glass, 405-408
  applied to subharmonic generation, 440
Replica
  broken symmetry, 387
  trick, 311, 330, 386, 399

Sample
  materials *listed* by chemical formula
  shape, 17, 28
Scaling
  asymptotic, 83
  axes, 1, 36, 64
  breakdown of dynamic, 100
  corrections to, 9, 330
  crossover form, 97, 310
  dynamic, 83, 102, 114, 120, 122
  extended, 50
  function, 7, 49, 117, 118, 321
  hypothesis, 221, 348
  linear, 1
  nonlinear, 1, 27
  nonuniversal, 2
  tricritical, 27
  variables, 7, 25, 91
Self-similarity, 245, 317, 447
Sherrington-Kirkpatrick model, 386
  non-ergodicity of, 390
Shift exponent, 39, 47, 144, 149, 152
Sierpinski gasket, 317, 412, 419
Singlet ground-state, 219
  in Praseodymium, 213
Sliding mode, *see* phason
Soliton, 234, 238
  density, 254
  of fractional spin, 247
  interaction energy, 255
  lattice, 252
  pinning 253-257
Specific heat, 105, 133, 134, 137, 194, 289, 338
  in disordered Ising systems, 329-332, 353
  exponent for random force constants, 202

Spectral dimension, 420
Spin
  clusters, 380
  dimensionality, 1
  flop transition, 2, 36
  freezing, 347, 364
  glass, 335, 363-408
    characteristic barrier height, 395
    dynamics of, 394-381
    field cooled, 379-381
    fluctuations in, 391-394
    freezing temperature, 385
    longitudinal, 400
    mean field theory, 386-391
    multicritical point in, 399-403
    order parameter, 386
    phase transition, 373-377
    specific heat, 367-369
    susceptibility, 370-371
    theoretical development, 383-408
    transverse, 400
Spiral phase, 53 (*see also* modulated phase)
SQUID magnetometer, 379
$SrTiO_3$, 2, 130, 133
Staggered field, 168
Staggered magnetisation, 24, 28, 31, 36, 47, 319, 340
Starfish, 258
Strange attractor, 427
Structural phase transition, 129-152, 162
Sulphur, 2
Superconductivity, 409-420
Supersymmetry, 321, 339
Surface phases, 289-291
Symmetry breaking, 130, 131, 144 156

Tetracritical point, 2, 38, 40, 46, 324
  decoupled, 324, 357
  in stressed $LaAℓO_3$, 143-152
  dynamics, 85, 97
$ThBr_4$, 179
Thiourea, 186-191, 210, 279, 286
Tight-binding model, 413, 417
TMATC-Zn, 191-194
Transfer matrix technique, 393
Tricritical exponents, 23, 90, 107
Tricritical line, 166
Tricritical
  point
    in $CsCoCℓ_3.2D_2O$, 20
    in DAG, 20
    dynamics, 85, 95, 104
    in $FeCℓ_2$, 20
    in fluid mixtures, 2
    in ^3He-^4He mixtures, 2, 13, 20, 103
    isotropic, 156
    in liquid crystals, 77
    in magnetic systems, 13-32
    at structural transitions, 134
Transition-metal dichalcogenides, 179, 185
TTF-TCNQ, 185

Ultrasonic measurements, 113-127
Umbilicus, 36
Uniaxial stress
  applied to Praseodymium, 213
  at structural transitions, 129-152, 162
Universal shape function, 83
Universality classes, 35, 56, 87
Upper critical dimension
  for Lifshitz point, 56
  in random systems, 321
  in spin glass, 392, 405

Variable
  conserved, 83, 92-93, 95
  dangerous irrelevant, 3
  marginal, 3
  relevant, 2
Verneuil crystals, 149
Vibrational absorption, 227-231

Wetting transition, 290, 301
Winding number, 251, 265
Wing critical point, 24

X-ray scattering, 75, 183, 202, 207